Soil Fertility: Role of Fertilizers for Plant Nutrition and Growth

Soil Fertility: Role of Fertilizers for Plant Nutrition and Growth

Editor: Eugene Perry

www.callistoreference.com

Callisto Reference,
118-35 Queens Blvd., Suite 400,
Forest Hills, NY 11375, USA

Visit us on the World Wide Web at:
www.callistoreference.com

ISBN: 978-1-63239-981-6 (Hardback)

Cataloging-in-Publication Data

Soil fertility : role of fertilizers for plant nutrition and growth / edited by Eugene Perry.
    p. cm.
Includes bibliographical references and index.
ISBN 978-1-63239-981-6
1. Soil fertility. 2. Fertilizers. 3. Plants--Nutrition. 4. Growth (Plants). I. Perry, Eugene.
S596.7 .S65 2018
631.422--dc23

Table of Contents

Permissions

List of Contributors

Index

Preface

Soil fertility is the capability of soil to supply essential plant nutrients. Soil fertility has a direct impact on crop yield. Some of the aspects which affect soil fertility are soil depth, internal drainage, soil pH, presence of microorganisms, etc. Fertilizers are either natural or synthetic substances which are added to soil to enhance its fertility. This book elucidates the concepts and innovative models around prospective developments with respect to soil fertility and fertilizers. As this field is emerging at a rapid pace, the contents of this book will help the readers understand the modern concepts and applications of the subject.

The researches compiled throughout the book are authentic and of high quality, combining several disciplines and from very diverse regions from around the world. Drawing on the contributions of many researchers from diverse countries, the book's objective is to provide the readers with the latest achievements in the area of research. This book will surely be a source of knowledge to all interested and researching the field.

In the end, I would like to express my deep sense of gratitude to all the authors for meeting the set deadlines in completing and submitting their research chapters. I would also like to thank the publisher for the support offered to us throughout the course of the book. Finally, I extend my sincere thanks to my family for being a constant source of inspiration and encouragement.

Editor

Testing the Growth Rate Hypothesis in Vascular Plants with Above- and Below-Ground Biomass

Qiang Yu[1,2]*, Honghui Wu[1,2], Nianpeng He[3], Xiaotao Lü[1,2], Zhiping Wang[2], James J. Elser[4], Jianguo Wu[4,5], Xingguo Han[1,2]*

1 State Key Laboratory of Forest and Soil Ecology, Institute of Applied Ecology, Chinese Academy of Sciences, Shenyang, China, 2 State Key Laboratory of Vegetation and Environmental Change, Institute of Botany, Chinese Academy of Sciences, Beijing, China, 3 Institute of Geographic Sciences and Natural Resources Research, Chinese Academy of Sciences, Beijing, China, 4 School of Life Sciences, Arizona State University, Tempe, Arizona, United States of America, 5 Sino-US Center for Conservation, Energy and Sustainability Science (SUCCESS), Inner Mongolia University, Hohhot, Inner Mongolia, China

Abstract

The growth rate hypothesis (GRH) proposes that higher growth rate (the rate of change in biomass per unit biomass, μ) is associated with higher P concentration and lower C:P and N:P ratios. However, the applicability of the GRH to vascular plants is not well-studied and few studies have been done on belowground biomass. Here we showed that, for aboveground, belowground and total biomass of three study species, μ was positively correlated with N:C under N limitation and positively correlated with P:C under P limitation. However, the N:P ratio was a unimodal function of μ, increasing for small values of μ, reaching a maximum, and then decreasing. The range of variations in μ was positively correlated with variation in C:N:P stoichiometry. Furthermore, μ and C:N:P ranges for aboveground biomass were negatively correlated with those for belowground. Our results confirm the well-known association of growth rate with tissue concentration of the limiting nutrient and provide empirical support for recent theoretical formulations.

Editor: Dorian Q. Fuller, University College London, United Kingdom

Funding: This work was supported by the National Natural Science Foundation of China (NSFC, 31170434) and the Key Project of NSFC (30830026). N. He acknowledges support from the Ministry of Education Key Laboratory for Biodiversity Science and Ecological Engineering (K0802). J. Elser and J. Wu acknowledge support from the National Science Foundation (DEB-0618193). The funders had no role in study design, data collection and analysis, decision to publish, or preparation of the manuscript.

Competing Interests: The authors have declared that no competing interests exist.

* E-mail: turfyu@gmail.com (QY); xghan@ibcas.ac.cn (XH)

Introduction

Carbon (C), nitrogen (N) and phosphorus (P) are very important elements for living organisms [1]. Their relative use in biomass (i.e. their C:N:P stoichiometry) reflects a complex interplay of evolutionary processes [2] coupled to phenotypic plasticity that is driven by patterns of element supply from the environment or diet. Thus, it is increasingly recognized that the values and ranges of C:N:P ratios in an organism are important determinants of the ecological niche. Indeed, C:N:P stoichiometry, and especially N:P ratio, is a powerful factor underlying diverse ecological processes [3], such as population stability [4], competitive interactions [5], community organization [6], trophic dynamics [7], litter decomposition [8,9], nutrient limitation [10,11], and biogeochemical cycling [12]. Thus, it is important to understand the underlying biological factors that drive observed variation in C:N:P ratios in organisms.

Considerable recent work has proposed specific connections between C:N:P stoichiometry and growth rate [1]. Growth rate is a central integrating parameter of overall life history strategy [13] and is closely linked to fitness [14]. Initiated from the study of crustacean zooplankton, the growth rate hypothesis (GRH) proposes that fast-growing organisms have low biomass C:P and N:P ratios [1,3] because of differential allocaiton to P-rich ribosomal RNA. By integrating ecological consequences with cellular and genetic mechanisms, the GRH broadened the use of

stoichiometric concept in evolutionary studies [3,15,16], providing a unifying thread connecting genes to ecosystems. The GRH has been intensively tested and generally supported via both theoretical and empirical analysis in zooplankton, arthropods, and bacteria [3,16–20]. However, the applicability of the hypothesis to photoautotrophs is not entirely clear, especially given the fact that storage materials in plants may obscure the associations between C:N:P stoichiometry and growth rate [1,21,22]. So, it is not clear whether the relationships between growth rate and C:N:P observed in the world of bacteria and zooplankton would also be observed for plants.

Diverse comprehensive reviews have shown that foliar N content in vascular plants tends to increase less than proportionately with P content [23–27]; thus, nutrient-rich foliage tends to have low N:P ratio, suggesting that the GRH has validity in the realm of vascular plants. However, not all studies in plants provide consistent support for the GRH. For example, Matzek and Vitousek's data for pine species showed that it was plant protein:RNA ratio but not foliar N:P ratio that was significantly correlated (negatively) with growth rate [28]. Thus, the interactions between N:P stoichiometry and growth rate require further study.

Ågren proposed to adapt the GRH to plants via a quantitative model of relationship between growth rate (μ) and N:C ($R_{N:C}$), P:C ($R_{P:C}$), N:P ($R_{N:P}$) with the following four equations [21]:

$$R_{N:C} = \frac{\mu}{\theta_{CN}} + \beta_N \qquad (1)$$

$$R_{P:C} = \frac{\mu^2}{\theta_{CN}\theta_{NP}} + \beta_P \qquad (2)$$

$$R_{N:P} = \frac{\mu\theta_{NP} + \beta_N\theta_{CN}\theta_{NP}}{\mu^2 + \beta_P\theta_{CN}\theta_{NP}} \qquad (3)$$

$$\text{or } R_{N:P} = \frac{\beta_{Nri}}{\theta_{CN}\theta_{NP}}\mu^2 + \frac{1}{\theta_{CN}}\mu + \beta_N \qquad (4)$$

where θ_{CN} represents the rate of C assimilated by proteins; θ_{NP} represents the rate of proteins assimilated by ribosomes, please see more details for these equations in [21]. Equation 1 predicts that N:C ratio is a linear increasing function of μ. Equation 2 predicts that P:C changes quadratically with μ. Thus, the N:P ratio is predicted to be a unimodal function of μ, increasing for small values of μ, reaching a maximum, and then decreasing.

Previous studies have provided considerable evidence for positive relationships between N:C or P:C with growth rate of plants [21]. However, the relationship between N:P and growth rate is unclear. Only a few experiments have tested the GRH in vascular plants [21,22,28,29], especially under both N- and P-limited conditions [21,22]. Unfortunately, even among those limited studies, the results are mixed. N:P ratio of birch seedlings decreased with μ when P was limiting but increased with μ when N was limiting [21], suggesting the relationship between N:P and μ varies considerably under different nutrient conditions. Consistent with Ågren's theory, Cernusak et al. [29] found that seedlings of 13 tropical tree and liana species showed hump-shaped relationships between N:P ratio and the relative growth rate. However, Matzek and Vitousek [28] found no relationship between μ and N:P in greenhouse experiments across 14 species. Furthermore, most studies have focused only on foliage and above-ground biomass. To our knowledge, no study has been done to test the relationship between root C:N:P stoichiometry and μ. Thus, not only the GRH but also Ågren's model need more comprehensive testing in terrestrial vascular plants, especially for belowground tissues.

To test the GRH and Ågren's theory in vascular plants, here we conducted a sand culture experiment in the temperate steppe of Inner Mongolia. Three grassland plants were planted in sand pots with various N and P levels to examine the relationship between C:N:P and growth rates of aboveground, belowground and total biomass under the variation of N and P.

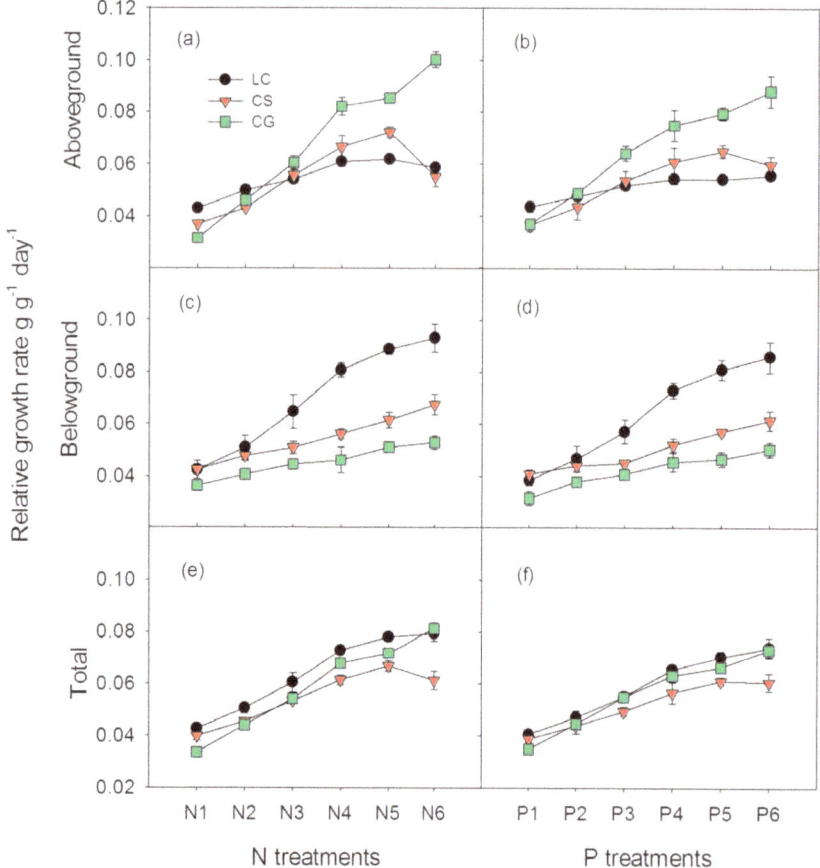

Figure 1. The comparison of relative growth rate of aboveground (a, b), belowground (c, d) and total biomass (e, f) along N and P enrichment levels across the three species. The three species are: *Leymus chinensis* (LC), *Cleistogenes squarrosa* (CS) and *Chenopodium glaucum* (CG).

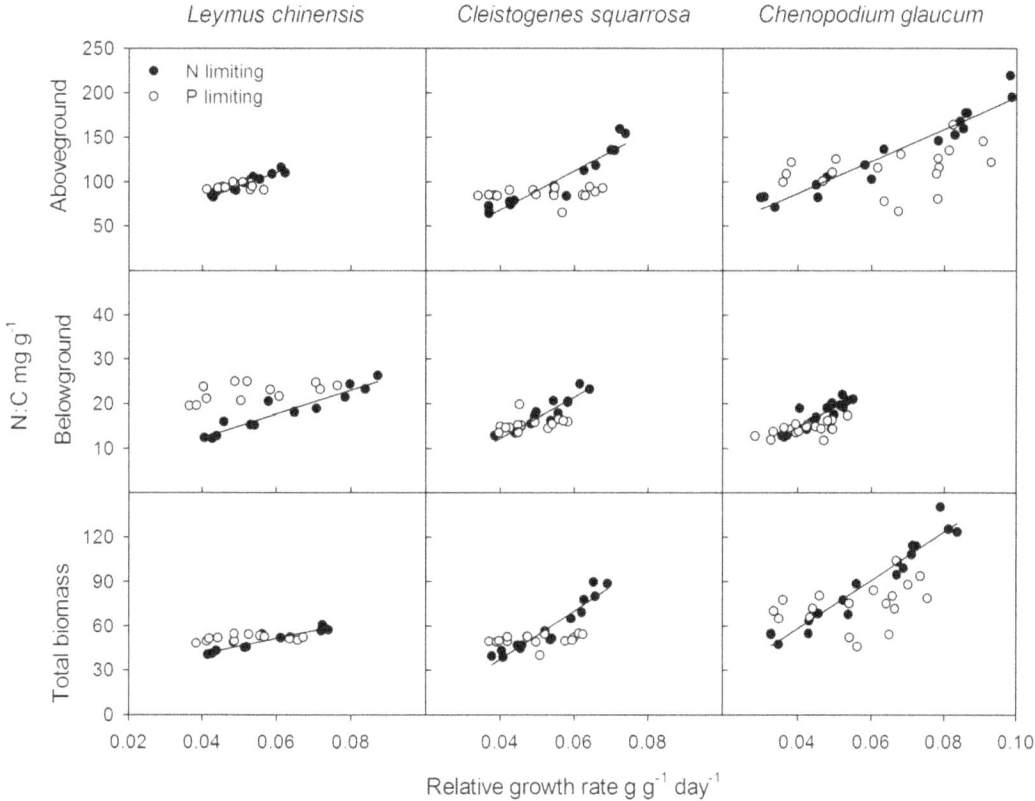

Figure 2. Relationships between relative growth rate and N:C for aboveground, belowground and total biomass. See analysis results in Table 1.

Results

Comparison of μ along N and P enrichment levels across the three species

Relative growth rate of each species increased significantly with increasing N and P availability for aboveground, belowground, and total biomass across low N and P fertilization levels (Fig. 1). However, for aboveground and total biomass at high levels of N and P, relative growth rate did not increase significantly or decreased with increasing N and P fertilization for *Leymus chinensis* and *Cleistogenes squarrosa*. The existence of N or P limitation was estimated by the changes of total biomass in response to N or P fertilization. If total biomass under a certain N or P treatment did not increase significantly compared to the lower N or P treatment, the treatment was delineated as an "excess treatment" and all the lower treatments were delineated as N or P limiting treatments. For *Leymus chinensis*, N5, N6, P5 and P6 were excess N and P levels respectively, i.e. N0–N4 were N limiting treatments while P0–P4 were P limiting treatments; for *Cleistogenes squarrosa*, N6 and P6 were excess N and P levels, i.e. N0–N5 were N limiting treatments while P0–P5 were P limiting treatments. For *Chenopodium glaucum* μ of aboveground, belowground, and total biomass all increased with increasing N and P availability across all the fertilization levels, indicating that all N treatments were N limiting and all P treatments were P limiting for *Chenopodium glaucum*.

Relationships between μ and N:C of the three species

Consistent with the predictions of equation 1, aboveground, belowground and total biomass N:C ratio increased linearly as a function of μ for each of the three species when N was limiting

(Fig. 2 and Table 1). However, when excess N treatments were included, no significant relationships were found for *Leymus chinensis* and *Cleistogenes squarrosa*.

No significant relationships were found between aboveground and total biomass N:C and μ for any of the three species when P was limiting (Fig. 2 and Table 1). For belowground biomass of *Cleistogenes squarrosa*, no significant correlation between N:C and μ was found in the P treatments (Table 1); however, significant relationships were found for the other two species.

Relationships between μ and P:C of the three species

All aboveground, belowground, and total biomass P:C ratios increased as a nonlinear function of μ either when N or P was limiting (with the only exception being *Leymus chinensis* aboveground biomass in the N treatments) (Fig. 3 and Table 1). These results are qualitatively consistent with the predictions of equation 2, with a lower Akaike information criterion (AIC) value than linear regression when P was limiting. The increase in P:C with growth was considerably larger in the P fertilization series than in the N fertilization series. As for aboveground data, if the data for excess N and P treatments were included, significant relationships between P:C and μ disappeared for *Leymus chinensis* and *Cleistogenes squarrosa*.

Relationships between μ and N:P of the three species

Almost all relationships between μ and N:P of the three species for aboveground, belowground and total biomass were not significant based on equation 3 with the only exception of total biomass of *Leymus chinensis* ($P = 0.0136$, $R^2 = 0.72$). However, all relationships of the three species were significant based on

Table 1. Results of regression analyses of the relationships relative growth rate and C:N:P of three species in the sand culture experiment.

Plant parts	Experiment	Species	N:C		P:C	
			P	R^2	P	R^2
aboveground	N limiting	LC	<0.0001	0.93	n.s.	0.1
		CS	<0.0001	0.90	<0.0001	0.84
		CG	<0.0001	0.94	<0.0001	0.88
	P limiting	LC	n.s.	0.02	=0.0001	0.87
		CS	n.s.	0.02	<0.0001	0.82
		CG	n.s.	0.10	<0.0001	0.78
belowground	N limiting	LC	<0.0001	0.94	<0.0001	0.91
		CS	<0.0001	0.90	=0.0059	0.50
		CG	<0.0001	0.82	=0.0232	0.39
	P limiting	LC	=0.0013	0.49	<0.0001	0.95
		CS	n.s.	0.18	<0.0001	0.90
		CG	=0.0113	0.34	<0.0001	0.84
Total	N limiting	LC	<0.0001	0.88	<0.0001	0.88
		CS	<0.0001	0.91	=0.0091	0.49
		CG	<0.0001	0.94	<0.0001	0.83
	P limiting	LC	n.s.	0.09	<0.0001	0.93
		CS	n.s.	0.15	<0.0001	0.95
		CG	n.s.	0.13	<0.0001	0.88

For each of the three species, analysis was conducted without the data in excess N or P treatments. No significant difference (n.s.) represents $P>0.05$. *Leymus chinensis* (LC), *Cleistogenes squarrosa* (CS) and *Chenopodium glaucum* (CG).

equation 4 (Fig. 4). AIC values of quadratic regressions were lower than those of linear regressions. In most cases, the N:P ratio was a unimodal function of μ, increasing for small values of μ, reaching a maximum, and then decreasing. Although the relationship for all belowground biomass and total biomass of *Leymus chinensis* showed the same trends as predicted by equation 4, we did not find the maximum of N:P. No significant relationships were found for aboveground biomass of *Leymus chinensis* and *Cleistogenes squarrosa* when data from the excess N and P treatments were included based on equation 4.

Variation of relative growth rate, N:C, P:C and N:P across the three species

The rank order of aboveground μ for the three species reversed between low and high nutrient conditions (Fig. 1). Growth rate of *Leymus chinensis* was the highest in both the low N and P levels ($P<0.05$, ANOVA; 2 and 0.15 mmol L^{-1}, respectively), whereas it was the lowest in high N and P treatments ($P<0.05$, 32 and 4.8 mmol L^{-1} respectively). While the mean value of μ of *Chenopodium glaucum* was the highest in high fertilization treatments for both N and P, it was the lowest in low N and P treatments (Fig. 1). Interestingly, the species rankings for μ for belowground biomass were opposite with that for aboveground.

There were two main trends for the ranges of variation in μ, N:C, P:C and N:P for the three species and aboveground versus belowground. First, the ranges of μ were generally consistent with those of C:N:P stoichiometry. Both when N and P were limiting, *Chenopodium glaucum* exhibited the highest range of aboveground μ

(Fig. 5), while *Leymus chinensis* showed the lowest range. There were consistent patterns for the three species for ranges of foliar N:C, P:C, and N:P ratios (except P:C when N was limiting) in both N and P treatments; i.e., *Chenopodium glaucum*>*Cleistogenes squarrosa*>''' *Leymus chinensis* (Fig 5), which was the same as the rank order of ranges of μ. The highest foliar N:P and the lowest P:C occurred in *Leymus chinensis*, while the lowest N:P and the highest P:C were found in *Cleistogenes squarrosa*. Second, the patterns for belowground biomass were almost completely opposite those just discussed for aboveground biomass (Fig. 5). Overall, belowground biomass of *Leymus chinensis* had the highest ranges of μ, N:C, P:C and N:P, while *Chenopodium glaucum* had the lowest ranges. However, no significant differences were found for N:P among the three species when N was limiting.

Discussion

Our results from the sand culture experiment clearly demonstrate strong positive associations of aboveground, belowground and total biomass N:C and P:C ratios with μ for these Inner Mongolia grassland vascular plants, which are consistent with numerous previous findings [1,21,22,25,29–31]. Furthermore, when N or P was in excess, positive correlations were lost in *Leymus chinensis* and *Cleistogenes squarrosa*, which is consistent with other research about excess supply of N under strong P limitation [32–34], suggesting that excess uptake will confound the physiological need for elements at different relative growth rates with the capacity of excess uptake. For example, if both N:C and P:C are taken from P-limited conditions, P:C describes the physiological need for P but N:C describes the capacity of N excess uptake. Overall, N:C ratio was positively related with μ when N was limiting, and P:C ratio was positively correlated with μ when P was limiting.

As predicted by equation 4, N:P ratio was a unimodal function of μ, increasing for small values of μ, reaching a maximum, and then decreasing. However, in contrast with Ågren's study, the data did not fit equation 3. Our study suggests that, in vascular plants, N in ribosomes may need to be considered when analyzing the relationship between μ and N:P [21], at least for grasses in Inner Mongolia grassland. Most studies have shown a negative association between μ and N:P, not only among plant species but also within populations or cultivars of a given species [1,10,27,29,33,34] consistent with the results of belowground and the results of aboveground and total biomass with high μ in our study. Results from this study also provide further support for the suggestion that the negative relationship between μ and N:P ratio may not hold for plants when P is not limiting [27], likely due to effects of P storage under N limitation. Thus, for vascular plants, μ is positively correlated with N:P when μ is low while negatively correlated with N:P when μ is high.

To our knowledge, our data represent the first documentation of the relationships between belowground C:N:P stoichiometry and μ, which is consistent with aboveground biomass. In the derivation of equations 1–4, the N:C is based on the need for proteins for C capture and P:C is based on the need for ribosomes for protein production [21]. However, roots do not take up C (except for limited uptake of organic compounds). So, the question is why should belowground biomass fit the equations? Likely the answer is that translocation of carbon and nutrients among tissues is a whole-plant process. As μ and C:N:P ratios of aboveground biomass were positively related with those of belowground (Fig. 1 and Yu et al unpublished data), belowground biomass should exhibit similar relationships as those seen aboveground.

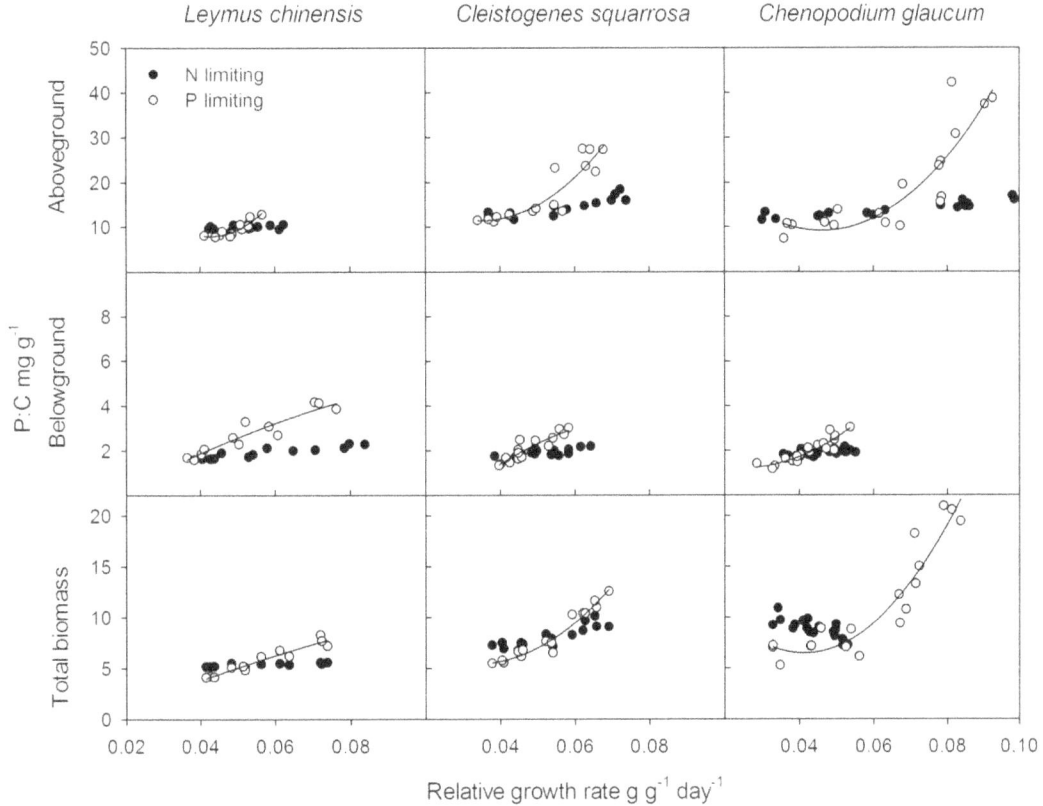

Figure 3. Relationships between relative growth rate and P:C for aboveground, belowground and total biomass. See analysis results in Table 1.

Across all N treatments, the annual species, *Chenopodium glaucum*, had the highest aboveground μ but had an intermediate N:P ratio (Fig. 5). *Leymus chinensis* had the highest N:P while *Cleistogenes squarrosa* had the lowest N:P but they had similar μ. Similarly, it was difficult to establish clear ranking patterns when we considered N:C, P:C and μ both for aboveground and belowground biomass at the inter-specific level. Thus, while data for additional species are clearly needed, across all the nutrient levels, μ was not clearly associated with C:N:P ratios at the inter-specific level in our data, consistent with the results from a study focused on 14 *Pinus* species growing at high and low nutrient levels [28]. In contrast, Elser *et al.* [35] found strong and consistent rank order relationships between growth rate, RNA content, and P content (and thus likely N:P ratio) across five species of *Drosophila*. It is likely that the increased nutrient storage capacities of plants relative to metazoans may underpin this contrast between animal and plant studies.

The reversal in the rank order of aboveground μ of the three species between low and high nutrient treatments (Fig. 1) is notable and suggests an overall performance tradeoff that is manifested when environmental nutrient conditions fluctuate. *Chenopodium glaucum*, an annual species considered to be a fast grower [36], performed well under high nutrient conditions but grew very slowly under low ambient nutrient supply. In contrast, *Leymus chinensis*, a perennial rhizomatous species usually considered as a slow grower [36], grew relatively slowly in highly fertilized conditions but could tolerate poor nutrient conditions. Thus, "fast growers" are not always growing faster than "slow growers" and instead there may be important trade-offs between growth and tolerance that are mediated by above- and below-ground

allocation differences (see below). These responses likely reflect inter-specific differences in adaptive strategies, i.e. the "fast growth" species has an advantage in fertile habitats while "slow growth" species are better suited to infertile soils and other stresses related to the efficient husbanding of limiting nutrients [37].

Related to these trade-offs, another notable pattern is that almost all the species ranges for μ, C:N:P ratios were opposite between aboveground and belowground biomass, which corresponds well with our previous observations for above- vs below-ground stoichiometric homeostasis (an index of variability in C:N:P stoichiometry) [38]. The functional equilibrium model predicts that fertilization will cause reduced allocation to roots [39–41] and species that rapidly shift their biomass allocation to aboveground tend to outcompete species with a less responsive root : shoot ratio after N eutrophication [41]. In contrast, slow growers (for aboveground biomass) tend to dominate in arid or low-fertility ecosystems [37]. Our results indicate that these aboveground "slow growers" are likely "fast growers" belowground, a response that may allow them to more effectively exploit below-ground nutrient resources in infertile soils. This corresponds well with the observation that these aboveground "slow growers" have high root : shoot ratios [39,42].

Consistent with these arguments, the taxon with the highest ranges of μ (*Chenopodium glaucum* for aboveground, *Leymus chinensis* for belowground) tended to also have the highest ranges in N:C, P:C and N:P ratios across treatments, while the species with the lowest ranges of μ tended to have the lowest ranges in C:N:P stoichiometry. While more data are clearly needed, these data suggest a pattern in which species-level variation of growth rate is positively correlated with the variation of C:N:P stoichiometry.

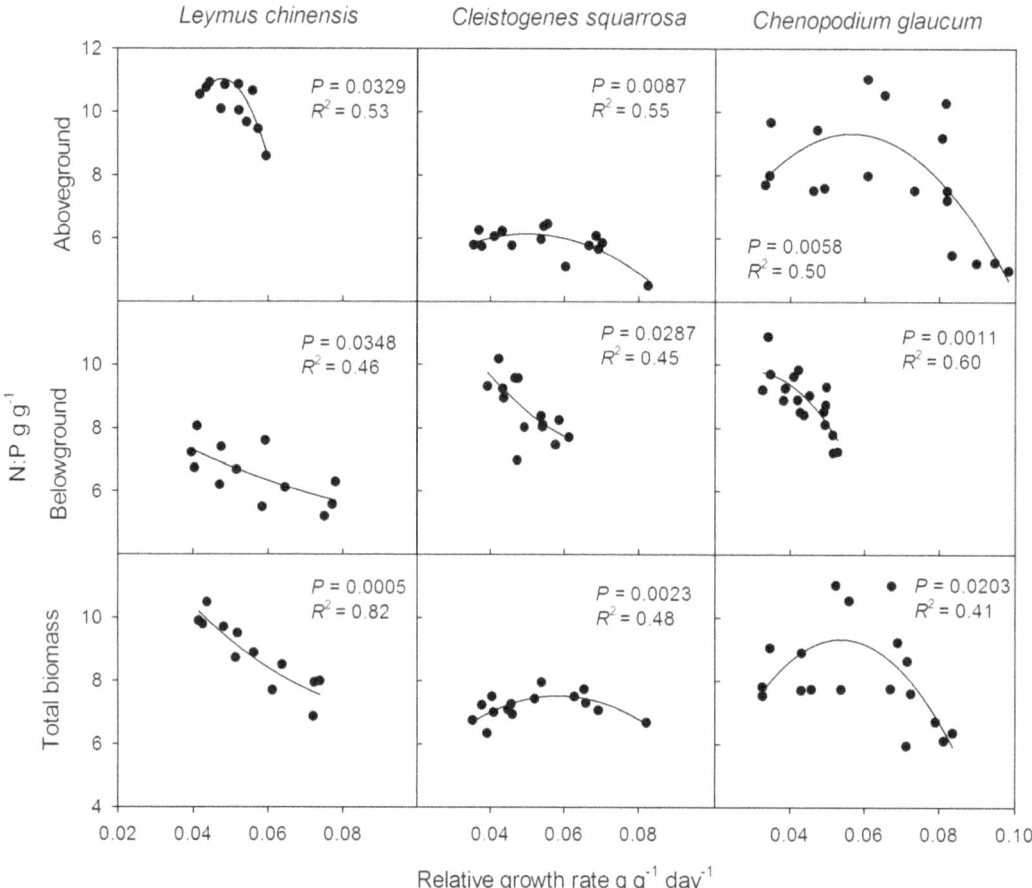

Figure 4. Relationships between relative growth rate and N:P for aboveground, belowground and total biomass.

The rank order of stoichiometric homeostasis both for aboveground and belowground biomass of these three species shown in our previous work [38] was the opposite of the rank of variation of relative growth rate in this study, indicating that stoichiometrically homeostasic species are also "homeostatic" in growth rate. Stoichiometric homeostasis in autotrophs, as well as in other organisms, is known to depend on relative growth rate [22,43,44]. These studies have shown that, the closer to its maximum capacity a plant grows, the more constrained is its elemental composition. As stoichiometrically homeostatic species tend to be dominant in the community, and homeostatic ecosystems are more productive and stable [45], the ability of species to control their variation of growth rate may be another mechanism responsible for important ecological properties, such as ecosystem structure, functioning and stability.

There are two main limitations to our study. First, because we only studied three species and because newly-germinated perennial plants may not accurately reflect above- and below-ground growth rates and allocation patterns, the relationships between μ and biomass C:N:P ratios and the contrasting patterns between above- and below-ground that we document need more testing in future studies. Second, it is preferable to study plant growth as a function of a range of steady-state N and P supplies rather than the pulsed nutrient supply regime in the sand culture approach used in our study [46,47], because relative growth rates are more closely associated with uptake rates of nutrients rather than with external concentrations supplied under relatively static conditions. If possible, techniques for steady state should be adopted in the future studies testing the growth rate hypothesis for vascular plants.

This study is one of the few evaluations of the GRH in vascular plants and the first to evaluate patterns in belowground biomass. The results showed that, for aboveground, belowground and total biomass, μ was positively correlated with N:C and P:C ratios for each of the three species but μ was positively correlated with N:P ratio in N treatments and negatively in P treatments. No clear associations among μ and C:N:P ratios were found at the interspecific level. Thus, we suggest that the GRH need to be refined for application to vascular plants, likely due to the effects of storage of non-limiting nutrient in plants. More specifically, we propose a revised GRH in which: 1) Plant nutrient contents (N:C and P:C) are positively correlated with μ; and 2) There are disproportionate increases in the content of the limiting nutrient relative to the non-limiting nutrient in a cell-quota dependent manner such that the relationship between μ and N:P ratio is context-dependent.

Materials and Methods

No specific permits were required for the described field studies and we confirmed that the location is not privately-owned and the field studies did not involve endangered or protected species.

Sand culture experiment

The sand culture experiment was conducted at the Inner Mongolia Grassland Ecosystem Research Station in 2006. Three plant species, *Leymus chinensis* (a C-3 perennial rhizome grass),

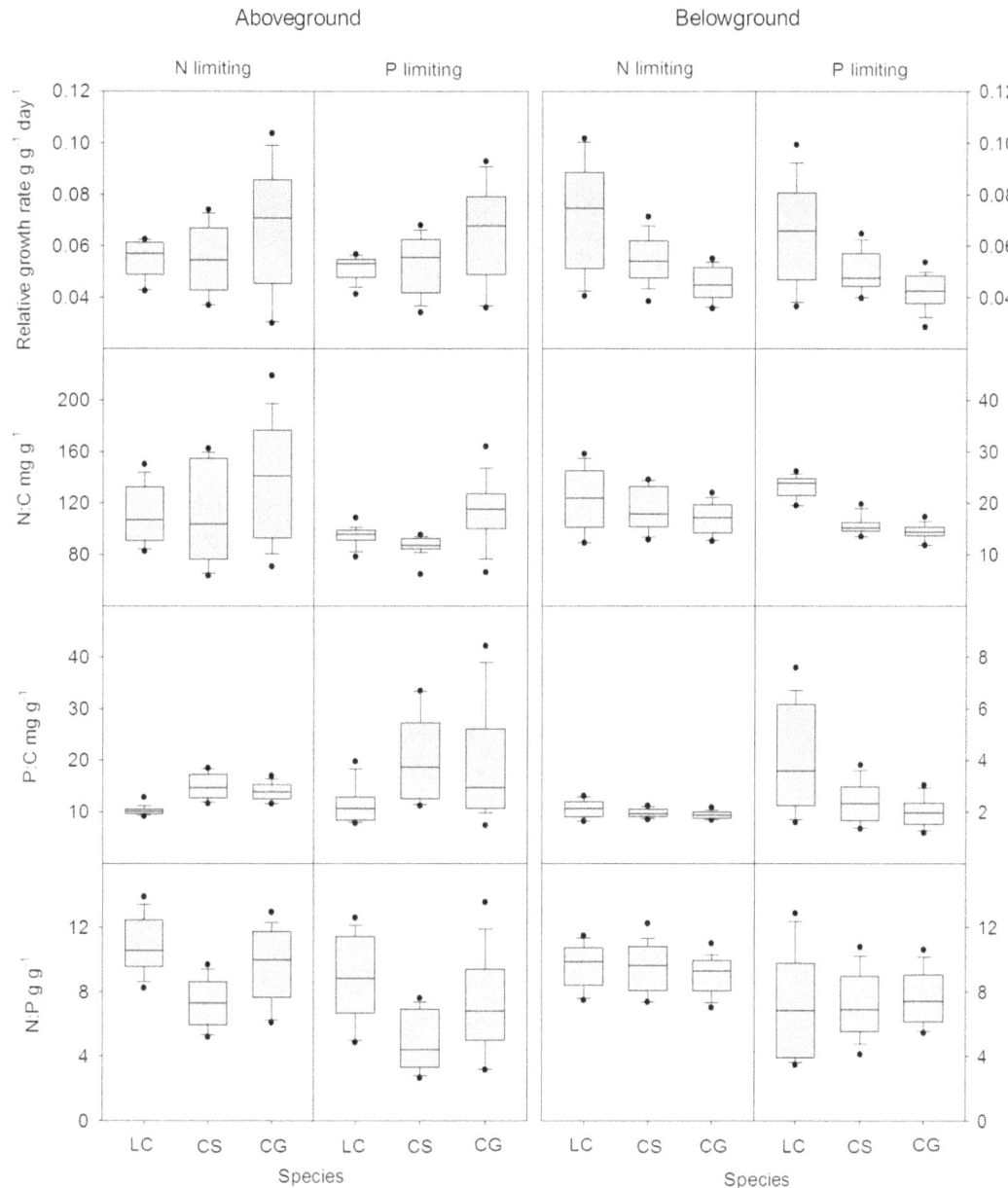

Figure 5. Variation of relative growth rate, N:C, P:C and N:P across the three species.

Cleistogenes squarrosa (a C-4 perennial bunchgrass) and *Chenopodium glaucum* (a C-3 annual species) were selected as our target species, representing the dominant, subdominant, and the annual minor species respectively. Their seeds were planted in pots filled with sand on May 1 and watered with nutrient solutions each day. Sand between 0.2 and 2.0-mm was screened with mesh, and, to minimize the soil nutrient content, washed five times prior to filling plastic pots (30 cm diameter, 35 cm height). For each plant species, we applied treatments consisting of 6 N levels (2, 4, 8, 16, 24, 32 mmol N L^{-1}, added as NH_4NO_3) and 6 P levels (0.15, 0.3, 0.6, 1.2, 2.4, 4.8 mmol P L^{-1}, added as KH_2PO_4,) respectively. P concentration was held constant at 1 mmol P L^{-1} in N treatments and N concentration was held constant at 15 mmol N L^{-1} in P treatments. Each experimental pot received the same amounts of macro- and micronutrients except for N and P. The macroelement composition of the solution followed the formula developed by

Hoagland and Arnon [48] and the microelement composition was based on Jensen's formula [49]. Each level had three replicates, with three pots randomly allocated to a replicate and a total of 18 pots (3 replicates * 3 pots * 2 harvests) for each species. Each experimental pot received 250-mL of solution every day to maintain a relatively constant macro- and micronutrient concentration (excess solutions were drained through the four holes at the bottom). The pots were washed with 500-ml water twice every 10 days followed by 250-mL solutions immediately to avoid ionic toxicity. All pots were covered when it rained and, if rainwater entered pots, additional 250-mL nutrient solutions were added. There were 10 to 30 individuals in each pot depending on the plant size. The density was controlled to ensure that the plant individuals did not shade each other. The above- and below-ground biomass of thirty plants (within 3 pots) of each plant species for each treatment replicate was harvested on 10 July and 10

August 2006. Healthy, fully expanded leaves and roots were oven-dried at 60°C, powdered and screened with 0.1-mm mesh for chemical analysis (total N and total P).

Elemental analysis

Total N (% of dry mass) were analyzed with the micro-Kjeldahl method [50] using 2300 Kjeltec Analyzer Unit (FOSS, Sweden). Total P concentrations (% of dry mass) were measured by the ammonium molybdate method after persulfate oxidation [51]. The total C concentration (%) was measured using a modified Mebius method [52]. Briefly, 0.01 g samples were digested with 10 ml 0.50 mol·L^{-1} K$_2$Cr$_2$O$_7$ at 180°C for 5 minutes followed by titration of the digests with standardized FeSO$_4$.

Estimation of relative growth rate

The relative growth rates were calculated by

$$\mu = Ln(M_t/M_0)/t$$

where M$_0$ is the initial biomass, M$_t$ is the final biomass, and t is the time interval. In this study, M$_0$ is the aboveground, belowground

or total biomass of each species on 10 July and M$_t$ is the biomass on 10 August.

Statistical analysis

Regression analysis was used to assess the relationships between μ and C:N:P ratios (linear regression for μ with N:C and quadratic regression for μ with P:C and N:P). ANOVA was used to test the difference of μ among species. All analysis was performed by SAS (version 9.0, SAS Inst., Cary, NC, USA).

Acknowledgments

We thank Dan Flynn, Andrew Hector, as well as two anonymous reviewers, for providing helpful comments on the manuscript. We are grateful to the Inner Mongolia Grassland Ecosystem Research Station (IMGERS) for providing the experimental sites.

Author Contributions

Conceived and designed the experiments: QY XH. Performed the experiments: QY HW NH. Analyzed the data: QY. Contributed reagents/materials/analysis tools: QY. Wrote the paper: QY HW NH ZW XL JE JW XH.

References

1. Sterner RW, Elser JJ (2002) Ecological Stoichiometry: *The Biology of Elements from Moleculaes to the Biosphere*. Princeton: Princeton University Press.
2. Hessen DO, Ågren GI, Anderson TR, Elser JJ, de Ruiter P (2004) Carbon sequestration in ecosystems: the role of stoichiometry. Ecology 85: 1179–1192.
3. Elser JJ, Sterner RW, Gorokhova E, Fagan WF, Markow TA, et al. (2000) Biological stoichiometry from genes to ecosystems. Ecol Lett 3: 540–550.
4. Loladze I, Kuang Y, Elser JJ (2000) Stoichiometry in producer-grazer systems: linking energy flow with element cycling. Bull Math Biol 62: 1137–62.
5. Olde Venterink H, Güsewell S (2010) Competitive interactions between two meadow grasses under nitrogen and phosphorus limitation. Funct Ecol 24: 877–886.
6. Sterner RW, Hessen DO (1994) Algal nutrient limitation and the nutrition of aquatic herbivores. Annu Rev Ecol Evol S 25: 1–29.
7. Elser JJ, Chrzanowski TH, Sterner RW, Mills KH (1998) Stoichiometric constraints on food-web dynamics: A whole-lake experiment on the Canadian shield. Ecosystems 1: 120–136.
8. Güsewell S, Gessner MO (2009) N : P ratios influence litter decomposition and colonization by fungi and bacteria in microcosms. Funct Ecol 23: 211–219.
9. Manzoni S, Trofymow JA, Jackson RB, Porporato A (2010) Stoichiometric controls dynamics on carbon, nitrogen, and phosphorus in decomposing litter. Ecol Monogr 80: 89–106.
10. Güsewell S (2004) N : P ratios in terrestrial plants: variation and functional significance. New Phytol 164: 243–266.
11. Elser JJ, Bracken MES, Cleland EE, Gruner DS, Harpole WS, et al. (2007) Global analysis of nitrogen and phosphorus limitation of primary producers in freshwater, marine and terrestrial ecosystems. Ecolo Lett 10: 1135–1142.
12. Elser JJ, Andersen T, Baron JS, Bergstrom A-K, Jansson M, et al. (2009) Shifts in lake N:P stoichiometry and nutrient limitation driven by atmospheric nitrogen deposition. Science 326: 835–837.
13. Arendt JD (1997) Adaptive intrinsic growth rates: An integration across taxa. Q Rev of Biol 72: 149–177.
14. Sterner RW, Elser JJ, Hessen DO (1992) Stoichiometric relationships among producers, consumers and nutrient cycling in pelagic ecosystems. Biogeochemistry 17: 49–67.
15. Gorokhova E, Dowling TE, Weider LJ, Crease TJ, Elser JJ (2002) Functional and ecological significance of rDNA intergenic spacer variation in a clonal organism under divergent selection for production rate. Proc R Soc B 269: 2373–2379.
16. Makino W, Cotner JB, Sterner RW, Elser JJ (2003) Are bacteria more like plants or animals? Growth rate and resource dependence of bacterial C : N : P stoichiometry. Funct Ecol 17: 121–130.
17. Elser JJ, Acharya K, Kyle M, Cotner J, Makino W, et al. (2003) Growth rate-stoichiometry couplings in diverse biota. Ecolo Lett 6: 936–943.
18. Acharya K, Kyle M, Elser JJ (2004) Biological stoichiometry of Daphnia growth: An ecophysiological test of the growth rate hypothesis. Limnol and Oceanogr 49: 656–665.
19. Watts T, Woods HA, Hargand S, Elser JJ, Markow TA (2006) Biological stoichiometry of growth in *Drosophila melanogaster*. J Insect Physiol 52: 187–193.
20. Hessen DO, Jensen TC, Kyle M, Elser JJ (2007) RNA responses to N- and P-limitation; reciprocal regulation of stoichiometry and growth rate in Brachionus. Funct Ecol 21: 956–962.

21. Ågren GI (2004) The C : N : P stoichiometry of autotrophs - theory and observations. Ecol Lett 7: 185–191.
22. Ågren GI (2008) Stoichiometry and nutrition of plant growth in natural communities. Annu Rev Ecol Evol S 39: 153–170.
23. Wright IJ, Reich PB, Cornelissen JHC, Falster DS, Garnier E, et al. (2005) Assessing the generality of global leaf trait relationships. New Phytol 166: 485–496.
24. Kerkhoff AJ, Enquist BJ (2006) Ecosystem allometry: the scaling of nutrient stocks and primary productivity across plant communities. Ecol Lett 9: 419–427.
25. Niklas KJ (2006) Plant allometry, leaf nitrogen and phosphorus stoichiometry, and interspecific trends in annual growth rates. Ann Bot 97: 155–163.
26. Reich PB, Oleksyn J, Wright IJ, Niklas KJ, Hedin L, et al. (2010) Evidence of a general 2/3-power law of scaling leaf nitrogen to phosphorus among major plant groups and biomes. Proc R Soc B 277: 877–883.
27. Elser JJ, Fagan WF, Kerkhoff AJ, Swenson NG, Enquist BJ (2010) Biological stoichiometry of plant production: metabolism, scaling and ecological response to global change. New Phytol 186: 593–608.
28. Matzek V, Vitousek PM (2009) N : P stoichiometry and protein : RNA ratios in vascular plants: an evaluation of the growth-rate hypothesis. Ecol Lett 12: 765–771.
29. Cernusak LA, Winter K, Turner BL (2010) Leaf nitrogen to phosphorus ratios of tropical trees: experimental assessment of physiological and environmental controls. New Phytol 185: 770–779.
30. Niklas KJ, Owens T, Reich PB, Cobb ED (2005) Nitrogen/phosphorus leaf stoichiometry and the scaling of plant growth. Ecol Lett 8: 636–642.
31. Poorter H, Bergkotte M (1992) Chemical composition of 24 wild species differing in relative growth rate. Plant, Cell & Environment 15: 221–229.
32. McJannet CL, Keddy PA, Pick FR (1995) Nitrogen and phosphorus tissue concentrations in 41 wetland plants: a comparison across habitats and functional-groups. Funct Ecol 9: 231–238.
33. Limpens J, Berendse F, Klees H (2003) N deposition affects N availability in interstitial water, growth of Sphagnum and invasion of vascular plants in bog vegetation. New Phytol 157: 339–347.
34. Tomassen HBM, Smolders AJP, Limpens J, Lamers LPM, Roelofs JGM (2004) Expansion of invasive species on ombrotrophic bogs: desiccation or high N deposition? J Appl Ecol 41: 139–150.
35. Elser JJ, Watts T, Bitler B, Markow TA (2006) Ontogenetic coupling of growth rate with RNA and P contents in five species of *Drosophila*. Funct Ecol 20: 846–856.
36. Bai YF, Wu JG, Clark CM, Naeem S, Pan QM, et al. (2010) Tradeoffs and thresholds in the effects of nitrogen addition on biodiversity and ecosystem functioning: evidence from inner Mongolia Grasslands. Global Change Biolo 16: 358–372.
37. Chapin FS (1980) The mineral nutrition of wild plants. Annu Rev Ecol Evol S 11: 233–260.
38. Yu Q, Elser JJ, He NP, Wu HH, Chen QS, et al. (2011) Stoichiometric homeostasis of vascular plants in the Inner Mongolia grassland. Oecologia 166: 1–10.
39. Tilman D (1988) Plant strategies and the dynamics and structure of plant communities. Princeton University Press, Princeton, New Jersey, USA.
40. Ericsson T (1995) Growth and shoot:root ratio of seedlings in relation to nutrient availability. Plant Soil 168–169: 205–214.

41. Johnson NC, Rowland DL, Corkidi L, Allen EB (2008) Plant winners and losers during grassland N-eutrophication differ in biomass allocation and mycorrhizas. Ecology 89: 2868–2878.

42. Bai YF, Han XG, Wu JG, Chen ZZ, Li LH (2004) Ecosystem stability and compensatory effects in the Inner Mongolia grassland. Nature 431: 181–184.

43. Elrifi IR, Turpin DH (1985) Steady-state luxury consumption and the concept of optimum nutrient ratios: a study with phosphate and nitrate limited *Selenastrum minutum* (Chlorophyta). J Phycol 21: 592–602.

44. Persson J, Fink P, Goto A, Hood JM, Jonas J, et al. (2010) To be or not to be what you eat: regulation of stoichiometric homeostasis among autotrophs and heterotrophs. Oikos 119: 741–751.

45. Yu Q, Chen QS, Elser JJ, He NP, Wu HH, et al. (2010) Linking stoichiometric homeostasis with ecosystem structure, functioning, and stability. Ecol Lett 13: 1390–1399.

46. Ingestad T, Lund AB (1986) Theory and techniques for steady state mineral nutrition and growth of plants. Seand J For Res 1: 439–453.

47. Ingestad T (1982) Relative addition rate and external concentration - driving variables used in plant nutrition research. Plant Cell and Environment 5: 443–453.

48. Hoagland R, Arnon DI The water culture method for growing plants without soil, Circular 347, California Agricultural Experiment Station, College of Agriculture, University of California, Berkeley.

49. Jensen MH, Collins WL (1985) Hydroponic Vegetable Production. Hortic Rev 7: 483–558.

50. Bremner JM (1996) Nitrogen: total. in Methods of Soil Analysis Part 3: Chemical Methods Sparks DL, Page AL, Loeppert PA, Soltanpour PN, Tabatabai MA, Johnston CT, et al., editor. Soil Science Society of America and American Society of Agronomy, Madison. pp 1085–1123.

51. Kuo S (1996) Phosphorus. in Methods of Soil Analysis Part 3: Chemical methods Sparks DL, Page AL, Loeppert PA, Soltanpour PN, Tabatabai MA, Johnston CT, et al., editor. Soil Science Society of America and American Society of Agronomy, Madison. pp 869–920.

52. Nelson DW, Sommers LE (1982) Total carbon, organic carbon, and organic matter. In: Methods of Soil Analysis Page AL, Miller RH, Keeney DR, eds. American Society of Agronomy and Soil Science Society of American, Madison, WI. pp 1–129.

2

Distinct Microbial Limitations in Litter and Underlying Soil Revealed by Carbon and Nutrient Fertilization in a Tropical Rainforest

Nicolas Fanin[1,2], Sandra Barantal[1], Nathalie Fromin[1]*, Heidy Schimann[3], Patrick Schevin[1], Stephan Hättenschwiler[1]

1 Centre d'Ecologie Fonctionnelle et Evolutive (CEFE), CNRS, Montpellier, France, 2 Université of Montpellier II, Montpellier, France, 3 UMR Ecologie des Forêts de Guyane (EcoFoG), Campus Agronomique, Kourou, French Guiana

Abstract

Human-caused alterations of the carbon and nutrient cycles are expected to impact tropical ecosystems in the near future. Here we evaluated how a combined change in carbon (C), nitrogen (N) and phosphorus (P) availability affects soil and litter microbial respiration and litter decomposition in an undisturbed Amazonian rainforest in French Guiana. In a fully factorial C (as cellulose), N (as urea), and P (as phosphate) fertilization experiment we analyzed a total of 540 litterbag-soil pairs after a 158-day exposure in the field. Rates of substrate-induced respiration (SIR) measured in litter and litter mass loss were similarly affected by fertilization showing the strongest stimulation when N and P were added simultaneously. The stimulating NP effect on litter SIR increased considerably with increasing initial dissolved organic carbon (DOC) concentrations in litter, suggesting that the combined availability of N, P, and a labile C source has a particularly strong effect on microbial activity. Cellulose fertilization, however, did not further stimulate the NP effect. In contrast to litter SIR and litter mass loss, soil SIR was reduced with N fertilization and showed only a positive effect in response to P fertilization that was further enhanced with additional C fertilization. Our data suggest that increased nutrient enrichment in the studied Amazonian rainforest can considerably change microbial activity and litter decomposition, and that these effects differ between the litter layer and the underlying soil. Any resulting change in relative C and nutrient fluxes between the litter layer and the soil can have important consequences for biogeochemical cycles in tropical forest ecosystems.

Editor: Caroline P. Slomp, Utrecht University, The Netherlands

Funding: This research was funded through CNRS "PIR Amazonie II" and "EC2CO" grants. N. Fanin received a Ph.D. grant from the French Research and Education Ministry and from the University of Montpellier II. The funders had no role in study design, data collection and analysis, decision to publish, or preparation of the manuscript.

Competing Interests: The authors have declared that no competing interests exist.

* E-mail: nathalie.fromin@cefe.cnrs.fr

Introduction

Over the last two decades, considerable efforts were made towards a better understanding of the effects of global change factors such as climate change or nutrient deposition on the quality of plant litter, its subsequent decomposition and the consequences on ecosystem carbon (C) dynamics [1–4]. Whether or not C is sequestered in forest ecosystems depends on the often small difference between photosynthetic C fixation and ecosystem respiration, with soil respiration representing between half to two thirds of the total ecosystem respiration [5–8]. Nutrient availability is a key factor in the regulation of soil respiration, and anthropogenic alterations of the nitrogen (N) and phosphorus (P) cycles can have important consequences for the net CO_2 exchange between the biosphere and the atmosphere, and thus for the global C budget [9,10].

Tropical forests are a particularly important component in the terrestrial C budget and even small changes in tropical CO_2 fluxes may modify the global C cycle [11–14]. With a share of about 55% to 76% of total soil CO_2 efflux from tropical soils [6,15,16] microbial-driven heterotrophic soil respiration is a critical CO_2 flux to the atmosphere in tropical ecosystems. Microbial hetero-trophs in the litter layer and the underlying soil are highly responsive to altered nutrient availability (*e.g.* [17–20]). N and P inputs in particular modify the soil C:N:P stoichiometry and that of plant residues, which in turn affect decomposer activity and growth [21,22], and the processes of litter decomposition and organic matter mineralization [23,24]. In addition to nutrient deposition, global change-induced shifts in plant tissue C quality (*e.g.* secondary metabolites, non-structural carbohydrates) [25–27] may also affect heterotrophic soil organisms. Such C-quality changes may have important consequences in some tropical forests where the poor C quality of leaf litter has been proposed to impose energy starvation on decomposers [28].

External resource supply through fertilization provides a straightforward experimental test of ecosystem nutrient limitation and the response of ecosystem processes to altered resource availability [10,29,30]. Despite a large diversity in geology, soil characteristics, climatic factors and biological diversity of tropical rainforests [31,32], only a relatively small number of fertilization experiments have been performed in this biome. Some of these experiments reported positive effects of P fertilization on decomposition and on CO_2 release into the atmosphere [33–

36], while others have found an increase of soil C stocks associated with lower soil respiration following N additions [37–39], or contrasting effects of these resources as a result of site age related soil fertility gradients [40–42]. Additionally, studies that included a combination of different resources in their fertilization protocol observed interacting effects between resources such as N×P on litter mass loss or microbial activity during decomposition [41–43], suggesting that some limiting resources can influence the fate and the impact of other resources on soil processes [29,30]. However, only few studies have simultaneously manipulated the availability of all three key elements C, N, and P and none of them were performed in a lowland tropical forest. By adding a labile C source (*i.e.* cane sugar, glucose monohydrate) in factorial CNP fertilization designs, important and significant interactions between C×P or C×N on microbial responses and soil C dynamics have been reported in a tropical montane forest of Ecuador [44] and along a successional gradient in a temperate system [45]. Although these studies provide clear evidence of interactions between C and nutrients, sugars used as C fertilization may produce immediate responses by favoring opportunistic soil organisms. Other, more complex, but still relatively easily accessible C sources - such as cellulose - should be tested to provide a more detailed understanding of the potential regulation of soil processes by multiple elements.

The distinction between microbial processes in the litter layer and the underlying soil are rarely made clearly [46,47], and the effects of resource addition on litter and soil heterotrophs are seldom addressed in the same study [17]. Although the litter layer and the underlying soil are intimately connected through the exchange of energy and matter, microclimatic and physical conditions as well as the chemical composition, with notably a stark contrast in organic matter quality and C:N:P stoichiometry, differ strongly. Leaf litter material display much wider C to nutrient ratios as well as distinct C quality compared to that of soil organic matter, with especially more labile compounds in freshly fallen leaf litter. These different qualities of organic substrates available for microbial heterotrophs might result in distinct constraints for litter and soil communities. In addition, the very high tree species richness, typical for most tropical forests, results in chemically diverse leaf litter inputs at small spatial scales [48,49]. These distinct litter substrates decompose at different rates [50,51] and affect the respiration and structure of microbial communities in the underlying soil differently [52,53]. It is important to account for this chemical heterogeneity of tropical leaf litter when assessing the effects of increased resource supply, because nutrient enrichment effects are likely to be modified by differences in initial litter quality. In fact, in a companion paper, Barantal and coll. [43] demonstrated that combined fertilization with N and P increasingly stimulated leaf litter decomposition with decreasing initial litter P concentration and increasing initial litter N:P ratios. Moreover, these positive NP fertilization effects were enhanced when soil fauna had access to decomposing litter [43]. It is widely accepted that soil fauna are important decomposers in tropical rainforests [51,54–57], and McGlynn and coll. [58] showed that soil C:P stoichiometry controls soil fauna abundance in a Costa Rican rainforest. However, the importance of litter identity in the response of decomposition and associated microbial processes (especially in the underlying soil) to fertilization, as well as the role of fauna in modulating this response, remains little explored.

In this study, we addressed the question of how multiple resource fertilizations affect heterotrophic processes within decomposing leaf litter and in the underlying soil, and investigated how these effects are influenced by species differences in litter quality and the presence of soil fauna. Data were collected from an ongoing fertilization experiment in a low-fertile lowland rainforest in French Guiana where C (cellulose), N (urea) and P (phosphate) are added in a fully factorial fertilization experiment since 2009 [43]. We specifically addressed the following hypotheses: (i) external supply of readily available C, N and P alleviates resource limitation and consequently stimulates the overall microbial capacity (estimated by substrate induced respiration, SIR); (ii) the stimulating effects of external resource supply increase with decreasing initial litter quality; (iii) the previously reported fauna-induced stimulation of fertilization effects on decomposition [43] translates into increased consumption of microorganisms by litter-feeding fauna and consequently decreases microbial respiration; (iv) the response of SIR to external resources differ between litter and the underlying soil, with stronger effects of nutrient addition on litter SIR because soil organic matter has a lower C to nutrient ratio compared to leaf litter, and stronger effects of C fertilization on soil microbial respiration compared to litter because litter is richer in labile C substrates than soil.

Materials and Methods

All necessary permits were obtained for the described field studies (fertilization), and no specific permits were required for the described measurements in the field (sampling of soil and leaf litter), in agreement with the owner, the French research center CIRAD. We confirm that the field studies did not involve endangered or protected species.

Study site

The study site is located within the undisturbed Amazonian rainforest of Paracou near Sinnamary, French Guiana (5°15′N, 53°′W). The mean annual air temperature is 25.5°C (10-year average, 1995–2005) with only slight intra annual variations. Total annual rainfall is approximately 2575 mm (10-year average, 1995–2005), with two distinct rainy seasons (a moderate one from December to February and a stronger one from April to July) with an associated range in relative air humidity between 70 and 90% [59]. Tree species richness is around 150 species per hectare with a mean density of 620 individual trees ha^{-1} (individuals of a diameter >0.1 m at breast height) [60]. Soils in the study area are classified as acrisol, developed over a Precambrian metamorphic formation called the Bonidoro series. The soil is nutrient-poor with 24% clay, 7% silt and 69% sand, and a pH (water extract) of 4.7 in the top 0.2 m [57]. Average soil C:N is 14.7 with a total C of 2.21 g kg^{-1}, a total N of 0.15 g kg^{-1} soil and a total P of 0.010 g kg^{-1} soil (for more details on soil composition and texture see [53]).

Plant material

For the construction of litterbags we used leaf litter from the six tree species *Carapa procera* (Aublet), *Goupia glabra* (Aublet), *Platonia insignis* (Martius), *Hymenaea courbaril* (Linnaeus), *Simarouba amara* (Aublet) and *Vochysia tomentosa* (G. Mey.) (Table 1). A representative pool of fresh fallen leaf litter of each species was obtained from a tree plantation close to our study site. These more than 25-year-old tree stands have been established using local seed sources and have a fully closed canopy composed of about 40 individuals of each of a total of 16 tree species growing in monocultures [61]. Litter was collected twice a month during the year 2009 in suspended 25 m^2 litter traps and pooled across sampling dates. Leaves with obvious signs of damage (*e.g.* herbivory, galls, fungal attacks) and green leaves were excluded (typically <15% of total collected leaves). Leaf litter was air-dried, weighed (8.0±0.1 g air-dry to oven-dry corrected mass per litterbag) and enclosed in plastic mesh bags for each species individually. We used coarse-mesh (8 mm) and fine-mesh (0.06 mm) bags in order to allow or not the access of soil and litter macrofauna. The initial quality of pooled leaf litter differed significantly among the six species (Table 1). For example, the C:N ratio varied between 34.5 (*P. insignis*) and 51.5 (*C. procera*), and the N:P ratio varied between 21.8 (*H. courbaril*) and 78.9 (*P. insignis*).

Experimental design

A full-factorial fertilization experiment (control, C, N, P, CN, CP, NP, CNP) plus one additional fertilization treatment (called +other nut. throughout the paper) with major cations (K, Ca, Mg) and micronutrients (i.e. B, Cu, Fe, Mn, Mo, S, Zn) was set up in the field using a total of five blocks. Each of the five blocks measured approximately 3000 m^2 and was situated within a 2.5 ha zone of rather homogeneous flat topography. Each of the nine treatment plots within blocks measured 5.5 m×5.5 m and was separated from neighbor plots by a buffer zone of at least 5 m. Fertilization was applied twice a year during the two dry periods in order to limit potential wash-off of fertilizer just after application. The fertilization was started in April 2009 and is ongoing since then. Based on preliminary microcosm tests of different fertilizer concentrations (Barantal, *unpublished data*) and the concentrations used in other tropical fertilization experiments [34,36,41], we used the annual doses of 1405 kg C ha^{-1} year^{-1} provided as cellulose (Waterspare, celliob industry, France), 130 kg N ha^{-1} year^{-1} as coated urea [(NH$_2$)$_2$CO] and 69 kg P ha^{-1} year^{-1} as mono-potassium phosphate [KH$_2$PO$_4$] corresponding to C:N of 10.8, C:P of 20.4 and N:P of 1.9. The cations and micronutrients in the +other nut. treatment was equivalent to 22 kg ha^{-1} year^{-1} of a mixture of H$_3$BO$_3$ (1150 ppm), CuSO$_4$ (1150 ppm), Fe-EDTA (2%), MnSO$_4$ (1150 ppm), ZnSO$_4$ (600 ppm) and (NH$_4$)$_2$MoO$_4$ (600 ppm), plus 87 kg K ha^{-1} year^{-1} as K$_2$SO$_4$, 92 kg Mg ha^{-1} year^{-1} as MgSO$_4$, and 50 kg Ca ha^{-1} year^{-1}as Ca-EDTA. Twelve 15 cm×15 cm large litterbags (6 litter species×2 mesh

sizes) were randomly placed directly on the soil surface (natural litter was removed prior to litterbag placement), fixed on the forest floor with wire and exposed in each of the 45 plots for a total of five months from September 2009 (just before the second fertilization event) to February 2010.

Sample collection

After 158 days of exposure in the field, the litterbags were retrieved and the underlying soil underneath each litterbag was collected, resulting in a total of 540 pairs of litterbag-soil samples (5 blocks×9 treatments×6 species×2 mesh size). The underlying soil was sampled in the center of the litterbag using a stainless steel cylinder (diameter of 5 cm) to a depth of 8 cm. All sampling was done from 9th to 14th February 2010 during the wet season, approximately two months after peak litter fall [59]. In the laboratory, litter from the litterbags was weighed for total fresh mass and an aliquot (2 g fresh weight) was dried at 65°C to determine litter dry mass and litter mass loss. The remaining litter material of each litterbag was air-dried and stored dry until further analyses. Soil samples were air-dried, passed through a 2 mm sieve to remove roots and stones, homogenized and stored dry until further analyses.

Determination of soil and litter SIR

Substrate induced respiration, SIR, as a measurement of potential activity, encompasses several aspects of the microbial community, and is often used as a proxy of the soil respiration process [62]. It was used as an indicator of the overall capacity of the litter and soil microbial communities [53]. Soil SIR was measured according to Beare and coll. [63]. For each sample, 10 g of soil (dry weight) were placed in a sealed plasma flask of 150 ml. A solution of glucose (1.5 mg C g^{-1} of dry soil) was added to reach 80% of field capacity. The flasks were incubated at 25°C for 6 h, a time span that is considered short enough to avoid *de novo* enzyme synthesis. Two hundred μl air samples from the headspace of each flask were analyzed for CO$_2$ concentration after 2 and 6 h

Table 1. Initial litter quality parameters measured for leaf litter from the six different tree species used in our study.

Litter characteristics[+]	C. procera	G. glabra	H. courbaril	P. insignis	S. amara	V. tomentosa
Litter elements (%DM)						
Carbon	48.4±0.2	49.7±0.2	49.7±0.1	49.0±0.2	49.1±0.1	42.9±0.4
Nitrogen	0.94±0.04	1.21±0.13	1.22±0.03	1.42±0.03	1.11±0.07	0.87±0.04
Phosphorus	0.019±0.012	0.033±0.004	0.056±0.002	0.018±0.001	0.032±0.002	0.029±0.001
Litter stoichiometry						
C:N	51.5±2.1	41.1±4.1	40.7±1.1	34.5±0.6	44.2±2.9	49.3±2.9
C:P	2547±147	1507±168	888±36	2722±154	1534±111	1479±87
N:P	49.5±2.9	36.7±1.5	21.8±0.4	78.9±5.6	34.7±0.7	30±2.9
Carbon compounds (%DM)						
Dissolved organic carbon	0.59±0.09	1.93±0.24	0.56±0.02	1.46±0.16	1.07±0.07	0.74±0.03
Water soluble compounds	32.4±0.3	36.6±0.4	31.0±1.0	29.3±0.3	45.4±0.4	34.6±1.1
Hemicellulose	7.5±0.5	16.2±0.7	10.3±0.1	23.5±0.7	11.7±0.2	20.1±1.1
Cellulose	22.7±0.4	18.8±0.3	22.3±0.6	22.5±0.7	20.0±0.3	19.7±0.4
Lignin	37.5±0.5	28.4±0.8	36.3±0.7	24.7±1.1	22.8±0.7	25.6±0.4
Soluble phenolics	2.8±0.2	1.1±0.2	1.0±0.1	1.0±0.1	4.4±0.2	0.6±0.1
Total phenolics	7.9±0.8	2.8±0.3	4.2±0.4	12.5±0.5	11.0±0.8	4.4±0.4
Condensed tannin	7.7±0.7	0.6±0.1	3.8±0.4	0.4±0.1	6.3±0.3	3.9±0.3

incubation with a gas chromatograph using a microcatharometer (VARIAN GC 4900; Varian, Walnut Creek, USA). From the amount of CO_2 released during this time we calculated SIR expressed in μg of C-CO_2 per g of soil per hour. Litter SIR was measured in the same way with the exception that we used 2 g of litter material (dry weight) and 2 ml of a solution of glucose to supply 20 mg C g^{-1} of dry litter mass. For some of the most rapidly decomposing litter types (19% of all litter samples collected in the field) there was not enough litter material left for these measurements, but each combination (litter species×fertilization treatment) was replicated at least three times, allowing robust statistical analyses.

Data analysis

Normality of the distribution of data was assessed for all variables using Shapiro-Wilkinson's test and the homogeneity of variance using the Fisher (F) test. When data were not normally distributed, transformations of variables were performed in order to meet the assumptions before any further statistical tests. In particular, litter mass loss, litter SIR and soil SIR rates were log-transformed.

The effect of species and mesh size on litter mass loss, litter SIR and soil SIR without fertilization was assessed with linear mixed models, LMM, in control plots only (using the "nlme" R package [64,65]). Blocks were considered as a random factor while litter species, mesh size and their interactions were fixed factors. To evaluate the relationship in control plots between litter species-specific initial quality and litter mass loss, litter SIR and soil SIR, we performed stepwise regression to select the best litter quality predictor when soil fauna were included or not. We divided the data into two sets based on mesh size before running the statistical analysis. The results of stepwise regression should be interpreted with caution because this method leads to several biases such as errors in parameter estimation, inconsistencies among model selection algorithms or reliance on a single best model [66].

We analyzed the effect of fertilization in two steps. First, to test for the effect of any of the major resources C, N or P added, the effect of fertilization was assessed with full factorial LMM (for these tests the +other nut. treatment was excluded in order to keep a balanced design). In these analyses, we compared all plots receiving C-, N- or P- fertilization and all plots with no addition of this particular resource (C, N or P presence/absence in each combination). Blocks were considered as random factor while C, N, P supply, litter species, mesh size and their interactions were fixed factors. Second, to test for the effect of each external resource singly or in combination with each other, a "net fertilization effect" was calculated within each block as the difference for response variables between the plot receiving a given fertilization treatment and the control plot. A positive net fertilization effect denoted higher mass loss or SIR with fertilization. When significant effects were found, we ran post-hoc means separation tests using Tukey-HSD ($\alpha = 0.05$).

Mathematical correlations between litter mass loss and litter and soil SIR rates were explored with simple linear or non-linear regressions. Regression analyses were also used to assess potential relationships between the "net fertilization effect" and initial litter quality in order to evaluate whether the effect size depended on specific initial litter quality traits. Levels of significance are indicated as * ($p<0.05$), ** ($p<0.001$), and *** ($p<0.0001$). All statistical tests were performed with the R software (version 2.11.1).

Results

Litter mass loss and SIR without fertilization

In the unfertilized control plots we observed a mesh size effect on litter mass loss, but not on litter SIR and soil SIR (Table 2). Fauna access to litterbags increased mass loss, and this effect depended on litter species identity (significant species×mesh size interaction, Table 2). Fauna access also led to variation in litter mass loss (from 30.3% in *C. procera* to 68.3% in *G. glabra*), and to a higher variation (CV = 30%) compared to small mesh width litterbags (range between 23.5% in *V. tomentosa* and 37.2% in *H. courbaril* with a CV of 12%, Table 3, Table S1).

Similar to litter mass loss, litter SIR and soil SIR were also significantly different among litter species. Litter SIR was highest in decomposing *S. amara* litter (23.3 μg g^{-1} h^{-1}) and lowest in *V. tomentosa* litter (12.9 μg g^{-1} h^{-1}). In contrast, soil SIR was highest underneath *G. glabra* litter (1.83 μg g^{-1} h^{-1}) and lowest underneath *H. courbaril* litter (1.03 μg g^{-1} h^{-1}). However, in contrast to litter mass loss, litter SIR and soil SIR showed no significant mesh size×litter species interaction (Table 2).

The observed litter species effects on litter mass loss and SIR were related to initial litter carbon quality (Table 1, 3). The best predictor for litter mass loss when fauna had access to litterbags was the initial concentration of dissolved organic carbon (DOC) in leaf litter with increasing mass loss when DOC concentrations increased ($r^2 = 0.89$, $p = 0.05$). In contrast, the concentration of condensed tannins (CT) showed a negative correlation with litter mass loss when fauna was present ($r^2 = 0.74$, $p = 0.02$). Similar to litter mass loss, litter SIR showed a trend for a positive correlation with initial litter DOC in the presence of fauna ($r^2 = 0.58$, $p = 0.07$), while soil SIR tended to correlate negatively with initial litter lignin content ($r^2 = 0.59$, $p = 0.07$). Without macrofauna, litter mass loss correlated best and positively with initial concentrations of total carbon ($r^2 = 0.76$, $p = 0.02$) (Table 3). The same trend was found for litter SIR, while soil SIR without fauna access to litterbags was best predicted with the initial litter DOC concentration (increasing soil SIR with increasing DOC, $r^2 = 0.69$, $p = 0.04$).

Fertilization effects

In a first analysis of fertilization effects we identified how absolute litter mass loss and rates of litter and soil SIR differed with C, N, and P amendment compared to when these respective fertilizers were not added (*e.g.* all plots receiving C compared to all plots without C addition). The significant effects of mesh size and litter species identity on litter mass loss reported in control plots persisted in fertilized plots and explained a higher amount of variation in mass loss compared to that of C, N and P supply (Table 4). On average, N and P fertilization increased litter mass loss by 17% and by 12%, respectively (Figure 1). In contrast, C fertilization showed no significant effect on litter mass loss with a trend for negative effects (Figure 1, 2, Table 4). The effects of N and P fertilization both depended on litter species (significant N and P×litter species interactions, Table 4). The P fertilization was further influenced by mesh size with a stronger effect when fauna had access (significant P×mesh size interaction).

The litter species-specific differences in litter SIR remained essentially the same across fertilization treatments as those observed in control plots. However the mesh size effect was significant when fertilized plots were included in the analysis (Table 4). With fauna access, litter SIR was on average 17% lower than that measured in litter without fauna access. N and P fertilization explained a higher amount of variation in litter SIR than mesh size and litter species identity (Table 4). Overall, N and

Table 2. Results of mixed linear models to test for the effects of litterbag mesh size and litter species identity on (a) litter mass loss, (b) SIR litter and (c) soil SIR within control plots only (no fertilization).

(a) Litter mass loss	Num. d.f.†	Den. d.f.†	F value	p-value
mesh size	1	43	**19.1**	**<0.0001**
species	5	43	**5.9**	**0.0003**
mesh size × species	5	43	**4.1**	**0.004**
(b) SIR litter	Num. d.f.	Den. d.f.	F value	p-value
mesh size	1	34	2.0	0.17
species	5	34	**14.7**	**<0.0001**
mesh size × species	5	34	1.3	0.28
(c) SIR soil	Num. d.f.	Den. d.f.	F value	p-value
mesh size	1	44	0.6	0.43
species	5	44	**2.9**	**0.02**
mesh size × species	5	44	1.5	0.20

†Num d.f., numerator degrees of freedom; Den d.f., denominator degrees of freedom.

P fertilization increased litter SIR by 29% and by 20%, respectively (Figure 1). Carbon fertilization also significantly increased litter SIR by 6%. The positive C effect, however, was influenced by mesh size with a weaker C fertilization effect when fauna had access to the litter. Moreover, a positive interaction was observed between C and N addition and N and P addition (higher litter SIR when C or P was added with N simultaneously).

The significant litter species effect on soil SIR observed in control plots (Table 2) disappeared with fertilization, and mesh size still had no significant effect on soil SIR (Table 4). While the addition of N and P and their interaction significantly changed soil SIR, C addition had no impact (Figure 1, Table 4). On average, N fertilization decreased soil SIR by 15%. In contrast, P fertilization showed an average increase of 16% (Figure 1). Moreover, a negative interaction between N and P additions was observed (on average lower SIR than P alone when N was added in combination to P).

Treatment specific net fertilization effects

In a second analysis we explored in more detail how litter mass loss and rates of litter and soil SIR changed in the eight different fertilization treatments compared to the control treatment (net fertilization effect = absolute difference between treatment and control). Litter mass loss and litter SIR both showed the highest net fertilization effects with a combined addition of N and P supply (Figure 2). These net NP fertilization effects were highest for mass loss when fauna had access (on average 35% higher than in control plots), and highest for litter SIR when fauna was excluded (on average 96% higher than in control plots). We observed broadly similar patterns for the net effects of the different fertilizer combinations on litter mass loss and litter SIR (Figure 2). However, the net fertilization effects on litter SIR were stronger than those on litter mass loss. Also, the presence of fauna tended to decrease litter SIR, and to rather increase litter mass loss, respectively. As a result, the overall positive relationship between litter SIR and litter mass loss across all fertilization treatments when fauna was absent disappeared in the presence of fauna (Figure S1).

Table 3. Means (± SE) and CV (in %) of litter mass loss, litter SIR, and soil SIR measured in control plots (no fertilization) with or without fauna access.

Variable	Mean	CV	Best Predictor	Effect	r^2	p-value
With fauna access						
Litter mass loss	44.2±13.9	30	Dissolved Organic Carbon	+	**0.89**	**0.005**
			Condensed Tannins	−	**0.74**	**0.02**
SIR litter	18.2±3.9	23	Dissolved Organic Carbon	+	0.58	0.07
SIR soil	1.44±0.72	47	Lignin	−	0.59	0.07
Without fauna						
Litter mass loss	31.6±3.9	12	Total Carbon	+	**0.76**	**0.02**
SIR litter	19.5±4.3	21	Total Carbon	+	0.49	0.12
SIR soil	1.33±0.76	59	Dissolved Organic Carbon	+	**0.69**	**0.04**

When several litter quality traits significantly explained litter mass loss or SIR, all the corresponding models from stepwise regression analysis are displayed (in bold). When no litter trait significantly explained the variable ($p>0.05$), the best model is shown.

Table 4. Results from mixed linear models to test for the effects of fertilization (addition or not of either one of C, N, and P), litterbag mesh size, litter species identity, and their interactions on (a) litter mass loss, (b) litter SIR and (c) soil SIR.

(a) Litter mass loss	Num. d.f.†	Den. d.f.†	F value	p-value
C (Carbon)	1	436	1.7	0.19
N (Nitrogen)	1	436	35.5	<0.0001
P (Phosphorus)	1	436	47.6	<0.0001
mesh size	1	436	87.3	<0.0001
species	5	436	69.8	<0.0001
mesh size × species	5	436	26.6	<0.0001
P × mesh size	1	436	5.2	0.023
N × species	5	436	2.6	0.026
P × species	5	436	4.1	0.0012
(b) SIR litter	**Num. d.f.**	**Den. d.f.**	*F value*	*p-value*
C	1	358	13.6	0.0003
N	1	358	195.1	<0.0001
P	1	358	64.2	<0.0001
mesh size	1	358	24.4	<0.0001
species	5	358	62.6	<0.0001
C×N	1	358	4.89	0.028
N×P	1	358	4.5	0.034
mesh size × species	5	358	4.9	0.0002
P × species	5	358	3.6	0.0037
C × mesh size	1	358	5.3	0.022
(c) SIR soil	**Num. d.f.**	**Den. d.f.**	*F value*	*p-value*
C	1	461	1.1	0.30
N	1	461	21.0	<0.0001
P	1	461	20.9	<0.0001
mesh size	1	461	0.01	0.93
species	5	461	1.6	0.17
N×P	1	461	5.0	0.026

Only significant interaction terms are shown.
†Num d.f., numerator degrees of freedom; Den d.f., denominator degrees of freedom.

Soil SIR responded distinctly to fertilization compared to litter SIR or litter mass loss (Figure 2). Most fertilization treatments showed no significant net effect on soil SIR rates, notably the combined addition of N and P that induced the strongest response on litter mass loss and litter SIR. The combined C and P fertilization was the only treatment showing a positive net effect on soil SIR when fauna was excluded from the litterbags (on average 46% higher than in control plots). This CP fertilization effect, however, was not statistically significant when fauna had access to the litterbags on top of the sampled soil. The net fertilization effects on soil SIR were negative when plots were either fertilized with CN or with cations and micronutrients (Figure 2). Soil SIR showed no correlation with litter SIR or litter mass loss.

Litter species-specific responses to fertilization

The effects of nutrient fertilization on litter mass loss and litter SIR differed among litter species (significant nutrient×litter species interactions, Table 4) apparently as a result of distinct initial litter quality. For example, N fertilization effects increased with decreasing litter initial N concentration. Likewise, the P fertilization effect was particularly strong in litter of low initial P

concentrations (*e.g. P. insignis*). Interestingly though, N and P fertilization effects on litter SIR were strongest in litter species with the highest initial DOC concentrations, *i.e.* in *G. glabra* (+37.9% litter SIR) and *S. amara* (+30.7% litter SIR) for N addition and in *P. insignis* (+34.2% litter SIR) and *G. glabra* (+26.2% litter SIR) for P addition. The strong effect of combined N and P fertilization observed for litter mass loss (Figure 2) depended on initial litter P concentrations, *i.e.* the net NP fertilization effect increased with decreasing litter P concentration. In contrast, the NP fertilization effect on litter SIR correlated best with initial concentrations of DOC, *i.e.* the net NP fertilization effect increased with increasing litter DOC concentration (Figure 3). This relationship was positive independently of the mesh-size. Additional fertilization with cellulose did not change these relationships between initial litter quality and the net NP fertilization effects.

Discussion

Decomposition and litter SIR in response to fertilization

In our first hypothesis we stated that in the studied low fertile Amazonian rainforest, an increased availability of the key

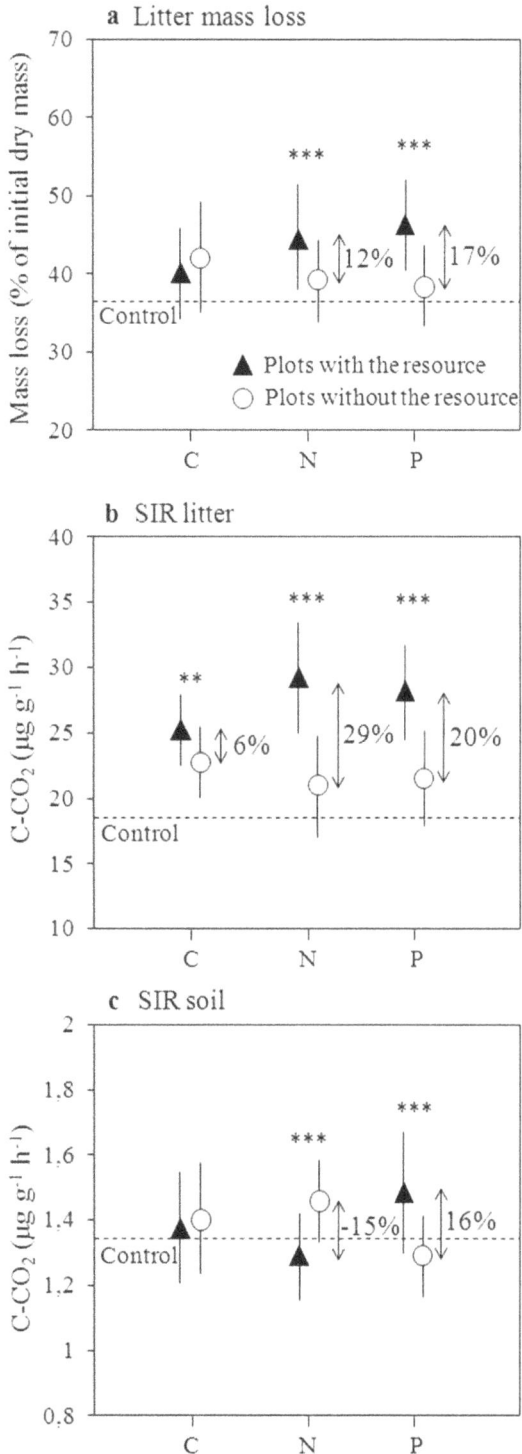

Figure 1. Effects of C, N, and P fertilization (alone or in any combination with the other resources) on (a) litter mass loss, (b) litter SIR and (c) soil SIR, without distinction of litter species and mesh size. These effects were analyzed using linear mixed models (dashed lines indicate the mean values of control plots). Black triangles represent the mean values (± SE) for all plots receiving C, N or P fertilization, and open circles the values for all plots receiving no addition of C, N or P, respectively (*e.g.* C, CN, CP and CNP vs control, N, P and NP for the C resource). Stars denote significant differences between plots with or without the addition of C, N or P as follows: * ($p<0.05$), ** ($p<0.01$), *** ($p<0.001$).

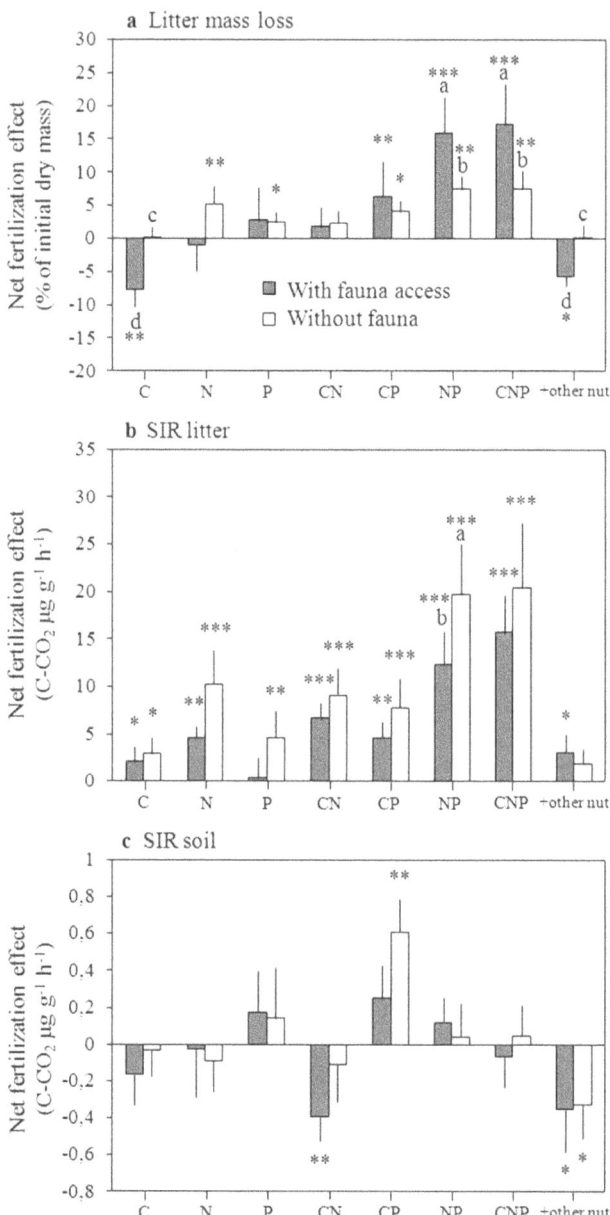

Figure 2. Net fertilization effects (mean ± SE) on (a) litter mass loss, (b) litter SIR and (c) soil SIR. Net fertilization effects are defined as the absolute difference between values measured on control plots and those measured on the plots of the respective fertilization treatment. Gray bars represent treatments with fauna access (coarse mesh litterbags) and open bars represent treatments without fauna access (fine mesh litterbags). Different letters indicate significant differences between coarse and fine mesh litterbags for a given treatment. Stars denote net treatment effects that are significantly different from zero using paired Student's *t* tests: * ($p<0.05$), ** ($p<0.001$), *** ($p<0.0001$).

resources C, N and P should increase leaf litter SIR in parallel to faster decomposition. We tested this hypothesis with a fully factorial fertilization experiment that was the first to our knowledge to use cellulose addition, a less labile C form than the commonly used highly labile sugars. In support of our hypothesis we found that litter mass loss and litter SIR both increased on P- and N-fertilized plots compared to plots that were not amended with P or N (Figure 1). Contrary to the predicted

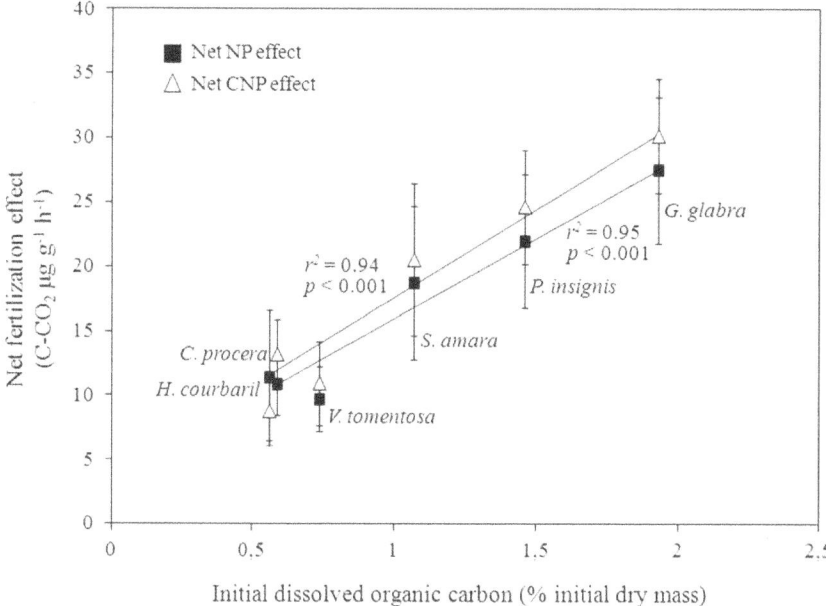

Figure 3. Net effects of NP (black squares) and CNP (open triangles) fertilization (mean ± SE) on litter SIR (data pooled across mesh-size) as a function of the initial litter species-specific DOC concentration. Net fertilization effects are defined as the absolute difference between values measured on control plots and those measured on the plots of the respective fertilization treatment.

strong effect of P fertilization at our study site of particularly low soil P availability, our results suggest that both P and N limit litter microbial decomposers and decomposition simultaneously. In fact, the net fertilization effect of a combined P and N addition was clearly stronger than the effects of separate P or N fertilization, especially for SIR. The microbial communities in decomposing litter respired on average 85% more with a combined N and P supply compared to an increase of 11% and 31% with P and N supplied singly, respectively. Except for a small positive effect on litter SIR, C fertilization showed little effects. Litter mass loss was actually slower in plots fertilized with just C compared to control plots. Barantal and coll. [43] previously argued that decomposers might prefer cellulose to leaf litter that contains large quantities of recalcitrant C compounds. Such cellulose preference may explain the somewhat higher litter SIR and slower litter decomposition with cellulose fertilization. However, we can not exclude the possibility that this potential initial cellulose preference followed by enhanced litter SIR may lead to an increased consumption of litter C *via* a priming effect in the longer term [67,68].

Our results are in line with those from previous studies in montane forests of Hawai'i showing that N and P together can constrain litter mass loss and microbial functioning during decomposition [41,42], and support the increasing evidence that ecosystem processes are more often than not co-limited by N and P [10,69]. Mineralization and acquisition of N from decomposing leaf litter material requires the breakdown of the C skeleton of rather complex organic compounds [70,71]. In contrast, P is less strongly bound and may be lost from decomposing leaf litter at higher rates than N [72]. In addition, more than half of the total litter P may be readily available mineral phosphate in contrast to organic N that dominates the total litter N pool [49]. Therefore, P is more easily accessible than N in the early stages of litter decomposition, and N may initially be relatively more limiting. The relative importance of N availability should shift with increasing age and decreasing C:N ratios of organic matter. Accordingly, N fertilization is expected to have less of an effect

within the soil than P, which is in line with the observed positive P and a negative N effect on soil SIR in our study. A second, not mutually exclusive, explanation is that bacterial and fungal communities differ in their resource limitation. Indeed, in experimental manipulations of C (as glucose), N and P in a tropical montane rainforest in southern Ecuador, Krashevska and coll. [44] showed that fungi predominantly responded to N whereas bacteria responded to P. Consequently, such differences in primary limitation between fungi and bacteria may also explain the particularly strong effect on microbial functioning in the litter with a combined addition of N and P that stimulates both fungal and bacterial communities.

Litter species-specific resource limitation

Large variation in green leaf quality and stoichiometry was observed among tropical tree species at regional scales [73], and a similar large variation in litter quality and stoichiometry has also been documented at small local scales [48,50]. Such high interspecific variation in litter quality results in a spatially highly variable organic matter input to the soil, creating a mosaic of diverse resources for heterotrophic microbial communities [52,53]. This variability in chemical quality of tree leaf litter was taken into account here by selecting litter from six tree species with contrasting stoichiometry and C quality. According to our second hypothesis we expected the relative effect of external resource supply to increase with decreasing initial litter quality.

In the litter layer, N and P fertilization interacted with litter species identity, suggesting that the response to nutrient addition were dependent on initial litter quality (Table 3). With a combined N and P fertilization, litter mass loss correlated negatively with initial litter P concentration and with decreasing initial litter N:P ratios. Such relationship was also observed across various mixtures of the studied six litter species [43]. This influence of initial litter P status suggests a pivotal role of litter P availability in determining the strength of NP fertilization effects on litter decomposition. On the other hand, litter SIR correlated positively with initial litter

DOC concentration under a combined N and P fertilization, indicating that microbial activity increased more with NP fertilization in litter with more labile C substrates (Figure 3). In a different NP fertilization experiment in a Costa Rican rainforest, the NP effect was strongest at the beginning of the wet season, when labile C content was maximal in litter leachates [34]. The results from the Costa Rican study and our own study both suggest that labile C compounds in leaf litter provide the microorganisms with the required energy to efficiently use external nutrients. Since the DOC from the litter used in our study should be available, particularly at the beginning of litter decay, the persistent interactive effect with fertilization after 158 days of litter exposure in the field may suggest that DOC primed litter SIR responses to increased nutrient availability. Apparently, the addition of cellulose had a different effect compared to litter inherent DOC. Cellulose decomposition requires particular enzymatic activities while DOC is a cocktail of various and easily accessible C-compounds that are likely used by a more diverse microbial community and may be also more quickly mineralized by opportunistic microorganisms. In that sense, litter inherent DOC may have similar effects like sugar fertilization [44–45], confirming the importance of the C quality in determining heterotrophic responses to C additions [74].

Nutrient fertilization has previously been shown to stimulate mass loss of litter from one tree species in a lowland Costa Rican rainforest [34] and of the original site-specific litter mixture from the fertilized plots in montane forests of Hawai'i [41] and in a lowland Panamanian rainforest [36]. Here we additionally highlighted that fertilization effects on decomposers depended on litter species-specific initial quality at our study site in an Amazonian rainforest. These results underline the importance of tree species-specific litter input to the forest floor for the understanding of how decomposers respond to changes in external N and P availability. Consequently, potential shifts in tree species composition and/or losses of tree species diversity in the Amazon [75,76], global change induced changes in litter quality [27] and changes in anthropogenic nutrient inputs [9,77] may interactively affect decomposer communities, litter decomposition and organic matter turnover.

Fauna effect on heterotrophic microbial functioning

The contribution of fauna, especially that of macrofauna (*e.g.* millipedes, isopods, termites), to the decomposition process in tropical wet forests is disproportionately higher compared to forests ecosystems at higher latitudes [57]. The fauna impact on decomposition was shown to be influenced by litter stoichiometry in the rainforest of French Guiana [72] and the abundance and composition of soil fauna communities depended on soil C:P stoichiometry in a tropical Costa-Rican rainforest [58]. Accordingly, changes in the relative availability of nutrients and/or of substrate C quality are likely to affect the composition and activity of fauna communities, with potential indirect effects on heterotrophic microbial functioning as well. In our third hypothesis, we expected that increased fauna activity with fertilization would decrease SIR rates as a result of increased predation on microorganisms by litter-feeding fauna.

In line with our hypothesis we observed that fauna increased and decreased the positive NP effect on litter mass loss and on litter SIR, respectively (Figure 2). The stronger net NP effect on litter mass loss in presence of fauna may suggest enhanced litter-feeding with a higher availability of nutrients. Alternatively, this fauna response may result from intensified detritivore foraging on litter that is more heavily colonized by microorganisms. This hypothesis is supported by the higher SIR rates measured in litter

fertilized with NP in the absence of fauna. Higher detritivore feeding may then have reduced microbial biomass by direct consumption and indirect physical disruption of the microbial communities, possibly explaining the lower NP effect on SIR in presence of fauna.

Are heterotrophic processes in the litter layer and underlying soil distinctively affected by fertilization?

We stated in our fourth hypothesis that nutrient fertilization stimulates the SIR rates more in litter than in the soil because of wider C:nutrient stoichiometries in litter compared to soil. We actually observed an overall positive effect of N fertilization on litter SIR, but a negative N effect on soil SIR (Figure 1) that is broadly in agreement with our hypothesis. Such negative N fertilization effect was not associated with potential fertilizer induced changes in soil pH (*data not shown*). These contrasting effects of N fertilization in litter and soil are in line with Berg & Matzner's [2] reasoning of distinct effects of N fertilization on organic matter breakdown depending on the stage of decomposition. During the initial stage of decomposition when mostly soluble compounds and cellulose are broken down, N fertilization should have positive effects and during later stages of decomposition when more recalcitrant lignin-like compounds dominate the remaining organic matter, N fertilization is expected to have rather negative effects [2,68]. Accordingly, Neff and coll. [78] showed that labile C fractions, present during the early stages of decomposition, are consumed more rapidly when N is added. Suppression of soil respiration in tropical forests following N fertilization has been repeatedly reported [37–39], but the mechanisms underlying this response are not yet clarified. Slower decomposition of organic matter *via* the decrease of oxidative enzyme production [38,79,80], decrease in labile C pools [39], inhibition of microbial biomass [37,81], or changes in microbial community structure [82] have been proposed as potential mechanisms.

In contrast to N fertilization, P fertilization stimulated litter SIR (+20%) and soil SIR (+16%, Figure 1) in similar ways. A positive P effect on soil microbial activity was expected at our site with P-poor soils indicating P-deficient conditions [28]. Similarly, Cleveland and coll. [33] reported that P availability constrained the total respiratory CO_2-flux in a Costa Rican tropical forest, and that P fertilization increased the proportion of added dissolved organic matter that was converted to CO_2 [34]. Much of the positive P effect on soil SIR observed here was driven by the combined fertilization with P and C (Figure 2) resulting in an overall higher stimulation of soil SIR than when fertilized with P only. This positive interaction of a combined P and C fertilization was the only indication of a stimulating C effect on soil microbial activity as we initially hypothesized. The positive response in SIR may indicate that soil microorganisms are simultaneously limited by low soil P and by the access to labile C. Collectively, our data suggest that altered nutrient inputs in the studied Amazonian rainforest distinctly affect decomposer communities in the litter layer and the underlying soil with contrasting effects on organic matter turnover that is further modified by the quality of organic C sources.

Conclusions

Taken together, our data show that increasing inputs of N and P, and in particular of both of them together can considerably change microbial activity and litter decomposition in a low fertile Amazonian rainforest. These effects are modified by soil fauna and depend on the quality of plant litter, especially on its quantities of labile C compounds and P. Moreover, soil and litter microorganisms are distinctly affected by increasing N inputs that may change relative C and nutrient fluxes between the litter layer and the soil. In a context of strong rise of N and P deposition predicted for tropical regions [9,77], our results suggest important consequences for biogeochemical cycles in tropical forest ecosystems, and that simultaneous global change-induced shifts in the quality of leaf litter input could modulate these effects. However, in order to compare our data with previous fertilization experiments we used nutrient concentrations that exceed predictions for nutrient depositions in tropical rainforests. The effects of anthropogenic depositions might thus be lower than those observed here and studies utilizing more realistic levels of nutrients will be needed to estimate their true impact.

Supporting Information

Table S1   Mean values of litter mass loss, litter SIR, and soil SIR for each of the six different litter species and each individual fertilization treatment separated into fine and coarse mesh litterbags.

Figure S1   Litter SIR as a function of litter mass loss across all litter species and fertilization treatments but separated into coarse (grey circles) and fine (open circles) mesh litterbags. Lines indicated fitted exponential (solid line) or linear (dashed line) regressions for the two fauna treatments separately.

Acknowledgments

We thank the CIRAD institute for access to the Paracou forest research station. We are grateful to thank Audin Patent, Elianne Louisiana, Lindon Yansen and Frits Kwasie (UMR EcoFoG) for litter sampling and management of the fertilization experiment, Bruno Buatois and Raphaëlle Leclerc for analyses, Margaret Amui-Vedel, Flavien Branchereau and Jessica Kok for technical help and Elisabeth Hättenschwiler for litter bag construction. SIR measurements were performed at the Plate-Forme d'Analyses Chimiques en Ecologie, Structure Fédérative de Recherche «Montpellier Environnement Biodiversité».

Author Contributions

Conceived and designed the experiments: SH HS N. Fromin. Performed the experiments: N. Fanin SB HS PS. Analyzed the data: N. Fanin SB N. Fromin SH. Contributed reagents/materials/analysis tools: HS PS. Wrote the paper: N. Fanin N. Fromin SB SH HS.

References

1. Coûteaux M-M, Bottner P, Berg B (1995) Litter decomposition, climate and litter quality. Trends Ecol Evol 10: 63–66.
2. Berg B, Matzner E (1997) Effect of N deposition on decomposition of plant litter and soil organic matter in forest systems. Environ Rev 5: 1–25.
3. Henry HAL, Cleland EE, Field CB, Vitousek PM (2005) Interactive effects of elevated CO_2, N deposition and climate change on plant litter quality in a California annual grassland. Oecologia 142: 465–473.
4. Davidson EA, Janssens IA (2006) Temperature sensitivity of soil carbon decomposition and feedbacks to climate change. Nature 440: 165–173.
5. Janssens IA, Lankreijer H, Matteucci G, Kowalski AS, Buchmann N, et al. (2001) Productivity overshadows temperature in determining soil and ecosystem respiration across European forests. Global Change Biol 7: 269–278.
6. Chambers JQ, Tribuzy ES, Toledo LC, Crispim BF, Higuchi N, et al. (2004) Respiration from a tropical forest ecosystem: Partitioning of sources and low carbon use efficiency. Ecol Appl 14: S72–S88.
7. Epron D, Nouvellon Y, Roupsard O, Mouvondy W, Mabiala A, et al. (2004) Spatial and temporal variations of soil respiration in a Eucalyptus plantation in Congo. Forest Ecol Manag 202: 149–160.
8. Yuste JC, Nagy M, Janssens IA, Carrara A, Ceulemans R (2005) Soil respiration in a mixed temperate forest and its contribution to total ecosystem respiration. Tree Physiol 25: 609–619.
9. Galloway JN, Dentener FJ, Capone DG, Boyer EW, Howarth RW, et al. (2004) Nitrogen cycles: past, present, and future. Biogeochemistry 70: 153–226.
10. Vitousek PM, Porder S, Houlton BZ, Chadwick OA (2010) Terrestrial phosphorus limitation: mechanisms, implications, and nitrogen-phosphorus interactions. Ecol Appl 20: 5–15.
11. Phillips OL, Malhi Y, Higuchi N, Laurance WF, Nunez PV, et al. (1998) Changes in the carbon balance of tropical forests: Evidence from long-term plots. Science 282: 439–442.
12. Malhi Y, Grace J (2000) Tropical forests and atmospheric carbon dioxide. Trends Ecol Evol 15: 332–337.
13. Grace J, Malhi Y, Higuchi N (2001) Productivity and carbon fluxes of tropical rain forests. . In: Mooney HA, Roy J., Saugier B, editors. Terrestrial Global Productivity: Past, Present, and Future. San Diego: Academic Press. pp. 401–428.
14. Luyssaert S, Schulze ED, Borner A, Knohl A, Hessenmoller D, et al. (2008) Old-growth forests as global carbon sinks. Nature 455: 213–215.
15. Sayer EJ, Tanner EVJ (2010) A new approach to trenching experiments for measuring root-rhizosphere respiration in a lowland tropical forest. Soil Biol Biochem 42: 347–352.
16. Silver WL, Thompson AW, McGroddy ME, Varner RK, Dias JD, et al. (2005) Fine root dynamics and trace gas fluxes in two lowland tropical forest soils. Global Change Biol 11: 290–306.
17. Gallardo A, Schelsinger WH (1994) Factors limiting microbial biomass in the mineral soil and forest floor of a warm-temperate forest. Soil Biol Biochem 26: 1409–1415.
18. Joergensen RG, Scheu S (1999) Response of soil microorganisms to the addition of carbon, nitrogen and phosphorus in a forest Rendzina. Soil Biol Biochem 31: 859–866.
19. Ilstedt U, Singh S (2005) Nitrogen and phosphorus limitations of microbial respiration in a tropical phosphorus-fixing acrisol (ultisol) compared with organic compost. Soil Biol Biochem 37: 1407–1410.
20. Gnankambary Z, Stedt U, Nyberg G, Hien V, Malmer A (2008) Nitrogen and phosphorus limitation of soil microbial respiration in two tropical agroforestry parklands in the south-Sudanese zone of Burkina Faso: The effects of tree canopy and fertilization. Soil Biol Biochem 40: 350–359.
21. Sterner RW, Elser JJ (2002) Ecological stoichiometry: The biology of elements from molecules of the biosphere. Princeton: Princeton University Press. 439 p.
22. Cherif M, Loreau M (2009) When microbes and consumers determine the limiting nutrient of autotrophs: a theoretical analysis. Proc Roy Soc B-Biol Sci 276:487–497
23. Enriquez S, Duarte CM, Sand-Jensen K (1993) Patterns in decomposition rates among photosynthetic organisms: the importance of detritus C:N:P content. Oecologia 94: 457–471.
24. Manzoni S, Trofymow JA, Jackson RB, Porporato A (2010) Stoichiometric controls on carbon, nitrogen, and phosphorus dynamics in decomposing litter. Ecol Mono 80: 89–106.
25. De Angelis P, Chigwerewe KS, Mugnozza GES (2000) Litter quality and decomposition in a CO_2-enriched Mediterranean forest ecosystem. Plant Soil 224: 31–41.
26. Körner C, Miglietta F (1994) Long-term effects of naturally elevated CO_2 on Mediterranean grassland and forest trees. Oecologia 99: 343–351.
27. Hättenschwiler S, Schafellner C (2004) Gypsy moth feeding in the canopy of a CO_2-enriched mature forest. Global Change Biol 10: 1899–1908.
28. Hättenschwiler S, Coq S, Barantal S, Handa IT (2011) Leaf traits and decomposition in tropical rainforests: revisiting some commonly held views and towards a new hypothesis. New Phytol 189: 950–965.
29. Cleveland CC, Townsend AR, Taylor P, Alvarez-Clare S, Bustamante MMC, et al. (2011) Relationships among net primary productivity, nutrients and climate in tropical rain forest: a pan-tropical analysis. Ecol Lett 14: 939–947.
30. Townsend AR, Cleveland CC, Houlton BZ, Alden CB, White JWC (2011) Multi-element regulation of the tropical forest carbon cycle. Front Ecol Environ 9: 9–17.
31. Townsend AR, Asner GP, Cleveland CC (2008) The biogeochemical heterogeneity of tropical forests. Trends Ecol Evol 8: 424–431.
32. Quesada CA, Lloyd J, Schwarz M, Patiño S, Baker TR, et al. (2010) Variations in chemical and physical properties of Amazon forest soils in relation to their genesis. Biogeosciences 7: 1515–1541.
33. Cleveland CC, Townsend AR, Schmidt SK (2002) Phosphorus limitation of microbial processes in moist tropical forests: Evidence from short-term laboratory incubations and field studies. Ecosystems 5: 680–691.
34. Cleveland CC, Reed SC, Townsend AR (2006) Nutrient regulation of organic matter decomposition in a tropical rain forest. Ecology 87: 492–503.
35. Cleveland CC, Townsend AR (2006) Nutrient additions to a tropical rain forest drive substantial soil carbon dioxide losses to the atmosphere. P Natl Acad Sci USA 103: 10316–10321.
36. Kaspari M, Garcia MN, Harms KE, Santana M, Wright SJ, et al. (2008) Multiple nutrients limit litterfall and decomposition in a tropical forest. Ecol Lett 11: 35–43.

37. Mo J, Zhang W, Zhu W, Gundersen P, Fang Y, et al. (2008) Nitrogen addition reduces soil respiration in a mature tropical forest in southern China. Global Change Biol 14: 403–412.

38. Cusack DF, Torn MS, McDowell WH, Silver WL (2010) The response of heterotrophic activity and carbon cycling to nitrogen additions and warming in two tropical soils. Global Change Biol 16: 2555–2572.

39. Cusack DF, Silver WL, Torn MS, McDowell WH (2011) Effects of nitrogen additions on above- and belowground carbon dynamics in two tropical forests. Biogeochemistry 104: 203–225.

40. Vitousek PM, Farrington H (1997) Nutrient limitation and soil development: Experimental test of a biogeochemical theory. Biogeochemistry 37: 63–75.

41. Hobbie SE, Vitousek PM (2000) Nutrient limitation of decomposition in Hawaiian forests. Ecology 81: 1867–1877.

42. Reed S, Vitousek P, Cleveland C (2011) Are patterns in nutrient limitation belowground consistent with those aboveground: results from a 4 million year chronosequence. Biogeochemistry 106: 323–336.

43. Barantal S, Schimann H, Fromin N, Hättenschwiler S. (2012) Nutrient and carbon limitation on decomposition in an Amazonian rainforest. Ecosystems 15: 1039–1052.

44. Krashevska V, Maraun M, Ruess L, Scheu S (2010) Carbon and nutrient limitation of soil microorganisms and microbial grazers in a tropical montane rain forest. Oikos 119: 1020–1028.

45. Milcu A, Heim A, Ellis RJ, Scheu S, Manning P (2011) Identification of general patterns of nutrient and labile carbon control on soil carbon dynamics across a successional gradient. Ecosystems 14: 710–719.

46. Priess JA, Fölster H (2001) Microbial properties and soil respiration in submontane forests of Venezuelian Guyana: characteristics and response to fertilizer treatments. Soil Biol Biochem 33: 503–509.

47. Brown ALP, Garland JL, Day FP (2009) Physiological profiling of soil microbial communities in a Florida scrub-oak ecosystem: spatial distribution and nutrient limitations. Micro Ecol 57: 14–24.

48. Hättenschwiler S, Aeschlimann B, Coûteaux MM, Roy J, Bonal D (2008) High variation in foliage and leaf litter chemistry among 45 tree species of a neotropical rainforest community. New Phytol 179: 165–175.

49. Schreeg L, Mack M, Turner B (2012) Nutrient-specific patterns in leaf litter solubility across 41 lowland tropical woody species. Ecology, in press.

50. Wieder RW, Cleveland CC, Townsend AR (2009) Controls over leaf litter decomposition in wet tropical forests. Ecology 90: 3333–3341.

51. Coq S, Souquet JM, Meudec E, Cheynier V, Hättenschwiler S (2010) Interspecific variation in leaf litter tannins drives decomposition in a tropical rain forest of French Guiana. Ecology 91: 2080–2091.

52. Ushio M, Wagai R, Balser TC, Kitayama K (2008) Variations in the soil microbial community composition of a tropical montane forest ecosystem: Does tree species matter? Soil Biol Biochem 40: 2699–2702.

53. Fanin N, Hättenschwiler S, Barantal S, Schimann H, Fromin N (2011) Does variability in litter quality determine soil microbial respiration in an Amazonian rainforest? Soil Biol Biochem 43: 1014–1022.

54. Gonzalez G, Seastedt TR (2001) Soil fauna and plant litter decomposition in tropical and subalpine forests. Ecology 82: 955:964.

55. Wall DH, Bradford MA, St John MG, Tofymow JA, Behan Pelletier VM, et al. (2008) Global decomposition experiment shows soil animal impacts on decomposition are climate-dependent. Global Change Biol 14: 2661–2677.

56. Yang X, Chen J (2009) Plant litter quality influences the contribution of soil fauna to litter decomposition in humid tropical forests, southwestern China. Soil Biol Biochem 41: 910–918.

57. Makkonen M, Berg MP, Handa T, Hättenschwiler S, van Ruijven J, et al. (2012) Highly consistent effects of plant litter identity and functional traits on decomposition across a latitudinal gradient. Ecol Lett 15: 1033–1041.

58. McGlynn TP, Salinas DJ, Dunn RR, Wood TE, Lawrence D, et al. (2007) Phosphorus limits tropical rain forest litter fauna. Biotropica 39: 50–53.

59. Bonal D, Bosc A, Ponton S, Goret JY, Burban B, et al. (2008) Impact of severe dry season on net ecosystem exchange in the Neotropical rainforest of French Guiana. Global Change Biol 14: 1917–1933.

60. Gourlet-Fleury S, Houllier F (2000) Modelling diameter increment in a lowland evergreen rain forest in French Guiana. Forest Ecol Manag 131: 269–289.

61. Roy J, Hättenschwiler S, Domenach AM (2005) Tree diversity and soil biology: a new research program in French Guiana. In: Binkley D, Menyailo O, editors. Tree species effects on soils: implications for global change. Kluwer Academic, Dordrecht. pp. 337–348.

62. Nannipieri P, Ascher J, Ceccherini MT, Landi L, Pietramellara G, et al. (2003) Microbial diversity and soil functions. Eur J Soil Sci 54: 655–670.

63. Beare MH, Neely CL, Coleman DC, Hargrove WL (1990) A substrate-induced respiration (SIR) method for measurement of fungal and bacterial biomass on plant residues. Soil Biol Biochem 22: 585–594.

64. Pinheiro JC, Bates D, DebRoy S, Deepayan S (2007) nlme: Linear and nonlinear mixed effects models. R Package Version 31. 339 p.

65. Bates D (2010) Linear mixed model implementation in lme4. R Package Version 3. 32 p.

66. Whittingham MJ, Stephens PA, Bradbury RB, Freckleton RP (2006) Why do we still use stepwise modelling in ecology and behaviour? J Anim Ecol 75: 1182–1189.

67. Fontaine S, Barot S, Barre P, Bdioui N, Mary B, et al. (2007) Stability of organic carbon in deep soil layers controlled by fresh carbon supply. Nature 450: 277–U210.

68. Talbot JM, Treseder KK (2012) Interactions among lignin, cellulose, and nitrogen drive litter chemistry-decay relationships. Ecology 93: 345–354.

69. Elser JJ, Bracken MES, Cleland EE, Gruner DS, Harpole WS, et al. (2007) Global analysis of nitrogen and phosphorus limitation of primary producers in freshwater, marine and terrestrial ecosystems. Ecol Lett 10: 1135–1142.

70. Schimel JP, Bennett J (2004) Nitrogen mineralization: Challenges of a changing paradigm. Ecology 85: 591–602.

71. Craine JM, Morrow C, Fierer N (2007) Microbial nitrogen limitation increases decomposition. Ecology 88: 2105–2113.

72. Hättenschwiler S, Bracht-Jorgensen HB (2010) Carbon quality rather than stoichiometry controls litter decomposition in a tropical rain forest. J Ecol 98: 754–763.

73. Townsend AR, Cleveland CC, Asner GP, Bustamante MMC (2007) Controls over foliar N : P ratios in tropical rain forests. Ecology 88: 107–118.

74. Orwin KH, Wardle DA, Greenfield LG (2006) Ecological consequences of carbon substrate identity and diversity in a laboratory study. Ecology 87:580–593.

75. Feeley KJ, Davies SJ, Perez R, Hubbell SP, Foster RB (2011) Directional changes in the species composition of a tropical forest. Ecology 92: 871–882.

76. Feeley KJ, Malhi Y, Zelazowski P, Silman MR (2012) The relative importance of deforestation, precipitation change, and temperature sensitivity in determining the future distributions and diversity of Amazonian plant species. Global Change Biol 18: 2636–2647.

77. Okin GS, Mahowald N, Chadwick OA, Artaxo P (2004) Impact of desert dust on the biogeochemistry of phosphorus in terrestrial ecosystems. Global Biogeochem Cy DOI:10.1029/2003GB002145

78. Neff JC, Townsend AR, Gleixner G, Lehman SJ, Turnbull J, et al. (2002) Variable effects of nitrogen additions on the stability and turnover of soil carbon. Nature 419: 915–917.

79. Carreiro MM, Sinsabaugh RL, Repert DA, Parkhurst DF (2000) Microbial enzyme shifts explain litter decay responses to simulated nitrogen deposition. Ecology 81: 2359–2365.

80. DeForest JL, Zak DR, Pregitzer KS, Burton AJ (2004) Atmospheric nitrate deposition, microbial community composition, and enzyme activity in northern hardwood forests. Soil Sci Soc Am J 68: 132–138.

81. Treseder KK (2008) Nitrogen additions and microbial biomass: a meta-analysis of ecosystem studies. Ecol Lett 11: 1111–1120.

82. Cusack DF, Silver WL, Torn MS, Burton SD, Firestone MK (2011) Changes in microbial community characteristics and soil organic matter with nitrogen additions in two tropical forests. Ecology 92: 621–632.

3

Plastic Traits of an Exotic Grass Contribute to Its Abundance but Are Not Always Favourable

Jennifer Firn[1]*, Suzanne M. Prober[2], Yvonne M. Buckley[3,4]

1 School of Earth, Environment and Biological Sciences, Queensland University of Technology, Brisbane, Queensland, Australia, **2** Ecosystem Sciences, CSIRO, Wembley, Western Australia, Australia, **3** School of Biological Sciences, The University of Queensland, St. Lucia, Queensland, Australia, **4** Ecosystem Sciences CSIRO, Dutton Park, Queensland, Australia

Abstract

In herbaceous ecosystems worldwide, biodiversity has been negatively impacted by changed grazing regimes and nutrient enrichment. Altered disturbance regimes are thought to favour invasive species that have a high phenotypic plasticity, although most studies measure plasticity under controlled conditions in the greenhouse and then assume plasticity is an advantage in the field. Here, we compare trait plasticity between three co-occurring, C_4 perennial grass species, an invader *Eragrostis curvula*, and natives *Eragrostis sororia* and *Aristida personata* to grazing and fertilizer in a three-year field trial. We measured abundances and several leaf traits known to correlate with strategies used by plants to fix carbon and acquire resources, i.e. specific leaf area (SLA), leaf dry matter content (LDMC), leaf nutrient concentrations (N, C:N, P), assimilation rates (*Amax*) and photosynthetic nitrogen use efficiency (PNUE). In the control treatment (grazed only), trait values for SLA, leaf C:N ratios, *Amax* and PNUE differed significantly between the three grass species. When trait values were compared across treatments, *E. curvula* showed higher trait plasticity than the native grasses, and this correlated with an increase in abundance across all but the grazed/fertilized treatment. The native grasses showed little trait plasticity in response to the treatments. *Aristida personata* decreased significantly in the treatments where *E. curvula* increased, and *E. sororia* abundance increased possibly due to increased rainfall and not in response to treatments or invader abundance. Overall, we found that plasticity did not favour an increase in abundance of *E. curvula* under the grazed/fertilized treatment likely because leaf nutrient contents increased and subsequently its' palatability to consumers. *E. curvula* also displayed a higher resource use efficiency than the native grasses. These findings suggest resource conditions and disturbance regimes can be manipulated to disadvantage the success of even plastic exotic species.

Editor: Justin Wright, Duke University, United States of America

Funding: Thank you to CSIRO Ecosystem Sciences and Australian Research Council (YB, DP0771387) for funding. The funders had no role in study design, data collection and analysis, decision to publish, or preparation of the manuscript.

Competing Interests: The authors have declared that no competing interests exist.

* E-mail: jennifer.firn@qut.edu.au

Introduction

Exotic plant species can establish and dominate sites despite lacking evolutionary familiarity with local conditions and having small founder populations (i.e. the invasion paradox [1,2]). Substantial evidence suggests disturbances such as changed grazing regimes and nutrient addition increase opportunities for invasive species to establish [3,4,5,6,7,8], particularly if disturbances are novel to an ecosystem [9,10]. Disturbance favours the growth and survival of some species over others depending on the characteristics of the disturbance itself including the type, frequency, duration and intensity [11], but also on the traits of species present [12,13]. Despite extensive research, evidence for a generic set of traits that favour exotic over native species remains inconclusive [14,15,16,17].

Evidence suggests invasive species tend to display traits of fast growing species that are resource acquisition specialists and native species tend to display traits of slow-growing species that are conservation specialists [18,19,20,21,22]. The leaf economic spectrum proposes a fundamental trade-off in the traits held by fast- and slow-growing plant species [23,24,25]. Fast growing species, better at resource capture, tend to dominate disturbed

ecosystems where resource availability is not limited. These fast growing species have generally higher specific leaf area (SLA, mm^2/mg, fresh leaf area/oven-dry mass), lower leaf dry matter content (LDMC, mg/g, oven dry mass/water-saturated fresh mass), higher nutrient contents and higher rates of assimilation (*Amax*) [23,24,25]. Slower growing plant species generally occupying low resource and less disturbed sites are better at resource conservation and to tend to hold opposite traits—lower SLAs, higher LDMCs, lower nutrient contents and lower rates of *Amax* [23,24,25]. Studies comparing the leaf traits of exotics and natives have consistently found evidence for this trade-off, with exotics showing better resource acquisition strategies and natives better resource conservation strategies [26]. However, recent findings by Leishman et al. [27] suggest that exotic and native plant species can hold similar strategies for capturing resources, with exotic and native species at disturbed sites possessing similar traits, but different traits to natives at pristine sites.

To date, most studies investigating plant traits focused on differences between species (interspecific variability) and across sites affected by different disturbances and environmental conditions, but recent research has highlighted the importance of intraspecific variability in traits or phenotypic plasticity [28,29].

Table 1. General characteristics of the invasive exotic lovegrass, and the native grasses purple wiregrass and woodlands lovegrass.

Characteristics	*Eragrostis curvula* **lovegrass, Exotic grass**	*Aristida personata* **purple wiregrass Native grass**	*Eragrostis sororia* **woodlands lovegrass Native grass**
Mean abundance at site (± S.E.) at time 0	47.56%±3.98	22.66±5.12	1.61%±0.67
Growth Habit	Tufted perennial	Tufted perennial	Tufted perennial
Photosynthetic Pathway	C_4	C_4	C_4
Height	Up to 120 cm	Up to 120 cm	Up to 70 cm
Growth season	Summer	Summer	Summer
Flowering time	Spring to Autumn	Summer to Autumn	Summer
Palatability to livestock	Low	Low	Moderate
Native continental distribution	Africa	Australia	Australia

[61,62,63,64,65].

Evidence suggests that invasive exotic species display higher phenotypic plasticity than natives—the potential of each individual genotype to produce different traits/phenotypes in response to disturbance and fluctuating environmental conditions [14,30,31]. This capacity to change morphological or physiological traits may allow genotypes of a species to thrive across a wider range of environmental conditions (genotype-level plasticity), and/or allow individuals within a population to thrive at sites during and after disturbance or resource pulses (species-level plasticity) [14,17,30,32,33].

Invasive species have shown higher trait plasticity in response to increased resources, e.g soil nutrients and water, in comparison to phylogenetically-related non-invasive species from high resource environments [34,35], and phylogenetically-related native species from low resource environments [33]. Although a recent greenhouse study comparing 20 phylogenetically-related invasive and native trees and shrubs found similar levels of trait plasticity in response to nutrient and light treatments, but enhanced performance by invasive species measured as mean trait values [36]. Studies have also shown individuals of the same species sampled from both introduced and native sites have a higher trait plasticity at introduced sites [37]. However, studies measuring trait plasticity have generally grown species over short-periods of time under controlled greenhouse conditions. Adults growing in the field may display different morphological and physiological traits when subjected to a wider range of resource conditions and biotic interactions in comparison to controlled greenhouse experiments [38].

Here, we use a factorial field trial to compare trait plasticity between an invasive exotic grass (*Eragrostis curvula* (Schrad.) Nees, hereafter lovegrass), and two native grasses (*Aristida personata* Henrard, hereafter purple wiregrass, and *Eragrostis sororia* Domin, hereafter woodlands lovegrass). Our study is unique as we measure how traits of key species in a community change in response to treatments, and measure these changes under 'realistic conditions' to increase the reliability of the results for explaining invasion success. We measured how grazing and fertilizer addition treatments altered abundances and several leaf traits. We hypothesised that under the existing site conditions (grazing) that the invader would display traits consistent with faster growth than the natives. We also hypothesised that under different experimental treatments the invader would exhibit higher phenotypic plasticity than the natives, evidenced by predictable changes in traits based on trends identified in the leaf economic spectrum. We then relate these results to differences in abundance of all three species between the treatments. This invasion scenario is a model

system to compare plasticity because these species share life-history traits, co-exist at the same site, and native woodlands lovegrass is a congener of the invader lovegrass (Table 1).

Results

After three years of treatments, the abundance of all species was significantly correlated with abundance prior to the start of the treatments, time 0 (Table 2). Lovegrass (exotic) abundance was best explained by the additive effects of grazing and fertilizer, but not the interaction (Table 2 a). Lovegrass abundance increased across all treatments except the grazed/fertilized treatment where its abundance decreased (Fig. 1 a, 53.11%±27.80 reduction in comparison to time 0). Grazing treatments had the strongest effect ($F_{1, 2} = 139.91$, $P<0.008$); but fertilizer treatments also had a significant effect ($F_{1, 54} = 6.14$, $P<0.02$). Purple wiregrass abundance was best explained by the effect of the grazing treatment ($F_{1, 2} = 44.51$, $P<0.025$). After three years, the abundance of purple wiregrass was reduced across all treatments, but most significantly in the grazing exclusion treatments (>35% decrease, Fig. 1 b), which was also the treatment where lovegrass abundance increased the most (>20% increase Fig. 1a). Woodlands lovegrass was low in abundance pre-treatment (Table 1), and increased across all treatments when compared to its abundance at year 0. It increased in abundance in the grazing exclusion treatments to more than 2% and in the grazed treatments to more than 10% (Fig. 1 c), but the difference between treatments was not significant.

In year 3, the availability of soil nutrients also varied significantly depending on the treatments (Fig. S1 and Table S1). Soil nitrate (NO_3) levels varied marginally by the interaction of grazing and fertilizer treatments ($F_{1, 58} = 3.00$, $P<0.09$, Fig. 1 a), although overall nitrate levels were higher in the grazing exclusion treatments and highest in the grazing exclusion and unfertilized treatment. This decreasing trend in soil nitrate levels, despite the application of fertilizer, is likely reflective of increased leaching and/or use by plants and soil fauna. Soil ammonium (NH_4) levels did not vary significantly between the treatments (Fig. S1b and Table S1b). Soil phosphate (PO_4) levels increased significantly with the grazing treatments ($F_{1, 2} = 30.53$, $P<0.03$, Fig. 1 c) and the fertilizer treatments ($F_{1, 58} = 11.67$, $P<0.001$, Fig. 1 c), but not the interaction.

Traits differed between species in the control treatment

In the control treatment (grazed/no fertilizer), mean LDMC values did not vary significantly between species ($F_{2, 60} = 0.89$,

Table 2. Results from an ANOVA conducted to assess the significance of the fixed effects for LMEMs of abundance (arc-sine transformed) in year 3, with a fixed effects structure of grazing and fertilizer treatments and a co-variate of abundance in time 0, and a random effects structure of block/plot.

abundance$_{time\ 3}$	Fixed effects	F values (df as subscript), P value
a) Lovegrass	**grazing**	$F_{1,\ 2} = 139.91$, $P < 0.008$
	fertilizer	$F_{1,\ 54} = 6.14$, $P < 0.02$
	abundance$_{time\ 0}$	$F_{1,\ 54} = 48.79$, $P < 0.0002$
	grazing ×fertilizer	$F_{1,\ 54} = 1.96$, $P < 0.20$
	grazing ×abundance$_{time\ 0}$	$F_{1,\ 54} = 1.98$, $P < 0.20$
	fertilizer ×abundance$_{time\ 0}$	$F_{1,\ 54} = 0.38$, $P < 0.60$
	grazing ×fertilizer ×abundance$_{time\ 0}$	$F_{1,\ 54} = 0.26$, $P < 0.65$
b) Purple wiregrass	**grazing**	$F_{1,\ 2} = 44.51$, $P < 0.025$
	fertilizer	$F_{1,\ 54} = 0.25$, $P < 0.70$
	abundance$_{time\ 0}$	$F_{1,\ 54} = 33.25$, $P < 0.002$
	grazing ×fertilizer	$F_{1,\ 54} = 0.02$, $P < 0.90$
	grazing ×abundance$_{time\ 0}$	$F_{1,\ 54} = 0.16$, $P < 0.70$
	fertilizer ×abundance$_{time\ 0}$	$F_{1,\ 54} = 1.16$, $P < 0.30$
	grazing ×fertilizer ×abundance$_{time\ 0}$	$F_{1,\ 54} = 0.31$, $P < 0.60$
c) Woodlands lovegrass	grazing	$F_{1,\ 2} = 4.18$, $P < 0.20$
	fertilizer	$F_{1,\ 32} = 4.18$, $P < 0.20$
	abundance$_{time\ 0}$	$F_{1,\ 32} = 8.00$, $P < 0.04$
	grazing ×fertilizer	$F_{1,\ 32} = 1.18$, $P < 0.30$
	grazing ×abundance$_{time\ 0}$	$F_{1,\ 32} = 2.29$, $P < 0.20$
	fertilizer ×abundance$_{time\ 0}$	$F_{1,\ 32} = 0.91$, $P < 0.40$
	grazing ×fertilizer ×abundance$_{time\ 0}$	$F_{1,\ 32} = 0.13$, $P < 0.80$

$P < 0.50$, Fig. 2 a). Mean SLA values differed marginally between species, but contrary to expectations, with lovegrass showing a lower mean SLA value than Purple wiregrass ($F_{2,\ 60} = 2.69$, $P < 0.08$, Fig. 2 b), a trait indicative of a slower growing species. In agreement with expectations that lovegrass would display characteristics of a faster growing species under the grazing treatment (control), lovegrass had a significantly higher assimilation rate (Amax; $F_{2,\ 30} = 10.1$, $P < 0.002$, Fig. S2 "grazing") and photosynthetic nitrogen use efficiency (PNUE = Amax/leaf nitrogen; $F_{2,\ 30} = 8.78$, $P < 0.001$, Fig. 3 "grazing") than the two native grasses. Leaf nutrient concentrations in the control treatment did not vary significantly between species, except in the case of Leaf C:N ratios where lovegrass showed a significantly higher ratio than woodlands lovegrass ($F_{2,\ 30} = 3.13$, $P < 0.05$, Fig. 4 b), again contrary to expectations as this is a trait indicative of a slower growing species.

Trait plasticity in response to the treatments differed amongst species

The traits of lovegrass varied predictably with the treatments, with adult individuals showing significant differences in LDMC, SLA, PNUE, Amax and leaf nutrients. The traits of purple wiregrass also changed with the nutrient treatments, but woodlands lovegrass showed little change (Fig. 3–5, Fig. S2, Table 3–5 and Table S1). Differences in LDMC and SLA values for lovegrass were best explained by the interaction of grazing and fertilizer treatments (LDMC: $F_{1,\ 58} = 10.10$, $P < 0.002$ and SLA: $F_{1,\ 58} = 3.89$, $P < 0.05$, Table 3). In agreement with expectations, LDMC decreased and SLA increased for lovegrass with increasing

amounts of disturbance from grazing exclusion treatments to the grazed/fertilized treatments (Fig. 5 a and b). The highest LDMC and lowest SLA values were found in both grazing exclusion treatments, whereas the lowest LDMC and highest SLA values were shown in the grazed/fertilized treatments. For purple wiregrass, differences in LDMC were not explained by the grazing or fertilizer treatments, whereas differences in SLA were explained by fertilizer treatments ($F_{1,\ 58} = 4.25$, $P < 0.05$, Table 3).

PNUE varied depending on the interaction between species and treatments ($F_{6,\ 42} = 2.38$, $P < 0.05$). Lovegrass showed a three-fold increase in PNUE between the grazed and exclusion treatments (Fig. 3). Woodlands lovegrass overall had a lower PNUE than the other grasses, but rates did not vary between treatments (Fig. 3). Purple wiregrass had a higher PNUE rate than woodlands lovegrass, but did not show a significant difference between treatments (Fig. 3). Amax varied similarly to PNUE depending on the interaction between species and treatments ($F_{6,\ 42} = 2.84$, $P < 0.02$, Fig. S2).

Differences in total nitrogen concentration were marginally significant and C:N ratio were significant for lovegrass leaves collected from fertilized and unfertilized treatments (Fig. 4 a and b, Table 4 a and Table S2 a). Lovegrass leaves showed a marginally significant increase in leaf nitrogen concentration in the fertilized plots, with the highest increase occurring in the grazed/fertilized treatment (Fig. 4 a). Consistent with this increase in N, leaf C:N ratios for lovegrass decreased when fertilizer was added (Fig. 4 b and Table S2 a). The total phosphorus concentration of lovegrass leaves varied significantly with the grazing treatment, with the highest phosphorus concentration occurring in treatments where grazing was maintained (Fig. 4 c, Table 4 a and Table S2 a). Total

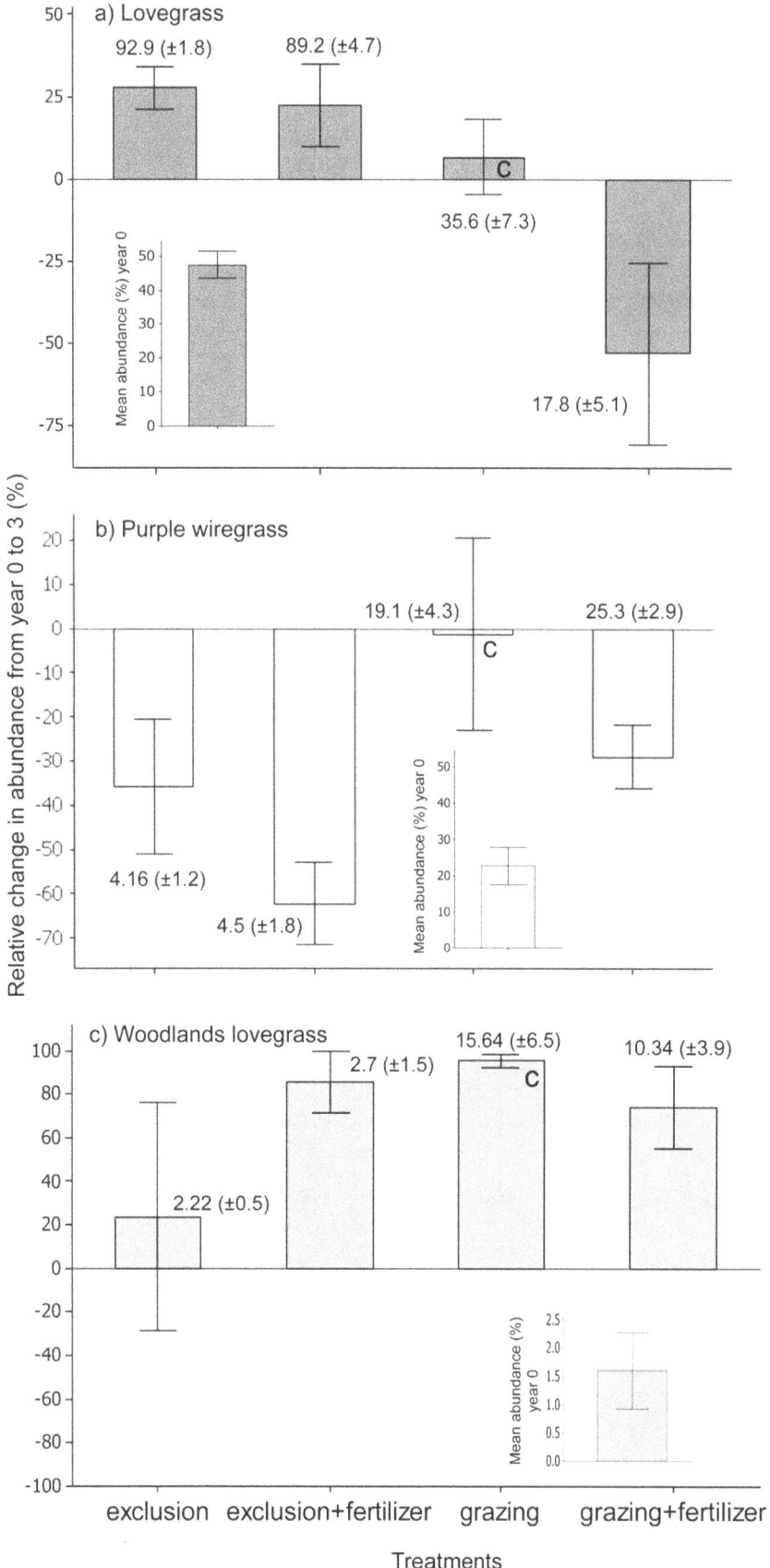

Figure 1. Relative change in abundance of each species from measurements taken prior to the start of the field trial and measurements taken again after three years of treatments (calculated as ((abundance$_{T3}$-abundance$_{T0}$)/abundance$_{T0}$)×100). The

insets show the mean abundance values (± S.E.) at time 0 and the values shown next to each bar are the mean abundance values (± S.E.) after three years. **C** indicates the control treatment grazing/no fertilizer.

leaf phosphorus for purple wiregrass did not vary significantly between treatments, but total nitrogen and C:N ratios differed marginally with the grazing treatment (Table 4 b and Table S2 b). The total nitrogen concentration of purple wiregrass leaves was lower in the grazed treatment, and C:N ratios higher in the grazed treatment for purple wiregrass. The nitrogen concentration of woodlands lovegrass leaves did not vary significantly, but did vary depending on the grazing treatments for both C:N ratios and total phosphorus concentration (Table 4 c and Table S2 c). In both cases, woodlands lovegrass leaves, collected from the exclusion treatment, had the highest C:N ratios and the highest total phosphorus concentration (Fig. 4 c).

Table 5 summarises the response of each trait to the grazing and fertilizer treatments for each of the three grass species and

indicates whether the change followed or was contrary to expectations.

Discussion

Overall we found the invasive exotic grass displayed higher trait plasticity in response to the treatments than the two native grasses (Table 5). Lovegrass changed its traits according to predictions based on trends from the Leaf Economic Spectrum for all six traits, compared with only one trait for the native grasses [23]. A recent meta-analysis comparing 75 invasive/non-invasive pairs of plant species found invaders were more plastic in their response to increased resource availability than non-invaders, but plasticity was only a fitness advantage for the invasive species when resource conditions were high [39]. Increased resource availability is widely

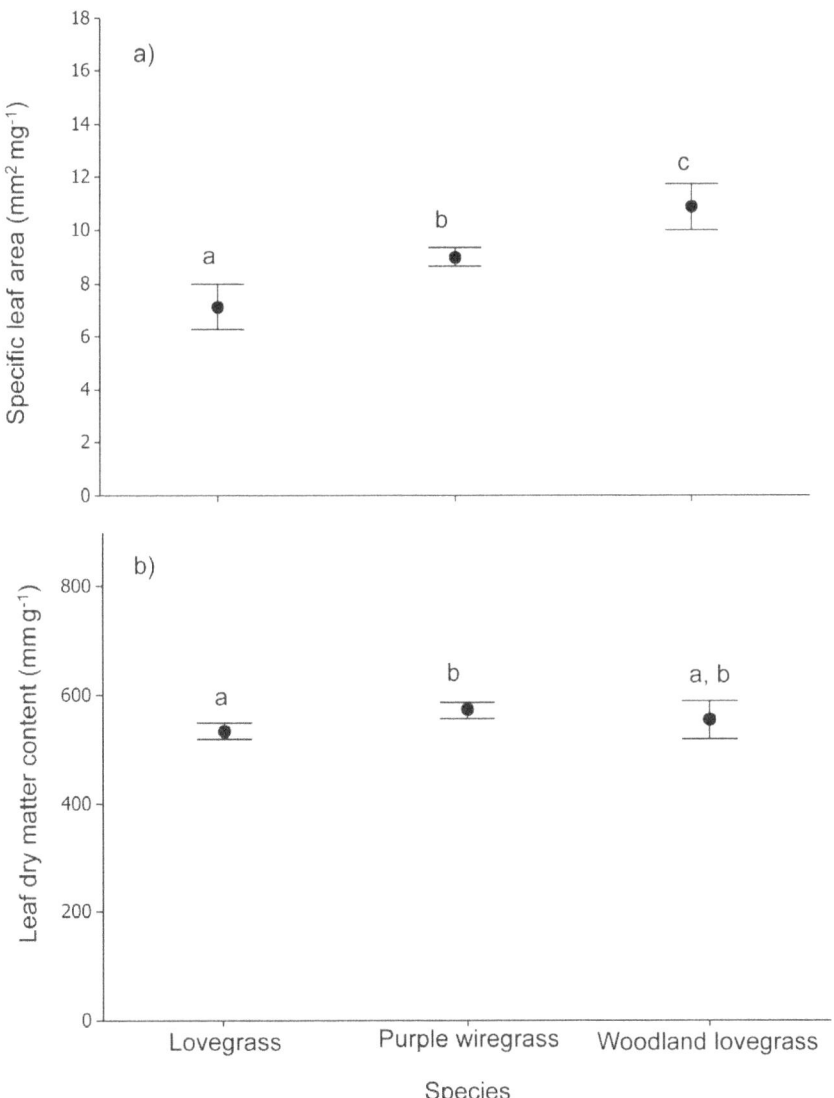

Figure 2. Comparison of mean trait values between species at the site level. Mean LDMC and SLA values (± S.E.) for each species for the grazing only treatment, which was the original disturbance at this site and therefore represents a control. Different letters indicate means are significantly different at p<0.05.

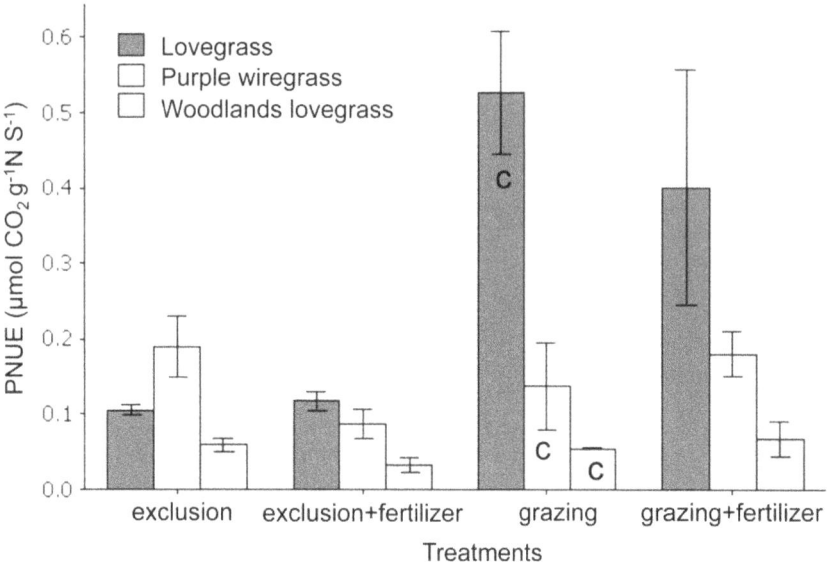

Figure 3. Mean photosynthetic nitrogen use efficiency (± SE) for each species depending on the grazing and fertilizer treatments. C indicates the control treatment grazing/no fertilizer.

agreed to promote invasion [40,41,42]. Because most studies in this meta-analysis were pot trials growing plants in the absence of competition and other biotic interactions such as grazing, it is difficult to extrapolate these findings to field conditions [38]. We found the plastic response of Lovegrass was not an advantage in the field when resources are high as increased soil nutrients, led to increased resource uptake by the exotic but also increased selective grazing pressure.

Using a three year field study, we found increased nutrients coupled with grazing decreased the abundance of the exotic, a trend also measured in the first two years of the study and published in Firn et al. [43]. Under these conditions lovegrass leaves increased in SLA, decreased in LDMC and increased in leaf total nitrogen and phosphorus concentration in response to fertilizer, but this response likely also increased its palatability to grazing livestock. An alternative explanation for these results to trait plasticity may be increased genetic diversity prior to the start of the experiment within the lovegrass population, and the treatments led to differential survival or 'filtering' of phenotypes better adapted to the different experimental conditions. This explanation is, however, unlikely as the grasses are long-lived perennials and we were careful to measure traits from large mature tussocks. Also, if genetic diversity were the explanation, genotypes in the experimental treatments would likely be subsets of those in the control; therefore, we would have expected higher trait variation in the control treatment (grazing only).

The leaf traits of lovegrass also changed in the exclusion treatments showing lower SLA and higher LDMC suggesting it has at least a comparable capacity to conserve resources as the native grasses. Using a greenhouse study, Funk [33] compared the response of several related exotic and native species from resource limiting environments and also found exotics were equally or more efficient at resource conservation. Lovegrass and both native species showed similar mean traits under the control treatment of grazing only, except the exotic had a lower SLA (indicative of a slower growing species) than purple wiregrass and a higher PNUE and $Amax$ than both native species. This result suggests lovegrass is more efficient at resource capture than the native species. A study comparing traits of exotics to native species in the same region of

Australia, found several C_4 exotic grass species (including lovegrass), had higher LDMC than native species [44], similarly suggesting successful exotic grasses in this region may be resource conservation specialists [44].

We also found evidence that lovegrass has a higher resource use efficiency (RUE, carbon assimilation per unit of resource, measured as PNUE) than the native grasses. Funk and Vitousek [45] compared RUE between related and co-occurring exotic and native species within Hawaii, and also found exotics had a higher RUE. PNUE increased more than three-fold in the grazed versus exclusion treatments. Leaf C:N ratios decreased in the treatments that were fertilized, but this same response was not shown by the native grasses. Higher RUE would be an advantage at the field site, as rainfall is highly variable and soil nutrients low.

Lovegrass was the dominant species at the site in year 0 and displayed the highest plasticity in response to grazing and fertilizer after three years of treatment, and this plasticity correlated with changes in its abundance in the short-term. Grime's mass ratio hypothesis [46] describes dominant species as having the highest impact on ecosystem functions. Dominant species may then be the most plastic in response to changed conditions. While intermediate/subordinate species abundance may be most influenced by the abundance of the dominant species and transient species (a species whose abundance fluctuates depending on resources) abundance responsive to environmental fluctuations [46].

Purple wiregrass, a subordinate species, was reduced in abundance across the treatments with the highest reductions occurring where grazing was excluded. Purple wiregrass did show some trait plasticity. Although it was the least disturbed treatments where purple wiregrass showed some plasticity, including an increased SLA in exclusion/fertilised treatment and an increased leaf nitrogen concentration in the exclusion treatments. In accordance with Grime's mass ratio hypothesis, increased abundance of lovegrass (>85%) in the exclusion treatments may account for the significant reduction of purple wiregrass abundance (reduced by >35%).

Woodlands lovegrass increased in abundance across the treatments. Although related to lovegrass, woodlands lovegrass did not show similar trait plasticity. This finding suggests that the

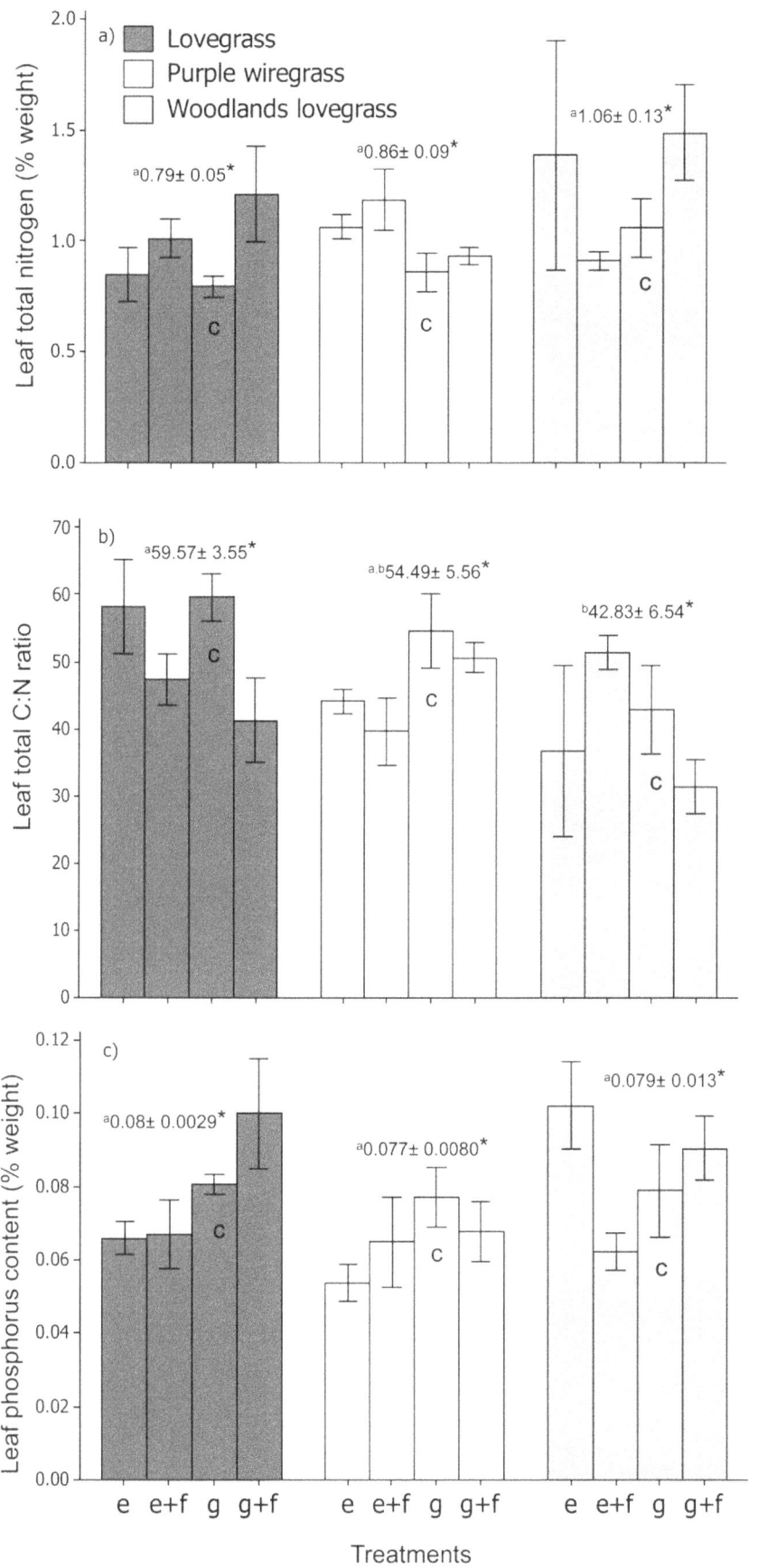

Figure 4. Mean leaf nutrient concentrations (± SE) for each species depending on the grazing and fertilizer treatments. Panel a) shows leaf total nitrogen concentration (% weight), b) leaf carbon to nitrogen ratios and c) leaf phosphorus concentration (% weight). Values shown in each panel are the mean leaf nutrient concentrations (± SE) for each species at the site regardless of treatment. Different letters indicate means are significantly different at p<0.05. **C** indicates control treatment, grazing/no fertilizer.

increased abundance of woodlands lovegrass may be driven by other factors such as increased rainfall in year 3 of the study. Mean rainfall in year 0 was 215 mm, which was lower than the local 20 year average of 600 mm [47]; while mean rainfall in year 3 was higher than the local average at 652 mm.

At disturbed sites, invasive exotic species may be successful because of traits that allow quick growth in response to increased resource conditions [40,41], but in generally low resource environment these species likely also need traits that temper

growth to survive lulls between resource pulses [48]. Pursuit of a tangible set of generic traits that distinguish exotics from natives may not be plausible or meaningful [49]; instead, we suggest the pursuit should focus on plasticity, as this may be the trait that leads to characteristically dominant plant species whether native or exotic. Lovegrass may have replaced a more plastic and characteristically dominant native species, and future studies should compare invasive and native species that are all generally considered to hold a similar hierarchical role in a community (i.e.

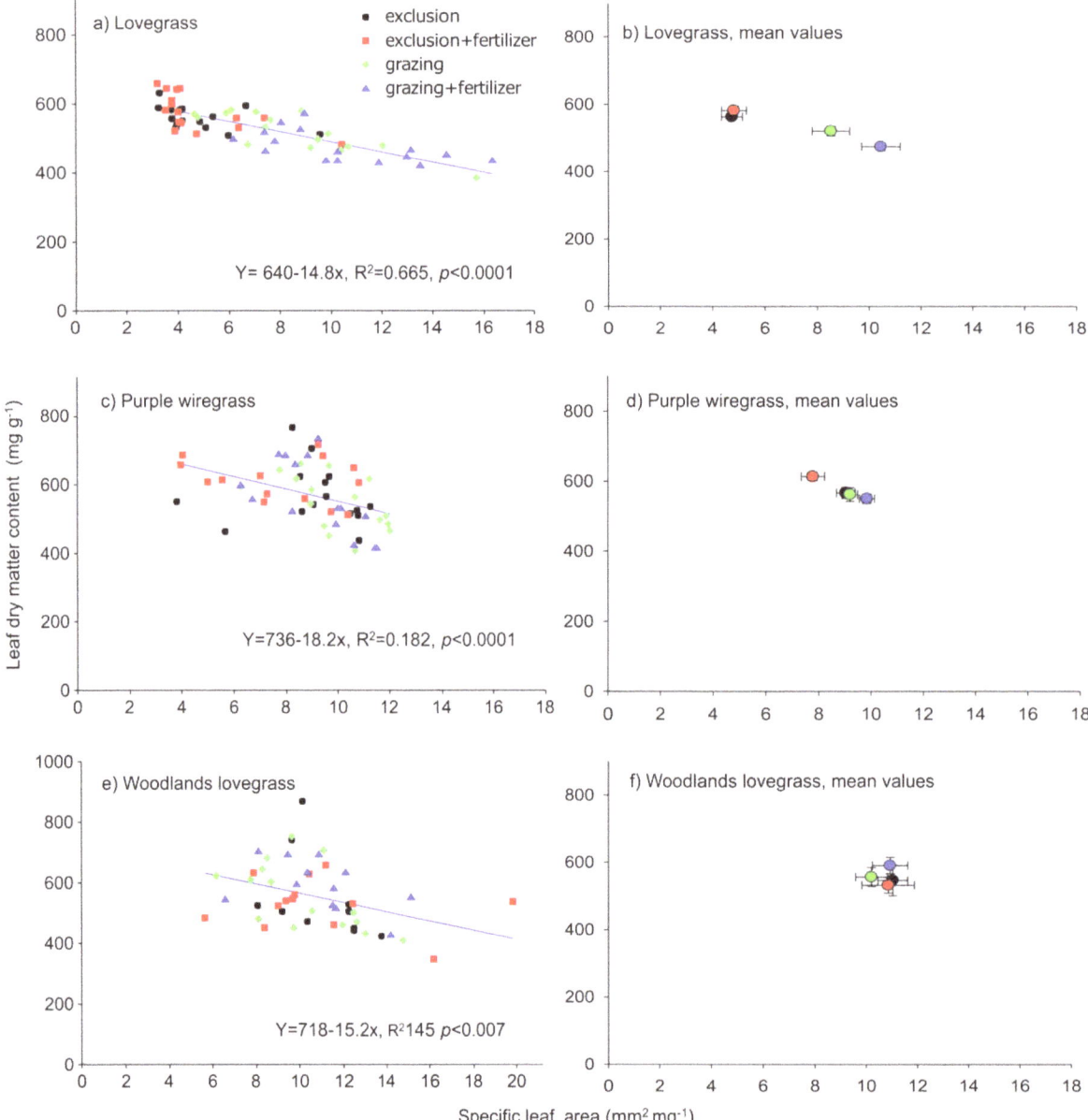

Figure 5. Correlations between LDMC and SLA values for each species depending on the four grazing and fertilizer treatments. Panel a), c) and e) show the mean LDMC and SLA values for each species collected from each plot and b), d), f) show the mean values for LDMC and SLA (±) for each treatment.

Table 3. Results from an ANOVA of LMEMs of leaf dry matter content (LDMC) and specific leaf area (SLA) for each of the grass species, with a fixed effects structure of grazing and fertilizer treatments and a random effects structure of block/plot.

Species & response variable	Fixed effects	F values (dfs as subscript), P values
a) Lovegrass LDMC (mg g^{-1})	grazing	$F_{1,2} = 5.60, P<0.20$
	fertilizer	$F_{1,58} = 2.58, P<0.10$
	grazing×fertilizer	$\mathbf{F_{1,58} = 10.10, P<0.002}$
SLA (mm^2 mg^{-1})	grazing	$F_{1,2} = 8.85, P<0.10$
	fertilizer	$\mathbf{F_{1,58} = 4.55, P<0.04}$
	grazing×fertilizer	$\mathbf{F_{1,58} = 3.89, P<0.05}$
b) Purple wiregrass LDMC (mg g^{-1})	grazing	$F_{1,2} = 0.40, P<0.60$
	fertilizer	$F_{1,58} = 1.91, P<0.20$
	grazing×fertilizer	$F_{1,58} = 1.00, P<0.30$
SLA (mm^2 mg^{-1})	grazing	$F_{1,2} = 0.64, P<0.50$
	fertilizer	$\mathbf{F_{1,58} = 4.25, P<0.05}$
	grazing×fertilizer	$F_{1,58} = 0.28, P<0.60$
c) Woodlands lovegrass LDMC (mg g^{-1})	grazing	$F_{1,2} = 0.08, P<0.80$
	fertilizer	$F_{1,58} = 0.17, P<0.70$
	grazing×fertilizer	$F_{1,58} = 0.51, P<0.50$
SLA (mm^2 mg^{-1})	grazing	$F_{1,2} = 0.10, P<0.80$
	fertilizer	$F_{1,58} = 0.54, P<0.50$
	grazing×fertilizer	$F_{1,58} = 0.35, P<0.60$

compare dominants to dominants, subordinates to subordinates and rare to rare).

Overall, our results show that plasticity at the species level, however, does not necessarily equate to a 'super invader'; instead, if plasticity of an undesirable species is understood, biotic interactions and resource availability can be manipulated to limit abundance. Exotic species with high species-level phenotypic plasticity may then be vulnerable to changed resource conditions as a direct result of this plasticity.

Materials and Methods

Study species

Lovegrass was introduced into Australia in the early 1900s for pasture improvement and soil conservation [50], and is now found in every Australian state, spreading into many regions where it was never intentionally introduced [51]. The increased dominance of lovegrass poses a significant threat to native biodiversity because of its ability to dominant communities, and the sustainability of production in farming communities because it is not palatable (low nutrients and crude protein content) to grazing livestock in the low

Table 4. For each of the three grass species, treatments that significantly predicted differences in total leaf nitrogen concentration, leaf carbon to nitrogen ratio and total leaf phosphorus concentration.

	Predictor variables	F values (df as subscript), P value
a) Lovegrass leaves		
Total nitrogen (% weight)	Fertilizer treatment	$F_1 = 4.67, P<0.06$
C:N ratio	Fertilizer treatment	$F_1 = 7.26, P<0.02$
Total phosphorus (% weight)	Grazing treatment	$F_1 = 6.16, P<0.03$
b) Purple wiregrass leaves		
Total nitrogen (% weight)	Grazing treatment	$F_1 = 5.33, P<0.08$
C:N ratio	Grazing treatment	$F_1 = 7.17, P<0.06$
Total phosphorus (% weight)	NS	
d) Woodlands lovegrass leaves		
Total nitrogen (% weight)	NS	
C:N ratio	Grazing treatment	$F_1 = 7.82, P<0.06$
Total phosphorus (% weight)	Grazing treatment	$F_1 = 9.33, P<0.02$

Table 5. Summary of the traits of each grass species that showed a plastic response to the treatments according to expectations "√", contrary to expectations "X" or traits that did not change in response to the treatments "-".

Trait	Eragrostis curvula lovegrass, Exotic grass	Aristida personata purple wiregrass Native grass	Eragrostis sororia woodlands lovegrass, Native grass
LDMC (mg g^{-1})	✓ (grazing & fertilizer)	-	-
SLA (mm^2 mg^{-1})	✓ (grazing & fertilizer)	✓ (fertilizer)	-
Amax (µmol co$_2$ g^{-1} s^{-1})	✓ (grazing)	-	-
PNUE (µmol co$_2$ g^{-1}N s^{-1})	✓ (grazing)	-	-
Leaf total Nitrogen (%)	✓ (fertilizer)	X (grazing)	-
Leaf total C:N	✓ (fertilizer)	X (grazing)	X (grazing)
Leaf Phosphorus (%)	✓ (grazing)	X (grazing)	X (grazing)

productivity regions where it is spreading, and difficult to control [51].

In June 2006, we established a large field trial on a private cattle grazing property in the Millmerran region of south-western Queensland, Australia [43]. The field site had been grazed by cattle with a low stocking rate since 1980 and has never been cultivated or fertilized. Lovegrass was first identified on the property in 1998 by the landholder. Average rainfall of this area is 600 mm p.a. [47] with two-thirds of the rain occurring during the summer months from October to April. The soil is a yellow sodosol derived from sandstone, which is characteristically low in nutrients, slightly acidic (ranging from pH 4.8 to 5.9), low in water holding capacity, and highly susceptible to compaction [47].

Data collection and sampling design

In this experiment, we measured traits and abundances from a subset of treatments and plots from a larger field trial with a randomized split-plot design [43]. In June 2006, four large blocks (35×40 m) were established randomly in a pasture dominated by lovegrass. Two blocks were fenced to exclude grazing by cattle and limit access by other native and exotic gazers (e.g. kangaroos, wallabies, hares and rabbits). The other two blocks were left open to grazing. In each block, we established 48 plots (each was 9 m^2 in size with an additional 4 m^2 buffer between each plot). In this study, we sampled 16 plots in each block, 8 fertilized plots and 8 unfertilized plots. We applied a slow-release fertilizer to half of the plots in a pellet form (N 21.6%, P 1.1%, K 4.1%) at a low application rate of 2 kg/ha at the start of each growing season from 2006 to 2009, which can begin anytime between October and December depending on rainfall. In this experiment, the grazed/-unfertilized treatment is considered the control treatment, because this was the disturbance acting on the site prior to the start of the experiment. Firn *et al.* (2010) contains species abundance results from 2006 to 2008.

In December 2009, prior to applying the fourth year of treatments, we measured species abundance and leaf traits (specific leaf area, leaf dry matter content, *Amax* and leaf nitrogen, phosphorus, and carbon to nitrogen levels) from the grazing and fertilizer treatment combinations. The abundances of all species were recorded within each plot in the central 9 m^2 section using the point-intercept method (modified from [52]). A 4 mm dowel was placed vertically on set points along a grid of 100 points. Relative abundance was measured by identifying and counting each leaf, stem and inflorescence that touched the dowel at each point along the grid.

To measure SLA, LDMC, LN traits, we collected five young but fully expanded leaves from three mature individuals of each species within each of the 64 plots, using the standardised protocols detailed by Cornelissen et al. [53] and the rehydration methods proposed by Garnier et al. [54]. Because of the low abundance of woodlands lovegrass, we did not find individuals of this species in all plots, but we were able to collect samples from 42 plots. The leaves collected from each species in each plot were combined (leaves of most species were small), weighed and scanned for area, using a flat bed scanner (Epson perfection V300) and image analysis software (ImageJ, [55]). Leaf samples were then dried in an oven for 72 hours at 60°C and re-weighed.

These leaf samples were then bulk sampled by species and treatment and analysed for total nitrogen and carbon concentration using a LECO CNS 2000 combustion analyser set at 1100°C. Total leaf phosphorus concentration was measured using a Varian Vista Pro ICPOES on samples digested in 5:1 nitric:perchloric acid (six samples were analysed per species per treatment) [56]. We measured leaf nitrogen and phosphorus concentration because extensive research has shown a stronger relationship between these nutrients within the leaf economic spectrum than other nutrients [57]. Six soil samples (core radius of 5 cm, 10 cm deep) were collected from each plot at the same time as the botanical surveys. Available soil nitrate, ammonium and phosphorus were analysed with colorimetric methods using a SEAL AQ2 [58]. Soil and leaf nutrient analyses were conducted by the Analytical Services Unit, School of Land and Food Sciences, the University of Queensland.

In September 2010, we measured assimilation rates (A_{max}) of eight individuals of lovegrass and purple wiregrass within four fertilised plots and four unfertilized plots in each of the four blocks. We were very careful with our leaf selection, choosing young, intact leaves with a healthy appearance and growing in full sun. We took measurements between 6:00 am and 10:00 am over five days to standardize measurements. Because of the smaller size of woodlands lovegrass leaves and its low abundance, we were only able to measure four individuals within each of the treatments. We used a LI-COR LI-6400 photosynthesis system and the narrow leaf chamber LI-COR LI-6400-11. For assimilation rates ambient CO_2 conditions were maintained at 400 µmol L^{-1}, relative humidity at 40–50%, leaf temperature at 22–23°C and PAR at 1300–1530 µL L^{-1}. To fill the chamber, multiple leaves growing in full sun were selected from each individual and leaf area was measured with a LI-COR, LI-3000c Portable Area Meter. We calculated photosynthetic nitrogen use efficiency (PNUE) as the ratio of *Amax* to leaf nitrogen.

Data analysis

To analyse the effects of different treatments on species abundance and leaf traits, we developed Linear Mixed Effects Models (hereafter LMEM), using R 2.12.1 (R Foundation for Statistical Computing©) and the nlme package. We modelled the abundance (arc-sine transformed) of each study species in time 3 as a function of grazing and fertilizer treatments and abundance at time 0 as a covariate with a nested random effects structure of block/plot. We also modelled each of the leaf traits and soil nutrient levels as a function of the grazing and fertilizer treatments with a nested random effects structure of block/plot. Maximum likelihood was used when comparing nested models to simplify the model for fixed effects [59,60]. We used diagnostic plots to check model assumptions [59]; there was no evidence of correlation of observations within groups and we assumed that within group errors were normally distributed. Finally, we used ANOVAs to assess the significance of the fixed effects within the LMEMs (Pinheiro and Bates, 2000). To analyse leaf nutrient concentrations, we used ANOVAs as opposed to LMEMs because these values were measured from bulked samples.

Supporting Information

Table S1 Effect of treatments on Soil NO_3, NH_4 and PO_4 levels in year 4. Results of an ANOVA conducted to assess the significance of the fixed effects for LMEMs of soil nutrient levels, with a fixed effects structure of grazing and fertilizer treatments, and a random effects structure of block/plot.

Table S2 Effect of treatments on leaf nitrogen concentration, leaf carbon to nitrogen ratio and total leaf phosphorus concentration. Results of an ANOVA conducted for each of the three grass species.

Figure S1 Soil nitrate, ammonium and phosphate levels taken across the treatments in year 3 of the field trial. e = grazing exclusion treatment, e+f = grazing exclusion and fertilized treatment, g = grazing treatment, and g+f = grazing and fertilized treatment.

Figure S2 Mean assimilation rates (± SE) for each species depending on the grazing and fertilizer treatments. **C** indicates the control treatment grazing/no fertilizer.

Acknowledgments

Thank you to the land-owner for generously permitting us to set-up this experiment on the property and Katrina Cousins, Karri Hartley and Richard Unwin for help in the field and laboratory. Thank you to Dr. S. McIntyre and Dr. H. Murphy from CSIRO Ecosystem Sciences for commenting on an earlier draft of this manuscript.

Author Contributions

Conceived and designed the experiments: JF SMP YMB. Performed the experiments: JF. Analyzed the data: JF. Wrote the paper: JF SMP YMB.

References

1. Sax DF, Brown JH (2000) The paradox of invasion. Global Ecology and Biogeography 9: 363–371.
2. Rout ME, Callaway RM (2009) An Invasive Plant Paradox. Science 324: 734–735.
3. MacDougall AS, Turkington R (2005) Are invasive species the drivers or passengers of change in degraded ecosystems? Ecology 86: 42–55.
4. Hobbs RJ, Atkins L (1988) Effect of disturbance and nutrient addition on native and introduced annuals in plant communities in the Western Australian wheatbelt. Australian Journal of Ecology 13: 171–179.
5. HilleRisLambers J, Yelenik SG, Colman BP, Levine JM (2010) California annual grass invaders: the drivers or passengers of change? Journal of Ecology 98: 1147–1156.
6. Seabloom EW, Harpole WS, Reichman OJ, Tilman D (2003) Invasion, competitive dominance, and resource use by exotic and native California grassland species. Proceedings of the National Academy of Science 100: 13384–13389.
7. Suding KN, Goldberg D (2001) Do disturbances alter competitive hierarchies? Mechanisms of change following gap creation. Ecology 82: 2133–2149.
8. Firn J, Rout T, Possingham HP, Buckley YM (2008) Managing beyond the invader: manipulating disturbance of natives simplifies control efforts. Journal of Applied Ecology 45: 1143–1151.
9. Hobbs RJ, Arico S, Aronson J, Baron JS, Bridgewater P, et al. (2006) Novel Ecosystems: theoretical and management aspects of the new ecological world order. Global Ecology and Biogeography 15: 1–7.
10. Zavaleta ES, Hobbs RJ, Mooney HA (2001) Viewing invasive species removal in a whole-ecosystem context. Trends in Ecology and Evolution 16: 454–459.
11. Pickett STA, Cadenasso ML, Meiners SJ (2008) Ever since Clements: from succession to vegetation dynamics and understanding intervention. Applied Vegetation Science 12: 9–21.
12. McIntyre S, Lavorel S (2006) A conceptual model of land use effects on the structure and function of herbaceous vegetation. Agriculture, Ecosystems and Environment 2007: 11–21.
13. Suding KN, Lavorel S, Chapin FS, Cornelissen JHC, Diaz S, et al. (2008) Scaling environmental change through the community-level: a trait-based response-and-effect framework for plants. Global Change Biology 14: 1125–1140.
14. Daehler CC (2003) Performance comparisons of co-occurring native and alien invasive plants: implications for conservation and restoration. Annual Review of Ecology and Systematics 34: 183–211.
15. Bazzaz FA (1986) Life history of colonizing plants: some demographic, genetic and physiological features. In: Mooney HA, Drake JA, eds. Ecology of Biological Invasion of North America and Hawaii. New York: Springer-Verlag. pp 96–110.
16. Rejmanek M, Richardson DM (1996) What attributes make some plants species more invasive? Ecology 77: 1655–1661.
17. Baker HG (1964) Characteristics and modes of origin of weeds. In: Baker HG, Ledyard Stebbins G, eds. The Genetics of Colonizing Species. New York: Academic Press Inc. pp 147–172.
18. Leishman MR, Haslehurst T, Ares A, Baruch Z (2007) Leaf trait relationships of native and invasive plants: community- and global-scale comparisons. New Phytologist 176: 635–643.
19. Baruch Z, Goldstein G (1999) Leaf contruction cost, nutrient concentration, and net CO_2 assimilation of native and invasive species in Hawaii. Oecologia 121: 183–192.
20. Durand LZ, Goldstein G (2001) Growth, leaf characteristics and spore production in native and invasive tree ferns in Hawaii. American Fern Society 91: 25–35.
21. Gulias J, Flexas J, Mus M, Cifre J, Lefi E, et al. (2003) Relationship between maximum leaf photosynthesis, nitrogen content and specific leaf area in Balearic endemic and non-endemic Mediterranean species. Annals of Botany 92: 215–222.
22. Grotkopp E, Rejmanek M, Rost TL (2002) Toward a causal explanation of plant invasiveness: seedling growth and life-history strategies of 29 pine (*Pinus*) species. American Naturalist 159: 396–419.
23. Wright IJ, Reich PB, Westoby M, Ackerly DD, Baruch Z, et al. (2004) The worldwide leaf economics spectrum. Nature 428: 821–827.
24. Westoby M, Wright IJ (2006) Land-plant ecology on the basis of functional traits. Trends in Ecology & Evolution 21: 261–268.
25. Diaz S, Hodgson JG, Thompson K, Cabido M, Cornelissen JHC, et al. (2004) The plant traits that drive ecosystems: Evidence from three continents. Journal of Vegetation Science 15: 295–304.
26. Moles AT, Gruber MAM, Bonser S (2008) A new framework for predicting invasive plant species. Journal of Ecology 96: 13–17.
27. Leishman MR, Thomson VP, Cooke J (2010) Native and exotic invasive plants have fundamentally similar carbon capture strategies. Journal of Ecology 98: 28–42.
28. Albert CH, Grassein F, Schurr FM, Vieilledent G, Violle C (2011) When and how should intraspecific variability be considered in trait-based plant ecology? Perspectives in Plant Ecology, Evolution and Systematics doi:10.1016/j.ppees.2011.04.003. pp 1–9.
29. Albert CH, Thuiller W, Yoccoz NG, Soudant A, Boucher F, et al. (2010) Intraspecific functional variabillity: extent, structure and sources of variation. Journal of Ecology 98: 604–613.
30. Richards CL, Bossdorf O, Muth NZ, Gurevitch J, Pigliucci M (2006) Jack of all trades, master of some? On the role of phenotypic plasticity in plant invasions. Ecology Letters 9: 981–993.

31. DeWitt TJ, Sih A, Wilson DS (1998) Costs and limits of phenotypic plasticity. Trends in Ecology and Evolution 13: 77–81.

32. Elton CS (1958) The Ecology of Invasions by Animals and Plants. London: Methuen and CO Ltd.

33. Funk JL (2008) Differences in plasticity between invasive and native plants from a low resource environment. Journal of Ecology 96: 1162–1173.

34. Burns JH, Winn AA (2006) A comparison of plastic responses to competition by invasive and non-invasive congeners in Commelinaceae. Biological Invasions 8: 797–807.

35. Muth NZ, Pigliucci M (2007) Implementation of a novel framework for assessing species plasticity in biological invasions: responses of Centaurea and Crepis to phosphorus and water availability. Journal of Ecology 95: 1001–1013.

36. Godoy O, Valladares F, Castro-Diez P (2011) Multispecies comparison reveals that invasive and native plants differ in their traits but not in their plasticity. Functional Ecology 25: 1428–1259.

37. Zou J, Rogers WE, Siemann E (2007) Differences in morphological and physiological traits between native and invasive populations of *Sapium sebiferum*. Functional Ecology 21: 721–730.

38. Hulme PE (2008) Phenotypic plasticity and plant invasions: is it all Jack? Functional Ecology 22: 3–7.

39. Davidson AM, Jennions M, Nicotra AB (2011) Do invasive species show higher phenotypic plasticity than native species and, if so, is it adaptive? A meta-analyses. Ecology Letters 14: 419–431.

40. Melbourne BA, Cornell HV, Davies KF, Dugaw CJ, Elmendorf S, et al. (2007) Invasion in a heterogeneous world: resistance, coexistence or hostile takeover? Ecology Letters 10: 77–94.

41. Davis MA, Grime JP, Thompson K (2000) Fluctuating resources in plant communities: a general theory of invasibility. Journal of Ecology 88: 528–534.

42. Pysek P, Richardson DM (2007) Traits associated with invasiveness in alien plants: where do we stand? In: Nentwig W, ed. Biological Invasions. Berlin Heidelberg: Springer-Verlag.

43. Firn J, House APN, Buckley YM (2010) Alternative states models provide an effective framework for invasive species control and restoration of native communities. Journal of Applied Ecology 47: 96–105.

44. McIntyre S, Martin TG, Heard KM, Kinloch J (2005) Plant traits predict impact of invading species: an analyses of herbaceous vegetation in the subtropics. Australian Journal of Botany 53: 757–770.

45. Funk JL, Vitousek PM (2007) Resource-use efficiency and plant invasion in low-resource systems. Nature 446: 1079–1081.

46. Grime JP (1998) Benefits of plant diversity to ecosystems: immediate, filter and founder effects. Journal of Ecology 86: 902–910.

47. Biggs A, Coutts A, Harris PS, eds. (1999) Central Darling Down Land Management Manual: Department of Primary Industries.

48. Chesson P, Gebauer RLE, Schwinning S, Huntly N, Wiegand K, et al. (2004) Resource pulses, species interactions, and diversity maintenance in arid and semi-arid environments. Oecologia 141: 236–253.

49. Thompson K, Davis MA (2011) Why research on traits of invasive plants tells us very little. Trends in Ecology and Evolution 26: 155–156.

50. Leigh JH, Davidson RL (1968) *Eragrostis curvula* (Schrad.) Nees and some other African lovegrasses. Plant Introduction Review 5: 21–46.

51. Firn J (2009) African lovegrass in Australia: a valuable pasture species or embarrassing invader. Tropical Grasslands 43: 86–97.

52. Everson CS (1987) A comparison of six methods of botanical analysis in the montane grasslands of Natal. Vegetatio 73: 47–51.

53. Cornelissen JHC, Lavorel S, Garnier E, Diaz S, Buchmann N, et al. (2003) A handbook of protocols for standardised and easy measurement of plant functional traits worldwide. Australian Journal of Botany 51: 335–380.

54. Garnier E, Shipley B, Roumet C, Laurent G (2001) A standardized protocol for the determination of specific leaf area and leaf dry matter content. Functional Ecology 15: 688–695.

55. Rasband WS (1997–2011) ImageJ website, U.S. National Institutes of Health, Betheseda Maryland, U.S.A. http://imagejnihgov/ij/. Accessed 2012 March 29.

56. Reuter DJ, Robinson JB (1986) Plant Analysis and Interpretation Manual. Melbourne, Australia: Inkata Press.

57. Wright IJ, Reich PB, Cornelissen JHC, Falster DS, Garnier E, et al. (2005) Assessing the generality of global leaf trait relationships. New Phytologist 166: 485–496.

58. Raymont GE, Higginson FR (1992) The Australian Handbook of Soil and Water Chemical Methods. Melbourne: Inkata Press.

59. Pinheiro JC, Bates DM (2000) Mixed-Effects Models in S and S-Plus; Chambers J, Eddy W, Hardle W, Sheather S, Tierney L, eds. New York: Springer Verlag.

60. Ives AR, Zhu J (2006) Statistics for correlated data: phylogenies, space, and time. Ecological Applications 16: 20–32.

61. Voight PW, Kneebone WR, McIlvain EH, Shoop MC, Webster JE (1968) Palatability, chemical composition, and animal gains from selection pf Weeping Lovegrass, Eragrostis curvula (Schrad.) Nees. Agronomy Journal 62: 673–676.

62. Mitchell M (2002) Native Grasses: an identification handbook for Temperate Australia. Collingwood: Landlinks Press 42 pages p.

63. Anderson E (2003) Plants of Central Queensland. Brisbane: Department of Primary Industries. 271 p.

64. Henry DR, Hall TJ, Jordan DJ, Milson JA, Schefe CM, et al. (1995) Pasture Plants of Southern Inland Queensland; Pavasaris S, ed. Brisbane: Queensland Department of Primary Industries.

65. Sharp D, Simon BK (2002) AusGrass: Grasses of Australia Australian Biological Resources Study, Canberra and Environmental Protection Agency, Queensland.

4

Biofortification and Bioavailability of Rice Grain Zinc as Affected by Different Forms of Foliar Zinc Fertilization

Yanyan Wei[⁹], M. J. I. Shohag[⁹], Xiaoe Yang*

Ministry of Education (MOE) Key Laboratory of Environmental Remediation and Ecosystem Health, College of Environmental and Resources Science, Zhejiang University, Hangzhou, People's Republic of China

Abstract

Background: Zinc (Zn) biofortification through foliar Zn application is an attractive strategy to reduce human Zn deficiency. However, little is known about the biofortification efficiency and bioavailability of rice grain from different forms of foliar Zn fertilizers.

Methodology/Principal Findings: Four different Zn forms were applied as a foliar treatment among three rice cultivars under field trial. Zinc bioavailability was assessed by *in vitro* digestion/Caco-2 cell model. Foliar Zn fertilization was an effective agronomic practice to promote grain Zn concentration and Zn bioavailability among three rice cultivars, especially, in case of Zn-amino acid and $ZnSO_4$. On average, Zn-amino acid and $ZnSO_4$ increased Zn concentration in polished rice up to 24.04% and 22.47%, respectively. On average, Zn-amino acid and $ZnSO_4$ increased Zn bioavailability in polished rice up to 68.37% and 64.43%, respectively. The effectiveness of foliar applied Zn-amino acid and $ZnSO_4$ were higher than Zn-EDTA and Zn-Citrate on improvement of Zn concentration, and reduction of phytic acid, as a results higher accumulation of bioavailable Zn in polished rice. Moreover, foliar Zn application could maintain grain yield, the protein and minerals (Fe and Ca) quality of the polished rice.

Conclusions: Foliar application of Zn in rice offers a practical and useful approach to improve bioavailable Zn in polished rice. According to current study, Zn-amino acid and $ZnSO_4$ are recommended as excellent foliar Zn forms to ongoing agronomic biofortification.

Editor: Keqiang Wu, National Taiwan University, Taiwan

Funding: This research work was financially supported by the HarvestPlus-China Program (8234), International Cooperation Program from the Ministry of Science and Technology, China (2006AA06Z386), Project from the Zhejiang Provincial Department of Education, Zhejiang, China (20100339) and Fundamental Research Funds for the Central Universities (2012FZA6008). The funders had no role in study design, data collection and analysis, decision to publish, or preparation of the manuscript.

Competing Interests: The authors have declared that no competing interests exist.

* E-mail: xyang571@yahoo.com

⁹ These authors contributed equally to this work.

Introduction

Zinc (Zn) deficiency is a well documented global health problem, affecting nearly half of the world population, particular in developing countries, where high proportion of cereal crops, such as rice and wheat, consumed as a staple food [1,2]. The reliance on cereal based food induce Zn deficiency related health problem, such as impairments in physical growth, immune system and brain function [3,4]. Among the cereals, Rice (*Oryza stavia* L.), being one of the leading staple crop for half of the world's population and, hence, is the main source of Zn to human [5]. Rice, however unfortunately, is a poor source of metabolizable Zn, due to inherently low in Zn content and the bioavailable Zn [6]. Enrichment of rice with high bioavailable Zn is, therefore, suggested as a way to generate major health benefits for a large number of susceptible people.

Zinc biofortification, which aims to enhance Zn concentration as well as bioavailability of rice grain, is considered as the more sustainable and economical solution to address human Zn deficiency [7]. Genetic biofortification and agronomic biofortifica-

tion are two important agricultural tools to improve rice grain Zn concentration [2,8]. However, yield factor, interactions between genotype and environment, lack of sufficient genetic diversity in current cultivars for breeding program, consumer resistance and safety of genetically modified crops are the main bottlenecks of genetic biofortification [2,9,10,11]. The traditional and efficient strategy of agronomic biofortification, such as Zn fertilization is, therefore, urgent, essential and rapid solution for improving Zn concentration in rice grain to address the ongoing human Zn deficiency.

Three methods, including soil amendment, seed priming and foliar application, used in Zn fertilizations, have been extensively reviewed [2]. In recent years, a considerable progress has been made on the impact of foliar Zn fertilization on biofortification of Zn in rice grain [12,13,14], since it has the advantages of low application rates and avoiding Zn losses through soil fixation [15]. Furthermore, foliar applied Zn caused greater increases in brown rice Zn concentration than soil application [13,14]. There is evidence in literature demonstrating that foliar applied Zn can be absorbed by leaf epidermis, and remobilized and transferred into the rice grains through the phloem [16] and several members of the Zn-regulated transporters regulate this process [17]. In most of those literatures, the reported data are mostly based on brown rice. As polished rice is the main consumed portion by human, rare information was found on Zn concentration in polished rice after foliar Zn fertilizations. Moreover, time of foliar application and the different forms of foliar Zn fertilizers may differentially influence grain Zn concentration. In recent past, several studies have been conducted to adjust time of foliar Zn application in cereal crops [2,13,18]. It is now well established that foliar Zn application after flowering stage (e.g., at early milk plus dough stages) more distinctly increase the grain Zn concentration [13]. On the other hand, different Zn fertilizers such as inorganic and organic Zn salts play a fundamental role in the way in nutrient transport from leave to the grain [19]. Unfortunately, studies evaluating the effectiveness of foliar application of different Zn forms on rice grain Zn accumulation are still rare.

The metabolizable Zn from biofortified crop grain not only depends on net Zn concentration, but also a large extent on the bioavailability of Zn. Zinc bioavailability defined as the proportion of the total amount of Zn that is potentially absorbable in a metabolically active form [20]. Phytic acid, the naturally occurring anti-nutrient presents in the seed, reduces the bioavailability of Zn, because of its ability to form complex with Zn, and inhibits Zn solubility, digestibility and absorption in human body [21]. Although, it is assume that foliar Zn fertilization improved Zn bioavailability, but till now there are rare studies on the Zn bioavailability of rice grain deserved from different forms of foliar Zn application [2]. Hence, an in vivo approach to assess the potential benefits of different forms of foliar Zn application on grain Zn bioavailability is required.

Ideally, Zn bioavailability in crop grains should be evaluated through in vivo human study. However, complexity to perform large-scale screening of sample and cost limit their applicability [22]. In vitro digestion/Caco-2 cell model has been proposed as an alternative to in vivo method for estimating mineral bioavailability in diets. In recent years, in vitro digestion/Caco-2 cell culture model is being utilized for absorption studies involving Zn. This in vitro model is currently considered as the best approach, in term of cost and time, to investigate the bioavailability of different food components as a prelude to in vivo study [23]. The present study used this model to assess the bioavailability of Zn from polished rice grain fortified with different forms of foliar Zn fertilization.

Viewing the above circumstances, the current study were aimed: (i) to assess the effect of different forms of foliar Zn fertilizer on Zn concentration in brown rice and polished rice, (ii) to assess the effect of different forms of foliar Zn fertilizer on Zn bioavailability in polished rice. The findings of the current study were used to design experiment to identify some useful foliar Zn fertilizer for increasing the level of bioavailable Zn in rice grain.

Materials and Methods

Field Experiment and Sampling

Field location. Experimental site was Longyou, Zhejiang province (29° 02′ N, 119° 11′ E), China. The climate of the experimental site is subtropical humid. The soil type of experimental field was periodical water logged paddy soil. Before the start of experiments, puddle layer (0–15 cm top soil) soil samples were taken from four random spots of the field and analyzed for various physico-chemical properties (Table1).

Experimental design and treatment. Experimental design was a split plot with four replications. Foliar Zn fertilization treatments were treated as main plot and rice cultivars as sub-plot.

Table 1. Selected physical and chemical properties of the soils.

Characteristics	Value
pH (H$_2$O, 20 °C)	5.8
Total N (g kg^{-1})	1.36
Organic matter (g kg^{-1})	13.70
Olsen P (mg kg^{-1})	42.6
CaCO$_3$ (%)	1.98
NH$_4$OAC-exchangeable K (mg kg^{-1})	90.35
DTPA-extractable Zn (mg kg^{-1})	3.84
DTPA-extractable Fe (mg kg^{-1})	198.45

Foliar Zn fertilization treatments comprised of four different forms of Zn fertilizer: (i) ZnNa$_2$EDTA (Zn-EDTA), Zn-EDTA was the common Zn fertilizer which contain 9% Zn (ii) Zn-Citrate, in which Zn content was 10% (iii) ZnSO$_4$·7H$_2$O (ZnSO$_4$), common Zn fertilizer, in which Zn content was 36% (iv) Zn-amino acids (Zn-AA), Zn-AA contains Zn as ZnSO$_4$ (10%) and amino acid (25%) and, (v) Control, sprayed with distilled water. Three rice (Oryza sativa L.) varieties differ in their grain Zn concentration namely Hai7, Bing91185 and Biyuzaonuo were selected according to our previous study [24]. Thus, there were 60 plots with each 4 m^2 (2×2 m).

Thirty days old seedlings of each cultivar were transplanted to the plot. Before transplanting, the standard recommended dose of NPK fertilizer was applied to all plots at rates of 187.5 kg N ha^{-1} (70% applied as basal dose and 30% as topdressing at panicle initiation stage), 70 kg P$_2$O$_5$ ha^{-1} and 93 kg K$_2$O ha^{-1}. Water management was the same as conventional rice farming practice. The foliar Zn was applied three times, one time at panicle initiation stage, two times at 7 days after flowering stage. Spray was applied after sunset. During spray, soil surface was covered to minimize the contamination of soil with foliar applied Zn. The concentration of Zn fertilizer was 0.2%. The amount of foliar Zn applied was equivalent to 2.5 kg Zn ha^{-1}. All foliar sprays contained 0.01% (v/v) Tween80 as a surfactant.

Rice sample preparation. Plants were harvested from the center of each plot at maturity and were manually threshed to separate grains. Rice grains were air dried; the brown rice was prepared by removing the husk using a laboratory de-husker (JLGJ4.5, Taizhou Cereal and Oil Instrument Co. Ltd., Zhejiang, China), the polished rice was prepared by polishing the bran by a laboratory polishing machine (JNMJ3, Taizhou Cereal and Oil Instrument Co. Ltd., Zhejiang, China). The rice samples were powdered to make flour by using a ball mill (Retsch, MM-301, Germany), then put in the plastic bag and keep at −20°C until analysis. A part of rice was cooked for 15 min with 1:2 rice/ deionized water (w/v). The cooked rice samples were then homogenized in a polytron homogenizer and then the homogenates were frozen and lyophilized before testing via the in vitro digestion/Caco-2 cell model.

Chemical Analysis

Mineral concentration determination. The ground rice samples (0.3 g) of each treatment were placed in to PTFE digestion tube and, digested with nitric acid (2 mL) and hydrogen peroxide (0.5 mL). After cooling, the digestion solution was transferred to a 25 mL volumetric flask, made up the volume with deionized water. The concentrations of Zn, iron (Fe), calcium (Ca) in sample were determined by inductively coupled plasma mass spectrometry (ICP-MS, Agilent 7500a, Agilent Technologies, CA, USA) following our previously described method [16].

Phytic acid determination. Phytic acid from the rice samples was determined by the method described by Dai et al., [25]. Briefly, 0.5 g of rice flour was extracted with 10 mL of 0.2 M HCl for 2 h by a rotary shaker and then centrifuged at 10000 g for 10 min. The clear supernatant was collected, and 2 mL of 0.2% $FeCl_3$ was added to 2.5 mL of supernatant. The resulting solution was mixed thoroughly, heated in a boiling water bath for 30 min, cooled in room temperature, and centrifuged at 10000 g for 15 min. Then supernatant was discarded and the residue in the tube washed three times with 5 mL of deionized water. The tube was then centrifuged again at 10000 g for 10 min after adding 3 mL of 1.5 M NaOH to it. The supernatant was discarded again, and 3 mL of 0.5 M HCl was added to the tube to dissolve the residue. Finally, deionized water was added to the solution made up to the volume of 10 mL. The Fe concentration in the solution was measured by ICP-MS (Agilent 7500a, Agilent Technologies, CA, USA). The phytic acid content was calculated by multiplying Fe content by the factor 4.2.

Protein content determination. Rice samples were analyzed for protein content by determination of total nitrogen. The Kjeldahl method was used to determine total nitrogen. Rice samples (0.5 g) were digested by 20 mL H_2SO_4, and then distilled in KjelFlex K-360 (Buchi, Flawil, Switzerland) with 40% (w/v) NaOH and 2% (w/v) boric acid (methyl red and bromcresol green used as an indicator solution), then titrated with 0.02 mM H_2SO_4. The protein content was subsequently calculated by multiplying nitrogen content by a conversion factor of 5.95 [26].

Zinc Bioavailability Assay

In vitro digestion of rice sample. The *in vitro* digestion method according to our previously described method [27]. Briefly, 5 g of cooked rice powder was mixed with 15 mL of saline buffer (140 mM NaCl, 5 mM KCl), and pH was adjusted to 2 with 6 M HCl. Then, the sample was mixed with 0.5 mL of pepsin solution (0.2 g pepsin in 5 mL of 0.1 M HCl) and incubated on a shaking water bath for 2 h at 37°C. After 2 h of gastric digestion, the pH of digest was adjusted to 5.0. The intestinal phase of digestion was then initiated with the addition of 2.5 mL of pancreatin-bile solution (0.45 g of bile salts and 0.075 g of pancreatin in 37.5 mL of 0.1 M $NaHCO_3$), and the digest samples were then incubated in a shaking water bath for 2 h. To stop the intestinal digestion, the digest sample was cooled in ice for 10 min, and then the pH was adjusted to 7.2 by adding 0.5 M NaOH. The obtained digests were heated at 100°C for 4 min to inhibit the proteases. The gastrointestinal digest were centrifuged 3500 g for 1 h at 4°C. Prior to addition of rice soluble fraction to the cells, glucose (5 mM) and HEPES (50 mM) were added in order to make it similar to culture media. Deionized water was used to adjust the osmolarity to 310 ± 10 mOsm kg^{-1} (Freezing point osmometer, Osmomat 030, Berlin, Germany). The supernatants (soluble fraction) were analyzed for Zn content and used in cell uptake assays.

Cell culture. The Caco-2 human intestinal cell line was purchased from Institute of Biochemistry and Cell Biology (SIBS, CAS, Shanghai, China) were used in experiments at passage 25–37. The cells were grown in 75 cm^2 flasks (Corning Inc., NY, USA) and maintained in high-glucose (4.5 g L^{-1}) Dulbecco's modified minimal essential medium (GIBCO, Grand Island, NY, USA), supplemented with 10% (v/v) heat-inactivated fetal bovine serum (GIBCO, Grand Island, NY, USA), 4 mM L-glutamine,

1% (v/v) non-essential amino acids and 1% (v/v) antibiotic solution (GIBCO, Grand Island, NY, USA). The cells were incubated in a cell culture incubator (Heraeus, BB15, Germany) set at 37°C, 5% CO_2 and 95% atmospheric air at constant humidity. After reaching 80% confluence, the cells were digested by using 0.25% trypsin (GIBCO, Grand Island, NY, USA), then cells were seeded at a density of 50000 cells cm^{-2} in 1.5 mL of complete DMEM in polyester membrane chamber inserts (24 mm diameter, 0.4 μm pore size; Costar Corp. NY, USA), the basal compartment contained 2.5 mL of complete Dulbecco's modified minimal essential medium, and the medium was changed every 2 days. Zinc bioavailability experiments were carried out 21 days after initial seeding. Subsequently, the integrity of Caco-2 cells monolayer was assessed and validated by transepithelial electrical resistance (TEER) using a Millicell-ERS meter (Millipore Corporation, Bedford, MA, USA). The method of TEER was according to the manufacturer's protocols. Only those filters that had TEER values >250 Ω cm^2 at the beginning and the end of the experiment were included. The monolayer used in this study exhibited adequate TEER values 500–680 Ω cm^2.

Zinc uptake experiment in Caco-2 cell. Prior to the Zn bioavailability experiment, the growth medium was removed from each culture well, and the cell layer was washed three times with Ca^{2+} and Mg^{2+} free HBSS at 37°C at pH 7.4. Then, 2.5 mL of transport solution (130 mM NaCl, 10 mM KCl, 1 mM $MgSO_4$, 5 mM glucose and 50 mM HEPES, pH 7.4) was added to the bottom chamber, and 1.5 mL of rice soluble fraction was added to the apical chamber. Cell cultures were incubated at 37°C under 5% CO_2, with relative humidity 95% for 2 h. After incubation, basolateral compartments were collected for determine of Zn transported across the monolayer. The cell monolayer was washed twice with ice cold HBSS at pH 7.4 to remove nonspecifically bound mineral and residual medium. The cell on filters were lysed by the addition of 1 mL of deionized water in the well, and then harvested. Cell viability after 2 h of exposure to the uptake solution was assessed by trypan blue exclusion and typically 85–95%.

The concentrations of Zn in cell retention (Zn fraction in the cell monolayer) and the solution of transport (Zn collected from basolateral compartment) were assessed by ICP-MS (Agilent 7500a, Agilent Technologies, CA, USA). Zinc uptake in Caco-2 cell model was calculated as the Zn content in cell retention plus the Zn content in transport. Solubility percentages were calculated by using following equation: solubility% = soluble fraction (μg of Zn g^{-1} sample) × 100/C, where C = total Zn content of sample; The following equation was used for Zn retention percentage: Zn retention% = Zn retention (μg $well^{-1}$) × 100/C, where C = mineral soluble added (μg); The following equation was used for Zn transport percentage: Zn transport% = Zn transport (μg $well^{-1}$) × 100/C, where C = mineral soluble added (μg); The following equation was used for Zn uptake percentage: Zn uptake% = (retention+ transport μg $well^{-1}$) × 100/C, where C = mineral soluble added (μg). Due to the differences among samples in terms of solubility of Zn after *in vitro* digestion, Zn uptake availability was expressed as Zn uptake efficiency, Zn uptake efficiency% = (% solubility × % uptake)/100. Bioavailable Zn (μg g^{-1} polished rice) = Zn concentration (mg kg^{-1}) × Zn uptake efficiency%.

Quality Control of Mineral Analysis

Standard reference material rice flour (SRM 1568a) from National Institute of Standards and Technology (Gaithersburg, MD, USA) was used to check the accuracy of Zn, Fe and Ca analysis. The measured value was 19.7 ± 0.2 mg kg^{-1} for Zn, 6.9 ± 0.3 mg kg^{-1} for Fe and 114.5 ± 1.2 mg kg^{-1} for Ca, which values were in accordance with the certified ranges of 19.4 ± 0.5 mg kg^{-1} for Zn, 7.4 ± 0.9 mg kg^{-1} for Fe and 118 ± 6 mg kg^{-1} for Ca.

Statistical Analysis

Statistical analysis of the data was performed using SPSS12.0 (SPSS, Inc., Chicago, IL, USA). The data were subjected to a separate analysis of variance (ANOVA) for each cultivar, and Fisher's least significant difference (LSD) at P<0.05 was used to determine differences between treatment means. The Pearson correlation procedure and linear regression model was used to evaluate the relationship between brown rice and polished rice Zn concentration.

Results

Biomass and Grain Yield

Biomass, grain yield, harvest index and thousand seed weight of rice did not different among the four different forms of foliar Zn treatments for all three rice cultivars (Table 2).

Zinc Concentration in Brown Rice and Polished Rice

Foliar Zn fertilization had significant (P<0.05) impact on Zn concentration in brown rice and polished rice (Fig. 1). Brown rice Zn concentration was significantly increased by foliar Zn fertilizations (Fig. 1A). Regardless of the three cultivars, Zn concentration in brown rice was increased from 30.28 mg kg^{-1} in the control, to 33.75 mg kg^{-1} by foliar Zn-EDTA application, to 35.07 mg kg^{-1} by foliar Zn-Citrate application, to 38.45 mg kg^{-1} by foliar ZnSO$_4$ application, to 39.84 mg kg^{-1} by foliar Zn-AA application, these represented increases of 11.46%, 15.81%, 27.26% and 31.58%, respectively. Zn concentration in polished rice was significantly increased by foliar Zn fertilizations (Fig. 1B). Regardless of the three cultivars, Zn concentration in polished rice

was increased from 22.92 mg kg^{-1} in the control, to 25.26 mg kg^{-1} by foliar Zn-EDTA application, to 26.09 mg kg^{-1} by foliar Zn-Citrate application, to 28.08 mg kg^{-1} by foliar ZnSO$_4$ application, to 28.67 mg kg^{-1} by foliar Zn-AA application, these represented increases of 10.22%, 13.82%, 22.47% and 24.04%, respectively. Thus, foliar Zn fertilization could increase Zn concentration in brown rice and polished rice depending on Zn form.

The concentration of Zn in brown rice and polished rice of all cultivars were significantly increased by different forms of foliar applied Zn (Fig. 2). In control, brown rice Zn concentrations were 24.71, 30.29 and 35.82 mg kg^{-1} in cultivar Hai7, Bing91185 and Biyuzaonuo, respectively. After application of Zn-EDTA through the foliage, Zn concentrations in brown rice were 27.98, 34.47 and 38.79 mg kg^{-1} in cultivar Hai7, Bing91185 and Biyuzaonuo, respectively. After foliar application of Zn-Citrate, Zn concentrations in brown rice were 28.15, 36.08 and 40.96 mg kg^{-1} in cultivar Hai7, Bing91185 and Biyuzaonuo, respectively. After foliar application of ZnSO$_4$, Zn concentrations in brown rice were 30.46, 38.94 and 46.20 mg kg^{-1} in cultivar Hai7, Bing91185 and Biyuzaonuo, respectively. After foliar application of Zn-AA, Zn concentrations in brown rice were 31.53, 40.62 and 47.38 mg kg^{-1} in cultivar Hai7, Bing91185 and Biyuzaonuo, respectively (Fig. 2A). Similar trends were found in polished rice (Fig. 2B), the cultivar Biyuzaonuo had the highest Zn concentration, while Hai7 had the lowest Zn concentration in all Zn treatments.

With respect to Zn content, a significant correlation was found between polished rice and brown rice (y = 0.619x +4.232, R^2 = 0.897, P<0.01).

Phytic Acid Content in Polished Rice

Foliar Zn fertilization reduced the phytic acid content in the polished rice (Table 3). Regardless of cultivar, phytic acid content in polished rice ranged from 2.25 mg g^{-1} in the control, to

Table 2. Effect of different forms of foliar Zn fertilization on the biomass, grain yield, harvest index and thousand seed weight of three rice cultivars.

Treatments	Cultivars[a]					
	Hai7	Bing91185	Biyuzaonuo	Hai7	Bing91185	Biyuzaonuo
	Biomass (t hm^{-1})			Grain yield (t hm^{-1})		
Control	19.50 a	18.61 a	18.35 a	7.81 a	7.98 a	7.50 a
Zn-EDTA	19.57 a	18.73 a	18.55 a	7.86 a	7.93 a	7.61 a
Zn-Citrate	19.79 a	18.66 a	18.43 a	8.00 a	8.03 a	7.53 a
ZnSO$_4$	19.63 a	18.39 a	18.61 a	7.97 a	7.92 a	7.64 a
Zn-AA	19.14 a	18.84 a	18.80 a	8.01 a	8.04 a	7.66 a
Zn effect by f-test[b]	NS	NS	NS	NS	NS	NS
	Harvest index (%)			Thousand seed weight (g)		
Control	40.16 a	42.95 a	40.91 a	22.01 a	22.96 a	20.92 a
Zn-EDTA	40.20 a	42.37 a	41.05 a	23.05 a	24.13 a	21.85 a
Zn-Citrate	40.44 a	43.11 a	40.87 a	22.12 a	24.52 a	22.02 a
ZnSO$_4$	40.59 a	43.06 a	41.13 a	22.34 a	23.68 a	22.45 a
Zn-AA	41.86 a	42.68 a	40.96 a	23.23 a	24.56 a	22.56 a
Zn effect by f-test[b]	NS	NS	NS	NS	NS	NS

[a]Different letters after number in the same column designated significant difference by LSD$_{P<0.05}$.
[b]Significant effects: NS = not significant at P>0.05.

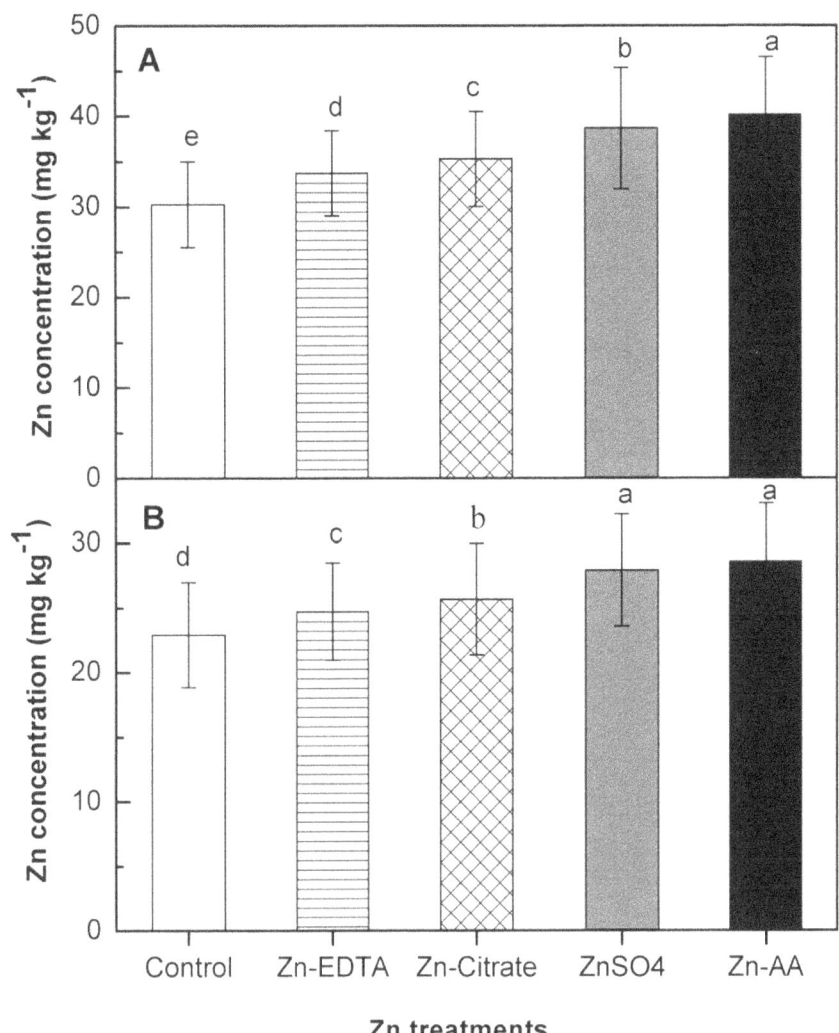

Figure 1. Effect of different forms of foliar Zn fertilization on Zn concentration in rice grain. (A) Zn concentration in brown rice. (B) Zn concentration in polished rice. Error bars indicate standard errors of the means (n = 4). Different letters indicate significant difference among Zn treatments according to LSD test (P<0.05).

2.09 mg g^{-1} by Zn-EDTA, to 2.06 mg g^{-1} Zn-Citrate, to 1.87 mg g^{-1} by foliar ZnSO$_4$ and 1.92 mg g^{-1} by Zn-AA, these showed that, the decreases of 6.97%, 8.42%, 16.91% and 14.57%, respectively.

Protein, Iron and Calcium Concentration in Polished Rice

Protein content in polished rice showed an increase trend by foliar Zn-AA, but not by other foliar Zn fertilizations (Table 3). Foliar application of Zn-AA had significant impact on grain protein content in cultivar Hai7 and Bing91185, except Biyuzao-nuo. Generally, compared to the control, foliar Zn-AA could increase protein content by 1.88–4.79% depending on cultivar. Grain Fe and Ca concentration did not change by different forms of Zn treatments in all three cultivars. Foliar Zn fertilization had little impact on Fe and Ca content (Table 3).

In vitro Zinc Solubility of Polished Rice

The amount of Zn solubilized after *in vitro* digestion is an indicator for bioavailability. Foliar Zn fertilization had significant (P<0.05) effect on Zn solubility (Table 4). Averaged across the cultivars, *in vitro* Zn solubility from polished rice was ranged from

28.48% in the control, to 29.34% by application of Zn-EDTA, 29.41% by application of Zn-Citrate, 31.15% by application of ZnSO$_4$ and 30.67% by application of Zn-AA, these represented increase of 2.99%, 3.24%, 9.36% and 7.65%, respectively. Foliar application of Zn-AA and ZnSO$_4$ significant improved *in vitro* Zn solubility in all cultivars.

Zinc Bioavailability of Polished Rice

The soluble fraction obtained from *in vitro* digestion was used to carry out the retention, transport and uptake experiments in Caco-2 cell (Table 4). Foliar Zn fertilization had significant (P<0.05) impact on the percentages of Zn retention, transport and uptake efficiency in polished rice grain for all cultivars. Generally, foliar application of Zn-EDTA, Zn-Citrate, ZnSO$_4$ and Zn-AA could increase the percentages of Zn retention, transport and uptake efficiency in polished rice grain, but only in case of foliar application ZnSO$_4$ and Zn-AA could reach significant level in most of the cultivars tested (P<0.05). Regardless of cultivar, compare to the control, after foliar application of ZnSO$_4$, the percentages of Zn retention, transport, and uptake efficiency from the polished rice increased by 12.96%, 31.36% and 34.98%,

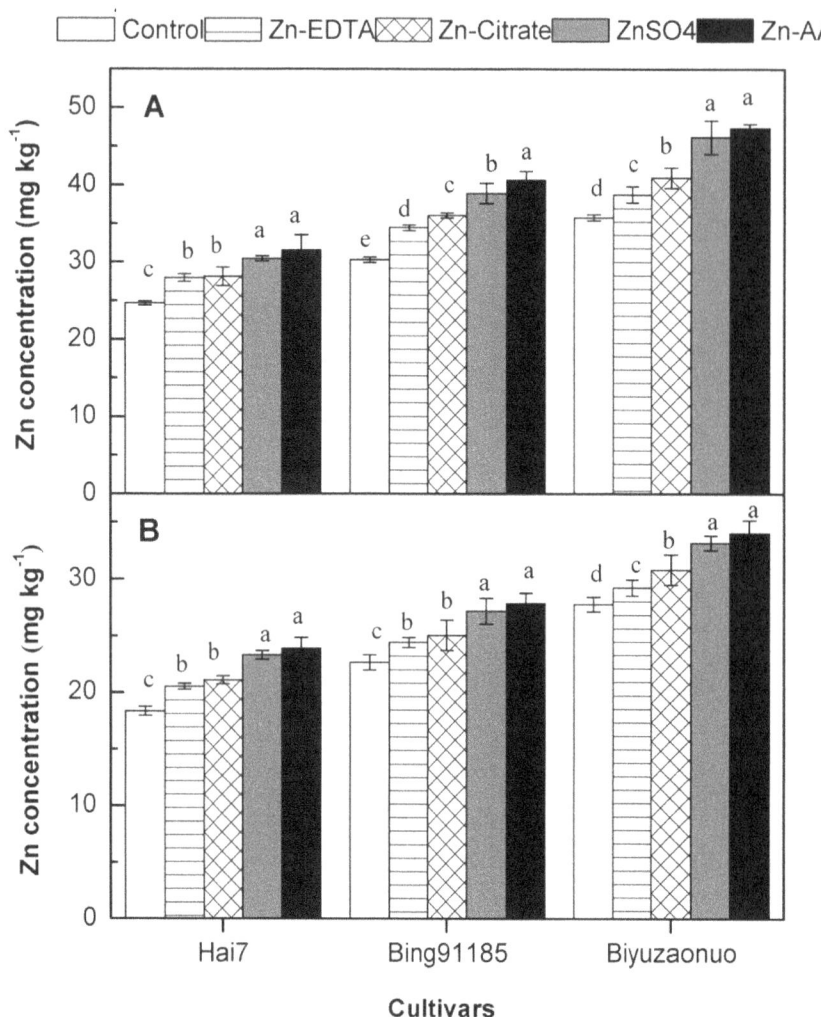

Figure 2. Zn concentration in rice grain among three cultivars. (A) Zn concentration in brown rice. (B) Zn concentration in polished rice. Error bars indicate standard errors of the means (n = 4). Different letters indicate significant difference among Zn treatments according to LSD test (P<0.05).

respectively; after foliar application of Zn-AA, the percentages of Zn retention, transport, and uptake efficiency from the polished rice increased by 19.4%, 29.47% and 35.25%, respectively. Regardless of cultivar, the amount of bioavailable Zn in the rice grain has the similar trend as the Zn uptake efficiency (Fig. 3). Regardless of cultivar, compared to the control, amount of bioavailable Zn was increased by 13.85%, 21.96%, 64.43% and 68.37% by Zn-EDTA, Zn-Citrate, $ZnSO_4$ and Zn-AA, respectively. Generally, the amount of bioavailable Zn could increased by the foliar Zn fertilizations (Zn-EDTA, Zn-Citrate, $ZnSO_4$ and Zn-AA), but only foliar $ZnSO_4$ and Zn-AA reach at significant level (P<0.05) in all cultivars tested.

Discussion

The DTPA extractable Zn (Table 1) in the soil of the current field study was higher than the critical level for rice (0.8 mg kg^{-1}) [28], thus the plant was in the sufficient Zn nutritional status. In the Zn sufficient soil, excess foliar application of Zn did not affect the biomass, grain yield, harvest index and thousand seed weight (Table 2). Similar results also found in previous reports [12,29].

In contrast to grain yield, foliar Zn fertilization significantly (P<0.05) increased the Zn concentration in brown rice (Fig. 1A), consisted with the previous studies [16,30]. Thus, foliar application Zn as an effective method could boost Zn level in rice grain. Here, in this study, particular attention should be given in the Zn concentration of polished rice, as this is the predominant fraction consumed by human. In current study, we found that regardless of cultivar, although polishing process decrease substantial amount Zn from mature grain, the polished rice obtained from foliar Zn applications was still contained 10.22–24.04% more Zn than those of control (Fig. 1B), and a significant positively correlation in Zn concentration between polished rice and brown rice was also found ($R^2 = 0.897$, P<0.01), indicating that Zn concentration in polished rice might be improved by increasing the Zn concentration in brown rice, consisted with previous studies [18,29], suggesting that excess foliar applied Zn could penetrate into the inner layers of rice endosperm. However, the mechanism of absorbed Zn from aleurone layer into rice endosperm still not clearly, some recent studies reported that nicotianamine and deoxymugineic acid play important role in this process [17,31].

Furthermore, the effectiveness of foliar Zn fertilization on Zn concentration of brown rice and polished rice varied with the

Table 3. Effect of different forms of foliar Zn fertilization on the grain protein, phytic acid, Fe and Ca contents of three rice cultivars.

Treatments	Cultivars[a]					
	Hai7	Bing91185	Biyuzaonuo	Hai7	Bing91185	Biyuzaonuo
	Protein (%)			Phytic acid (mg g^{-1})		
Control	9.06 b	9.32 b	9.98 a	1.78 a	1.64 a	3.32 a
Zn-EDTA	9.15 b	9.34 b	9.99 a	1.66 b	1.48 b	3.14 b
Zn-Citrate	9.16 b	9.66 ab	9.91 a	1.65 b	1.47 b	3.05 b
ZnSO$_4$	9.15 b	9.37 b	9.99 a	1.54 c	1.39 c	2.89 c
Zn-AA	9.43 a	9.76 a	10.17 a	1.50 c	1.36 c	2.67 d
Zn effect by f-test[b]	*	*	NS	***	***	***
	Fe (mg kg^{-1})			Ca (mg kg^{-1})		
Control	2.92 a	3.35 a	5.89 a	92.00 a	110.22 a	146.48 a
Zn-EDTA	3.12 a	3.74 a	5.84 a	83.67 a	107.13 a	150.49 a
Zn-Citrate	3.10 a	3.81 a	5.73 a	96.38 a	112.01 a	145.90 a
ZnSO$_4$	2.98 a	3.88 a	5.69 a	91.00 a	108.86 a	146.89 a
Zn-AA	2.98 a	3.86 a	5.79 a	94.12 a	107.43 a	143.80 a
Zn effect by f-test[b]	NS	NS	NS	NS	NS	NS

[a]Different letters after number in the same column designated significant difference by LSD$_{P<0.05}$.
[b]Significant effects: NS = not significant at $P>0.05$;
*at $P<0.05$;
**at $P<0.01$;
***at $P<0.001$.

forms of Zn fertilizer (Fig. 1). Among the forms tested for foliar application, foliar Zn-AA and ZnSO$_4$ were more effective than Zn-EDTA and Zn-Citrate in improving the Zn concentration in brown rice and polished rice, the results agreed with the previous study [32]. The reasons might be due to the different capacity of leaf penetration of different forms of foliar applied Zn fertilizer [19]. Foliar fertilizer with low molecular weight like Zn-AA and ZnSO$_4$ might be easily penetrate into the leaves than those of Zn-EDTA and Zn-Citrate with the high molecular weight, as a result, the more plant-available Zn from foliage might be transported and accumulated in rice grain [19,33]. The response of grain Zn concentration to foliar Zn fertilization was cultivar dependent (Fig. 2). The results consisted with the previous studies reported rice genotypes differ greatly in their response to foliar applied Zn to increase grain Zn concentration [14,16]. Thus, impact of foliar Zn application on grain Zn can be maximized by selecting genotypes with higher ability in leaf absorption and seed deposition of foliar applied Zn with low molecular weight.

Although the amount of grain Zn is important for Zn bioavailability, information about changes of anti-nutrients and other nutrients in rice grain especially polished rice during foliar Zn applications is crucial, because it related with Zn bioavailability and nutritional quality of rice, which impact on global human health. Phytic acid has long been known as a form of stored phosphorus in seeds, which was considered as an inhibitor of Zn bioavailability in rice grain [34]. In the current study, it was documented that foliar Zn applications could significantly reduced phytic acid content in polished rice, especially in case of Zn-AA and ZnSO$_4$ applications (Table 3), consisted with the previous studies [29,35]. Possible explanation was that foliar Zn application could inhibit the conversion of inorganic phosphorus to phytic acid in rice grain [2,35]. Iron and Ca in rice grain are also

important minerals because they are frequently deficient in human. In current study, the concentrations of Fe and Ca were varied with cultivar, but not by the foliar Zn application (Table 3). Foliar Zn fertilization had significant effect on grain protein content except the cultivar Biyuzaonuo, however, grain protein content showed increasing tread by foliar application of Zn-AA in all cultivars (Table 3), the results consisted with previous studies [12,36]. Consequently, above results indicated that foliar Zn application could reduce the phytic acid content, while could maintain the Fe, Ca and protein content in polished rice grain.

In vivo situation, Zn needs to be in soluble before it can be taken up by the enterocytes. In current study, we also determined the soluble Zn in polished rice obtained from the different foliar Zn treatments by *in vitro* digestion. The results showed that regardless of cultivar, the solubility of Zn in polished rice from foliar Zn application was higher than control, especially in the case of foliar applications of Zn-AA and ZnSO$_4$ (Table 4). To the best of our knowledge, no literature data on regarding foliar Zn fertilizers on the *in vitro* Zn solubility in rice grain are yet available. One possible reason was the *in vitro* Zn solubility in grain was increased by the reduction of phytic acid in grain [2,37]. The solubility method involves a simulation of gastrointestinal digestion followed by a measurement of soluble Zn in the digest and thus covers only the first phase of the overall Zn absorption process. The soluble Zn fraction obtained form *in vitro* digestion was used to carry out retention, transport and uptake experiment in Caco-2 cell model, which offer a more physiological tool for screening Zn bioavailability in food matrices [27,37,38,39]. In the current study, the percentage of Zn uptake efficiency of polished rice ranged from 6.01% to 12.33%, falling within the previous reported Zn uptake efficiency in Caoc-2 cell model from cereal foods (ranging from 4.1% to 48.1%) [37]. Compared to the control, foliar Zn-AA and

Table 4. Effect of different forms of foliar Zn fertilization on the percentages of solubility, retention, transported and uptake efficiency of Zn among three rice cultivars.

Treatments	Cultivars[a]					
	Hai7	Bing91185	Biyuzaonuo	Hai7	Bing91185	Biyuzaonuo
	Solubility (%)			Retention (%)		
Control	29.17 c	30.58 c	25.70 b	14.53 c	14.91 b	14.39 c
Zn-EDTA	30.75 c	31.21 bc	26.06 b	14.82 bc	14.96 b	15.04 bc
Zn-Citrate	30.90 bc	31.13 bc	26.20 b	14.93 bc	15.78 b	15.11 bc
ZnSO$_4$	32.68 a	32.24 a	28.54 a	16.68 ab	18.29 a	16.80 ab
Zn-AA	31.64 ab	32.62 a	27.73 a	17.54 a	18.80 a	17.53 a
Zn effect by f-test[b]	*	*	***	*	***	*
	Transport (%)			Uptake efficiency (%)		
Control	9.35 b	16.09 b	8.99 c	6.95 c	9.48 b	6.01 b
Zn-EDTA	9.61 b	16.23 b	9.57 c	7.52 bc	9.73 b	6.39 b
Zn-Citrate	11.08 ab	16.15 b	9.96 bc	8.02 b	9.94 b	6.57 b
ZnSO$_4$	13.27 a	18.43 ab	13.53 a	9.79 a	11.84 a	8.66 a
Zn-AA	13.05 a	19.11 a	12.42 ab	9.68 a	12.39 a	8.30 a
Zn effect by f-test[b]	**	*	*	***	***	***

[a]Different letters after number in the same column designated significant difference by LSD$_{P<0.05}$.
[b]Significant effects: NS = not significant at P>0.05;
*at P<0.05;
**at P<0.01;
***at P<0.001.

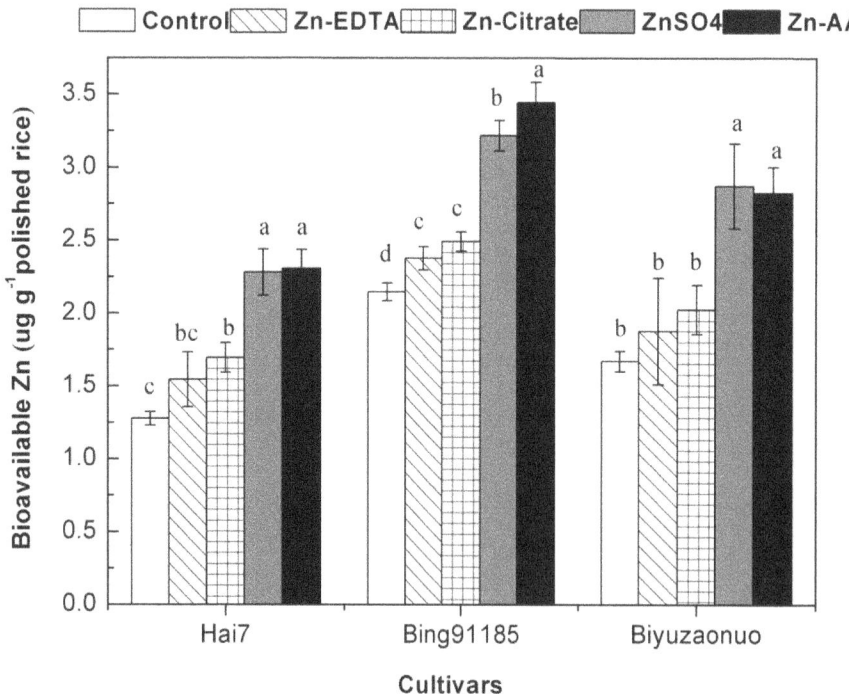

Figure 3. The amount of bioavailable Zn in the polished rice among three cultivars. Error bars indicate standard errors of the means (n = 4). Different letters indicate significant difference among Zn treatments according to LSD test (P<0.05).

$ZnSO_4$ could significantly increase the percentages of Zn retention, transport and uptake efficiency of polished rice in most of the cultivars tested (Table 4). Regardless of cultivar, the amount of bioavailable Zn has the similar trend as Zn uptake efficiency (Fig. 3), showing the foliar Zn could improve the amount of bioavailable Zn in the polished rice, especially foliar Zn-AA and $ZnSO_4$ could increased by 68.37% and 64.43%, respectively. The results indicated that foliar Zn application could improve the bioavailability of Zn from polished rice but depended on the forms of Zn application. Till now, few literatures on the Zn bioavailability of biofortified polished rice obtained from the foliar Zn fertilization are available. The possible explanation of our results might be foliar Zn-AA and $ZnSO_4$ have the higher efficiency to decrease the phytic acid, and improve the total amount of Zn than Zn-EDTA and Zn-Citrate, as a result, increase the amount of bioavailable Zn in the polished rice grain. In addition, in the current study, we also observed that the polished rice of cultivar Biyuzaonuo contained the highest amount of Zn, interestingly, it was not contained the highest amount of bioavailable Zn, this is might be due to presence of significant amount of phytic acid or other anti-nutrients in this cultivar [40]. Thus, it is suggested that not only the net grain Zn concentration but also the bioavailability of grain Zn should be considered in ongoing breeding program.

In conclusion, foliar Zn fertilization was an effective agronomic practice to promote grain Zn concentration and Zn bioavailability, especially, in case of Zn-AA and $ZnSO_4$. On average, Zn-AA and $ZnSO_4$ increased Zn concentration in polished rice up to 24.04% and 22.47%, respectively. On average, Zn-AA and $ZnSO_4$ increased Zn bioavailability in polished rice up to 68.37% and 64.43%, respectively. The effectiveness of foliar applied Zn-AA and $ZnSO_4$ were higher than Zn-EDTA and Zn-Citrate to improve the Zn concentration, and reduction of phytic acid, as a results higher accumulation of bioavailable Zn in polished rice. Moreover, foliar Zn application could maintain the protein and minerals (Fe and Ca) quality of the polished rice. Therefore, it's believed that foliar application of suitable Zn form is a feasible approach to improve the bioavailable Zn status in polished rice.

Acknowledgments

The authors gratefully thank Dr. Zed Rengel (The University of Western Australia) for critical comments on the manuscript.

Author Contributions

Conceived and designed the experiments: YW MJIS XY. Performed the experiments: YW MJIS. Analyzed the data: YW MJIS XY. Contributed reagents/materials/analysis tools: YW MJIS. Wrote the paper: YW MJIS.

References

1. Stein AJ (2010) Global impacts of human mineral malnutrition. Plant Soil 335: 133–154.
2. Cakmak I (2008) Enrichment of cereal grains with zinc: Agronomic or genetic biofortification? Plant Soil 302: 1–17.
3. Hambidge M (2000) Human zinc deficiency. J Nutr 130: 1344S–1349S.
4. Hotz C, Brown KH (2004) Assessment of the risk of zinc deficiency in populations and options for its control. Food Nutr Bull 25: S91 S204.
5. Fitzgerald MA, McCouch SR, Hall RD (2009) Not just a grain of rice: the quest for quality. Trends Plant Sci 14: 133–139.
6. Frossard E, Bucher M, Mächler F, Mozafar A, Hurrell R (2000) Potential for increasing the content and bioavailability of Fe, Zn and Ca in plants for human nutrition. J Sci Food Agric 80: 861–879.
7. Bouis HE (2002) Plant breeding: a new tool for fighting micronutrient malnutrition. J Nutr 132: 491S–494S.
8. Pfeiffer WH, McClafferty B (2007) HarvestPlus: Breeding crops for better nutrition. Crop Sci 47: S88 S105.
9. Joshi A, Crossa J, Arun B, Chand R, Trethowan R, et al. (2010) Genotype×environment interaction for zinc and iron concentration of wheat grain in eastern gangetic plains of India. Field Crops Res 116: 268–277.
10. Falk MC, Chassy BM, Harlander SK, Hoban TJ, McGloughlin MN, et al. (2002) Food biotechnology: Benefits and concerns. J Nutr 132: 1384–1390.
11. Palmgren MG, Clemens S, Williams LE, Krämer U, Borg S, et al. (2008) Zinc biofortification of cereals: problems and solutions. Trends Plant Sci 13: 464–473.
12. Fang Y, Wang L, Xin Z, Zhao L, An X, et al. (2008) Effect of foliar application of zinc, selenium, and iron fertilizers on nutrients concentration and yield of rice grain in China. J Agric Food Chem 56: 2079–2084.
13. Phattarakul N, Rerkasem B, Li L, Wu L, Zou C, et al. (2012) Biofortification of rice grain with zinc through zinc fertilization in different countries. Plant Soil: 1–11.
14. Wissuwa M, Ismail AM, Graham RD (2008) Rice grain zinc concentrations as affected by genotype, native soil-zinc availability, and zinc fertilization. Plant Soil 306: 37–48.
15. Nasri M, Khalatbari M, Farahani HA (2011) Zn-foliar application influence on quality and quantity features in phaseolous vulgaris under different levels of N and K fertilizers. Adv Environ Biol 5: 839–846.
16. Wu C, Lu L, Yang X, Feng Y, Wei Y, et al. (2010) Uptake, translocation, and remobilization of zinc absorbed at different growth stages by rice genotypes of different Zn densities. J Agric Food Chem 58: 6767–6773.
17. Bashir K, Ishimaru Y, Nishizawa NK (2012) Molecular mechanisms of zinc uptake and translocation in rice. Plant Soil: 1–13.
18. Cakmak I, Kalayci M, Kaya Y, Torun A, Aydin N, et al. (2010) Biofortification and localization of zinc in wheat grain. J Agric Food Chem 58: 9092–9102.
19. Colle C, Madoz-Escande C, Leclerc E (2009) Foliar transfer into the biosphere: review of translocation factors to cereal grains. J Environ. Radioactiv 100: 683–689.
20. House WA (1999) Trace element bioavailability as exemplified by iron and zinc. Field Crops Res 60: 115–141.
21. Lönnerdal B (2000) Dietary factors influencing zinc absorption. J Nutr 130: 1378S–1383S.
22. Van Campen DR, Glahn RP (1999) Micronutrient bioavailability techniques: accuracy, problems and limitations. Field Crops Res 60: 93–113.
23. Hotz C (2005) Evidence for the usefulness of in vitro dialyzability, Caco-2 cell models, animal models, and algorithms to predict zinc bioavailability in humans. Int J Vitam Nutr Res 75: 423–435.
24. Yang X, Ye Z, Shi CH, Zhu M, Graham R (1998) Genotypic differences in concentrations of iron, manganese, copper, and zinc in polished rice grains. J Plant Nutr 21: 1453–1462.
25. Dai F, Wang J, Zhang S, Xu Z, Zhang G (2007) Genotypic and environmental variation in phytic acid content and its relation to protein content and malt quality in barley. Food Chem 105: 606–611.
26. Ohtsubo K, Suzuki K, Yasui Y, Kasumi T (2005) Bio-functional components in the processed pre-germinated brown rice by a twin-screw extruder. J Food Compos Anal 18: 303–316.
27. Wei Y, Shohag M, Wang Y, Lu L, Wu C, et al. (2012) Effect of zinc sulfate fortification in germinated brown rice on seed zinc concentration, bioavailability and seed germination. J Agric Food Chem 60: 1871–1879.
28. Dobermann A, Fairhurst TH (2000) Rice: Nutrient disorders & nutrient management: Potash and Phosphate Institute, Potash and Phosphate Institute of Canada and International Rice Research Institute, Singapore.
29. Zhang Y, Shi R, Rezaul KM, Zhang F, Zou C (2010) Iron and zinc concentrations in grain and flour of winter wheat as affected by foliar application. J Agric Food Chem 58: 12268–12274.
30. Fang Y, Wang L, Xin Z, Zhao L, An X, et al. (2008) Effect of foliar application of zinc, selenium, and iron fertilizers on nutrients concentration and yield of rice grain in China. J Agric Food Chem 56: 2079–2084.
31. Lee S, Persson DP, Hansen TH, Husted S, Schjoerring JK, et al. (2011) Bio-available zinc in rice seeds is increased by activation tagging of nicotianamine synthase. Plant Biotechnol J 9: 865–873.
32. Cakmak I (2009) Enrichment of fertilizers with zinc: An excellent investment for humanity and crop production in India. J Trace Elem Med Biol 23: 281–289.
33. Stacey SP, Oosterhuis DM (2007) Effect of EDTA on the foliar absorption of trace element fertilizers. Soil Fertility Studies: 80–81.
34. Bohn L, Meyer A, Rasmussen S (2008) Phytate: impact on environment and human nutrition. A challenge for molecular breeding. J Zhejiang Univ Sci B 9: 165–191.
35. Yang X, Tian X, Lu X, Gale William J, Cao Y (2011) Foliar zinc fertilization improves the zinc nutritional value of wheat (Triticum aestivum L.) grain. Afr J Biotechnol 10: 14778–14785.
36. Liu D, Tian L, Shao J, Zhao H, Gao Z (2005) Effect of amino acid chelated microelement fertilizer on wheat and rice. J Soil Sci (in Chinese) 36: 917–920.
37. Frontela C, Scarino ML, Ferruzza S, Ros G, Martinez C (2009) Effect of dephytinization on bioavailability of iron, calcium and zinc from infant cereals assessed in the Caco-2 cell model. World J Gastroentero 15: 1977–1984.
38. Viadel B, Barberá R, Farré R (2006) Uptake and retention of calcium, iron, and zinc from raw legumes and the effect of cooking on lentils in Caco-2 cells. Nutr Res 26: 591–596.

39. Camara F, Barbera R, Amaro M, Farre R (2007) Calcium, iron, zinc and copper transport and uptake by Caco-2 cells in school meals: Influence of protein and mineral interactions. Food Chem 100: 1085–1092.

40. Rosado JL, Hambidge KM, Miller LV, Garcia OP, Westcott J, et al. (2009) The quantity of zinc absorbed from wheat in adult women is enhanced by biofortification. J Nutr 139: 1920–1925.

5

Tropical Andean Forests Are Highly Susceptible to Nutrient Inputs—Rapid Effects of Experimental N and P Addition to an Ecuadorian Montane Forest

Jürgen Homeier[1]*, Dietrich Hertel[1], Tessa Camenzind[2], Nixon L. Cumbicus[3], Mark Maraun[4], Guntars O. Martinson[5,6], L. Nohemy Poma[7], Matthias C. Rillig[2], Dorothee Sandmann[4], Stefan Scheu[4], Edzo Veldkamp[6], Wolfgang Wilcke[8], Hans Wullaert[9], Christoph Leuschner[1]

1 Albrecht von Haller Institute of Plant Sciences, Georg August University Göttingen, Göttingen, Germany, **2** Institute of Biology, Freie Universität Berlin, Berlin, Germany, **3** Instituto de Ecología, Universidad Técnica Particular de Loja, San Cayetano Alto, Loja, Ecuador, **4** J.F. Blumenbach Intitute of Zoology and Anthropology, Georg August University Göttingen, Göttingen, Germany, **5** Max Planck Institute for Terrestrial Microbiology, Marburg, Germany, **6** Buesgen Institute - Soil Science of Tropical and Subtropical Ecosystems, Georg August University Göttingen, Göttingen, Germany, **7** Universidad National de Loja, Ciudadela Universitaria Guillermo Falconí sector La Argelia, Loja, Ecuador, **8** Geographic Institute, University of Berne, Berne, Switzerland, **9** Geographic Institute, University of Mainz, Mainz, Germany

Abstract

Tropical regions are facing increasing atmospheric inputs of nutrients, which will have unknown consequences for the structure and functioning of these systems. Here, we show that Neotropical montane rainforests respond rapidly to moderate additions of N (50 kg ha^{-1} yr^{-1}) and P (10 kg ha^{-1} yr^{-1}). Monitoring of nutrient fluxes demonstrated that the majority of added nutrients remained in the system, in either soil or vegetation. N and P additions led to not only an increase in foliar N and P concentrations, but also altered soil microbial biomass, standing fine root biomass, stem growth, and litterfall. The different effects suggest that trees are primarily limited by P, whereas some processes—notably aboveground productivity—are limited by both N and P. Highly variable and partly contrasting responses of different tree species suggest marked changes in species composition and diversity of these forests by nutrient inputs in the long term. The unexpectedly fast response of the ecosystem to moderate nutrient additions suggests high vulnerability of tropical montane forests to the expected increase in nutrient inputs.

Editor: Han Y. H. Chen, Lakehead University, Canada

Funding: The study was funded by Deutsche Forschungsgemeinschaft (DFG RU 816). The funders had no role in study design, data collection and analysis, decision to publish, or preparation of the manuscript.

Competing Interests: The authors have declared that no competing interests exist.

* E-mail: jhomeie@gwdg.de

Introduction

Since the 1950/60s, anthropogenic changes to the cycling of the key nutrients nitrogen (N) and phosphorus (P) have dramatically altered the structure and functioning of many ecosystems in the world's industrialized regions [1–6]. Elevated N and P inputs affect virtually all components and processes of terrestrial and aquatic ecosystems, including plant growth, plant longevity and stress tolerance, plant community composition and diversity, biotic interactions (plant-plant, plant-fungus, plant-animal), the composition and activity of heterotrophic communities, and the storage and cycling of carbon, nutrients and water [7–9]. This is because primary production is limited by N or P, or both, in the vast majority of ecosystems around the globe [8,10–14].

In the past 30 years, research has focused on the structural and functional responses of temperate and boreal forests to atmospheric N inputs [15–18] because the bulk of fertilizer use worldwide was in the industrialized nations of the northern hemisphere. Furthermore, these regions had particularly pronounced gaseous NO_x emissions originating from the combustion of fossil fuels and NH_3 emissions from animal production [1]. However, this situation is changing rapidly. With the expansion of industrial agriculture into many tropical and southern hemispheric regions, the spread of N and P compounds to adjacent and more distant non-agricultural ecosystems in these regions has been greatly increased [1,19–20]. In the future, tropical forests will be increasingly exposed to airborne N and P inputs. Higher P inputs are mostly due to the deposition of dust [4,21–24], but sources of N can be varied, including oxidised and reduced N compounds emitted with farming, livestock breeding and the combustion of fossil fuels, and N released through biomass burning with the conversion of tropical forests [1,25–26].

Although tropical forests are likely to be sensitive to these changes, the size and direction of their responses are unclear [7,20,27–31]. A number of fertilization experiments in tropical forests have investigated responses to experimental high-dose treatments with N, or N and P (100–300 kg N ha^{-1} yr^{-1} and/or 50–100 kg P ha^{-1} yr^{-1}) [30,32–40]. These experiments typically focused on selected ecosystem properties, such as changes in tree growth, fine litter production or soil carbon pools, but did not provide comprehensive insight into ecosystem responses to elevated N and P loads. Here, we report data from a nutrient manipulation experiment (NUMEX) investigating the response of an old-growth montane forest ecosystem in the Andes of southern

Ecuador to moderate N (50 kg ha^{-1} yr^{-1}) and/or P (10 kg ha^{-1} yr^{-1}) additions considering a multitude of response variables. The study allows comprehensive insight into how highly diverse tropical montane forest ecosystems respond to moderate nutrient additions such as those predicted by climate change scenarios.

Results and Discussion

A large number of ecosystem properties and functions exhibited marked responses to the experimental nutrient additions after only one year.

Soil nutrient pools and soil biological activity

Nutrient addition did not result in a significant increase of the organic layer N pool while the P pool increased after combined addition of N and P (Fig. 1). The reason is likely the relatively high N stock in the thick organic layer. An adjacent micro-catchment between 1850 and 2150 m a.s.l. had 0.9–21 Mg ha^{-1} N in the organic layer while the P stock ranged from 30–700 kg ha^{-1} [41]. Soil microbial biomass decreased after adding N (Fig. 2a), whereas P addition had no effect. A decrease in microbial biomass after N addition suggests detrimental effects of nitrogen on soil microorganisms and has been seen in other studies [3,42], in particular with lignin decomposing fungi [43–44]. Conforming to the assumption of detrimental effects on microorganisms using complex organic compounds the metabolic activity of microorganisms significantly increased after N addition (Fig. 2b), indicating a shift towards microorganisms predominantly using more easily available carbon resources.

All treatments resulted in slightly higher net N mineralization rates, the increase after N addition was marginally significant (p = 0.08, Fig. 2c). N$_2$O emissions tended to increase after N+P addition (p = 0.05) but not after addition of N or P only (Fig. 2d). This suggests that N is needed as substrate for denitrification while P is limiting the respective organisms.

The assumption that P addition stimulated microorganisms which are responsible for N transformations is further supported by the finding that the combined addition of N and P had a significant effect on the NH$_4$-N/NO$_3$-N ratio in mineral soil solution (Fig. 2e–f). Although we added urea, which is primarily a source of NH$_4^+$, NO$_3^-$ concentrations increased after combined N and P addition while NH$_4^+$ concentrations did not. This was

probably because of the combination of stimulated nitrification and microbial NH$_4^+$ retention. Our finding that only N and P addition together stimulated N mineralization (and possibly also nitrification) is different than reports from other tropical sites where N addition alone stimulated nitrification and triggered NO$_3^-$ losses to the subsoil. This difference may be the result of no - or less pronounced - P limitation of the involved microorganisms [27,45]. In contrast to many temperate forests where NO$_3^-$ is the most abundant N form [46], NO$_3^-$ only accounted for 3–5% of total N in the soil solutions from our study site. Instead, the soil N pool was predominantly made up of dissolved organic nitrogen (DON) and NH$_4^+$ (contributing 50–70% and 27–43% respectively) [47]. This implies that small changes in N mineralization and nitrification rates can have a large impact on NO$_3^-$ concentrations and fluxes. Our results indicate that the addition of different nutrients may stimulate or inhibit different processes in the ecosystem resulting in a complex system response to nutrient deposition.

Nutrient cycling

Increased N and P contents in litterfall and throughfall (Figs. 3a–f) indicate that a large proportion of the added nutrients was taken up by trees and subsequently accelerated nutrient cycling through higher N and P return with litterfall and leachate [48] and through stimulated litter decomposition by higher N and P concentrations [49]. The annual increase in N and P fluxes in litterfall and throughfall after fertilization was equivalent to 25.4% and 26.7% of the applied N, after N addition and after N and P addition, respectively, and 3.8% and 6.1% of the applied P, after P addition and after N and P addition, respectively. Neither organic layer N and P pools nor N and P losses to the atmosphere or to the subsoil were significantly increased by N or P addition (Figs. 2d and 3i, j), suggesting that the bulk of N and P added was retained in the ecosystem [48]. The N and P effects on nutrient cycling were interrelated. Phosphorus increased the retention of N in the system since aboveground N losses were slightly reduced (Fig. 2d) and losses by leaching were negligible (Fig. 3i–j). We attribute this mainly to the stimulation of the N-mineralizing microbial community by P addition as reflected by the significantly increased net N mineralization rates [50]. Furthermore, N application increased the P return with litterfall and throughfall when added in combination with P (N+P treatment: Figs. 3d,f). The positve effects

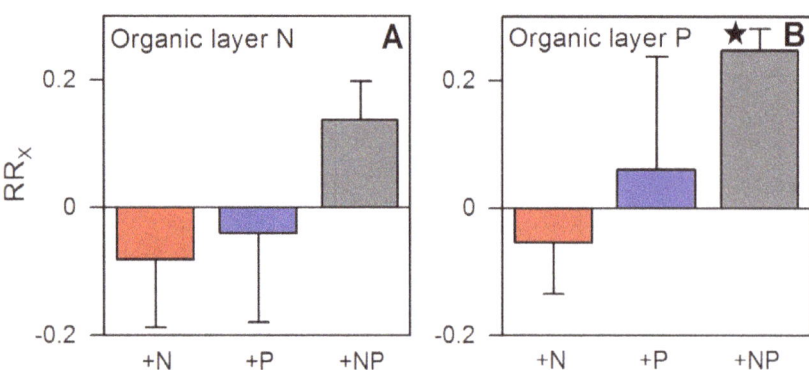

Figure 1. Effects of one year of experimental nutrient addition on various soil nutrient pools of a montane forest in Ecuador. Effects are presented as natural-log transformed response ratios (RR$_X$) in which the parameter in the enriched treatment is divided by its value in the control treatment and then ln-transformed. Hence, a value of 0.2 indicates a value in the manipulated treatment that is c. 23% higher than in the control, while a value of 0.5 indicates a 65% increase. Error bars indicate plus or minus one standard error. Data of the control treatment (mean ±1 SE) are given in parentheses below. Asterisks indicate significant differences to the control (P≤0.05). a. Organic layer nitrogen pool (3.79±0.31 Mg N ha^{-1}). b. Organic layer phosphorus pool (98.6±6.8 kg P ha^{-1}).

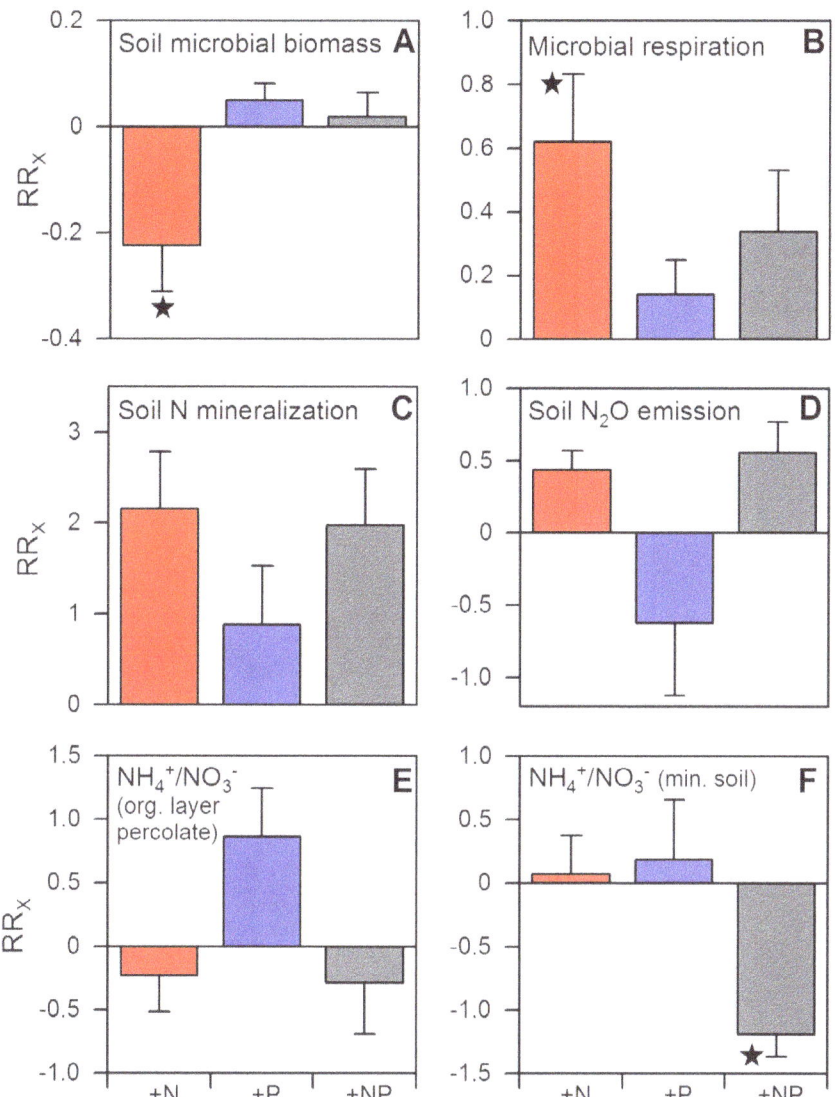

Figure 2. Effects of one year of experimental nutrient addition on biological soil activity of a montane forest in Ecuador. a. Soil microbial biomass in May 2009 (5881±25 µg C_{mic} g^{-1} soil dry mass). b. Respiration of soil microorganisms in May 2009 (5.066±0.36 µl O_2 mg C_{mic}^{-1} h^{-1}). c. Net N mineralization in September 2008 (23.4±10.5 ng N cm^2 h^{-1}). d. Annual emission of N_2O (0.25±0.03 kg N ha^{-1} yr^{-1}). e. Mean NH_4^+/NO_3^- ratio of organic layer percolate from February 2008 to January 2009 (15.7±6.7) and f. mean NH_4^+/NO_3^- ratio of mineral soil solution from February 2008 to January 2009 (8.9±2.0). Error bars indicate plus or minus one standard error. Data of the control treatment (mean ±1 SE) are given in parentheses in the legend. Asterisks indicate significant differences to the control (P≤0.05). For interpretation of graph see legend of Fig. 1.

of N addition on litter P concentration and P return with litterfall and throughfall might be a result of better soil P availability to plants by increased extracellular phosphatase activity after N addition. This has been observed in several temperate and tropical forest studies [51–53].

Tree biomass and forest productivity

Stand leaf area index (LAI) tended to increase after addition of N and P alone (although not significantly), while leaf litter production increased only after N+P addition (Figs. 4a,c). Since N fertilization tended to increase the specific leaf areas and foliar N concentrations of the four most common tree species (Table 1; difference significant only for *Myrcia* sp.), two foliage attributes generally associated with a shorter leaf lifespan [54], we infer that N addition has stimulated leaf production. The LAI increase after P addition probably resulted from an extension of mean leaf

lifespan as indicated by the observed slight reduction in litter production. This is supported by results from Hawaiian montane forests, where leaf production and leaf longevity were increased after P addition [55].

Stand basal area increment as a proxy of aboveground productivity tended to increase in all fertilization treatments (Fig. 4b), as reported from other Neotropical montane forests after addition of N [34,39,55], P [34,39] or N and P [35].

All fertilizer treatments resulted in a marked reduction in standing tree fine root biomass (by 15–28%) with the effect being strongest after P addition (Fig. 4d). This treatment also led to a strong increase in standing fine root necromass, while the N and N+P treatments had no effect on necromass (Fig. 4e). Presumably, the accumulation of fine root necromass in the P treatment resulted from reduced root litter decay due to an unfavorable litter N:P ratio for decomposers [56]. A decline of fine root biomass and

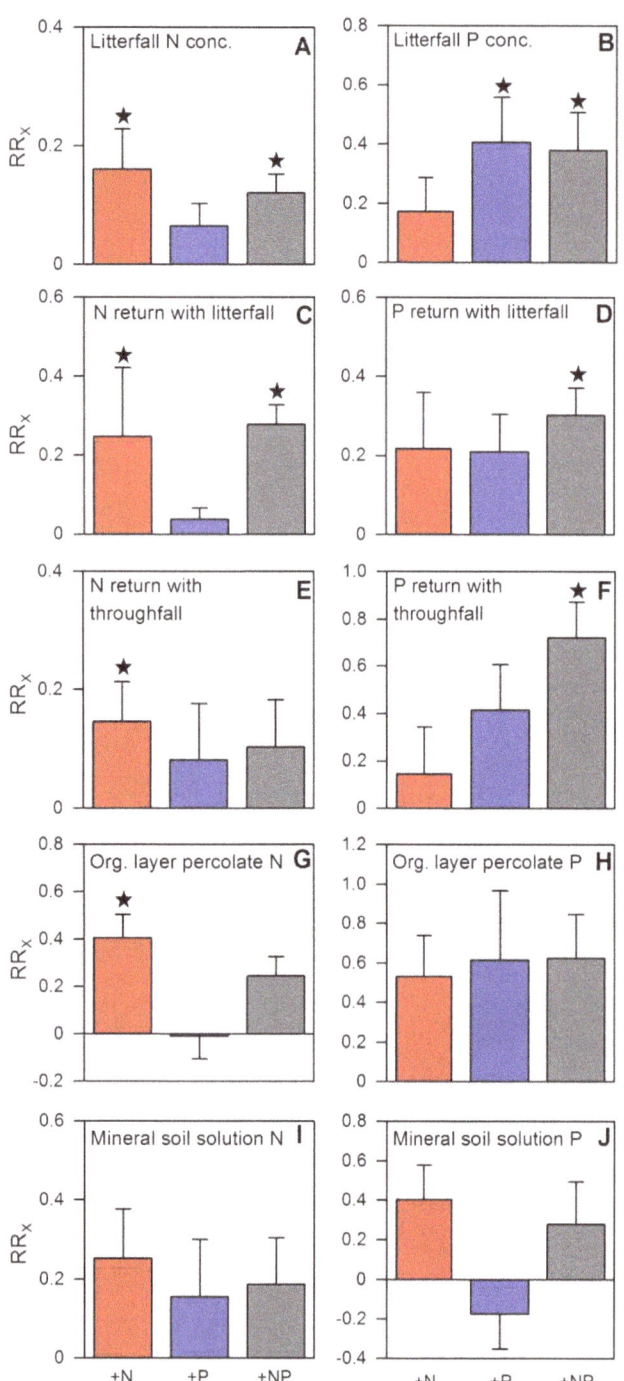

Figure 3. Effects of one year of experimental nutrient addition on nutrient cycling of a montane forest in Ecuador. a. Nitrogen and b. phosphorus concentrations of the litterfall in January 2009 after one year of nutrient addition (0.87±0.03% N and 0.025±0.006% P). Total annual return of c. nitrogen and d. phosphorus with litterfall (41.7±5.0 kg N ha^{-1} yr^{-1} and 1.49±0.24 kg P ha^{-1} yr^{-1}). Total annual return of e. nitrogen and f. phosphorus with throughfall (10.1±1.0 kg N ha^{-1} yr^{-1} and 0.15±0.03 kg P ha^{-1} yr^{-1}). g. Annual flux of nitrogen and h. phosphorus found in the organic layer percolate (14.65±1.21 kg N ha^{-1} yr^{-1} and 0.13±0.02 kg P ha^{-1} yr^{-1}). i. Annual flux of nitrogen and j. phosphorus found in the soil solution at 0.3 m soil depth (3.26±0.59 kg N ha^{-1} yr^{-1} and 0.03±0.004 kg P ha^{-1} yr^{-1}). Error bars indicate plus or minus one standard error. Data of the control treatment (mean ±1 SE) are given in parentheses in the legend. Asterisks indicate significant differences to the control (P≤0.05). For interpretation of graph see legend of Fig. 1.

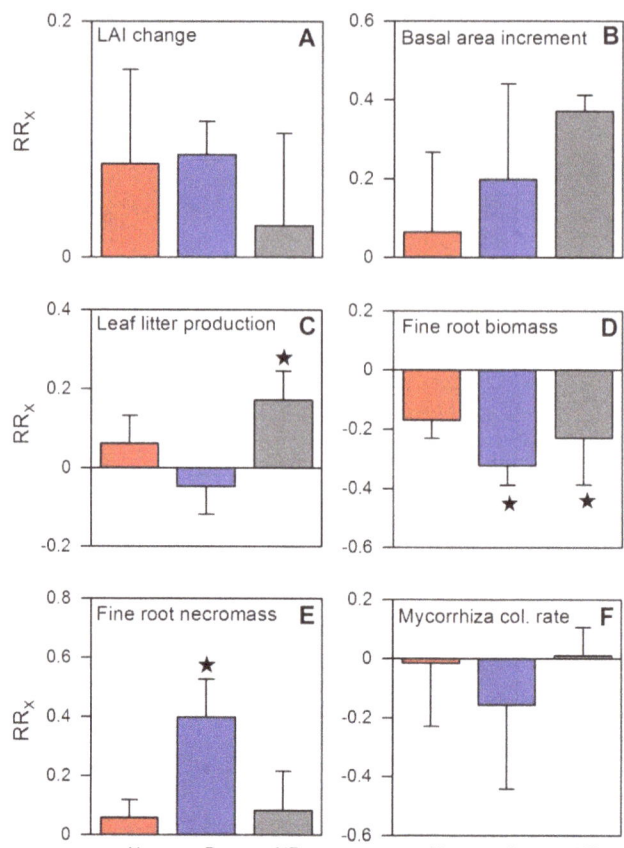

Figure 4. Effects of one year of experimental nutrient addition on vegetation related parameters of a montane forest in Ecuador. a. Relative change of leaf area index (LAI) after one year of nutrient addition (measurements from January 2009 were compared to measurements prior to nutrient addition in January 2008; control mean changed from 4.6±0.2 in 2008 to 4.7±0.4 in 2009). b. Plot basal area increment from February 2008 to January 2009 (0.111±0.018 m^2 ha^{-1}). c. Annual leaf litter production from February 2008 to January 2009 (3.46±0.46 Mg ha^{-1}). d. Fine root biomass in January 2009 (443±28 g m^{-2}), e. fine root necromass in January 2009 (426±29 g m^{-2}), and f. rate of fine root colonization by arbuscular mycorrhizal fungi in January 2009 (53.3±6.2%). Error bars indicate plus or minus one standard error. Data of the control treatment (mean ±1 SE) are given in parentheses in the legend. Asterisks indicate significant differences to the control (P≤0.05). For interpretation of graph see legend of Fig. 1.

a concurrent increase of dead roots after nutrient addition have also been shown in other montane forests after N addition [57: Puerto Rico] or after N, P and N+P addition [58: Hawaii].

The observed decrease of fine root biomass does not necessarily indicate a reduced fine root production. More likely is a stimulation of fine root production by nutrient addition and a concurrent increase of fine root turnover [59,60], the combination of both effects could result in a reduced standing fine root biomass.

The root colonization by arbuscular mycorrhizal fungi (AMF) was not significantly affected by nutrient addition, with values remaining more or less constant at about 50% (Fig. 4f). This result contrasts with the rapid response of fine root biomass and does not fit to the predictions of the functional equilibrium model of AMF [61]. Empirical data show root colonization by AMF to be reduced when N and particularly P availability is increased [62]. These results indicate that in our study area the addition of P (and to a lesser extent of N) relaxed the growth limitation imposed by P

Table 1. Effects of nutrient addition on foliar nutrient concentrations and leaf morphology of the four most common tree species of a montane forest in Ecuador.

	Tree species	control	Deviation from control (%)		
			+N	+P	+NP
Foliar N (mg g^{-1})	Graffenrieda	12.2	+5	±0	+4
	Myrcia sp.	11.3	**+9***	−6	**+11***
	Hieronyma	13.8	+1	−4	**+11***
	Alchornea	13.8	+7	**+15***	**+14***
Foliar P (mg g−1)	Graffenrieda	0.43	−2	**+37***	+21
	Myrcia sp.	0.43	**+28***	−5	**+33***
	Hieronyma	0.54	+2	+11	**+28***
	Alchornea	0.71	−11	**+34***	**+42***
Foliar N/P ratio	Graffenrieda	31	+7	**−31***	−17
	Myrcia sp.	27	−17	−4	−18
	Hieronyma	26	−3	**−14***	**−14***
	Alchornea	20	**+16***	**−15***	**−21***
Leaf area (cm^2)	Graffenrieda	178.2	+28	−1	+15
	Myrcia sp.	17.9	+3	−15	+12
	Hieronyma	26.8	−13	−29	−23
	Alchornea	30.5	−8	+1	+2
Specific leaf area (cm^2 g^{-1})	Graffenrieda	38.4	+9	+7	+10
	Myrcia sp.	40.2	+2	−5	+2
	Hieronyma	69.3	+6	+1	−7
	Alchornea	40.9	+13	+5	+12

Given are the absolute values for the control treatment and the percental effects of the treatments. Asterisks indicate significant differences to control (P<0.05). The number of sampled trees was for *Graffenrieda emarginata*: 5 (control), 6 (+N), 6 (+P) and 6 (+NP), for *Myrcia sp.*: 6, 5, 5 and 6, for *Hieronyma fendleri* 2, 5, 5 and 3 and for *Alchornea lojaensis*: 5, 4, 5 and 4.

(and N) scarcity and prompted the trees to allocate more carbon into aboveground structures and productivity. Our finding that fine root biomass was significantly reduced upon N and P addition, while the mycorrhizal infection of the trees remained unchanged, points at the importance of AMF functions other than nutrient acquisition in controlling the plant-fungus interaction [63], or resource uptake rates that are independent of infection rate.

The stand-level N and P use efficiencies in the control plots were in the upper range of values reported from other tropical forests [64] (Fig. 5). Addition of nitrogen, phosphorus, or both, led to a significant decrease (10–25%) in the N or P use efficiencies, respectively, within one year, suggesting rapid relaxation from growth limitation by P and N and most likely decreased nutrient resorption efficiencies of the vegetation. Lower nutrient resorption efficiencies with increasing green leaf nutrient status were also reported by Kobe et al. [65] for a global data set of perennial plant species.

Divergent tree species growth responses

We found a highly variable response of stem diameter growth upon N and/or P addition among the tree species in the fertilized plots. Depending on species, growth rates were either higher or lower relative to the control plots (Table 2). However, two of the most common species (*Hieronyma fendleri* and *Alchornea lojaensis*) showed increases in stem diameter growth rates after addition of N, P or N+P, while two other species (*Graffenrieda emarginata* and *Myrcia* sp.) tended to reduce growth upon N or P fertilization. The divergent growth response of different tree species is in agreement with fertilization studies from other tropical montane forests [33,34]. It appears that the responsiveness to N or P addition of stem diameter growth is highly species-specific and that while some species will increase, others will reduce their competitive strength with continued nutrient addition, likely resulting in species composition changes and diversity reductions in this species-rich forest over time [7,66–67]. Since the forest canopy will become denser in response to fertilization (increasing the LAI) and the trees will be relieved of growth constraints due to limited N and/or P availability, light competition will become more important with increasing input of nutrients. These changes may reduce the competitive ability of the seedlings and saplings of the currently abundant tree species, and will probably result in their eventual replacement by species adapted to more fertile soils. Changes in tree species composition (from slow-growing species adapted to nutrient-poor soils to faster growing species adapted to more fertile soils) will most likely accelerate the projected shifts in the C cycle by increasing the biomass turnover rate. The N:P ratio in leaf biomass of the unfertilized trees (means of 20–31 in the four most common species) provides support for the conclusion that tree growth in the studied forest is mainly limited by P [68]. Consequently, P was accumulated to a larger extent in the foliage than N after addition of P or N, and the N:P ratio responded more to P addition (decrease) than to N addition (no uniform effect; Table 1). The increased foliar N concentrations in three of the four studied common tree species after addition of N or N+P, respectively, should result in higher photosynthetic carbon gain, since photosynthetic capacity is closely related to foliar N [69].

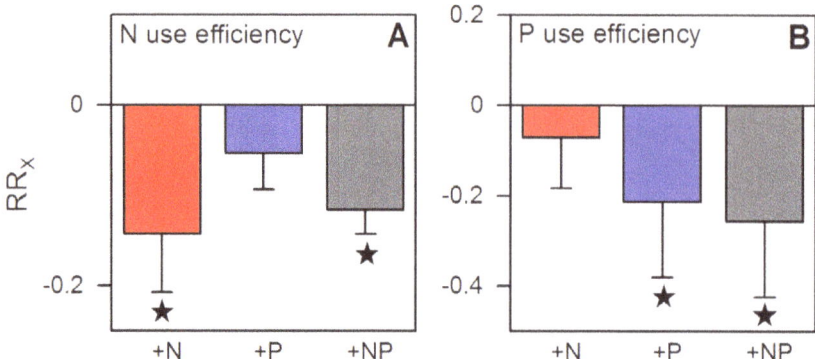

Figure 5. Nutrient use efficiencies and monthly nutrient return with litterfall. Nutrient use efficiencies (i.e. the ratio of total litterfall dry mass to nutrient content [64]) of the different treatments in the studied montane forest in Ecuador after one year of nutrient addition (samples from January 2009, means of N = 24 litter traps per treatment). a. N use efficiency (116.5±3.8 g g^{-1}), b. P use efficiency (4751±782 g g^{-1}). Error bars indicate plus or minus one standard error. Data of the control treatment (mean ±1 SE) are given in parentheses in the legend. Asterisks indicate significant differences to the control (P≤0.05). For interpretation of graph see legend of Fig. 1.

Conclusions

Overall, the strong and complex short-term response of the tropical montane forest ecosystem to moderate nutrient inputs suggests major consequences of expected future nutrient inputs into these ecosystems. This is particularly evident at our study site. The effects that we observed were larger than those reported from tropical lowland forests on more fertile soils, where only long-term nutrient addition resulted in significant effects [40]. Several of the responses to nutrient addition are similar to those known from other tropical montane forests, where they occurred either after chronic nutrient addition or after fertilization with higher amounts of N.

Provided that these initial trends persist, continued addition of substantial amounts of N and P will probably result in taller forests with a higher above-ground biomass but smaller below-ground biomass [60]. However, the below-ground response of the system to nutrient addition is still poorly understood. Given the large stocks of carbon in the organic layer, stimulated mineralization and soil respiration rates and less belowground C sequestration may turn these ecosystems into a significant future source of CO_2 to the atmosphere.

Further studies have to show how nutrient cycles and key ecosystem services such as carbon storage will adjust to continuing input of moderate amounts of N and P and how community composition will change in the long run.

Cross-study comparisons of nutrient manipulation experiments could contribute to a better understanding of ecosystem responses to increasing nutrient deposition, but the currently published studies are hard to compare due to different levels of fertilizer addition and methodological differences among the various studies. A network of coordinated experiments adding low amounts of nutrients to tropical forests, that covers a wide range of environmental conditions (climate, soil), would be the method choice of obtaining general patterns of tropical forest ecosystem responses to increasing nutrient availability.

Materials and Methods

Study area

The study was conducted at about 2000 m elevation, in a tropical montane moist forest of the San Francisco Reserve in the Andes of southern Ecuador (3°58'S, 79°04'W) (Fig. S1). This forest is in nearly pristine condition, and is one of the best-studied tropical montane forests worldwide, known for its extraordinary richness in tree species as well as other plant and animal groups [70–71]. The forest harbors more than 300 tree species with Lauraceae, Melastomataceae and Rubiaceae being the plant families with the highest species numbers. The study site has a mean annual precipitation of ~2200 mm and an annual mean temperature of ~15°C. The most abundant soil types at the study site are Cambisols that developed on paleozoic metamorphosed schists and sandstones. Soils are heterogeneous but usually nutrient-poor (thick organic layers can harbor locally high nutrient stocks, but often these are only slowly bioavailable) [41]. Estimates of total annual nutrient depositions (based on the monitoring of bulk and dry deposition between 1998 and 2010) range from 14–45 kg N ha^{-1} and 0.4–4.9 kg P ha^{-1} for the study area.

All necessary permits were required for the described field studies.

Experimental design

A full-factorial nutrient manipulation experiment (NUMEX) was conducted in 16 plots of 400 m^2 (20 m×20 m) consisting of four treatments (N, P, N+P, control) with four replicates in a

Table 2. Effects of nutrient addition on annual stem diameter growth of the four most common tree species of a montane forest in Ecuador.

| Tree species | control | Deviation from control (%) | | |
| | | +N | +P | +NP |
	(mm)			
Graffenrieda emarginata	0.84 (49)	−12 (49)	−21 (47)	+16 (54)
Myrcia sp.	0.80 (17)	−39 (15)	**−50*** (16)	−5 (11)
Hieronyma fendleri	0.07 (6)	+157 (22)	+71 (13)	**+414*** (13)
Alchornea lojaensis	0.12 (20)	**+33*** (18)	+17 (16)	**+42*** (9)
all other species pooled	0.28 (78)	−11 (94)	+82 (77)	+25 (89)

Given are the absolute values for the control treatment (February 2008–January 2009) and the percental effects of the treatments. Asterisks indicate significant differences to the control (P<0.05).

stratified random design in four blocks at 2020–2120m a.s.l. (Fig. S1). Minimum distance between two plots was 10 m.

Nitrogen and phosphorus were added at an annual rate of 50 kg N ha^{-1} as urea and 10 kg P ha^{-1} as monosodium phosphate. The fertilizer was dispersed homogeneously over the plots with two application dates per year (January 26 and July 26) starting in 2008.

The dominant tree species in the NUMEX plots, making up about one quarter of all stems (dbh≧10 cm), was *Graffenrieda emarginata* Ruiz & Pav. (Melastomataceae); other frequently found species were *Myrcia* sp. nov. (Myrtaceae), *Alchornea lojaensis* Secco and *Hieronyma fendleri* Briq. (both Euphorbiaceae). The mean number of trees, mean tree diameter and stem basal area per plot (pre-fertilization survey of trees ≥10 cm dbh) were 45.7, 15.0 cm and 0.91 m^2, respectively. Average stand height was 12 to 14 m.

Organic layer nutrient pools

In April 2009, the soil organic layer (including the Oi, Oe, and Oa horizons) was sampled with a 0.2×0.2 m frame to the depth of the organic layer/mineral soil boundary at five randomly selected points within each permanent plot. Samples were dried to constant mass at 40°C.

The organic horizons were separated from the underlying mineral soil at the point where bulk density abruptly increases from <0.2 g cm^3 in the organic layers to >1 g cm^3 in the mineral soil [41].

The N concentrations in the ground samples were determined with a CHNS-analyzer (Vario EL Cube, Elementar Analysensysteme GmbH). After microwave digestion with HNO_3 (Mars 5 Xpress, CEM Corporation, Matthews, NC), the total phosphate concentration was detected photometrically (Continuous Flow Analyser; Bran+Luebbe GmbH, Norderstedt, Germany). Soil bulk density was determined with two additional samples per plot which were dried at 105°C for 24 h. Detailed results are shown in Table S1.

Microbial biomass

In May 2009, three samples per plot were collected in the upper organic layer to a depth of 5 cm using a metal corer (5 cm diameter). The upper litter layer (1–2 cm) was removed and the three samples were pooled and stored at 5°C. Before measurements, roots >2 mm were removed from the soil and the remaining material was chopped to pieces of <25 mm^2, homogenized and pre-incubated at 20°C for five days.

Respiration of soil microorganisms was measured as O_2 consumption using an automated electrolytic O_2 microcompensation apparatus [72]. Respiration was measured at hourly intervals at 22°C for 24 h. Basal respiration (BR) of microorganisms was calculated as the mean oxygen consumption rates of hours 14–24 after the start of the measurements without addition of substrate. Microbial biomass carbon (C_{mic}) was calculated from substrate induced respiration measuring the respiratory response to D-glucose which activates the metabolism of living microorganisms in the soil [73]. After adding Glucose (80 mg g^{-1} dry mass in 300 μl deionized water) the mean of the lowest three readings within the first 5–10 h was taken as maximum initial respiratory response (MIRR; μg O_2 h^{-1} g^{-1} soil dry mass). Microbial biomass (μg C g^{-1} soil dry mass) was calculated as 38×MIRR [74].

Net rates of nitrogen cycling in the soil

Net rates of N cycling in the soil were measured in October 2008 using the buried bag method [75]. In each plot a soil sample was taken from 0 to 5 cm depth. One subsample was extracted immediately in the field with 0.5 mol L^{-1} K_2SO_4 to determine initial NH_4^+ and NO_3^- levels (T_0). The other sample was put into a plastic bag, reburied in the soil, incubated for ten days and afterwards extracted with 0.5 mol L^{-1} K_2SO_4 (T_1). The plastic bag was closed with a rubber band to prevent rain coming in but not too tight to permit air exchange. Net N mineralization and nitrification rates were calculated as the difference between T_1- and T_0- NH_4^+ and NO_3^- concentration.

Trace gas measurements

Nitrous oxide was measured monthly using static vented chambers. Four permanent chamber bases made of polyvinyl chloride (area 0.04 m^2, height 0.25 m, ~0.02 m inserted into the soil) were randomly placed in four of six subplots per plot at least four weeks before the first measurement, resulting in 16 chamber bases per block and 48 in total. Four gas samples (100 ml each) were removed at 2, 14, 26 and 38 min after chamber closure with an acrylonitrile butadiene styrene (ABS) lid and stored in pre-evacuated glass containers (60 mL) with stopcocks [76]. Gas samples were transported to the laboratory in Loja (Ecuador) within two days and analyzed using a gas chromatograph (Shimadzu GC-14B, Duisburg, Germany) equipped with an electron capture detector (ECD) and an autosampler [77]. Gas concentrations were determined by comparison of integrated peak areas of samples to standard gases (320, 501, 1001 and 3003 ppb N_2O; Deuste Steininger GmbH, Mühlhausen, Germany). Gas fluxes were calculated from the linear increase of gas concentration in the chamber vs. time, and were adjusted for air temperature and atmospheric pressure [76]. Zero fluxes were included.

Litterfall

Six litter traps (each 0.36 m^2 in surface area positioned 1 m above ground) were randomly placed in each plot. The litterfall was collected every four weeks starting on November 6th, 2007. The collected samples were oven-dried at 60°C before determining the dry weight.

Leaf morphology

Leaf samples from sun-exposed branches of each of 4–5 trees per treatment from four common species (*Alchornea lojaensis*, *Graffenrieda emarginata*, *Hieronyma fendleri* and *Myrcia* sp. nov.) were collected in January 2009 to quantify changes in leaf morphology and foliar nutrient concentrations one year after the onset of the experiment. For each sample 10–25 fresh leaves were scanned using a flat bed scanner (CanonScan LIDE 30, Canon). The images were analyzed subsequently with the WinFolia 2001a software (Regent Instruments Inc., Quebec, Canada) for calculation of leaf area. The leaves were then dried at 60°C to constant mass. Specific leaf area (SLA) was calculated as the ratio of leaf area and leaf dry weight.

Foliar and litter nutrient contents

The concentrations of total C and N in leaf and litter mass were determined with a C/N elemental analyzer (Vario EL III, elementar, Hanau, Germany). The concentrations of total P were analyzed using an Inductively Coupled Plasma Analyzer (Optima 5300DV ICP-OES, Perkin Elmer) after digesting the samples with concentrated HNO_3.

Throughfall and soil solutions

Throughfall was collected with 20 randomly distributed, fixed-positioned funnel gauges in each plot. The volume of throughfall water was measured in the field with a graduated cylinder, and the samples were then bulked according to their relative volume to result in a single sample per plot per collecting date.

Litter leachate was collected using three zero-tension lysimeters per plot, which consisted of plastic boxes with a collecting surface area (polyethylene net) of 0.15 m×0.15 m, installed below the organic layer. All collected litter leachate samples of a plot were bulked to yield a single sample per plot per collecting date.

In each plot, mineral soil solution was collected using three suction lysimeters (ceramic suction cups with 1 μm pore size) at 0.15 and 0.30 m depth, installed so that bulking of the soil solution per soil depth occurred *in situ*.

Throughfall, litter leachate and mineral soil solutions were sampled fortnightly. After collecting the mineral soil solution, a vacuum was applied to the suction lysimeters in order to collect sufficient sample for the next sampling period.

After field collection, throughfall, litter leachate, and soil solution samples were transported to our field laboratory where an aliquot was filtered (ashless filters with pore size 4–7 μm, folded filter type 389; Munktell & Filtrak GmbH, Bärenstein, Germany) and frozen until transport to Germany for further analysis. Samples were analyzed for concentrations of NH_4^+, NO_3^-, total dissolved N, and total dissolved P using continuous flow analysis (CFA, Bran+Luebbe GmbH, Norderstedt, Germany). Dissolved organic nitrogen concentrations were calculated as the difference between total dissolved N and the sum of total inorganic nitrogen ($NH_4^+ + NO_3^-$), assuming that NO_2^- concentrations were negligible.

Leaf area index

The LAI was quantified in the plots with two LAI-2000 plant canopy analyzers (LI-COR Inc., Lincoln, NE, USA). The LAI measurements were conducted in the remote mode, i.e. by synchronous readings below the canopy at 2 m height above the forest floor and in a nearby open area ("above-canopy" reading) using two devices. One measurement was done above each litter trap and a second at the same time outside the forest. Measurements were done in January 2008 (before 1st fertilization) and in January 2009 (one year after the 1st fertilization).

Stem diameter growth and basal area growth per plot

The stem diameter growth of all trees present with a dbh≥10 cm was monitored in the 16 plots every six weeks with permanent girth-increment tapes (D1 dendrometer, UMS, Munich; 713 stems in total). The cumulative increase in plot basal area per year was calculated as the sum of all tree basal area increments in a plot between February 15th, 2008 (after the first fertilization) and January 19th, 2009.

Fine root biomass

For measuring fine root biomass, we took six root samples per plot to a depth of 20 cm using a soil corer of 3.5 cm in diameter in January 2009. The soil samples were transferred to plastic bags and transported to the laboratory, where processing of the stored samples (4°C) took place within six weeks. In the lab, the samples were soaked in water and cleaned from soil residues using a sieve

with a mesh size of 0.25 mm. Only fine roots (roots <2 mm in diameter) of trees were considered for analysis. Live fine roots (biomass) were separated from dead rootlets (necromass) under the stereomicroscope based on color, root elasticity, and the degree of cohesion of cortex, periderm and stele [78–79]. The fine root biomass of each sample was dried at 70°C for 48 h and weighed.

Mycorrhiza

Root colonization by arbuscular mycorrhizal fungi was measured at 200× magnification following clearing and staining with 0.05% Trypan Blue according to [80], additionally including fungal structures as defined in [81].

Statistical analyses

The effects of N and/or P addition on the various investigated parameters were expressed by a response ratio metric ($RR_X = ln$ (measured value in nutrient addition treatment/measured value in the control)) [8] in order to compare the response of plant- and soil-related state variables or flux parameters in relative terms. Non-transformed data are shown in Table S2.

Effects of the addition of N and/or P on individual parameters were analyzed using linear mixed models (package lme4, R version 2.13.0) [82]. We included the fertilization treatments as fixed effects and the factor "block" as a random factor in the models, since in most parameters, samples were nested within plots. P-values for the fixed effects were calculated with the "cftest" function of the package "multcomp" (R version 2.13.0) [82].

Supporting Information

Figure S1 Location of the study area in southern Ecuador and outline of the Ecuadorian Nutrient Manipulation EXperiment (NUMEX).

Table S1 Soil nutrient status of the experimental plots in July 2007 prior to the first fertilization.

Table S2 Ranges and means of all parameters shown in Figures 1–5.

Acknowledgments

We thank the Ministerio de Ambiente del Ecuador for the research permit (0014-IC-FAUNA-DRLZCH-VS-MAE) for the Podocarpus National Park and the San Francisco reserve, Nature and Culture International (NCI) in Loja for granting access to the San Francisco reserve and the research station, and the Universidad Nacional de Loja and the Universidad Técnica Particular de Loja for cooperation.

Author Contributions

Conceived and designed the experiments: JH DH CL MM MCR SS EV WW. Performed the experiments: JH DH TC NLC GOM NP MCR DS HW. Analyzed the data: JH DH TC GOM DS HW. Wrote the paper: JH DH CL SS WW.

References

1. Galloway JN, Townsend AR, Erisman JW, Bekunda M, Cai Z, et al. (2008) Transformation of the nitrogen cycle: recent trends, questions, and potential solutions. Science 320: 889–892.
2. Gruber N, Galloway N (2008) An earth-system perspective of the global nitrogen cycle. Nature 451: 293–296.
3. Janssens IA, Dieleman W, Luyssaert S, Subke J-A, Reichstein M, et al. (2010) Reduction of forest soil respiration in response to nitrogen deposition. Nat Geosci 3: 315–322.
4. Mahowald NM, Artaxo P, Baker AR, Jickells TD, Okin GS, et al. (2005) Impacts of biomass burning emissions and land use change on Amazonian atmospheric phosphorus cycling and deposition. Global Biogeochem Cycles 19: GB4030.
5. Schlesinger WH (2009) On the fate of anthropogenic nitrogen. Proc Natl Acad Sci U S A 106: 203–208.

6. Tilman D, Fargione J, Wolff B, D'Antonio C, Dobson A, et al. (2001) Forecasting agriculturally driven global environmental change. Science 292: 281–284.

7. Bobbink R, Hicks K, Galloway J, Spranger T, Alkemade R, et al. (2010) Global assessment of nitrogen deposition effects on terrestrial plant diversity: a synthesis. Ecol Appl 20: 30–59.

8. Elser JJ, Bracken MES, Cleland EE, Gruner DS, Harpole WS, et al. (2007) Global analysis of nitrogen and phosphorus limitation of primary producers in freshwater, marine and terrestrial ecosystems. Ecol Lett 10: 1135–1142.

9. Reay DS, Dentener F, Smith P, Grace J, Feely RA (2008) Global nitrogen deposition and carbon sinks. Nat Geosci 1: 430–437.

10. Harpole WS, Ngai JT, Cleland EE, Seabloom EW, Borer ET, et al. (2011) Nutrient co-limitation of primary producer communities. Ecol Lett 14: 852–862.

11. Hedin LO, Brookshire ENJ, Menge DNL, Barron A (2009) The nitrogen paradox in tropical forest ecosystems. Annu Rev Ecol Evol Syst 40: 613–635.

12. LeBauer DS, Treseder KK (2008) Nitrogen limitation of net primary productivity in terrestrial ecosystems is globally distributed. Ecology 89: 371–379.

13. Vitousek PM, Porder S, Houlton BZ, Chadwick OA (2010) Terrestrial phosphorus limitation: mechanisms, implications, and nitrogen-phosphorus interactions. Ecol Appl 20: 5–15.

14. Xia J, Wan S (2008) Global response patterns of terrestrial plant species to nitrogen addition. New Phytol 179: 428–439.

15. Bedison JE, McNeill BE (2009) Is the growth of temperate forest trees enhanced along an ambient nitrogen deposition gradient? Ecology 90: 1736–1742.

16. Wright RF, Rasmussen L (1998) Introduction to the NITREX and EXMAN projects. For Ecol Manage 101: 1–7.

17. Stevens CJ, Dise NB, Mountford JO, Gowing DJ (2004) Impact of nitrogen deposition on the species richness of grasslands. Science 303: 1876–1879.

18. Högberg P, Fan H, Quist M, Binkley D, Tamm CO (2006) Tree growth and soil acidification in response to 30 years of experimental nitrogen loading on boreal forest. Glob Chang Biol 12: 489–499.

19. Bouwman L, Goldewijka KK, Van Der Hoek KW, Beusena AHW, Van Vuurena DP, et al. (in press) Exploring global changes in nitrogen and phosphorus cycles in agriculture induced by livestock production over the 1900–2050 period. Proc Natl Acad Sci U S A: doi: 10.1073/pnas.1012878108.

20. Phoenix G, Hicks WK, Cinderby S, Kuylenstierna JCI, Stocks WD, et al. (2006) Atmospheric nitrogen deposition in world biodiversity hotspots: the need for a greater global perspective in assessing N deposition impacts. Glob Chang Biol 12: 470–476.

21. Tamatamah RA, Hecky RE, Duthie HD (2005) The atmospheric deposition of phosphorus in Lake Victoria (East Africa). Biogeochemistry 73: 325–344.

22. Okin GS, Mahowald N, Chadwick OA, Artaxo P (2004) Impact of desert dust on the biogeochemistry of phosphorus in terrestrial ecosystems. Global Biogeochem Cycles 18: 1–14.

23. Pett-Ridge JC (2009) Contributions of dust to phosphorus cycling in tropical forests of the Luquillo Mountains, Puerto Rico. Biogeochemistry 94: 63–80.

24. Fabian P, Kohlpaintner M, Rollenbeck R (2005) Biomass burning in the Amazon – fertilizer for the mountaineous rain forest in Ecuador. Environ Sci Pollut Res Int 12: 290–296.

25. Boy J, Rollenbeck R, Valarezo C, Wilcke W (2008) Amazonian biomass burning-derived acid and nutrient deposition in the north Andean montane forest of Ecuador. Global Biogeochem Cycles 22: GB4011.

26. Matson PA, McDowell WH, Townsend AR, Vitousek PM (1999) The globalization of N deposition: ecosystem consequences in tropical environments. Biogeochemistry 46: 67–83.

27. Corre MD, Veldkamp E, Arnold J, Wright SJ (2010) Impact of elevated N input on N cycling and retention of soils under old-growth lowland and montane forests in Panama. Ecology 91: 1715–1729.

28. Hall SJ, Matson PA (2003) Nutrient status of tropical rain forests influences soil N dynamics after N addition. Ecol Monogr 73: 107–129.

29. Lewis SL, Malhi Y, Phillips OL (2004) Fingerprinting the impacts of global change on tropical forests. Phil Trans R Soc Lond B Biol Sci 359: 437–462.

30. Li Y, Xu M, Zou X (2006) Effects of nutrient additions on ecosystem carbon cycle in a Puerto Rican tropical wet forest. Glob Chang Biol 12: 284–293.

31. Wright SJ (2005) Tropical forests in a changing environment. Trends Ecol Evol 20: 553–560.

32. Adamek M, Corre MD, Hölscher D (2009) Early effect of elevated nitrogen input on above-ground net primary production of a lower montane rain forest, Panama. J Trop Ecol 25: 637–647.

33. Cavelier J, Tanner E, Santamaria J (2000) Effect of water, temperature and fertilizers on soil nitrogen net transformations and tree growth in an elfin cloud forest of Colombia. J Trop Ecol: 83–99.

34. Tanner EVJ, Kapos V, Freskos S, Healey JR, Theobald AM (1990) Nitrogen and phosphorus fertilization of Jamaican montane forest trees. J Trop Ecol 6: 231–238.

35. Tanner EVJ, Kapos V, Franco W (1992) Nitrogen and phosphorus fertilization effects on Venezuelan montane forest trunk growth and litterfall. Ecology 73: 78–86.

36. Mirmanto E, Proctor J, Green J, Nagy L, Suriantata (1999) Effects of nitrogen and phosphorus fertilization in a lowland evergreen rainforest. Phil Trans R Soc Lond B Biol Sci 354: 1825–1829.

37. Nomura N, Kikuzawa K (2003) Productive phenology of tropical montane forests: Fertilization experiments along a moisture gradient. Ecol Res 18: 573–586.

38. Ostertag R (2010) Foliar nitrogen and phosphorus accumulation responses after fertilization: an example from nutrient-limited Hawaiian forests. Plant Soil 334: 85–98.

39. Vitousek PM, Farrington H (1997) Nutrient limitation and soil development: Experimental test of a biogeochemical theory. Biogeochemistry 37: 63–75.

40. Wright SJ, Yavitt JB, Wurzburger N, Turner BL, Tanner EVJ, et al. (2011) Potassium, phosphorus, or nitrogen limit root allocation, tree growth, or litter production in a lowland tropical forest. Ecology 92: 1616–1625.

41. Wilcke W, Yasin S, Abramowski U, Valarezo C, Zech W (2002) Nutrient storage and turnover in organic layers under tropical montane rain forest in Ecuador. Europ J Soil Sci 53: 15–27.

42. Treseder KK (2008) Nitrogen additions and microbial biomass: a meta-analysis of ecosystem studies. Ecol Lett 11: 1111–1120.

43. Fog K (1988) The effect of added nitrogen on the rate of decomposition of organic matter. Biol Rev 63: 433–462.

44. Lu M, Yang Y, Luo Y, Fang C, Zhou X, et al. (2011) Responses of ecosystem nitrogen cycle to nitrogen addition: a meta-analysis. New Phytol 189: 1040–1050.

45. Lohse KA, Matson P (2005) Consequences of nitrogen additions for soil losses from wet tropical forests. Ecol Appl 15(5): 1629–1648.

46. Matzner E, Zuber T, Lischeid G (2004) Response of soil solution chemistry and solute fluxes to changing deposition rates. In: Matzner E, editor. Biogeochemstry of forested catchments in a changing environment - A German case study. Ecological Studies Vol 172. Berlin: Springer. pp. 339–360.

47. Goller R, Wilcke W, Fleischbein K, Valarezo C, Zech W (2006) Dissolved nitrogen, phosphorus, and sulfur forms in the ecosystem fluxes of a montane forest in Ecuador. Biogeochemistry 77: 57–89.

48. Wullaert H, Homeier J, Valarezo C, Wilcke W (2010) Response of the N and P cycles of an old-growth montane forest in Ecuador to experimental low-level N and P amendments. For Ecol Manage 260: 1434–1445.

49. Wieder WR, Cleveland CC, Townsend AR (2009) Controls over leaf litter decomposition in wet tropical forests. Ecology 90: 3333–3341.

50. Cleveland CC, Townsend AR (2006) Nutrient additions to a tropical rain forest drive substantial soil carbon dioxide losses to the atmosphere. Proc Natl Acad Sci U S A 103: 10316–10321.

51. Treseder KK, Vitousek PM (2001) Effects of soil nutrient availability on investment in acquisition of N and P in Hawaiian rain forests. Ecology 82: 946–954.

52. Gress SE, Nichols TD, Northcraft CC, Peterjohn W (2007) Nutrient limitation in soils exhibiting differing nitrogen availabilities: what lies beyond nitrogen saturation? Ecology 88: 119–130.

53. Olander LP, Vitousek PM (2000) Regulation of soil phosphatase and chitinase activity by N and P availability. Biogeochemistry 49: 175–190.

54. Poorter L, Bongers F (2006) Leaf traits are good predictors of plant performance across 53 rain forest species. Ecology 87: 1733–1743.

55. Vitousek PM, Walker LR, Whiteaker LD, Matson PA (1993) Nutrient limitations to plant growth during primary succession in Hawaii Volcanoes National Park. Biogeochemistry 23: 197–215.

56. Güsewell S, Gessner MO (2009) N:P ratios influence litter decomposition and colonization by fungi and bacteria in microcosms. Funct Ecol 23: 211–219.

57. Cusack DF, Silver WL, Torn MS, McDowell WH (2011) Effects of nitrogen additions on above- and belowground carbon dynamics in two tropical forests. Biogeochemistry 104: 203–225.

58. Gower ST, Vitousek PM (1989) Effects of nutrient amendments on fine root biomass in a primary successional forest in Hawai'i. Oecologia 81: 566–568.

59. Nadelhoffer KJ (2000) The potential effects of nitrogen deposition on fine-root production in forest ecosystems. New Phytol 147, 131–139.

60. Yuan ZY, Chen HYH (2012) A global analysis of fine root production as affected by soil nitrogen and phosphorus. Proc R Soc Lond B Biol Sci 279: 3796–3802.

61. Johnson NC, Rowland DL, Corkidi L, Allen EB (2008) Plant winners and losers during grassland N-eutrophication differ in biomass allocation and mycorrhizas. Ecology 89: 2868–2878.

62. Treseder KK (2004) A meta-analysis of mycorrhizal responses to nitrogen, phosphorus, and atmospheric CO_2 in field studies. New Phytol 164: 347–355.

63. Powell JR, Parrent JL, Hart MM, Klironomos JN, Rillig MC, et al. (2009) Phylogenetic trait conservatism and the evolution of functional tradeoffs in arbuscular mycorrhizal fungi. Proc R Soc Lond B Biol Sci 276: 4237–4245.

64. Vitousek PM (1984) Litterfall, nutrient cycling, and nutrient limitation in tropical forests. Ecology 65: 285–298.

65. Kobe RK, Lepczyk CA, Iyer M (2005) Resorption efficiency decreases with increasing green leaf nutrients in a global data set. Ecology 86: 2780–2792.

66. Güsewell S (2004) N:P ratios in terrestrial plants: variation and functional significance. New Phytol 164: 243–266.

67. Lu X, Mo J, Gilliam FS, Zhou G, Fang Y (2010) Effects of experimental nitrogen additions on plant diversity in old-growth tropical forest. Glob Chang Biol 16: 2688–2700.

68. Townsend AR, Cleveland CC, Asner GP, Bustamante MMC (2007) Controls over foliar N:P rations in tropical rain forests. Ecology 88: 107–118.

69. Wright IJ, Reich PB, Westoby M, Ackerly DD, Baruch Z, et al. (2004) The worldwide leaf economics spectrum. Nature 428: 821–827.

70. Beck E, Bendix J, Kottke I, Makeschin F, Mosandl R, editors (2008) Gradients in a tropical mountain ecosystem of Ecuador. Ecological Studies Vol. 198. Berlin: Springer. 525 p.

71. Homeier J, Breckle S-W Günter S, Rollenbeck RT, Leuschner C (2010) Tree diversity, forest structure and productivity along altitudinal and topographical gradients in a species-rich Ecuadorian montane rainforest. Biotropica 42: 140–148.

72. Scheu S (1992) Automated measurement of the respiratory response of soil microcompartments: Active microbial biomass in earthworm faeces. Soil Biol Biochem 24: 1113–1118.

73. Anderson JJPE, Domsch KH (1978) A physiological method for the quantitative measurement of microbial biomass in soil. Soil Biol Biochem 10: 215–221.

74. Beck T, Joergensen RG,, Kandeler E, Makeschin F, Nuss E, et al. (1997) An inter-laboratory comparison of ten different ways of measuring soil microbial biomass C. Soil Biol Biochem 29: 1023–1032.

75. Hart SC, Stark JM, Davidson EA & Firestone MK (1994) Nitrogen mineralization, immobilization and nitrification. In: Methods of Analysis Part 2. Microbiological and Biochemical Properties. Madison: Soil Science Society of America Book Series. pp 985–1018.

76. Koehler B, Corre MD, Veldkamp E, Wullaert H, Wright SJ (2009) Immediate and long-term nitrogen oxide emissions from tropical forest soils exposed to elevated nitrogen input. Glob Chang Biol 15: 2049–2066.

77. Loftfield N, Flessa H, Augustin J, Beese F (1997) Automated gas chromatographic system for rapid analysis of the atmospheric trace gases methane, carbon dioxide, and nitrous oxide. J Environ Qual 26: 560–564.

78. Persson H (1978) Root dynamics in a young Scots pine stand in Central Sweden. Oikos 30: 508–519.

79. Leuschner C, Hertel D, Coners H, Büttner V (2001) Root competition between beech and oak: a hypothesis. Oecologia 126: 276–284.

80. Rillig MC, Allen MF, Klironomos JN, Chiariello NR, Field CB (1998) Plant species-specific changes in root-inhabiting fungi in a California annual grassland: responses to elevated CO2 and nutrients. Oecologia 113: 252–259.

81. Beck A, Haug I, Oberwinkler F, Kottke I (2007) Structural characterization and molecular identification of arbuscular mycorrhiza morphotypes of Alzatea verticillata (Alzateaceae), a prominent tree in the tropical mountain rain forest of South Ecuador. Mycorrhiza 17:607–625.

82. R Development Core Team (2011) R: A language and environment for statistical computing. Vienna: R Foundation for Statistical Computing.

6

Groundwater Nitrogen Pollution and Assessment of Its Health Risks: A Case Study of a Typical Village in Rural-Urban Continuum, China

Yang Gao[1,3]*, Guirui Yu[1]*, Chunyan Luo[2], Pei Zhou[3]

1 Key Laboratory of Ecosystem Network Observation and Modeling, Institute of Geographic Sciences and Natural Resources Research, Chinese Academy of Sciences, Beijing, China, **2** Institute of Agricultural Resources and Regional Planning, CAAS, Beijing, China, **3** School of Agriculture and Biology, Shanghai Jiaotong University, Shanghai, China

Abstract

Protecting groundwater from nitrogen contamination is an important public-health concern and a major national environmental issue in China. In this study, we monitored water quality in 29 wells from 2009 to 2010 in a village in Shanghai city, whick belong to typical rural-urban continuum in China. The total N and NO_3-N exhibited seasonal changes, and there were large fluctuations in NH_4-N in residential areas, but without significant seasonal patterns. NO_2-N in the water was not stable, but was present at high levels. Total N and NO_3-N were significantly lower in residential areas than in agricultural areas. The groundwater quality in most wells belonged to Class III and IV in the Chinese water standard, which defines water that is unsuitable for human consumption. Our health risk assessments showed that NO_3-N posed the greatest carcinogenic risk, with risk values ranging from 19×10^{-6} to 80×10^{-6}, which accounted for more than 90% of the total risk in the study area.

Editor: Alex J. Cannon, Pacific Climate Impacts Consortium, Canada

Funding: This work was financially supported by National Key Basic Research Program (2010CB833504), and National Natural Science Foundation of China (40601097 and 30590381). The funders had no role in study design, data collection and analysis, decision to publish, or preparation of the manuscript.

Competing Interests: The authors have declared that no competing interests exist.

* E-mail: gaoyang0898@163.com (YG); yugr@igsnrr.ac.cn (GY)

Introduction

Groundwater is the major water supply for drinking and for the domestic, industrial, and agricultural sectors in the Shanghai region of China. One serious problem that affects the quality of the region's groundwater is leaching of nutrients from the soil, which is especially evident in areas dominated by agriculture [1–2]. Nitrogen percolates easily into the groundwater through the soil along with rainwater recharge or irrigation water. As a result, the shallow aquifers are more likely than deeper ones to initially suffer from contamination problems [3–4]. The application of large amounts of nitrogen fertilizers in regions of intensive agriculture contributes to excessive nitrogen accumulation in soils and excessive leaching into groundwater bodies [5–7]. Extensive irrigation and use of nitrogen (N) fertilizers together result in low N-use efficiency and high N loss [8]. Several studies have also reported increasing incidence of nitrogen pollution and dramatic increases in the nitrogen concentration in the groundwater of regions where intensive farming is practiced [9–11].

Because contaminated groundwater resources are often located in the vicinity of wells for drinking water, it is essential to determine how management practices in the area surrounding these wells will affect groundwater nitrogen concentrations, and particularly nitrate nitrogen (NO_3-N). Nitrate is formed from fertilizers, decaying plants, manure and other organic residues. It is found in the air, soil, water and food (particularly in vegetables) and is produced naturally within the human body. In many cases,

groundwater nitrate concentrations are currently approaching or exceeding the recommended 11.3 mg NO_3-N L^{-1} drinking water standard (e.g., [12]). Excess nitrates (levels >50 mg L^{-1}; [13]) in the drinking water cause health risks such as conversion of hemoglobin to methemoglobin, which depletes oxygen levels in the blood. Forman et al. [14] reported additional consequences among people who consumed drinking water containing high levels of nitrates: enlargement of the thyroid gland, increased incidence of 15 types of cancer and two kinds of birth defects, and even hypertension. In addition, increasing rates of stomach cancer caused by increasing nitrate intake have been reported [15].

In Shanghai, nitrogen pollution has become an increasingly serious problem. Villages in the Shanghai city are the main areas for developing urban agriculture, which can provide the main source of vegetables and fruits for many residents. Due to extensive irrigation and fertilizer use, non-point source pollution is the dominant form, and the non-point source nitrogen loading has substantially affected groundwater nitrogen concentrations [16]. Poinke and Urban [17] showed that the average nitrogen concentration in rural groundwater was five to seven times higher than that in adjacent forest-covered areas. Where groundwater is the main source of drinking, domestic, and agricultural water, potentially significant health risks are associated with the consumption of nitrate-rich groundwater. For this reason, it is important to study the nitrogen pollution problem in rural-urban continuum near Shanghai to determine the impact on food safety and health of the residents. The aim of the present study was therefore to investigate seasonal changes in levels of nitrate and other forms of nitrogen, and based on this data, to assess the health risk for a typical village in Shanghai, thereby providing a scientific basis for controlling nitrogen pollution and protecting groundwater safety.

Results

Changes in different types of nitrogen in the groundwater of agricultural areas

Figure 1 summarizes the results of the groundwater monitoring for the four types of nitrogen for wells in agricultural areas. The total N concentration was higher from June to August than during other months. The total N concentration exceeded 20 mg L^{-1} from June to August (Fig. 1a). From December to February, the total N concentration in groundwater reached its lowest value. Because NO_3-N accounted for 60 to 80% of total N, the seasonal changes in NO_3-N were similar to those for total N (Fig. 1b). According to the classification in Table 1, the groundwater quality for most wells from June to August was Class IV, with values ranging between 20 and 30 mg L^{-1}, although some wells were rated Class V, with NO_3-N exceeding 30 mg L^{-1}; in the other months, the groundwater quality was rated Class III or worse.

The degree of NH_4-N pollution was an order of magnitude lower than the NO_3-N pollution. The groundwater quality based on NH_4-N level was rated as Class III in most months, except from June to July, when the quality degraded to Class IV (Fig. 1c). From September to May of the following year, the NH_4-N concentration was relatively stable, decreasing to <0.2 mg L^{-1}. The NO_2-N concentration in groundwater was being in minimum from December to February in agricultural areas, but the change of NO_2-N was mostly rated Class IV (Fig. 1d).

Table 1. Classification standard for groundwater quality in China based on nitrogen levels. [18]

	Groundwater quality class				
	I	II	III	IV	V
Concentration (mg L^{-1})					
NH_4-N	≤0.02	≤0.02	≤0.20	≤0.50	>0.50
NO_2-N	≤0.001	≤0.010	≤0.020	≤0.100	>0.100
NO_3-N	≤2	≤5	≤20	≤30	>30

Note: if NO_3-N is class IV, it means that the concentration of NO_3-N is between 20 to 30 mg L^{-1}.

Changes in different types of nitrogen in the groundwater of residential areas

Total N concentrations in residential areas exhibited more gradual seasonal changes than in agricultural areas. In both the rainy season and the dry season, the total N concentration was significantly lower in residential areas than in agricultural areas (Fig. 2a). Although the NO_3-N levels were lower in residential than agricultural areas, none of the wells met the criteria for Class I water and the wells only met the Class II standard in December (Fig. 2b). The residential NO_3-N concentrations ranged from 5 to 20 mg L^{-1}, and were therefore graded Class III. They showed a similar pattern of change to that in the agricultural areas.

The fluctuation in NH_4-N concentrations in residential areas was large, but there was no significant seasonal change in NH_4-N (Fig. 2c). The groundwater quality based on NH_4-N levels ranged from Class IV to Class III, which was similar to the range in

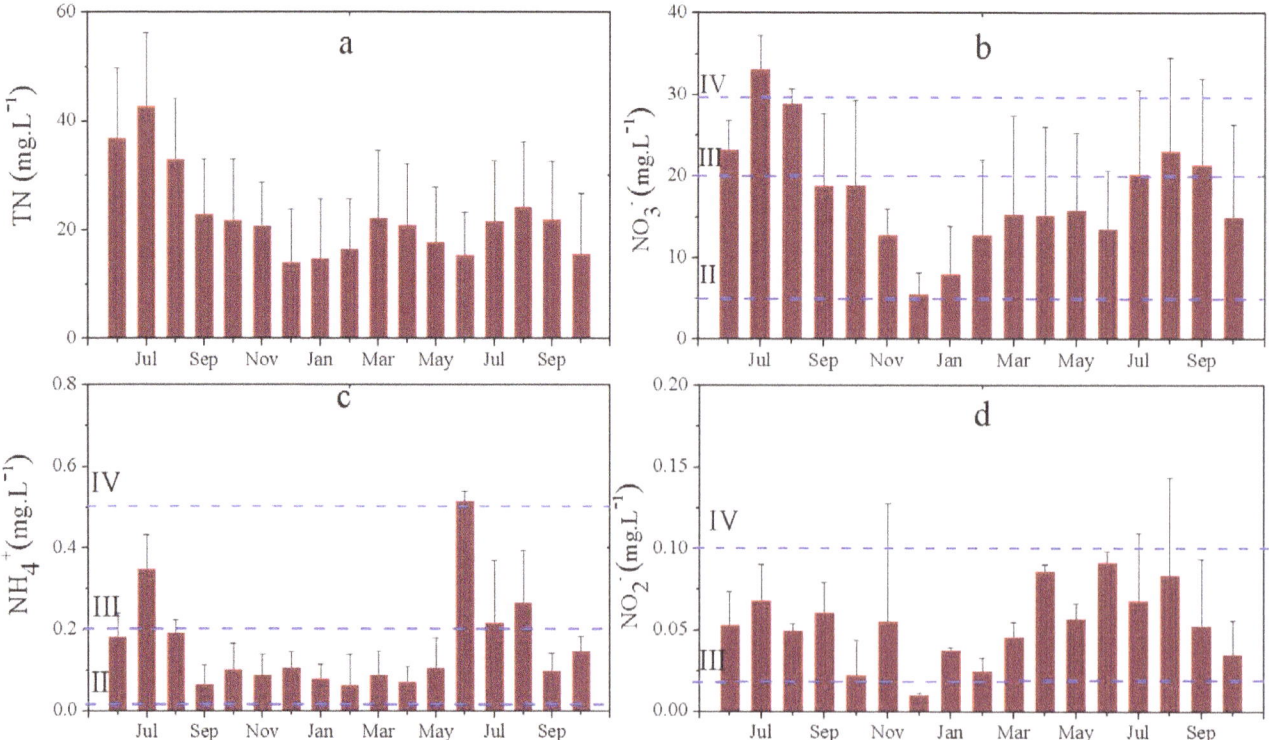

Figure 1. Seasonal changes in different types of nitrogen in the groundwater of agricultural areas of Xinchang village. Water quality grades are defined in Table 1. (a) total N; (b) nitrate nitrogen; (c) ammonia nitrogen; (d) nitrite nitrogen.

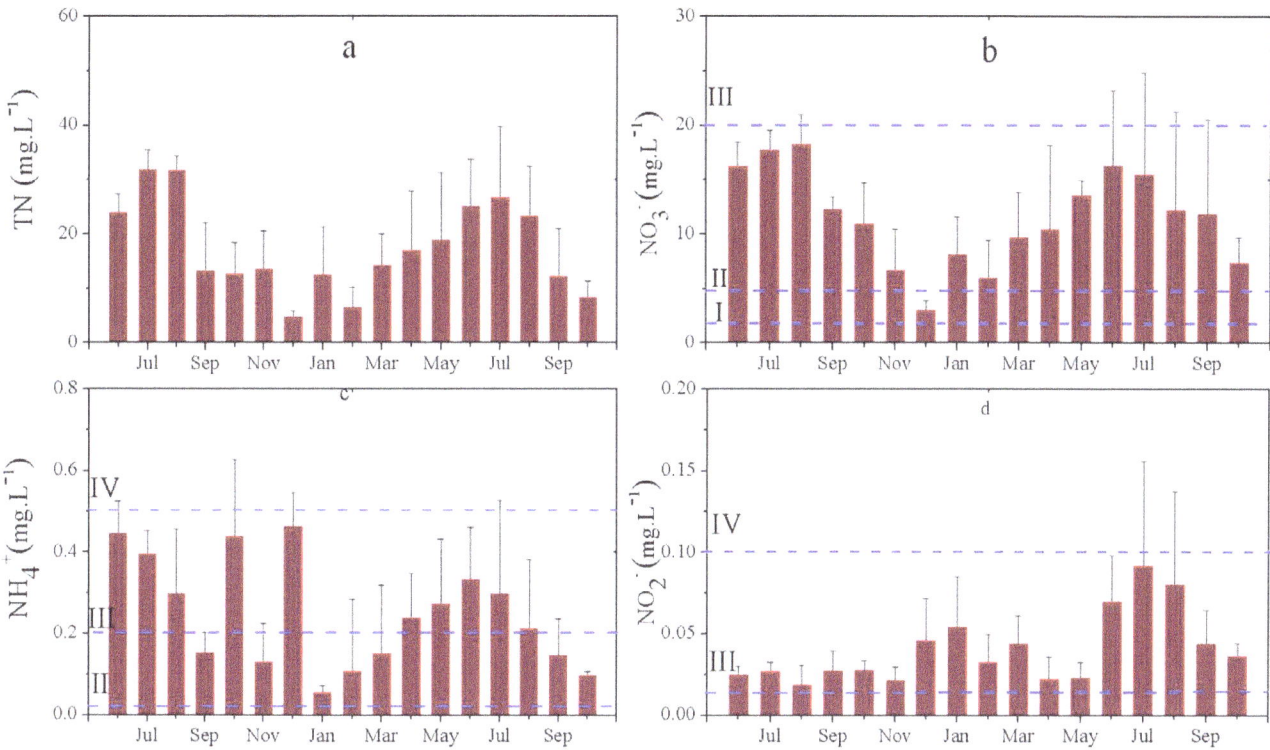

Figure 2. Seasonal changes in different types of nitrogen in the groundwater of residential areas of Xinchang village. Water quality grades are defined in Table 1. (a) total N; (b) nitrate nitrogen; (c) ammonia nitrogen; (d) nitrite nitrogen.

agricultural areas. The NO₂-N concentrations were lower than those in agricultural areas, but the groundwater quality based on this pollutant was still graded as Class III or Class IV throughout the year, with a large increase in June to August 2010 but no substantive differences during the rest of the study period (Fig. 2d).

Discussion

Nitrogen pollution in the groundwater

In the study area, the highest concentration of NO₃-N in the groundwater occurred from June to August, and NO₃-N was the most significant nitrogen contaminant. The NO₃-N concentrations were high from spring to summer and low from autumn to winter. During the jointing and booting stages of *Prunus persica* development, in March and April, most soil NO₃-N would be taken up by the trees and by other vegetables. As the rainy season began after May in the Shanghai region, the total N and NO₃-N levels in groundwater rapidly increased. From June to August, the NO₃-N concentration in the groundwater of Xinchang was close to the limit prescribed by the World Health Organization [13]. Because NO₃-N in solution is not adsorbed by soils, but NO₃-N can easily be absorbed by some tropical soils and leach into the subsurface soil and groundwater [1,5,8]. The changes in NH₄-N levels were similar in agricultural and residential areas. However, the concentration of NH₄-N was slightly lower in agricultural areas, indicating that the groundwater NH₄-N was affected both by agricultural practices such as fertilization and by human habitation. The peak values of NH₄-N content in the study area appeared from May to September, during the period of tree and vegetable growth and fertilization. Other inputs may come from agricultural production and domestic wastewater. Because nitrogenous fertilizers are applied to the soil, some of the NH₄-N, which

is a reactant for denitrification, can be transformed into NO₃-N through nitrification, and some is lost as a result of denitrification to produce volatile nitrogen gas [11].

Although the concentrations of NO₂-N were considerably lower than those of NO₃-N in the groundwater, the impact of pollution by NO₂-N was worse according to the Chinese groundwater quality criteria [18]; concentrations of NO₂-N in the groundwater throughout the study area exceeded the Class III water standard, whereas NO₃-N levels occasionally approached the Class II standard. NO₂-N is not stable in water or soil, and can easily be transformed into NO₃-N or into nitrogen gas through oxidation and denitrification. Therefore, the fluctuations of NO₂-N concentrations were irregular and did not appear to be associated with seasonal changes as a result of impact factors such as changes in fertilization, rainfall, and temperature.

Effects of rainfall and land use on nitrogen pollution

We found that land use patterns (here, residential vs. agricultural use) significantly affected NO₃-N concentrations in the groundwater. Enhanced agricultural activity is often accompanied by increased incorporation of organic matter into the soil. Nitrogen compounds in the fertilizer and organic matter are transported into the groundwater by percolating water from rainfall or from irrigation [3]. Hence, the nitrogen concentrations are typically high in agricultural areas [19–21]. Another reason for this phenomenon may be that in agricultural areas, the aquifer is typically shallow, and because it is relatively close to the surface, it receives direct inputs of NO₃-rich leachate from the agricultural soils. In residential areas, the nitrogen pollution was also serious, with levels close to those in agricultural areas. This can be explained by the high nitrogen content in groundwater around livestock and feedlot areas as well as near residences with septic tanks. Komar and Anderson [22] investigated the different nitrogen sources in a rural environment using nitrogen isotopes and obtained similar results to those in our study. Another reason for our observed results may be that the aquifers in the agricultural and residential areas are close to each other, so that leaching may transport pollutants between them; as a result, the magnitude of the difference in nitrogen contents in the groundwater would decrease.

The nitrogen concentrations in groundwater are affected by both rainfall and irrigation intensity [3], so we calculate the relationship between rainfall and the nitrogen concentration in groundwater (Fig. 3). The nitrogen concentrations in groundwater differed greatly between the rainy and dry seasons. The total N and NO_3-N in the groundwater were significantly correlated with rainfall in both agricultural and residential areas, but the correlations between rainfall and NH_4-N and NO_2-N concentrations was much weaker but still significant. This can be explained by the fact that the abundant rainfall in the study area is the most important impact factor responsible for nitrogen transport through subsurface runoff into the groundwater, and by the fact that NO_3-N accounted for 60 to 80% of total N. Soil nitrogen moves easily in water, especially during the first flush, when the runoff volume is high; internal and lateral solute movement in soils carried away nitrates even more intensively than surface runoff [23]. Zhu and Wen [24] showed that NH_4-N is strongly absorbed by soil particles and is more resistant to being detached or dissolved and transported by runoff waters; this is because the ammonium ion has a positive charge and can therefore be adsorbed to cation-exchange sites on soil particles. NH_4-N is easily oxidized or lost to denitrification, and NO_2-N is not stable in water or soil. The other

reason may be that soil pH in this area is 8.2, which easily affect NH_4-N and NO_2-N transformation via equilibrium. Therefore, NH_4-N and NO_2-N did not show a strong correlation with rainfall. The correlation between different types of nitrogen and rainfall was higher in agricultural areas than in residential areas. This is likely because in agricultural areas, groundwater nitrogen pollution was strongly influenced by agricultural activities such as irrigation and fertilization; human activities in residential area have less seasonal correlation than do activities in agricultural areas.

Health risk assessment

Based on data for Shanghai from 2009 to 2010, the main parameters used in our health risk assessment had the following values: $IR = 2$ L d^{-1}, $ED = 30$ years, $EF = 365$ d year^{-1}, $BW = 70$ kg, $AT = 70$ years, $Asd = 16\ 600$ cm^2, $FE = 0.5$ times d^{-1}, $f = 1$, $k = 1$ cm h^{-1}, $t = 1$ h, and $TE = 0.4$ h. The Rfd values for NO_3-N, NO_2-N, and NH_4-N were 34, 1.6, and 0.1, respectively [25]. Potential noncarcinogenic risks for exposure to contaminants of potential concern were evaluated by comparison of the estimated contaminant intakes from each exposure route (oral, dermal, inhalation) with the RfD. The HQ assumes that there is a level of exposure (i.e., RfD) below which it is unlikely for even sensitive populations to experience adverse health effects. There may be a concern arising for the potential noncarcinogenic effects if the HQ exceeds 1×10^{-6} (unity).

Figures 4 and 5 present the noncarcinogenic risk values for dermal and oral exposures to different type of nitrogen, respectively. Drinking and contact were assumed to be the main exposure routes of humans to nitrogen pollution in our risk

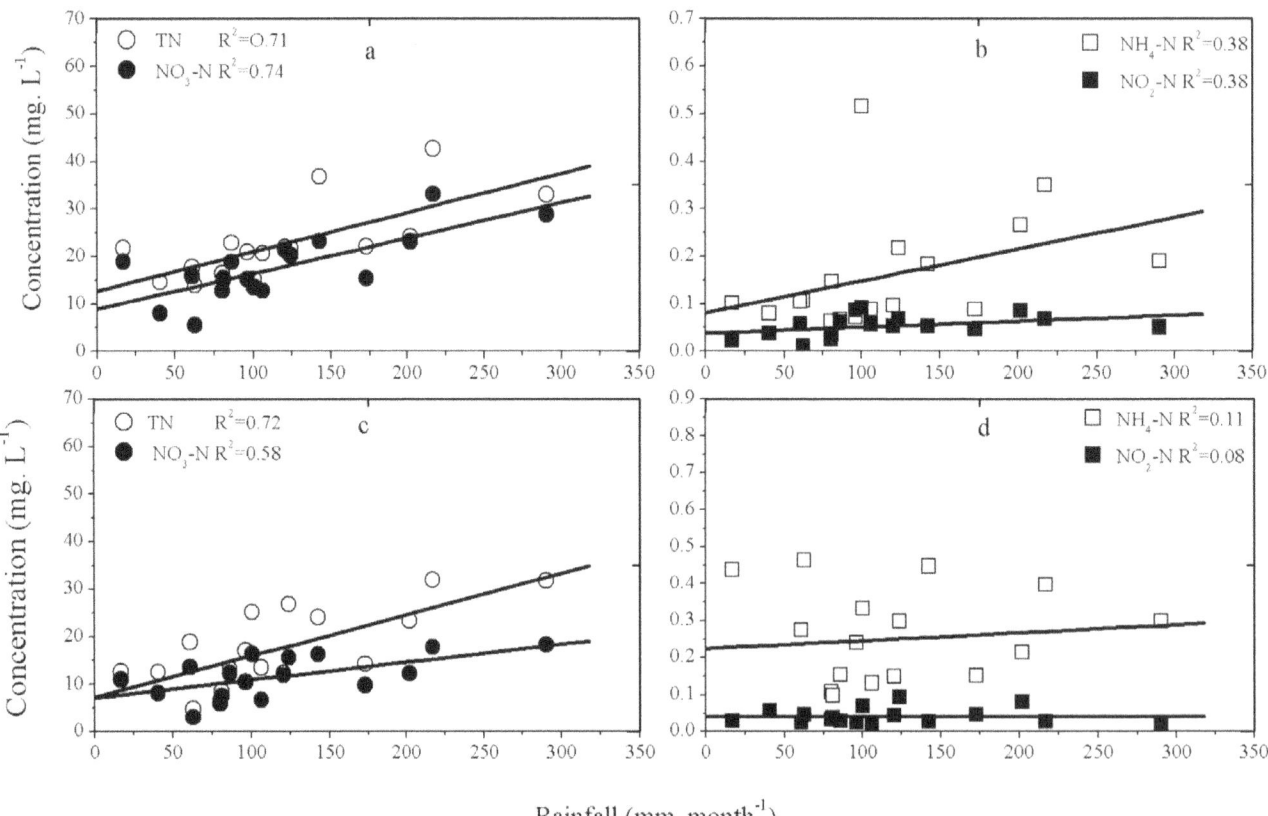

Figure 3. Relationships between the four types of nitrogen and rainfall. (a, b) agricultural areas; (c, d) residential areas.

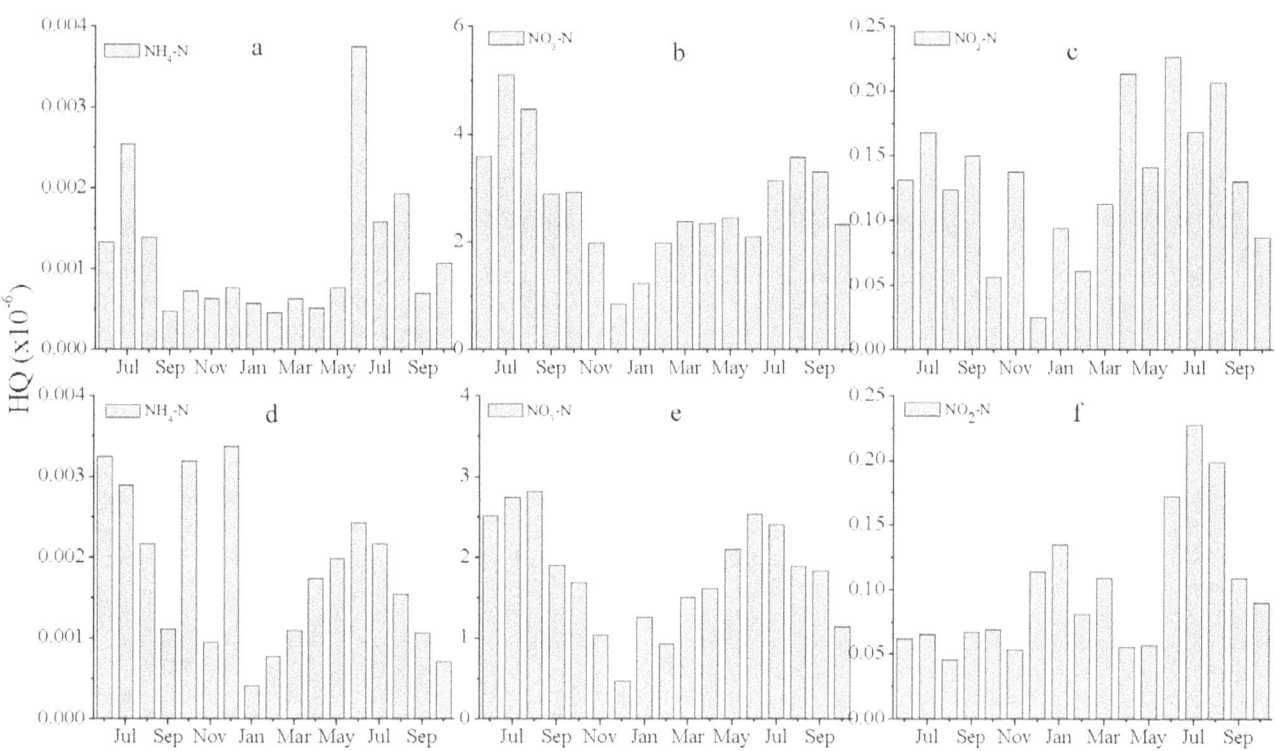

Figure 4. Noncarcinogenic dermal risk values for different types of nitrogen in the groundwater. (a, b, c) agricultural areas; (d, e, f) residential areas.

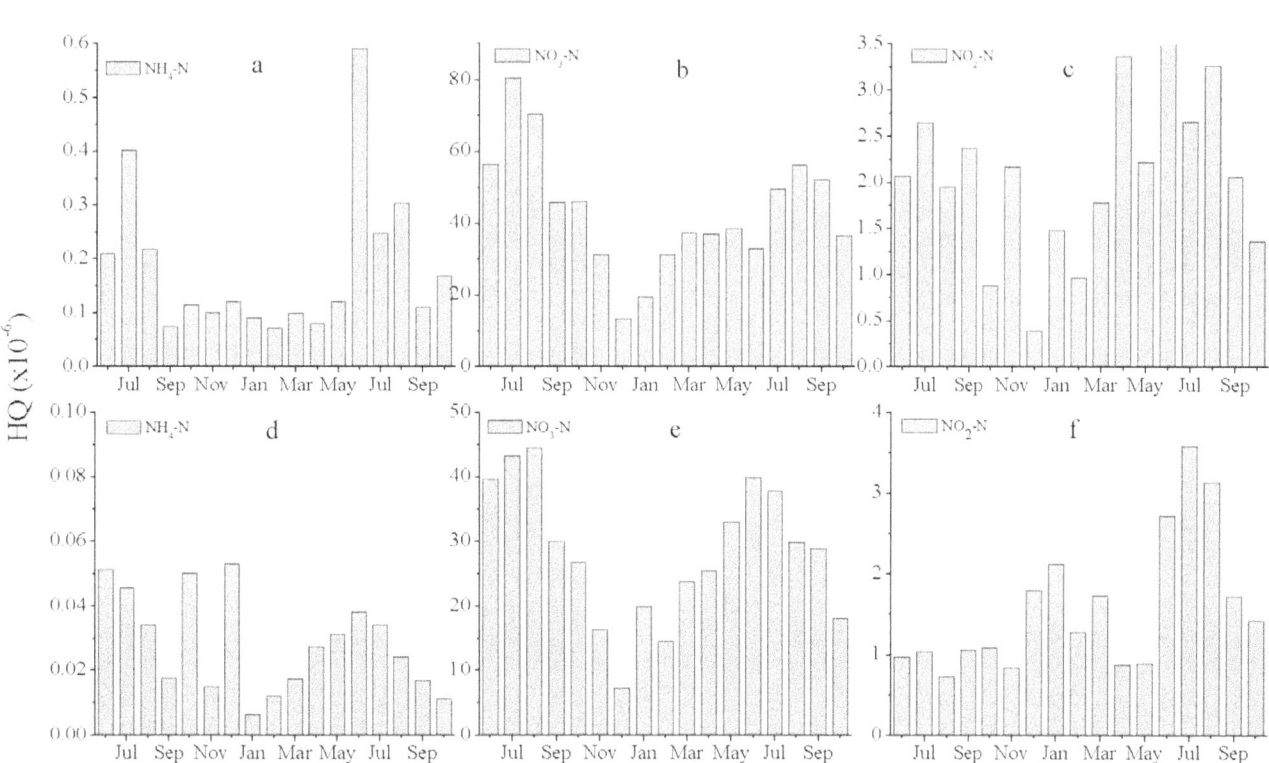

Figure 5. Noncarcinogenic oral risk values for different types of nitrogen in the groundwater. (a, b, c) agricultural areas; (d, e, f) residential areas.

assessment; we did not include inhalation as a source of exposure, which may mean that our risk estimates slightly underestimate the actual risk. The relative research showed that the levels of noncarcinogenic oral risk, toxic risk (HQ), ranged from 0.02 to 0.12×10^{-6} [26].The noncarcinogenic dermal risks due to NO_3-N and NO_2-N showed seasonal changes, ranging from 0.8×10^{-6} to 3.5×10^{-6} (3.0 to 23.1 mg L^{-1})and from 0.05×10^{-6} to 0.22×10^{-6} (0.01 to 0.35 mg L^{-1}), respectively. The HQ of the four types of nitrogen decreased in the following order: NO_3-N>NO_2-N>NH_4-N. NH_4-N represented the lowest noncarcinogenic dermal risk (Fig. 4). Noncarcinogenic dermal risk values were lower in residential areas than in agricultural areas.

The noncarcinogenic oral risk was two orders of magnitude higher than the noncarcinogenic dermal risk (Fig. 5). The levels of noncarcinogenic oral risk (HQ) ranged from 22×10^{-6} to 85×10^{-6}. NO_3-N posed the greatest risk, with HQ ranging from 19×10^{-6} to 80×10^{-6} (3.0 to 23.1 mg L^{-1}); this accounted for more than 90% of the total risk in the study area. Therefore, NO_3-N poses the greatest risk to human health. Daily intake of or contact with water by the local residents poses a potential health threat due to the cumulative impacts of long-term NO_3-N exposure. The difference between the dermal and oral values indicates that ingestion of water is a more critical exposure route for NO_3-N and NO_2-N. The spatial variation in oral risk was similar to the spatial patterns of the dermal risk. The noncarcinogenic oral risk values were lower in residential areas than in agricultural areas. Local farmers irrigated their crops using groundwater, which would enhance the noncarcinogenic risk for residents in Shanghai. Government statistics reported one case of cancer for every 100 woman living in Shanghai, which was the highest cancer incidence in any Chinese city, and the number of cases of cancer was double time in 2010 than that in 1960s [27].

Perspective

In China, there are many villages like Xinchang that are located in or near a big city, and which provide the main water resource and a supply of food and vegetables for city residents. Groundwater is the major source of water for drinking and for the domestic, industrial, and agricultural sectors in the Shanghai region. Therefore, protecting groundwater in this region has important implications for both food safety and human health. Because crops and animals take up nitrates from the soil and water, it will be important to quantify the nitrate contents of these foods and the quantities that are consumed in future research to determine how much this form of exposure increases the health risk to residents of Shanghai. In this study, our monitoring of the level of groundwater nitrogen pollution and our health risk assessment based on this data for Xinchang revealed that nitrogen pollution was a serious problem. Many wells exceeded the groundwater quality standard for human consumption for all forms of nitrogen, and particularly for nitrate, and higher levels of nitrogen contamination were significantly correlated with agricultural activity, human activity, and rainfall, especially in agricultural areas. The factors responsible for nitrogen pollution would be more complex in residential areas than in agricultural areas because of the greater diversity of activities. NO_3-N was the main form of nitrogen pollution of the groundwater and poses the greatest risk to the health of local residents. Long-term drinking of groundwater and irrigation using groundwater therefore pose a significant health risk for Shanghai's residents. Therefore, it is urgent to devise policy guidelines for efficient management of both the surface water and groundwater resources in this region to enhance groundwater recharge and minimize the pollution levels in both types of water to permit their safe use.

Materials and Methods

Study area

The study area was Xinchang village in the Nanhui District of Shanghai (31°03′N, 121°39′E), which is located in a typical alluvial plain of the region. The village covers an area of about 3564 ha and has an altitude ranging from 2 to 3 m asl. The farming, livestock, and agriculture in this area are well developed. The main type of land use in Xinchang is planting of peach trees (*Prunus persica*) and vegetables, which account for more than 50% of the area. The residential population is around 1000 people (Fig. 6). The village's water system belongs to the Huangpu River watershed, where there are number of crossed rivers with abundant fresh water resources. The main streams of Huangpu River include Huixin, Dazhi and Fengxin River. Groundwater storage condition in this area depends on the pore water of the unconsolidated rock, so the groundwater complement source in Xinchang village is abundant. The climate is a subtropical marine monsoon climate, with average annual rainfall of 1175 mm, an average annual temperature of 16.7°C, and 1932 h of annual sunshine. The rainfall variation is large, and 70% of the rain falls from June to August.

The soil type is yellow clay in FAO Soil Classification [28]. The soil properties are a pH of 8.2±0.2, a bulk density of 1.2±0.2 g cm^{-3}, a capillary porosity of 30.5±2.7%, a noncapillary porosity of 10.8±2.0%, a water content at field capacity of 13.6±1.0%, a total P of 0.8±0.2 g kg^{-1}, an available P of 44.7±5.9 mg kg^{-1}, an organic matter content of 20.4±0.7 g kg^{-1}, and a total N of 106.6±4.4 mg kg^{-1} (n = 29). The local combined annual application of inorganic fertilizer equals 0.018 kg N m^{-2} plus 0.011 kg P_2O_5 m^{-2}.

Water samples

There are many wells in Xinchang, and the depth to groundwater level of this area generally ranges from 1.2 to 1.5 m. We set groundwater sample point according to the land use of the village and the principle of uniform distributed points through GPS positioning, and then record the latitude, longitude and water level information for different well. We sampled water from 17 wells (red dot) surrounded by agricultural areas and 12 wells (blue dot) in residential areas (Fig. 1). The water from wells with red dot is mainly used in daily need for local resident in agricultural areas and crop irrigation, whereas the water from well with blue dot mainly meet daily need from resident in residential areas. The domestic wastewater in residential area contains large amounts of organic nitrogen, wherein part of the domestic wastewater can directly leak into groundwater from sewer, and then cause groundwater nitrogen pollution [29]. Water samples were collected from June 2009 to October 2010 at 1-month intervals period. Groundwater samples (each 500 mL in size) were collected from pumps connected to the wells. Rainfall data were obtained from the local meteorological station in Nanhui District. All data is reported as means ± S.D.

Analytical method

All water samples were passed through glass-fiber disks with a 0.70-mm pore size before analysis. To calculate total N, the water was digested in concentrated sulfuric acid using a $CuSO_4/Na_2SO_4$ mixture as a catalyst, followed by distillation of the resulting NH_4^+ into dilute boric acid and titration against a standardized 0.0025 M H_2SO_4 solution, as described by Rayment and Higginson [30]. NO_3-N was determined using an ultraviolet spectrometer [31], NO_2-N was determined by means of diazocoupling colorimetry [11], and NH_4^+-N was determined colorimetrically using the indophenol blue method [30].

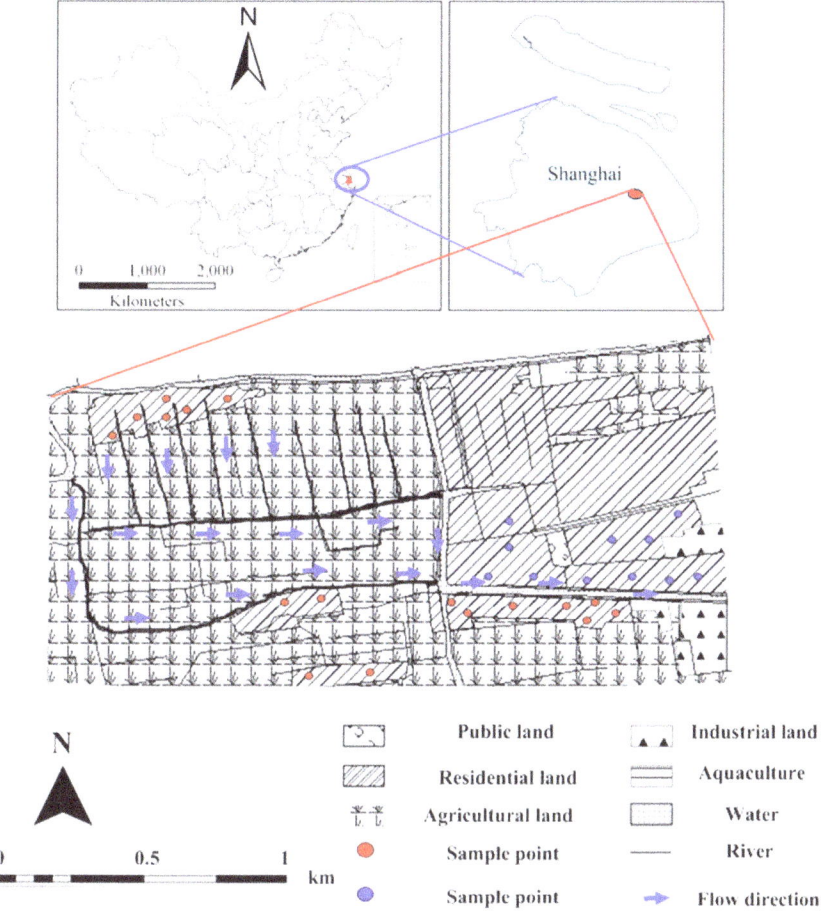

Figure 6. Location of the study area and sample points. 17 wells with red dot are in agricultural areas and 12 wells with blue dot are in residential areas.

Groundwater quality and health risk assessment

We selected NO_3-N, NO_2-N, and NH_4-N as the assessment index for groundwater nitrogen pollution. The national groundwater quality standard for nitrogen pollutants is presented in Table 1 [18]. Risk assessment is defined as the processes of estimating the probability of occurrence of an event and the probable magnitude of adverse health effects over a specified time period [32]. Human health risk assessment consists of four stages: (1) hazard identification, (2) toxicity (dose–response) assessment, (3) exposure assessment, and (4) risk characterization.

The estimated uptake of a potential toxin by the human body through contact with a contaminant is estimated using the chronic daily intake (*CDI*). The *CDI* value indicates the quantity of chemical substance ingested, inhaled, or absorbed through the skin per kilogram of body weight per day (mg kg^{-1} day^{-1}). The formulas for calculating intake are as follows:

Ingestion:

$$CDI_i = (C \times IR \times ED \times EF)/(BW \times AT) \qquad (1)$$

Dermal contact:

$$CDI_i = (C \times Asd \times EF \times FE \times ED)/(BW \times AT \times f) \qquad (2)$$

$$I = (2 \times 10^{-3} \times k \times C \times 6 \times t \times TE)/\pi \qquad (3)$$

where i represents a specific pollutant, C is that pollutant's concentration in water (mg L^{-1}), IR is the drinking rate (L d^{-1}), ED is the exposure duration (years), EF is the exposure frequency (d year^{-1}), BW is the average body weight (kg), AT is the average lifespan (years), Asd is the human body's surface area (cm^2), FE is the bathing frequency (number of times d^{-1}), f is the intestinal absorption rate (unitless, $= 1$), I is pollutant adsorption by the skin when bathing (mg cm^{-2} time^{-1}), k is the adsorption parameter for the skin (cm h^{-1}), t is the lag time (h), and TE is the bathing time (h).

Noncarcinogenic risks

We separately characterized the risk for carcinogenic and noncarcinogenic effects, and have discussed the factors that may result in either overestimation or underestimation of the risks for the residents of Xinchang. Potential noncarcinogenic risks for exposure to contaminants were evaluated by comparison of the estimated contaminant intakes from each exposure route (oral and dermal) with the reference dose (RfD, (mg kg^{-1} day^{-1}) to produce the hazard quotient (HQ, unitless), which is defined as follows [25]:

$$HQ = \text{CDI} \times 10^{-6} / Rfd \qquad (4)$$

where HQ is hazard quotient (unitless); RfD is reference dose (mg. kg^{-1} day^{-1}).

Carcinogenic risks

Carcinogenic risks were estimated as the incremental probability of an individual developing cancer over a lifetime as a result of exposure to a potential carcinogen. To do so, we used the following linear low-dose carcinogenic risk equation for each exposure route [25]:

$$CA = CDI \times slope\ factor \qquad (5)$$

where CA is the carcinogenic risk and "*slope factor*" is mg kg^{-1}

day^{-1}. Slope factor can be obtained from Risk Assessment Information System [33]. If a site has multiple carcinogenic contaminants, cancer risks for each carcinogen and each exposure route can be added (based on the assumption of additivity of effects) and compared with the accepted risk.

Acknowledgments

We thank Dr.Yafeng Wang in Research Center for Eco-Environmental Sciences, Dr. Liang Mao in Shanghai Jiaotong University, and Geoffrey Hart (Montréal, Canada) for his help in writing this paper. The authors would also like to thank the anonymous reviewers for their helpful remarks.

Author Contributions

Conceived and designed the experiments: YG. Performed the experiments: CL. Analyzed the data: YG. Contributed reagents/materials/analysis tools: PZ GY. Wrote the paper: YG GY.

References

1. Zhu B, Wang T, You X, Gao MR (2008) Nutrient release from weathering of purplish rocks in the Sichuan Basin, China. Pedosphere 18(2): 257–264.
2. Gao Y, Zhu B, Zhou P, Tang JL, Wang T, et al. (2009) Effects of vegetation cover on phosphorus loss from a hillslope cropland of purple soil under simulated rainfall: a case study in China. Nutrient Cycling in Agroecosystems 85: 263–273.
3. Naik PK, Tambe JA, Dehury BN, Tiwari AN (2008) Impact of urbanization on the groundwater regime in a fast growing city in central India. Environmental Monitoring and Assessment 146: 339–373.
4. Gao Y, Zhu B, Wang T, Wang YF (2012) Seasonal change of non-point source pollution-induced bioavailable phosphorus loss: a case study of Southwestern China. Journal of Hydrology 420–421: 373–379.
5. Zhu B, Wang T, Kuang FH, Luo ZX, Tang JL, et al. (2009) Measurements of nitrate leaching from a hillslope cropland in the central Sichuan Basin, China. Soil Science Society of America Journal 73(4): 1419–1426.
6. Akhavan S, Abedi-Koupai J, Mousavi SF, Afyuni M, Eslamian SS, et al. (2010) Application of SWAT model to investigate nitrate leaching in Hamadan–Bahar Watershed, Iran. Agriculture Ecosystems and Environment 139: 675–688.
7. Gao Y, Zhu B, Wang T, Tang JL, Zhou P, et al. (2010) Bioavailable phosphorus transport from a hillslope cropland of purple soil under natural and simulated rainfall. Environmental Monitoring and Assessment 171: 539–550.
8. Chen SF, Wu WL, Hu KL, Li W (2010) The effects of land use change and irrigation water resource on nitrate contamination in shallow groundwater at county scale. Ecological Complexity 7: 131–138.
9. Adhikary PP, Chandrasekharan H, Chakraborty D, Kamble K (2010) Assessment of groundwater pollution in West Delhi, India using geostatistical approach. Environmental Monitoring and Assessment 167: 599–615.
10. Hu KL, Huang YF, Li H, Li BG, Chen DL, et al. (2005) Spatial variability of shallow groundwater level, electrical conductivity and nitrate concentration, and risk assessment of nitrate contamination in North China Plain. Environment International 31: 896–903.
11. Chen XM, Wo F, Chen C, Fang K (2010) Seasonal changes in the concentrations of nitrogen and phosphorus in farmland drainage and groundwater of the Taihu Lake region of China. Environmental Monitoring and Assessment 169: 159–168.
12. Jackson BM, Browne CA, Butler AP, Peach D, Wade AJ, et al. (2008) Nitrate transport in chalk catchments: monitoring, modelling and policy implications. Environmental Science and Policy 11: 125–135.
13. WHO (2008) Guidelines for drinking-water quality- Third Edition Incorporating the first and second addenda. Volume 1. Recommendations. World Health Organization, Geneva website. Available: http://www.who.int/water_sanitation_health/dwq/GDWPRecomdrev1and2.pdf. Accessed 10 Sept 2010.
14. Forman D, Al-Dabbagh S, Doll R (1985) Nitrates, nitrites and gastric cancer in Great Britain. Nature 313: 620–625.
15. Payne MR (1993) Farm waste and nitrate pollution. In: Jones JG, ed. Agriculture and the environment 63–73. Horwood, New York.
16. Huang HB, Gao Y, Cao JJ, Huang HY, Zhang X, et al. (2010) Nonpoint source pollution of nitrogen in groundwater and risk assessment in urban agricultural region of Shanghai. Journal of Soil and Water Conservation 24(3): 56–70. (In Chinese).
17. Poinke HB, Urban JB (1985) Effect of agricultural land-use on groundwater quality in a small Pennsylvania watershed. Ground Water 23: 68–80.
18. State Environmental Protection Administration of China (1994) Chinese groundwater quality criteria of classification, (GB/T14848-93), (pp. 23–32). Beijing: China Environmental Science Press (in Chinese).
19. Vidal M, Melgar J, Opez AL, Santoalla MC (2000) Spatial and temporal hydrochemical changes in groundwater under the contaminating effects of fertilizers and wastewater. Journal of Environmental Management 60: 215–225.
20. Cepuder P, Shukla MK (2002) Groundwater nitrate in Austria: A case study in Tullnerfeld. Nutrient Cycling in Agroecosystems 64: 301–315.
21. Reddy AGS, Kumar KN, Rao DS, Rao SS (2009) Assessment of nitrate contamination due to groundwater pollution in north eastern part of Anantapur District, A.P. India. Environmental Monitoring and Assessment 148: 463–476.
22. Komar SC, Anderson HW (1993) Nitrogen isotopes as indicators of nitrate source in Minnesota sand plain aquifers. Groundwater 31(2): 250–270.
23. Yang JL, Zhang GL, Shi XZ, Wang HJ, Cao ZH, et al. (2009) Dynamic changes of nitrogen and phosphorus losses in ephemeral runoff processes by typical storm events in Sichuan Basin, Southwest China. Soil & Tillage Research 105: 292–299.
24. Zhu ZL, Wen QX (1992) Nitrogen of Chinese Soil. Jiangsu Science and Technology Press, Nanjing.
25. USEPA (1989) Risk assessment guidance for superfund, Vol. I, human health evaluation manual. Part A: (interim final), EPA/540/1–89/002. Washington, DC: Office of Emergency and Remedial Response, U.S. Environmental Protection Agency.
26. Li YL, Liu JL, Cao ZG, Lin C, Yang ZF (2010) Spatial distribution and health risk of heavy metals and polycyclic aromatic hydrocarbons (PAHs) in the water of the Luanhe River Basin, China. Environmental Monitoring and Assessment 163: 1–13.
27. Shanghai Municipal Center for Disease Control & Prevention (2010) website. Available: http://www.scdc.sh.cn/. Accessed 10 Sept 2010. (in Chinese).
28. Gong ZT Chinese Soil Taxonomy, Science Press, Beijing. (In Chinese).
29. Cao JJ, Gao Y, Huang HB, Huang HY, Mao L, Zhang X, et al. (2010) Output characteristics of non-point nitrogen from a typical village region in Yangtze Delta under an individual rainfall event. Environmental Science 31(11): 2587–2593. (In Chinese).
30. Rayment GE, Higginson FR (1992) Australian laboratory handbook of soil and water chemical methods. Inkata Press, Sydney.
31. Committee of Analytical Method of Water and Wastewater (1989) Analytical Method of Water and Wastewater. China Environmental Science Press, Beijing. (In Chinese).
32. Kolluru RV, Bartell SM, Pitblado RM, Stricoff RS (1996) Risk assessment and management handbook. McGraw-Hill, New York.
33. USEPA Risk Assessment Information System [EB/OL] website. Available: http://rais. ornl. gov/cgi-bin /tox / TOX- select ? select = nrad. Accessed 10 Sept 2010.

7

Nutrient Limitation on Ecosystem Productivity and Processes of Mature and Old-Growth Subtropical Forests in China

Enqing Hou[1,2,4], **Chengrong Chen**[2]*, **Megan E. McGroddy**[3], **Dazhi Wen**[1,4]*

1 Key Laboratory of Vegetation Restoration and Management of Degraded Ecosystems, South China Botanical Garden, Chinese Academy of Sciences, Guangzhou, China, **2** Environmental Futures Centre, Griffith School of Environment, Griffith University, Nathan, Queensland, Australia, **3** Department of Environmental Sciences, NASA/University of Virginia, Charlottesville, Virginia, United States of America, **4** University of Chinese Academy of Sciences, Beijing, China

Abstract

Nitrogen (N) is considered the dominant limiting nutrient in temperate regions, while phosphorus (P) limitation frequently occurs in tropical regions, but in subtropical regions nutrient limitation is poorly understood. In this study, we investigated N and P contents and N:P ratios of foliage, forest floors, fine roots and mineral soils, and their relationships with community biomass, litterfall C, N and P productions, forest floor turnover rate, and microbial processes in eight mature and old-growth subtropical forests (stand age >80 yr) at Dinghushan Biosphere Reserve, China. Average N:P ratios (mass based) in foliage, litter (L) layer and mixture of fermentation and humus (F/H) layer, and fine roots were 28.3, 42.3, 32.0 and 32.7, respectively. These values are higher than the critical N:P ratios for P limitation proposed (16–20 for foliage, ca. 25 for forest floors). The markedly high N:P ratios were mainly attributed to the high N concentrations of these plant materials. Community biomass, litterfall C, N and P productions, forest floor turnover rate and microbial properties were more strongly related to measures of P than N and frequently negatively related to the N:P ratios, suggesting a significant role of P availability in determining ecosystem production and productivity and nutrient cycling at all the study sites except for one prescribed disturbed site where N availability may also be important. We propose that N enrichment is probably a significant driver of the potential P limitation in the study area. Low P parent material may also contribute to the potential P limitation. In general, our results provided strong evidence supporting a significant role for P availability, rather than N availability, in determining ecosystem primary productivity and ecosystem processes in subtropical forests of China.

Editor: Sandra Maria Feliciano de Oliveira Azevedo, Federal University of Rio de Janeiro, Brazil

Funding: This study was supported by National Natural Science Foundation of China (No. 31070409), Strategic Priority Research Program - Climate Change: Carbon Budget and Relevant Issues of the Chinese Academy of Sciences (No. XDA05050205 and the Australian Research Council (FT0990547). The support from China Scholarship Council through an overseas joint doctoral fellowship to Enqing Hou is also kindly acknowledged. The funders had no role in study design, data collection and analysis, decision to publish, or preparation of the manuscript.

Competing Interests: The authors have declared that no competing interests exist.

* E-mail: c.chen@griffith.edu.au (CC); dzwen@scbg.ac.cn (DW)

Introduction

Nitrogen (N) and phosphorus (P) have both been shown to control the rates of ecosystem processes and primary productivity in both aquatic and terrestrial ecosystems [1–3]. Global pattern analysis of carbon (C):N:P stoichiometry in foliage and litter supports the hypothesis that N is the major limiting nutrient in temperate regions, while P tends to limit ecosystem productivity and processes in the tropical regions [4–7]. These analyses are generally consistent with the nutrient addition experiments or C:N:P stoichiometry studies at a local or regional scales [8–12], and well explained by variation in climate conditions (e.g. temperature) and soil types [7,12–14]. According to this global pattern, subtropical forests are likely to be co-limited by N and P. However, this supposition has rarely been tested.

Since the beginning of the industrial revolution, human activities (N fertilizer application and burning of fossil fuels) have doubled the N input into the terrestrial ecosystems [15,16]. Although anthropogenic P inputs (mainly as fertilizers) to the biosphere also increased fourfold in the period from 1950s to

1980s and remained more or less constant since 1989, the primary P inputs are mostly confined in agricultural soils and tend to remain and accumulate in crop soils [17]. The greater mobility and biological availability of N in the atmosphere are causing the imbalance supply between N and other mineral nutrients (especially P) in natural ecosystems [17,18], which is likely to transform N-limited ecosystems to P-limited ecosystems [16,19]. In a comprehensive study of nutrients on phytoplankton nutrient limitation in high- and low-N deposition lakes in Norway, Sweden, and Colorado, United States, Elser et al. (2009) found that continued anthropogenic N input increased the stoichiometric ratio of N and P in these lakes, resulting in a shift from N-limitation to P-limitation in high-N deposition lakes [16]. The imbalance of nutrient supply is likely to affect ecosystem productivity and processes and the carbon sequestration potential of terrestrial ecosystems [17,19,20].

China has 0.97 million km^2 of subtropical and tropical forests, which represent 62% of the country's total forested area, and play an important role in maintaining biodiversity and ecological equilibrium, sequestering atmospheric C, and providing important ecological services for social development [21,22]. However, these tropical and subtropical forests in the southern part of China, are generally close to or surrounded by large industrial and/or economic zones. Annual N deposition rate ranging from 18 to 53 kg N ha^{-1} yr^{-1} were reported at several long-term monitoring stations in tropical and subtropical forests [22], comparable to the highest levels of N deposition occurring in Europe [23,24]. Recent studies found that the understory plants generally showed no or even negative responses to experimental N additions (50, 100 and 150 kg N ha^{-1} yr^{-1}) in three mature and old-growth forests at the Dinghushan Biosphere Reserve, south China [25,26]. Nutrients other than N, were proposed as the primary constraint on plant growth at these forest sites with P being the mostly likely candidate [26]. However, direct evidence is still lacking, though one recent study reported a significant increase in litterfall production after experimental P addition (150 kg P ha^{-1} yr^{-1}) at these three forests [27].

While fertilization studies are the gold standard for determining the nature of nutrient limitation [28,29], they are difficult to do well in forest ecosystems [2,30,31] and, in many cases, after several years of study the results are still unclear [31–33]. It may be a question of how much fertilizer to add, as Chapin (1986) suggested [30], or if the nutrient limitation is ultimate, it may take decades or more for species replacement to happen in forest ecosystems and thus delaying measurable results [18,31]. The critical N:P ratio for biomass was shown to work well indicating the limiting nutrient in European wetland ecosystems [34], but is poorly supported in some other terrestrial ecosystems [35,36].

In this study, we investigated the N and P status of foliage, forest floors, fine roots and mineral soil, as well as microbial properties of the forest floors of eight forests in subtropical China. We used regression analysis, to study the relationships between rates of ecosystem productivity and nutrient cycling and N and P availability in these forests. We hypothesized that these selected parameters were more strongly related to P availability than N availability, due to the historically high rates of atmospheric N deposition [22].

Methods

Site Description

The research was conducted in the Dinghushan Biosphere Reserve, located in the middle of Guangdong province in southern China (112°31′ E to 112°34′ E, 23°09′ N to 23°12′ N; Figure 1). The Reserve covers an area of 1155 ha, and has a typical subtropical monsoon climate. The entire Reserve has 1843 plant species identified and documented [37]. Mean annual temperature at the site is 21°C, and mean annual precipitation is 1900 mm [38]. Nearly 80% of the precipitation falls in the wet season (from April to September) and 20% in the dry season (from October to March) [38]. Elevation ranges from 10 to 1000 m above sea level. The forest soil has developed from Devonian sandstone and shale during the Holocene (<15 kyr) [39]. Soils are Ferralsols according to the FAO classification, with a pH value ranging from 3.8 to 4.9 [40,41]. The annual rate of atmospheric N deposition was approximately 46 kg N ha^{-1} yr^{-1} between 1989 and 2007 [42].

The basic site information for the eight selected forest communities is summarized in Table S1. These eight communities cover all major forest types in DHSBR, representing five typical forest types in subtropical China [43]. They differ in tree species composition, stand age and topography (Table S1). Four of them (PF: pine forest; PBM1, 2 and 3, pine and broadleaved mixed forest 1, 2 and 3) are mature forests (about 80 years old), while the other four communities (REB1 and 2, ravine evergreen broad-

leaved forest 1 and 2; MEB, monsoon evergreen broadleaved forest; and MTEB, mountainous evergreen broadleaved forest) are old-growth forests (>100 year old) and were regional or topographical climax forests (Table S1) [44]. Seven of the forests have been protected from human disturbance since their establishment. The PF is the exception, it had been disturbed mainly by the harvest of understory vegetation and litter for fuel by local residents between 1950s and 1990s [45], whilst the community has remained and dominated by *Pinus massoniana*, of which biomass was about 90% of the total community biomass [46].

Sampling and Sample Preparation

During 15th – 18th October, 2010, we sampled foliage, forest floors, fine roots and soil at each of the eight study sites. At each site, mature and healthy foliage was sampled from three major tree species which were listed in Table S1. For each species, we collected foliar samples from four individuals.

At each location, four subplots (20×20 m^2) were randomly set up with a distance of at least 10 m between them. In each subplot, 3 small sampling areas (20×20 cm^2) were randomly located with the constraints that they were 1–2 m away from the nearest tree (diameter at breast height ≥5 cm) and at least 5 m from its nearest neighbor. All fine forest floor materials within the sampling area were collected, including leaf litter, and senesced branches, bark, flowers and fruits with diameters ≤1 cm. Forest floor materials were carefully separated into two layers (L layer, litter layer; F/H layer, mixture of fermentation layer and humus layer) in the field. After forest floor materials were sampled, a soil profile was excavated at the same area. Mineral soil from the 0–15 cm depth was sampled by 3 successive cutting rings (Height 5 cm, Volume 100 cm^3) from top to bottom (each 5 cm depth by one cutting ring). By using a cutting ring to sample soil, we also measured the bulk density at the same time of soil sampling. The three forest floor samples from the same layer in each subplot were bulked together as one composite sample (one composite L layer sample and one composite F/H layer sampler per subplot). Nine soils (3 cutting rings per area×3 areas) of each subplot were bulked together as one mineral soil sample. Both forest floor and soil samples were stored at 4°C in the refrigerator within 4 h after sampling.

Leaves were directly oven-dried at 65°C for 72 h prior to grinding for determination of total N and P concentrations. For forest floor materials, the fresh weight (w1, unit: g) was recorded and then the sample was mixed well. A subsample was oven-dried at 65°C for 72 h for the determination of dry weight transfer coefficient (t, proportion of dry weight over the fresh weight). The forest floor biomass was calculated by the equation followed:

L layer (or F/H layer) forest floor biomass (g/cm^2) = (w1 ×t)/ (400 cm^2×3).

The unit was converted later to Mg/ha and shown in the results. After t was determined, the oven-dried sample was ground for the determination of total N and P concentrations. Another subsample was taken and cut into 2–4 mm pieces and stored at 4°C prior to the determination of microbial biomass C concentration, respiration and β-glucosidase activity.

For soil samples, fresh weight (W1, unit: g) was recorded and the sample was mixed well. Stones with diameter >4 mm were picked out during the sieving (4 mm mesh) and weighed (W2, unit: g). A subsample of the sieved soil was air-dried for 2 weeks prior to grinding and determination of soil nutrient concentration. A subsample was used for the soil dry weight transfer coefficient (T; proportion of dry weight after over-dried in fresh weight)

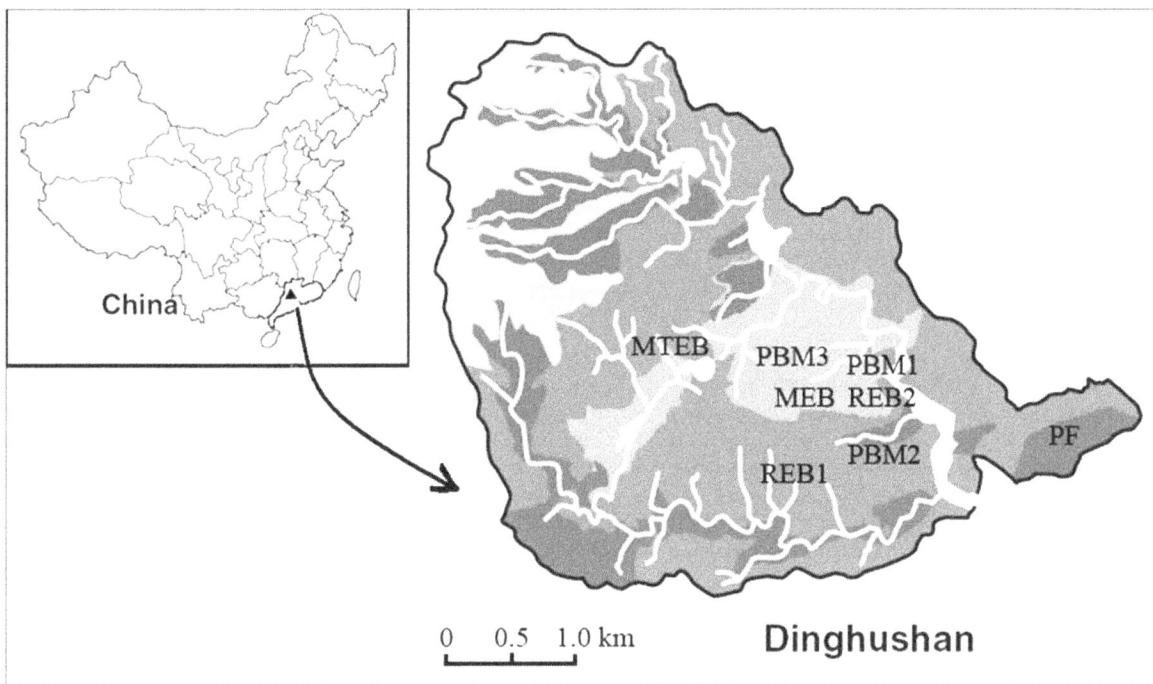

Figure 1. Location of eight study forest sites.

determination by oven-dried at 105°C for 72h. The bulk density was calculated by equations followed:

Bulk density $(g/cm^3) = W2 + (W1-W2) \times T/(100\ cm^3 \times 9)$.

The remaining soil was weighed and stored at 4°C for the determination of microbial properties (data not shown here) and fine roots (diameter ≤ 2 mm) collection. Fine roots retained on a 0.6 mm screen were collected and dried at 65°C for 72 h and then weighed for the fine root biomass calculation (data not shown here). Fine roots were finely ground for measurements of total N and P concentrations.

Analytical Methods

Microbial biomass C concentration in the L and F/H layers was determined by a fumigation-extraction (1:25) method using an E_C factor of 2.64 [47]. Microbial respiration was measured by using the incubation method. In brief, 2 g of fresh forest floor materials with moisture adjusted to 60% of the field capacity was incubated aerobically in a 1–L sealed plastic jar at room temperature (ranging from approximate 15°C at night time to 25°C at day time). All CO_2 evolved was trapped in 0.1 M NaOH and measured by acid titration (0.1 M HCl) after 1, 3, 7, 14, 21 and 28 days. The activity of β-glucosidase (EC 3.2.1.21) was analyzed following the procedure of Alef and Nannipieri (1995) [48], except that a fresh weight of 0.5 g was used for our forest floor samples.

Total N concentrations of foliage, L and F/H layers, fine roots and soil were all determined using an Isoprime isotope ratio mass spectrometer with a Eurovector elemental analyzer (Isoprime-Euro EA 3000). Total P concentrations were all measured using a nitric acid/perchloric acid digestion, followed by the molybdate blue method [49] using a UV–Vis spectrometer (UV1800, Shimadzu, Japan). Soil extractable N concentration was measured as the hot water extractable total N concentration according to method described by Sparling (1998) [50], which was found to be a simple and useful predictor of mineralizable N and plant available N [51,52]. In brief, 4.0 g air-dried soil was incubated

with 20 ml water in a capped test tube at 70°C for 18 h. The test tube was then shaken on an end-to-end shaker for 5 min, and filtered through Whatman 42 filter paper. Total N concentration of the extract was analyzed using a SHIMADZU TOC-$_{\text{VCPH/CPN}}$ analyser (Kyoto, Japan). Soil extractable P concentration was calculated as the sum of inorganic P concentration sequentially extracted by 1.0 M NH_4Cl, 0.1 M NH_4F and 0.1 M NaOH following the P fractionation scheme of McDowell and Condron (2000) [53]. Concentrations of inorganic P in the extracts of 0.1 M NH_4F and 0.1 M NaOH were determined by the molybdate blue method [49] using the same UV–Vis spectrometer mentioned above; while concentrations of inorganic P in the extracts of 1.0 M NH_4Cl were too low for the molybdate blue method and thus determined by the malachite green method [54], which works well for the determination of low concentrations of P in soil extracts [54], using the same UV–Vis spectrometer, too.

Community Biomass and Litterfall C, N and P Productions

Community biomass reported by Liu et al. (2007) [55] and litterfall production reported by Zhou et al. (2007) [44] and Yan et al. (2009) [56] from these forest communities were used to investigate the relationships of community biomass and litterfall C, N and P productions with soil nutrient pools in this study. Litterfall C, N and P productions were calculated by multiplying litterfall production by the C, N and P concentrations in the L layer. Community biomass includes dry weight of whole plant of all trees and shrubs with diameter at breast height ≥ 1 cm (Table S1). Litterfall production was available for seven of the eight study forests, while not for the PBM3 (Table S1). Methods of the data collection or calculation are described in details in Table S1.

Forest Floor Turnover Rate

Jenny et al. (1949) [57] and subsequently Olson (1963) [58] proposed that the rate of change in the forest floor biomass (or biomass C) could be used to determine nutrient transfers from the forest floor to the mineral soil in (near-) equilibrium forests. Forests selected in this study are all in or near an equilibrium status except for the PF site, as suggested by the study of long-term change of litterfall production [44]. Here, we calculated the forest floor turnover rate according to Olson (1963) [58]. The calculation formula is:

Forest floor turnover rate (yr^{-1}) = litterfall C production/forest floor biomass C.

Forest floor biomass C was the sum of L layer and F/H layer biomass C.

Statistical Analyses

Since the data are mostly (near-) normal distributed, Pearson correlation analysis and Pearson linear regression technique were used throughout the manuscript. All N:P ratios shown in this study were calculated on a mass basis. All analyses were performed using SPSS version 16.0. Pearson correlation was used to investigate the correlations between N and P concentration and N:P ratio for all plant and soil samples, and was also used to investigate the correlations between nutrient measures of the plant samples and those of the soil samples. Pearson linear regression technique was used to examine the relationships between community biomass, litterfall C, N and P productions, forest floor turnover rate and forest floor microbial properties with nutrient measures of the soil or the forest floors.

According to the foliar N concentration, foliage was divided into two groups (see Figure S1). One was the high N group with foliar N concentration higher than 25 mg/g (sample number = 20), including species of *Gironniera subaequalis* (from the REB2), *Ormosia fordiana* (REB1), *Caryota ochlandra* (REB2), *Euodia lepta* (PF) and *Sterculia lanceolata* (REB2); the other was low N group with foliar N concentration lower than 25 mg/g (sample number = 76), including the other eight species (Table S4 and Figure S1). For plants of the same species, only four species with a sample number ≥8 were selected for the correlation analysis. To be consistent with respect to units, when community biomass and litterfall C, N and P productions were regressed against the soil nutrients, pools were used; while for all other correlation and regression analysis, concentrations were used.

Results

Community Biomass and Litterfall Productions in Relation to the Nutrient Measures

Community biomass was positively related to both soil total P ($R^2 = 0.39$, $P<0.001$; Figure 2B) and extractable P pools ($R^2 = 0.59$, $P<0.001$; Figure 2E). The relationships with soil total N ($R^2 = 0.18$, $P<0.05$; Figure 2A) and extractable N ($R^2 = 0.20$, $P<0.05$; Figure 2D) pools were also significant but mainly caused by the inclusion of the prescribed PF site. When the PF site was excluded, community biomass was not related to either the soil total N ($R^2 = 0.04$, $P>0.05$; Figure 2A) or the extractable N pool ($R^2 = 0.02$, $P>0.05$; Figure 2D). Community biomass tended to increase with increasing soil total N:P ratio when community biomass was low (<300 Mg/ha), while increasing with decreasing soil total N:P ratio when community biomass was high (≥300 Mg/ha; Figure 2C). Overall, community biomass was not related to the soil total N:P ratio ($R^2<0.01$, $P>0.05$; Figure 2C). Community biomass was negatively related to soil extractable N:P ratio when the PF site was excluded from the analysis ($R^2 = 0.54$, $P<0.001$; Figure 2F).

Litterfall C, N and P productions were all positively related to the soil extractable P pool ($R^2 = 0.18$, 0.57 and 0.76, respectively, $P<0.05$; Figures 3B, D and F). Similar to the community biomass,

the relationships between litterfall C, N and P productions and the soil extractable N pool were significant for all sites (R^2 was 0.21, 0.33 and 0.23, respectively, $P<0.05$; Figures 3A, C and E), but none was significant if the PF site was excluded ($R^2 = 0.03$, 0.02 and 0.05, respectively, $P>0.05$; Figures 3A, C and E). The patterns for litterfall C, N and P productions with soil total fractions and soil extractable fractions were similar (data not shown). Moreover, litterfall C, N and P productions were not related to the soil total N:P ratio (data not shown).

Forest Floor Turnover Rates in Relation to the Nutrient Measures

Forest floor turnover rate (yr^{-1}) ranged from 0.10 to 1.75 among the seven study communities for which it was calculated (no data for PBM3), with an average value of 0.86 (Table S2). As summarized in the Table 1, forest floor turnover rate was positively related to the P concentration in both L and F/H layers (L layer: $R^2 = 0.35$, $P<0.001$; F/H layer: $R^2 = 0.40$, $P<0.001$), while only weakly related to the N concentration in the L layer ($R^2 = 0.18$, $P<0.05$). It was also negatively related to the N:P ratio in both L and F/H layers (L layer: $R^2 = 0.19$, $P<0.05$; F/H layer: $R^2 = 0.40$, $P<0.001$). When one outlier was excluded, the relationship of forest floor turnover rate with the F/H layer N:P ratio was even stronger ($R^2 = 0.53$, $P<0.001$; Figure 4).

Microbial biomass C concentration was not related to any nutrient measures in the L layer, but related to the N and P concentrations in the F/H layer, with a stronger relationship with the P concentration ($R^2 = 0.34$, $P<0.001$; Table 1) than with the N concentration ($R^2 = 0.16$, $P<0.05$; Table 1). Microbial respiration was positively related to the P concentration ($R^2 = 0.48$, $P<0.001$; Table 1) and also negatively related to the N:P ratio in the L layer ($R^2 = 0.37$, $P<0.001$; Table 1), while not related to any nutrient measures in the F/H layer (Table 1). In the L layer, β-glucosidase activity was more strongly related to the P concentration ($R^2 = 0.44$, $P<0.001$; Table 1) than the N concentration ($R^2 = 0.14$, $P<0.05$; Table 1), and negatively related to the N:P ratio ($R^2 = 0.23$, $P<0.01$; Table 1). In the F/H layer, β-glucosidase activity was poorly related to both N and P concentration ($R^2 = 0.17$ and 0.16, respectively, $P<0.05$; Table 1) while not related to the N:P ratio.

Correlations between Nutrient Measures

For both low N and high N groups of foliage, N and P concentrations were poorly correlated with each other (Table 2 and Figure S1A), and the P concentration was more strongly correlated with N:P ratio than the N concentration (Table 2 and Figures S1B and C). Within a species, N:P ratio was negatively correlated with the P concentration in all species ($P≤0.033$; Table 2), but was not correlated with the N concentration in any species ($P≥0.072$; Table 2). The relationship between foliar N and P concentration of plants of the same species was only significant for one species (*Castanea henryi*; Table 2). For the forest floors, fine roots and soil extractable fraction, N and P concentrations were all correlated with each other (r = 0.385–0.782, $P≤0.030$; Table 2), and N:P ratios were all negatively correlated with the P concentrations (r = −0.610– −0.798, $P<0.001$), while only poorly correlated with the N concentration in the F/H layer (r = 0.380, $P=0.032$; Table 2). Soil total N:P ratios were positively correlated with soil total N concentration (r = 0.450, $P=0.010$; Table 2) but not correlated with soil total P concentration (r = −0.201, $P=0.270$; Table 2). However, the plot of soil total N:P ratio against soil total P concentration showed that the points representing the PF site were distinct from other sites (Figure

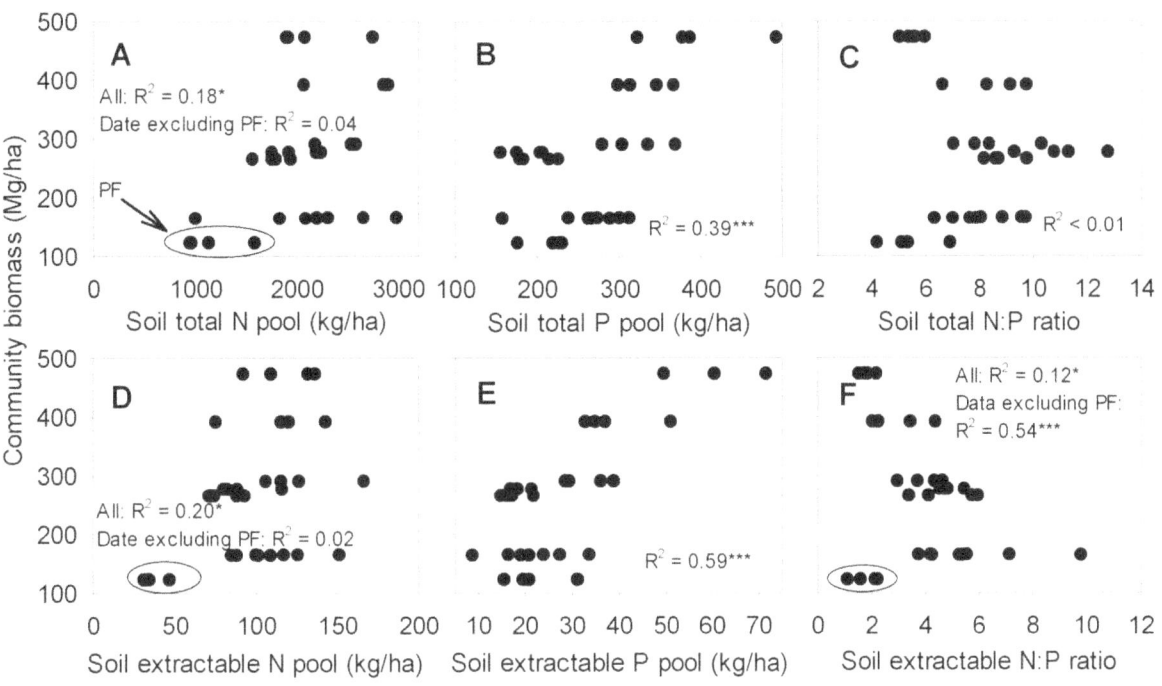

Figure 2. Relationships between community biomass and soil nutrient measures. PF indicates the pine forest. * $P<0.05$; *** $P<0.001$.

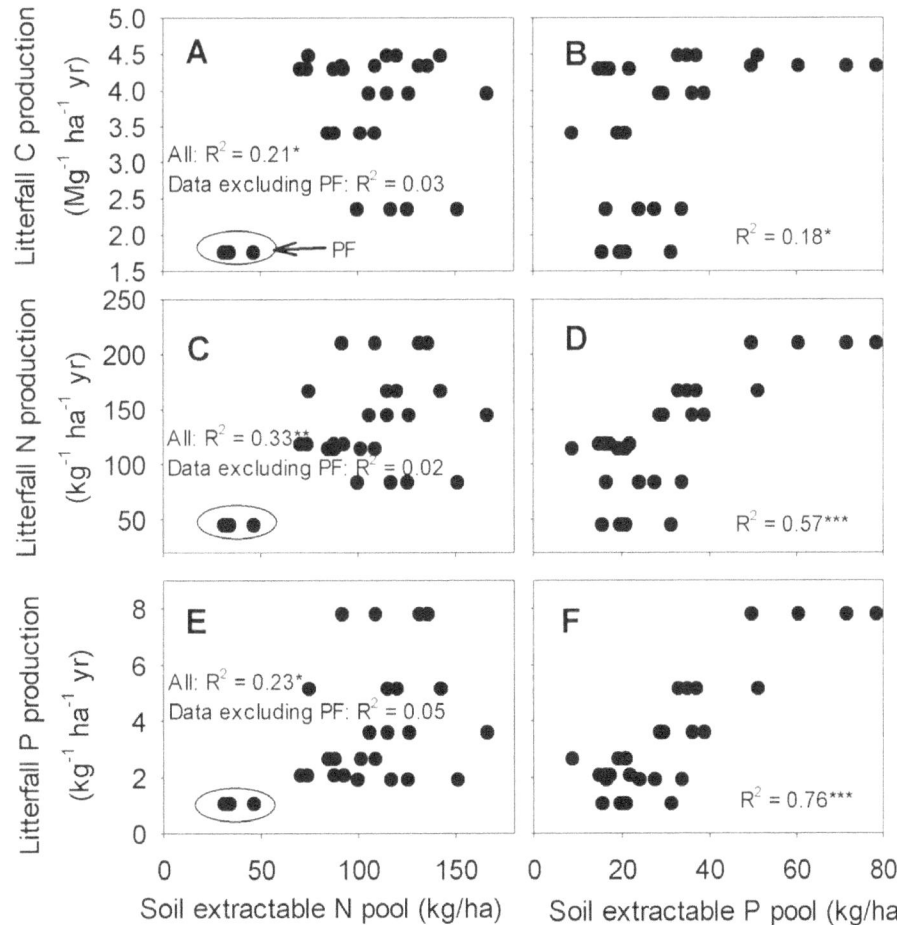

Figure 3. Relationships between litterfall C, N and P productions and soil nutrient measures. PF indicates the pine forest. ** $P<0.01$; *** $P<0.001$.

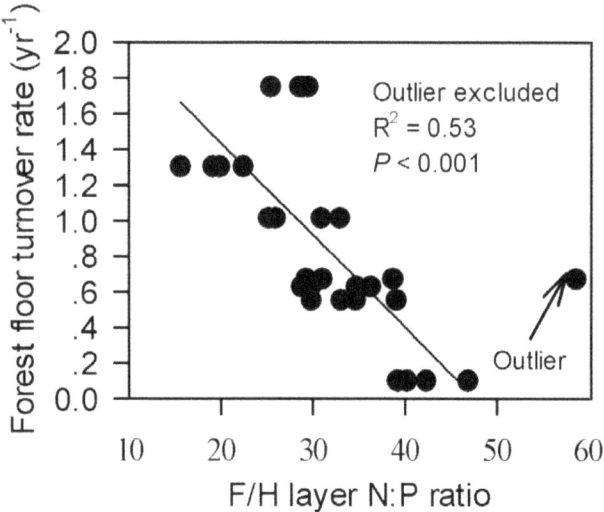

Figure 4. Relationships between forest floor turnover rate with F/H layer N:P ratio.

S2B). Excluding the PF site from analysis, the correlations between the soil total fractions were similar to those of forest floors, fine roots and soil extractable fraction (Table 3 and Figure S2).

As summarized in the Table 3, nutrient measures of all plant materials were generally more strongly correlated with the soil total P and extractable P concentrations than with soil total N and extractable N concentrations. Nitrogen: P ratios of forest floors and fine roots were all negatively correlated with soil total P and extractable P concentrations, while either not or negatively correlated with soil total N or extractable N concentration.

Nitrogen and P contents and N:P ratio

Soil total N:P ratios and extractable N:P ratios varied across the eight study sites, both were lower at the PF site (5.4 and 1.8, respectively) and REB1 site (5.5 and 1.8, respectively) than at other sites (7.2 to 11.0 and 3.0 to 6.1, respectively; Table 4). The PF site, with a long history of human disturbance, was distinct from other sites, with higher bulk density and lower organic C, total N and extractable N pools in the 0–15 cm mineral soil (Table 4). Soil total P and extractable P pools at the PF site were also lower than at the old-growth forest sites (REB1 and 2, MEB and MTEB), but

Table 1. Summary of regressions of forest floor turnover rates against nutrient measures of forest floors.

Parameter	Nutrient measure	Regression	R^2	Significance
Forest floor turnover rate (yr^{-1})				
L layer	N conc. (mg/g)	Rate = 0.067(N conc.) −0.226	0.18	<0.05
	P conc. (mg/g)	Rate = 1.729(P conc.) +0.012	0.35	<0.001
	N:P ratio	Rate = −0.019(N:P ratio) +1.638	0.19	<0.05
F/H layer	N conc. (mg/g)			NS
	P conc. (mg/g)	Rate = 2.70(P conc.) −0.36	0.4	<0.001
	N:P ratio	Rate = −0.036(N:P ratio) +2.025	0.4	<0.001
Microbial biomass C conc. (mg/g)				
L layer	N conc. (mg/g)			NS
	P conc. (mg/g)			NS
	N:P ratio			NS
F/H layer	N conc. (mg/g)	Microbial C = 0.94(N conc.) − 1.46	0.16	<0.05
	P conc. (mg/g)	Microbial C = 30.7(P conc.) − 2.25	0.34	<0.001
	N:P ratio			NS
Microbial respiration (µg CO$_2$-C g^{-1} h^{-1})				
L layer	N conc. (mg/g)			NS
	P conc. (mg/g)	Respiration = 24.6(P conc.) +21.1	0.48	<0.001
	N:P ratio	Respiration = −0.32(N:P ratio) +45.0	0.37	<0.001
F/H layer	N conc. (mg/g)			NS
	P conc. (mg/g)			NS
	N:P ratio			NS
β-glucosidase activity (µg p-nitrophenol g^{-1} h^{-1})				
L layer	N conc. (mg/g)	β-glucosidase = 19.3(N conc.) − 78.9	0.14	<0.05
	P conc. (mg/g)	β-glucosidase = 626.5(P conc.) − 29.5	0.44	<0.001
	N:P ratio	β-glucosidase = −6.8(N:P ratio) − 518.4	0.23	<0.01
F/H layer	N conc. (mg/g)	β-glucosidase = 8.1(N conc.) +45.0	0.17	<0.05
	P conc. (mg/g)	β-glucosidase = 176.5(P conc.) +75.8	0.16	<0.05
	N:P ratio			NS

n = 32. NS indicates statistically not significant at the level of $P < 0.05$.

Table 2. Correlations between nutrient measures in the plant and soil samples.

Sample	n	N conc. vs. P conc.		N:P ratio vs. N conc.		N:P ratio vs. P conc.	
		r	P	r	P	r	P
Foliage							
Low N group (<25 mg/g)	76	0.336	0.003	0.435	<0.001	−0.682	<0.001
High N group (≥25 mg/g)	20	−0.449	0.047	0.709	<0.001	−0.929	<0.001
C. henryi	16	0.709	0.002	0.085	0.753	−0.634	0.008
C. concinna	8	0.250	0.550	0.455	0.257	−0.747	0.033
P. massoniana	16	0.458	0.075	0.397	0.128	−0.627	0.009
S. superba	20	0.302	0.196	0.411	0.072	−0.740	<0.001
Forest floor							
L layer	32	0.630	<0.001	−0.111	0.545	−0.798	<0.001
F/H layer	32	0.385	0.030	0.380	0.032	−0.661	<0.001
Fine roots	32	0.782	<0.001	−0.094	0.610	−0.660	<0.001
0–15 cm mineral soil							
Total fraction	32	0.747	<0.001	0.450	0.010	−0.201	0.270
Total fraction excluding the PF site	28	0.617	<0.001	0.156	0.428	−0.647	<0.001
Extractable fraction	32	0.530	0.002	0.194	0.288	−0.610	<0.001

For soil total fraction, inclusion and exclusion of the pine forest (PF) site showed distinct results and thus, correlations for all data and data excluding PF were both shown.

comparable to those at other mature forest sites (PBM1, 2 and 3; Table 4).

Nitrogen and P concentrations varied widely in all plant materials, with average N concentrations of 21.0 mg/g, 16.1 mg/g, 13.2 mg/g and 13.8 mg/g, respectively in the foliage, L and F/H layers and fine roots, while corresponding values for P in these

materials were 0.77 mg/g, 0.41 mg/g, 0.43 mg/g and 0.44 mg/g, respectively (Table 5). Site averages of foliar N concentration were all above 15 mg/g (Table S3), and species averages were all higher than 15 mg/g as well, except for the *Pinus massoniana* (14.6 mg/g; from PF site) and *Rhododendron henryi* (11.2 mg/g; from MTEB site; Table S4). Although N:P ratio also varied widely in all plant

Table 3. Correlations of nutrient measures between the plant and soil samples.

Plant sample	0–15 cm mineral soil					
	Total N conc.	Total P conc.	Total N:P ratio	Extractable N conc.	Extractable P conc.	Extractable N:P ratio
Foliage						
N conc.	0.235	0.419	−0.313	0.126	0.548	−0.669
P conc.	0.277	0.357	−0.172	0.137	0.442	−0.538
N:P ratio	−0.253	−0.024	−0.399	−0.159	−0.105	−0.073
L layer forest floor						
N conc.	0.502**	0.668***	−0.074	0.607***	0.708***	−0.147
P conc.	0.218	0.661***	−0.426*	0.377*	0.833***	−0.562**
N:P ratio	0.040	−0.387*	0.482**	−0.076	−0.523**	0.578**
F/H layer forest floor						
N conc.	0.101	0.256	−0.377*	−0.018	0.042	−0.127
P conc.	0.349	0.782***	−0.468**	0.387*	0.787***	−0.604***
N:P ratio	−0.303	−0.550**	0.118	−0.381*	−0.646***	0.477**
Fine roots						
N conc.	0.632***	−0.614***	0.070	0.635***	0.618***	−0.286
P conc.	0.392*	0.697***	−0.326	0.484**	0.812***	−0.558**
N:P ratio	0.050	−0.442*	0.655***	−0.027	−0.489**	0.547**

Data are correlation coefficients. n = 8 for foliage; n = 32 for L and F/H layers and fine roots. * $P<0.05$; ** $P<0.01$; *** $P<0.001$.

Table 4. Selected characteristics of the 0–15 cm mineral soil.

Site	Bulk density (g/cm³)	Soil organic C pool (Mg/ha)	Soil total N pool (kg/ha)	Soil total P pool (kg/ha)	Soil total N:P ratio	Soil extractable N pool (kg/ha)	Soil extractable P pool (kg/ha)	Soil extractable N:P ratio
PF	1.35(0.02)	17(2)	1148(150)	213(13)	5.4(0.6)	36(3)	22(3)	1.8(0.3)
PBM1	0.92(0.03)	35(1)	1759(77)	200(11)	8.8(0.3)	81(5)	17(2)	4.8(0.6)
PBM2	1.02(0.07)	31(5)	1774(271)	243(29)	7.2(0.3)	96(6)	17(3)	6.1(1.2)
PBM3	0.98(0.04)	33(2)	2023(115)	185(12)	11.0(0.7)	91(8)	19(1)	4.9(0.2)
REB1	1.09(0.04)	28(2)	2156(201)	394(36)	5.5(0.2)	117(10)	65(6)	1.8(0.1)
REB2	0.89(0.04)	43(4)	2789(266)	330(15)	8.4(0.7)	113(14)	39(4)	3.0(0.5)
MEB	0.84(0.04)	40(4)	2682(268)	321(19)	8.4(0.7)	128(13)	33(3)	3.9(0.4)
MTEB	0.88(0.03)	44(3)	2533(178)	280(16)	9.0(0.4)	123(11)	25(4)	5.1(0.8)

All data are means (±1 SE), n = 4. The corresponding full names of eight study sites are listed in Table S1.

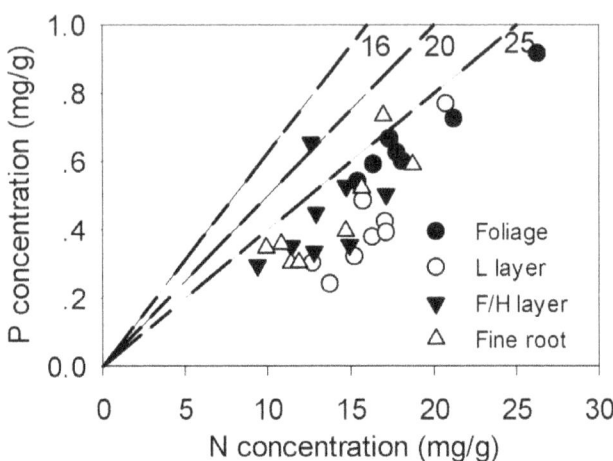

Figure 5. Mean N and P concentrations of plant materials of eight study forests. Dashed lines depict N:P ratios of 16, 20 and 25 on a mass basis. Ratios of 16 and 20 are P limitation thresholds of plant growth proposed by Koerselman and Meuleman (1996) [34] and Güsewell (2004) [31], respectively; ratio of 25 is the critcial N:P ratio that indicates P limitation on litter decomposition proposed by Güsewell and Verhoeven (2006) [61].

materials, the values (15.6–72.0) were generally high (Table 5), with 30 of the 32 site averages higher than 25 and other two values around 16 and 20 (Figure 5 and Table S3).

Discussion

Overall Relationship Patterns

Our results revealed consistent relationship patterns of community biomass, litterfall C, N and P productions, forest floor turnover rate and microbial properties with measures of N and P as well as N:P ratios (Figures 2, 3 and 4 and Table 1) that generally suggested that P availability played a more significant role than N availability in determining ecosystem primary productivity and nutrient cycling at the study sites. Considering that ecosystem productivity increases with increasing supply of a limiting nutrient, while showing no or even a negative response to the increased supply of a non-limiting nutrient [30,59,60], our results suggest that P availability was one of the limiting factors of plant growth and nutrient cycling at the study sites.

The idea is supported by the high N:P ratios of foliage, forest floors and fine roots at these sites. Site averages of the N:P ratios of these plant materials (30 of the 32 values higher than 25; Figure 5 and Table S3) were mostly higher than the proposed breakpoints of P limitation on plant growth by both Koerselman and Meuleman (1996) [34] (N:P ratio = 16) and Güsewell (2004) [31] (N:P ratio = 20), and also higher than the critical ratio that indicate P limitation on the decomposition of graminoid leaf litter (N:P ratio = 25) [61]. Moreover, average N:P ratios of foliage (28.3) and forest floors (L layer 42.3; F/H layer 32.0) in this study were much higher than the global averages of both temperature (broadleaf foliage 12.8, coniferous foliage 9.8; broadleaf litter 13.2, coniferous litter 11.8) and tropical regions (foliage 19.6, litter 28.4) [6]. These results generally suggest strong P limitation on primary productivity and litter decomposition at the study sites.

Our results are consistent with previous studies in which experimental N additions (50, 100 and 150 kg N ha^{-1} yr^{-1}) generally did not increase (and in some cases actually decreased) the understory vegetation biomass, litter decomposition rate and

Table 5. Statistical characteristics of nutrient measures of selected plant materials.

Material	N concentration (mg/g)		P concentration (mg/g)		N:P ratio	
	Mean	Range	Mean	Range	Mean	Range
Foliage	21.0(0.8)	10.7–43.9	0.77(0.04)	0.37–1.97	28.3(0.6)	15.8–41.4
L layer	16.1(0.5)	11.8–25.0	0.41(0.03)	0.21–0.89	42.3(2.0)	23.2–72.0
F/H layer	13.2(0.5)	8.2–19.5	0.43(0.02)	0.27–0.70	32.0(1.5)	15.6–58.7
Fine roots	13.8(0.7)	8.9–21.8	0.44(0.03)	0.24–0.88	32.7(1.2)	21.5–46.1

Data in the brackets are SE; n = 96 for foliage, = 32 for L and F/H layers and fine roots.

soil respiration rate at three (PF, PBM2 and MEB) of the eight study sites [25,26,62,63]. Experimental P addition (150 kg P ha^{-1} yr^{-1}) significantly increased soil microbial biomass C concentration, soil respiration and litterfall production at the old-growth MEB site, and also significantly increased the litterfall production at the other two sites (PF, PBM2) [27].

Different Relationship Patterns for Different Study Sites and Ecosystem Compartments

Although the relationship patterns suggested significant P limitation at the study sites in general, ecosystem productivity or processes may differ in the extent of P limitation or may be co-limited by N at different sites. Fertilization studies showed that at the old-growth MEB site, understory vegetation biomass, litter decomposition rate and soil respiration all responded negatively to experimental N additions [62,26], and soil microbial biomass and respiration and litterfall production showed positive responses to experimental P addition [27]. In contrast, at the PF site, understory vegetation biomass, litter decomposition and soil respiration at the PF site did not respond to the N additions [25,62], soil microbial biomass and respiration also did not respond to the P addition [27]. The responses of the PBM2 site to fertilization additions generally fell between the MEB site and PBM2 site [25–27,62].

The different nutrient limitation patterns of different sites were also reflected by our plots in this study. For example, plot of community biomass against soil extractable N:P ratio showed a distinct pattern at the PF site as compared to the other sites (Figure 2F). For sites with similar community biomass, soil extractable N:P ratio of the PF site was much lower than that of other sites. The distinct pattern was largely attributed to the markedly lower soil extractable N pool of the PF site than that of other sites (Table 4 and Figure 2D). Similarly, the plot of total N:P ratio against total P concentration in the soil also showed a distinct pattern of the PF site from other sites (Figure S2). For soils with similar total P concentration, total N:P ratio of the PF site was much lower than that of other sites due to its lower soil total N concentration compared with other sites. These distinct patterns of the PF site were due to its low soil N availability that probably because of the continuous removal of a large amount of nutrients, particular of N, by the prescribed over 40 years (1950s–1990s) of understory and litter harvest at this site [45].

Several recent studies have found that different ecosystem compartments or processes may differ in response to the addition of the same nutrient [27,64,65]. Nutrient limitation of one ecosystem compartment or process cannot be simply predicted from nutrient limitation of another ecosystem compartment or process [60,64,65]. In this study, community biomass was more strongly related to soil extractable P pool than litterfall C

production (Figures 2E and 3B), suggesting a greater constraint of P availability on community biomass than on litterfall C production. This was probably because of the indirect impact of P availability on litterfall C production via its impact on community biomass, as supported by significant relationship of litterfall C production with community biomass ($R^2 = 0.88$, $P = 0.014$, n = 7; data not shown).

Similarly, the relationships of litterfall C, N and P productions with soil P pools were also different. Stronger relationships of litterfall P and N productions with soil extractable P pool ($R^2 = 0.76$ and 0.47, respectively; Figures 3D and F) than litterfall C production ($R^2 = 0.18$; Figure 3B) suggest a stronger impact of soil P availability on litter chemistry than on litter quantity. Significant relationships between N and P concentrations and N:P ratio of forest floors and soil P concentrations (Table 2) also suggested the significant impacts of P availability on litter chemistry. These results are consistent with several other studies which revealed that forests on low-fertility soils tended to produce similar quantities of litters as forests on high-fertility soils nearby, but of lower quality [4,5,66–68].

Climate conditions (e.g. evapotranspiration) are likely to be the major factor affecting litter decomposition at a global scale, while at a local scale litter chemistry is always the major factor affecting litter decomposition [14,69,70]. In this study, forest floor turnover rate, microbial biomass C concentration, microbial respiration and β-glucosidase activity were all significantly related to the nutrient measures of the forest floors, with stronger relationships with the P concentration than with the N concentration and negative relationships with the N:P ratio in either L layer or F/H layer or both layers. As for the previous patterns these results suggest P availability rather than N availability plays an important role in controlling litter decomposition rates at all levels from ecosystem, to microbes and enzymes.

Constraint of Ecosystem Development by Physical Environment

The species composition of natural plant communities develops over a long period of succession. During succession, the community interacts with the physical environment (e.g. light, temperature and soil fertility) and finally reaches a relative steady status [71–73]. Physical environments may determine how far a community goes [71,73,74]. At the global or climate scale, primary productivity of forest ecosystems is largely determined by the climate conditions (e.g. precipitation) [75,76]. While at a local scale, if light, temperature and rainfall are relatively consistent across different terrestrial communities, soil fertility may be the major physical condition determining the mature stage of succession [1,2,14]. In this study, the close relationships of community biomass with soil nutrient pools (both total and extractable) did support this hypothesis. Soil total and extractable P pool explained 39% and 59% of the community biomass variation, respectively (Figures 2B and E).

Despite strong theoretical predictions suggesting the existence of relationship between community biomass and soil fertility [14,71,74], the relationships of community biomass and productivity with soil nutrients have been frequently poor in natural ecosystems [77,78], particular in forest ecosystems with high plant diversity as our study forests [14,79]. We proposed that the strong relationship of community biomass and litterfall N and P productions with soil P pools were probably because of markedly P limitation and N enrichment at the study sites compared with many other areas.

In addition to the high foliar and litter N:P ratios, mean N concentrations of the foliage (21.0 mg/g) and L layer (16.1 mg/g) were both ca. 2 times higher than the global averages for evergreen trees and shrubs (foliage 13.7 mg/g; litter 7.3 mg/g) [80], while mean P concentrations (foliage 0.77 mg/g; L layer 0.41 mg/g) were lower than reported global averages (foliage 1.02 mg/g and litter 0.50 mg/g) [80]. Moreover, foliar N concentrations of 11 of the 13 species in this study (range 17.1–42.1 mg/g, the other two values were 11.2 mg/g and 14.6 mg/g; Table S4) were in the upper range of species-specific foliar N concentrations of the tropical rain forests (mostly in the range of 10–30 mg/g) [35], and L layer N concentrations at six of the eight study sites (range of 15.2–20.7 mg/g, the other two values were 12.7 mg/g and 13.7 mg/g; Table S3) were in the upper ranges of site averages of litter N concentration of the tropical rain forests (mostly in the range of 10–19 mg/g) [67]. These results all suggest relatively high N availability and low P availability at our study sites compared with many other areas, which may underlie the strong relationships found between community biomass and litterfall N and P productions and soil P pools in this study.

How do these Consistent Patterns Occur?

As proposed above, strong limitation of P and enrichment of N might be the main causes of the consistent patterns observed in this study. The causes of P limitation in terrestrial ecosystems can be complicated. Here we only address three of six pathways proposed by Vitousek et al. (2010). First is the pathway of depletion that is likely to occur during millions of years of soil development [18,81,82]. Soils at the study sites developed during the Holocene (<15 ky) [39]. The soil ages are comparable to the Laupahoehoe site (20 kyr) in Hawaii that was found to be co-limited by N and P in fertilization studies [82]. The Laupahoehoe site is at comparable latitude (20°N), with similar mean annual temperature (16?), precipitation (2500 mm) and elevation (1170 m) [83] as our study sites (latitude 23°N; mean annual temperature 21?; precipitation 1900 mm; and elevation 50–600 m). Therefore, soil age is not likely to be the major cause of P limitation at the study sites. The second proposed pathway is low-P parent material that can cause P limitation developed quickly and persist over a long timescale [18]. Phosphorus concentrations in the C layer of the study area are approximately 0.40 mg/g [84], which are lower than the average in continental crust (0.70 mg/g, range of 0.04–3.00 mg/g) [18]. Therefore, low P concentration in the parent rock appears to contribute to the P limitation at the study sites but is not likely to be a major cause either.

The third possible pathway is anthropogenic P limitation by enhanced supply of other resources (especially N) [16–18]. This pathway is probably the most important cause of P limitation at the study sites, since N availability at the study sites was high compared with many other areas in the world as discussed above. The high N availability may be mainly attributed to two pathways:

accumulation of N through biological fixation during long-term (>80 years old) forest development [28,72], and high N deposition at the study sites [42,85]. Given an annual atmospheric N input of 40 kg N ha^{-1} yr^{-1} [42,85], atmospheric N input have accounted for 16% to 47% (mean 27%) of the annual total N inputs to the soils (atmospheric N input plus litterfall N production) at the study sites during the last two decades. A recent study reported the gradual increase of vegetation N:P ratios from 21 to 28 at the MEB site during the last three decades and a significant increase of foliar N:P ratio after 100 and 150 kg ha^{-1} yr^{-1} N added [42]. Together these lines of evidence suggested the possibility that the sites were driven to P limitation by excessive N inputs at the study sites.

Conclusions

Our study has revealed consistently stronger relationships between measures of P availability and community biomass, litterfall C, N and P productions, forest floor turnover rates and litter chemistry than we found for measures of N availability, indicating a significant role of P in determining ecosystem primary productivity and processes at the study sites. Vegetation N:P ratios indicated strong P limitation at all sites in this study. The results also showed that different ecosystem compartments or processes may differ in the extent of the nutrient limitation at different sites that depend on the soil nutrient status and land use history of these sites. In general, these results suggested constraint of ecosystem development by soil P availability at the study sites. We proposed that markedly N enrichment was probably a significant driver of the strong P limitation at these study sites. Low P parent material may also partly contribute to the P limitation. Further study is warranted on the mitigation of P limitation of ecosystem productivity and processes in the tropical and subtropical China.

Supporting Information

Figure S1   Correlations between N and P concentrations and N:P ratio of the foliage samples.

Figure S2   Correlations between total N and total P concentrations and total N:P ratio of the 0–15 cm mineral soil.

Table S1   Characteristics of eight study forest sites at Dinghushan Biosphere Reserve, China.

Table S2   Litterfall C production, forest floors biomass C, and forest floor turnover rate of eight study forests.

Table S3   Site averages of N and P concentrations and N:P ratio of foliage, L and F/H layers and fine roots of eight study forests.

Table S4   Species averages of foliar N and P concentrations and mass-based N:P ratio of 13 tree species selected from eight study forests.

Acknowledgments

The authors thank Marijke Heenan, Xuejin Wang, and Yujin Zhang for assistance in field samples collection and some laboratory analyses.

Author Contributions

Conceived and designed the experiments: EH CC DW MEM. Performed the experiments: EH CC DW. Analyzed the data: EH CC DW MEM.

Contributed reagents/materials/analysis tools: DW CC. Wrote the paper: EE CC DW MEM.

References

1. Chapin FS (1980) The mineral nutrition of wild plants. Annu Rev Ecol Syst 11: 233–260.

2. Aerts R, Chapin FS (2000) The mineral nutrition of wild plants revisited: A re-evaluation of processes and patterns. In: Fitter AH, Raffaelli DG, editors. Advances in Ecological Research 30. 1–67.

3. Elser JJ, Bracken MES, Cleland EE, Gruner DS, Harpole WS, et al. (2007) Global analysis of nitrogen and phosphorus limitation of primary producers in freshwater, marine and terrestrial ecosystems. Ecol Lett 10: 1135–1142.

4. Vitousek PM (1984) Litterfall, nutrient cycling, and nutrient limitation in tropical Forests. Ecology 65: 285–298.

5. Vitousek PM, Sanford RL (1986) Nutrient cycling in moist tropical forest. Annu Rev Ecol Syst 17: 137–167.

6. McGroddy ME, Daufresne T, Hedin LO (2004) Scaling of C:N:P stoichiometry in forests worldwide: Implications of terrestrial redfield-type ratios. Ecology 85: 2390–2401.

7. Reich PB, Oleksyn J (2004) Global patterns of plant leaf N and P in relation to temperature and latitude. Proc Natl Acad Sci 101: 11001–11006.

8. Kenk G, Fischer H (1988) Evidence from nitrogen fertilisation in the forests of Germany. Environ Pollut 54: 199–218.

9. Herbert DA, Fownes JH (1995) Phosphorus limitation of forest leaf area and net primary production on a highly weathered soil. Biogeochemistry 29: 223–235.

10. Tanner EVJ, Vitousek PM, Cuevas E (1998) Experimental investigation of nutrient limitation of forest growth on wet tropical mountains. Ecology 79: 10–22.

11. Cleveland CC, Townsend AR, Schmidt SK (2002) Phosphorus limitation of microbial processes in moist tropical forests: evidence from short-term laboratory incubations and field studies. Ecosystems 5: 680–691.

12. Tian HQ, Chen GS, Zhang C, Melillo JM, Hall CAS (2010) Pattern and variation of C:N:P ratios in China's soils: a synthesis of observational data. Biogeochemistry 98: 139–151.

13. Aerts R (1989) The effect of increased nutrient availability on leaf turnover and aboveground productivity of two evergreen ericaceous shrubs. Oecologia 78: 115–120.

14. Cleveland CC, Townsend AR, Taylor P, Alvarez-Clare S, Bustamante MMC, et al. (2011) Relationships among net primary productivity, nutrients and climate in tropical rain forest: a pan-tropical analysis. Ecol Lett 14: 939–947.

15. Vitousek PM, Mooney HA, Lubchenco J, Melillo JM (1997) Human domination of earth's ecosystems. Science 277: 494–499.

16. Elser JJ, Andersen T, Baron JS, Bergström AK, Jansson M, et al. 2009. Shifts in lake N:P stoichiometry and nutrient limitation driven by atmospheric nitrogen deposition. Science 326: 835–837.

17. Peñuelas J, Sardans J, Rivas A, Janssens IA (2012) The human induced imbalance between C, N and P in earth's life-system. Glob Chang Biol 18: 3–6.

18. Vitousek PM, Porder S, Houlton BZ, Chadwick OA (2010) Terrestrial phosphorus limitation: mechanisms, implications, and nitrogen-phosphorus interactions. Ecol Appl 20: 5–15.

19. Vitousek PM, Aber JD, Howarth RW, Likens GE, Matson PA, et al. (1997) Human alteration of the global nitrogen cycle: Sources and consequences. Ecol Appl 7: 737–750.

20. Gruber N, Galloway JN (2008) An earth-system perspective of the global nitrogen cycle. Nature 451: 293–296.

21. Zhao TQ, Ouyang ZY, Zheng H, Wang XK, Miao H (2004) Forest ecosystem services and their valuation in China. J Nat Resour 19: 480–491.

22. Liu XJ, Duan L, Mo JM, Du EZ, Shen JL, et al. (2011) Nitrogen deposition and its ecological impact in China: An overview. Environ Pollut 159: 2251–2264.

23. MacDonald JA, Dise NB, Matzner E, Armbruster M, Gundersen P, et al. (2002) Nitrogen input together with ecosystem nitrogen enrichment predict nitrate leaching from European forests. Glob Chang Biol 8: 1028–1033.

24. Aber JD, Goodale CL, Ollinger SV, Smith ML, Magill AH, et al. (2003) Is nitrogen deposition altering the nitrogen status of northeastern forests? BioScience 53: 375–389.

25. Lu XK, Mo JM, Gilliam FS, Yu GR, Zhang W, et al. 2011. Effects of experimental nitrogen additions on plant diversity in tropical forests of contrasting disturbance regimes in southern China. Environ Pollut 159: 2228–2235.

26. Lu XK, Mo JM, Gilliam FS, Zhou GY, Fang YT (2010) Effects of experimental nitrogen additions on plant diversity in an old-growth tropical forest. Glob Chang Biol 16: 2688–2700.

27. Liu L, Gundersen P, Zhang T, Mo JM (2012) Effects of phosphorus addition on soil microbial biomass and community composition in three forest types in tropical China. Soil Biol Biochem 44: 31–38.

28. Vitousek PM, Howarth RW (1991) Nitrogen limitation on land and in the sea - how can it occur. Biogeochemistry 13: 87–115.

29. Van Duren IC, Pegtel DM (2000) Nutrient limitations in wet, drained and rewetted fen meadows: evaluation of methods and results. Plant Soil 220: 35–47.

30. Chapin FS, Vitousek PM, Vancleve K (1986) The nature of nutrient limitation in plant-communities. Am Nat 127: 48–58.

31. Güsewell S (2004) N : P ratios in terrestrial plants: variation and functional significance. New Phytol 164: 243–266.

32. Lipson DA, Bowman WD, Monson RK (1996) Luxury uptake and storage of nitrogen in the rhizomatous alpine herb, *Bistorta bistortoïdes*. Ecology 77: 1277–1285.

33. Campo J, Dirzo R (2003) Leaf quality and herbivory responses to soil nutrient addition in secondary tropical dry forests of Yucatán, Mexico. J Trop Ecol 19: 525–530.

34. Koerselman W, Meuleman AFM (1996) The vegetation N:P Ratio: a new tool to detect the nature of nutrient limitation. J Appl Ecol 33: 1441–1450.

35. Townsend AR, Cleveland CC, Asner GP, Bustamante MMC (2007) Controls over foliar N : P ratios in tropical rain forests. Ecology 88: 107–118.

36. von Oheimb G, Härdtle W, Friedrich U, Power SA, Boschatzke N, et al. (2010) N:P Ratio and the nature of nutrient limitation in Calluna -Dominated Heathlands. Ecosystems 13: 317–327.

37. Peng SL, Zhang ZP (1995) Biomass, productivity and energy use efficiency of climax vegetation on Dinghu Mountains, Guangdong, China. Sci China B 38: 67–73.

38. Liu JC, Zhou GY, Zhang DQ (2007) Simulated effects of acidic solutions on element dynamics in monsoon evergreen broad-leaved forest at Dinghushan, China. Part 1: Dynamics of K, Na, Ca, Mg and P. Environ Sci Pollut Res 14: 123–129.

39. Shen CD, Yi WX, Sun YM, Xing CP, Yang Y, et al. (2001) Distribution of C-14 and C-13 in forest soils of the Dinghushan Biosphere Reserve. Radiocarbon 43: 671–678.

40. Lu XK, Mo JM, Gundersern P, Zhu WX, Zhou GY, et al. (2009) Effect of simulated N deposition on soil exchangeable cations in three forest types of subtropical China. Pedosphere 19: 189–198.

41. Zhou GY, Wei XH, Wu YP, Liu SG, Huang YH, et al. (2011) Quantifying the hydrological responses to climate change in an intact forested small watershed in Southern China. Glob Chang Biol 17: 3736–3746.

42. Huang WJ, Zhou GY, Liu JX (2011) Nitrogen and phosphorus status and their influence on aboveground production in three successional forests. Acta Oecol (in press).

43. Wang ZH, He DQ, Song SD, Chen SP, Chen DR, et al. (1982) The vegetation of Ding hu shan Biosphere Reserve. In: Ding Hu Shan Ecosystem Stationary Academia Sincia, editor. Tropical and subtropical forest ecosystem. Beijing: Science Press. 77–141.

44. Zhou GY, Guan LL, Wei XH, Zhang DQ, Zhang QM, et al. (2007) Litterfall production along successional and altitudinal gradients of subtropical monsoon evergreen broadleaved forests in Guangdong, China. Plant Ecol 188: 77–89.

45. Mo JM, Brown S, Lenart M, Kong GH (1995) Nutrient dynamics of a human-impacted pine forest in a MAB reserve of subtropical China. Biotropica 27: 290–304.

46. Fang YT, Mo JM (2002) Study on carbon distribution and storage of a pine forest ecosystem in Dinghushan Biosphere Reserve. Guihaia 22: 305–310.

47. Vance ED, Brookes PC, Jenkinson DS (1987) An extraction method for measuring soil microbial biomass-C. Soil Biol Biochem 19: 703–707.

48. Alef K, Nannipieri P (1995) β-Glucosidase activity. In: Alef K, Nannipieri P, editors. Methods in applied soil microbiology and biochemistry. London: Academic Press. 350–351.

49. Murphy J, Riley JP (1962) A modified single solution method for the determination of phosphate in natural waters. Anal Chim Acta 27: 31–36.

50. Sparling G, Vojvodic-Vukovic M, Schipper LA (1998) Hot-water-soluble C as a simple measure of labile soil organic matter: The relationship with microbial biomass C. Soil Biol Biochem 30: 1469–1472.

51. Chen CR, Xu ZH, Keay P, Zhang SL (2005) Total soluble nitrogen in forest soils as determined by persulfate oxidation and by high temperature catalytic oxidation. Aust J Soil Res 43: 515–523.

52. Curtin D, Wright CE, Beare MH, McCallum FM (2006) Hot water-extractable nitrogen as an indicator of soil nitrogen availability. Soil Sci Soc Am J 70: 1512–1521.

53. McDowell RW, Condron LM (2000) Chemical nature and potential mobility of phosphorus in fertilized grassland soils. Nutr Cycl Agroecosys 57: 225–233.

54. Ohno T, Zibilske LM (1991) Determination of low concentrations of phosphorus in soil extracts using malachite green. Soil Sci Soc Am J 55: 892–895.

55. Liu S, Luo Y, Huang YH, Zhou GY (2007) Studies on the community biomass and its allocations of five forest types in Dinghushan Nature Reserve. Ecol Sci 26: 387–393.

56. Yan JH, Zhang DQ, Zhou GY, Liu JX (2009) Soil respiration associated with forest succession in subtropical forests in Dinghushan Biosphere Reserve. Soil Biol Biochem 41: 991–999.

57. Jenny H, Gessel SP, Bingham FT (1949) Comparative study of decomposition rates of organic matter in temperate and tropical regions. Soil Sci 68: 419–432.

58. Olson JS (1963) Energy storage and the balance of producers and decomposers in ecological systems. Ecology 44: 322–331.

59. Güsewell S, Koerselman M (2002) Variation in nitrogen and phosphorus concentrations of wetland plants. Perspect Plant Ecol Evol Syst 5: 37–61.

60. Wright SJ, Yavitt JB, Wurzburger N, Turner BL, Tanner EVJ, et al. (2011) Potassium, phosphorus, or nitrogen limit root allocation, tree growth, or litter production in a lowland tropical forest. Ecology 92: 1616–1625.

61. Güsewell S, Verhoeven JTA (2006) Litter N:P ratios indicate whether N or P limits the decomposability of graminoid leaf litter. Plant Soil 287: 131–143.

62. Mo JM, Brown S, Xue JH, Fang YT, Li ZA (2006) Response of litter decomposition to simulated N deposition in disturbed, rehabilitated and mature forests in subtropical China. Plant Soil 282: 135–151.

63. Mo JM, Zhang W, Zhu WX, Gundersen P, Fang YT, et al. (2008) Nitrogen addition reduces soil respiration in a mature tropical forest in southern China. Glob Chang Biol 14: 403–412.

64. Hobbie SE, Vitousek PM (2000) Nutrient limitation of decomposition in Hawaiian forests. Ecology 81: 1867–1877.

65. Sundareshwar PV, Morris JT, Koepfler EK, Fornwalt B (2003) Phosphorus limitation of coastal ecosystem processes. Science 299: 563–565.

66. Edwards PJ (1982) Studies of mineral cycling in a montane rain forest in New Guinea: V. rates of cycling in throughfall and litter fall. J Ecol 70: 807–827.

67. Wood TE, Lawrence D, Clark DA (2006) Determinants of leaf litter nutrient cycling in a tropical rain forest: soil fertility versus topography. Ecosystems 9: 700–710.

68. Wood TE, Lawrence D, Clark DA, Chazdon RL (2009) Rain forest nutrient cycling and productivity in response to large-scale litter manipulation. Ecology 90: 109–121.

69. Aerts R (1997) Climate, leaf litter chemistry and leaf litter decomposition in terrestrial ecosystems: A triangular relationship. Oikos 79: 439–449.

70. Wieder WR, Cleveland CC, Townsend AR (2009) Controls over leaf litter decomposition in wet tropical forests. Ecology 90: 3333–3341.

71. Odum EP (1969) The strategy of ecosystem development. Science 164: 262–270.

72. Vitousek PM, Reiners WA (1975) Ecological succession and nutrient budgets - a hypothesis. BioScience 25: 376–381.

73. Wang SQ, Zhou L, Chen JM, Ju WM, Feng XF, et al. (2011) Relationships between net primary productivity and stand age for several forest types and their influence on China's carbon balance. J Environ Manage 92: 1651–1662.

74. Kerkhoff AJ, Enquist BJ, Elser JJ, Fagan WF (2005) Plant allometry, stoichiometry and the temperature-dependence of primary productivity. Glob Ecol Biogeogr 14: 585–598.

75. Schuur EA, Matson PA (2001) Net primary productivity and nutrient cycling across a mesic to wet precipitation gradient in Hawaiian montane forest. Oecologia 128: 431–442.

76. Schuur EA (2003) Productivity and global climate revisited: The sensitivity of tropical forest growth to precipitation. Ecology 84: 1165–1170.

77. DeLaune RD, Buresh RJ, Patrick WH (1979) Relationship of soil properties to standing crop biomass of *Spartina alterniflora* in a Louisiana marsh. Estuar Coast Mar Sci 8: 477–487.

78. Vermeer JG, Berendse F (1983) The relationship between nutrient availability, shoot biomass and species richness in grassland and wetland communities. Plant Ecol 53: 121–126.

79. Laurance WF, Fearnside PM, Laurance SG, Delamonica P, Lovejoy TE, et al. (1999) Relationship between soils and Amazon forest biomass: a landscape-scale study. Forest Ecol Manage 118: 127–138.

80. Aerts R (1996) Nutrient resorption from senescing leaves of perennials: are there general patterns? J Ecol 84: 597–608.

81. Walker TW, Syers JK (1976) The fate of phosphorus during pedogenesis. Geoderma 15: 1–19.

82. Vitousek PM, Farrington H (1997) Nutrient limitation and soil development: Experimental test of a biogeochemical theory. Biogeochemistry 37: 63–75.

83. Crews TE, Kitayama K, Fownes JH, Riley RH, Herbert DA, et al. (1995) Changes in soil phosphorus fractions and ecosystem dynamics across a long chronosequence in Hawaii. Ecology 76: 1407–1424.

84. Liu JX, Chu GW, Yu QF, Zhang DQ, Zhou GY (2002) The responses of soil chemical properties in different forest types to altitude at Dinghushan. In: Ding Hu Shan Forest Ecoysystem Stationary Academia Sinica, editor. Tropical and subtropical forest ecosystem. Beijing: Science Press. 125–131.

85. Fang YT, Gundersen P, Mo JM, Zhu WX (2008) Input and output of dissolved organic and inorganic nitrogen in subtropical forests of South China under high air pollution. Biogeosciences 5: 339–352.

8

Modeling Impacts of Alternative Practices on Net Global Warming Potential and Greenhouse Gas Intensity from Rice–Wheat Annual Rotation in China

Jinyang Wang[1], **Xiaolin Zhang**[1], **Yinglie Liu**[1], **Xiaojian Pan**[1], **Pingli Liu**[1,2], **Zhaozhi Chen**[1], **Taiqing Huang**[1], **Zhengqin Xiong**[1]*

1 Jiangsu Key Laboratory of Low Carbon Agriculture and GHGs Mitigation, College of Resources and Environmental Sciences, Nanjing Agricultural University, Nanjing, China, 2 Hebi Academy of Agricultural Sciences, Hebi, Henan, China

Abstract

Background: Evaluating the net exchange of greenhouse gas (GHG) emissions in conjunction with soil carbon sequestration may give a comprehensive insight on the role of agricultural production in global warming.

Materials and Methods: Measured data of methane (CH_4) and nitrous oxide (N_2O) were utilized to test the applicability of the Denitrification and Decomposition (DNDC) model to a winter wheat – single rice rotation system in southern China. Six alternative scenarios were simulated against the baseline scenario to evaluate their long-term (45-year) impacts on net global warming potential (GWP) and greenhouse gas intensity (GHGI).

Principal Results: The simulated cumulative CH_4 emissions fell within the statistical deviation ranges of the field data, with the exception of N_2O emissions during rice-growing season and both gases from the control treatment. Sensitivity tests showed that both CH_4 and N_2O emissions were significantly affected by changes in both environmental factors and management practices. Compared with the baseline scenario, the long-term simulation had the following results: (1) high straw return and manure amendment scenarios greatly increased CH_4 emissions, while other scenarios had similar CH_4 emissions, (2) high inorganic N fertilizer increased N_2O emissions while manure amendment and reduced inorganic N fertilizer scenarios decreased N_2O emissions, (3) the mean annual soil organic carbon sequestration rates (SOCSR) under manure amendment, high straw return, and no-tillage scenarios averaged 0.20 t C ha^{-1} yr^{-1}, being greater than other scenarios, and (4) the reduced inorganic N fertilizer scenario produced the least N loss from the system, while all the scenarios produced comparable grain yields.

Conclusions: In terms of net GWP and GHGI for the comprehensive assessment of climate change and crop production, reduced inorganic N fertilizer scenario followed by no-tillage scenario would be advocated for this specified cropping system.

Editor: Ben Bond-Lamberty, DOE Pacific Northwest National Laboratory, United States of America

Funding: This research was jointly supported by the National Science Foundation of China (41171238 and 40971139), the National Basic Research Program of China (2009CB118603), the Nonprofit Research Foundation for Agriculture (200903003), the Program for New Century Excellent Talent in Universities (NCET-10-0475), the Doctoral Program of Higher Education of China (20110097110001), the Fundamental Research Funds for the Central Universities (KYZ201110) and the PAPD (Priority Academic Program Development of Jiangsu Higher Education Institutions). The funders had no role in study design, data collection and analysis, decision to publish, or preparation of the manuscript.

Competing Interests: The authors have declared that no competing interests exist.

* E-mail: zqxiong@njau.edu.cn

Introduction

Agricultural activities are responsible for approximately 50% of global atmospheric methane (CH_4) emissions, and agricultural soils account for 75% of global nitrous oxide (N_2O) emissions [1]. Rice paddies have been identified as one of the major sources of atmospheric CH_4 and N_2O emissions [2–4]. China is one of the most important rice producing countries, rice planting area accounts for 20% of the world total and occurs on 23% of all cultivated land in China [5]. Flooded rice and upland crop, such as winter wheat and rice annual rotation system dominates in Chinese rice paddies [6]. The total CH_4 emissions from Chinese

rice paddies were estimated to be 6–10 Tg yr^{-1} in the 1990s [7,8], while N_2O emissions accounted for 25–35% of the total N_2O emissions from Chinese croplands [3,9]. These facts indicate that there is great potential for greenhouse gas (GHG) mitigations from Chinese rice agriculture [10].

Over the past decades, management practices affecting CH_4 and N_2O emissions from rice paddies have been well documented [4,11–14]. Shifting water regimes from continuous flooding to midseason drainage can significantly reduce CH_4 emissions, however, during the same period, N_2O emissions can increase due to trade-off among the emissions of CH_4 and N_2O [2,12,15]. Pronounced differences in CH_4 emissions from the rice season between straw returning time (i.e., the time of straw incorporation into soil after harvest) of on rice season (i.e., before soil flooded for rice transplanting) and off-rice season (i.e., after rice harvest and before wheat sowing), have also been demonstrated [13]. Altering the applications of inorganic fertilizer could either increase or decrease N_2O emissions, depending on the amounts and the timing applied. Conversion of conventional tillage to reduced or no-tillage may benefit soil carbon (C) stocks, but this conversion can also lead to anaerobic zones and thereby stimulate N_2O emissions [16,17].

A systematic approach for comprehensive assessment of GHG mitigation potential in agriculture is urgently needed [18]. Integrating the net exchanges of GHG with changes in the surface layer of soil organic carbon (SOC) has been proposed to analyze the effect of management practices on the net GWP of ecosystems [18,19]. The status of soil C pool plays an important role in regulating terrestrial ecosystem processes through the dynamic equilibrium of C gains and losses and is strongly dependent on current anthropogenic activities [20]. Although soil C sequestration is a separate issue from increasing crop productivity and protecting environmental health, the great potential of increasing SOC, to offset fossil fuel emissions and thus retard global warming, should be highlighted.

A process-based biogeochemical model – the Denitrification and Decomposition (DNDC) model– was originally developed to simulate N_2O emissions and SOC levels in US crop systems [21–23]. It has been widely used to simulate N_2O, N_2, nitric oxide (NO), CH_4, and carbon dioxide (CO_2) emissions for a wide range of ecosystems, such as cropland, grassland, and forests around the world [24–30]. Recently, the DNDC model was employed to estimate GHG emissions from uplands [27,29] and rice paddies [31–34] after validation with field measurement data. Via integrations of remote maps of soil and climate information with changing the alternative practice scenarios, or scaling up site-specific results to regional scale, DNDC simulations provided better understanding of the effect of site-specific management on global warming potential (GWP) at regional or large scales. However, large uncertainties still existed when estimating GHG emissions under certain managements on regional scale due to the spatial heterogeneity of soil properties such as texture, SOC content and pH [35]. Moreover, available evidence suggested that certain calibrations in the DNDC default parameters were essential for site-specific systems or scaling up to regional or large scales [36,37]. Due to the highly temporal and spatial variability of GHG emissions and their complex relationship to climatic and soil conditions, short-term field measurements may not capture the long-term effects of management practices on GHG emissions [38,39]. Although a number of field measurements have been conducted on GHG emissions and SOC change [2,40–42], the long-term impacts of the alternative management practices are poorly understood for the winter wheat – single rice cropping system.

The objectives of this study were to assess the applicability of the DNDC model tested against the field measurement data for the emissions of CH_4 and N_2O, and to utilize the validated model to evaluate the long-term (45-year) effects of alternative management practices on net GWP and GHGI for this specific rotation system of winter wheat – single rice in southern China.

Materials and Methods

Field Experiment

A field trial was carried out in Nanjing (31°52′N, 118°50′E), southern China for a winter wheat – single rice rotation system since 2008. The field studies did not involve endangered or protected species and the location is not protected in any way, No specific permits were required for the described field studies due to the local typical cropping and ambient air sampling.

The experimental soil was classified as *Stagnic Anthrosols* [43]. The texture of this studied soil was silt loam, consisting of 14% clay, 6% sand, and 80% silt with an initial pH of 5.7. Total organic C and N in the surface cultivated layer (0–20 cm) were 14.7 and 1.32 g kg^{-1}, and soil bulk density was 1.28 g cm^{-3} [44]. The annual mean temperature and total precipitation were 16.9°C and 136.5 cm, respectively in 2009, and 16.8°C and 130.4 cm, respectively in 2010, which were listed as baseline in Table 1.

During the period of 2009–2010, field measurements of the emissions of CH_4 and N_2O from this rotation system were conducted and adopted for the model test. Three treatments, each with three replicates, including a control treatment without N fertilization or straw return (CK), and the N fertilized treatments without straw (N) or with straw return (NS), were utilized to test the DNDC model (Table 2). For the straw return treatment, air-dried rice straw at the amount of 3 t ha^{-1} (C:N = 52:1) was applied to the surface before rice seedling transplantation, whereas no straw was returned during the winter wheat season. After crops were harvested, no above-ground residues were left *in situ*. The area of each plot was 20 m^2 (4 m×5 m), and cement bulkheads were placed between plots. The periods of crop planting and harvesting were November 13, 2009 and June 6, 2010, respectively, for winter wheat, and June 20, 2010 and October 10, 2010, respectively, for rice. The soil was conventionally tilled twice with a plough at a depth of approximately 10 cm, on November 8, 2009 and June 18, 2010, before crops were planted during the rotation cycle. For the fertilized treatment, urea was used as N fertilizer at the rate of 250 kg N ha^{-1} per crop with a split ratio of 4:3:3 for both crops. For winter wheat, the basal fertilization date was on November 13, 2009, and the two top dressings occurred on February 21, 2010 and March 18, 2010. The corresponding dates for single rice were June 18, July 6, and August 11, 2010. For each treatment, calcium superphosphate, used as phosphorus fertilizer, was applied at the local rate of 120 kg P_2O_5 ha^{-1}, and potassium chloride was applied at the local rate of 60 kg K_2O ha^{-1} as a basal fertilizer for each crop season. The common water management strategy of flooding – midseason drainage (June 29, 2010– August 7, 2010) – reflooding – final drainage (starting October 3, 2010) was employed for the rice season in this study, and no additional irrigation was used, with plots receiving only precipitation during the winter wheat season (Figure 1).

The fluxes of CH_4 and N_2O were measured using the static opaque chamber method [45]. The chamber was made of PVC, and consisted of two parts, one sized 45 cm×45 cm×50 cm and the other with an extended height of 60 cm to accommodate plant growth. The chamber was equipped with a circulating fan to ensure complete gas mixing and was wrapped with sponge and aluminum foil to minimize temperature changes inside the chamber during gas sampling. During gas sampling, these chambers were placed on permanently installed PVC collars of the same size and fitted into the groove of the collar sealed by water. The frequency of gas sampling was approximately once a week from November 13, 2009 to October 10, 2010, intensive gas sampling once every two days was performed after N fertilization and during drainage. For each plot, four gas samples were withdrawn from the chamber through a three-way stopcock using a 25-ml airtight syringe at 10 min intervals (0, 10, 20, and 30 min after the chamber closure). Gas samples were taken from 8:00 am through 11:00 am since the soil temperature during this period was close to the mean daily soil temperature.

Table 1. Input values utilized to the validated DNDC model for baseline scenario and the sensitivity tests.

Parameter	Baseline	Range tested
Environmental factors		
Annual mean temperature (°C)	16.9 (2009)/16.8 (2010)	Decrease by 2°C and 4°C and increase by 2°C and 4°C
Total annual precipitation (cm)	136.5 (2009)/130.4 (2010)	Decrease by 20% and increase by 20%
Soil texture	Silt loam	Loamy sand, sandy clay loam and sandy clay
SOC content (0–5 cm)	0.125%	0.05%, 0.1%, 0.15% and 0.2%
Soil pH	5.7	4.7, 6.7 and 7.7
Management alternatives		
Tillage	Conventional tillage (ploughed about 10 cm)	No-tillage and reduced tillage
Total annual N input (kg N ha^{-1} yr^{-1})	500 (250 for each crop season)	300 (150, 150) and 700 (350, 350)
Straw return (rice straw) (t ha^{-1})	3	1.5 and 6
Manure amendment (kg N ha^{-1} yr^{-1})	No amendment	250 and 500 applied as basal fertilizer instead of the equivalent annual rate of inorganic N

The gas samples were analyzed using a gas chromatograph (Agilent 7890A, USA) that was equipped with two detectors [45]. Methane was detected using a hydrogen flame ionization detector (FID), and N_2O was detected using an electron capture detector (ECD). Argon-CH_4 (5%) and N_2 were used as the carrier gas at a flow rate of 40 ml min^{-1} for N_2O and CH_4 analysis, respectively. The temperatures of the column and the ECD were maintained at 40°C and 300°C, respectively. The oven and the FID were operated at temperatures of 50°C and 300°C, respectively. The concentrations of CH_4 and N_2O were quantified by comparing their peak areas with those of reference gases (Nanjing Special Gas Factory). The fluxes were determined from the change in the slope of the mixing ratio of the collected samples after the chamber was closed. The seasonal or cumulative amounts of CH_4 and N_2O emissions were sequentially accumulated from the emissions between every two adjacent intervals of the measurements.

DNDC Model and the Sensitivity Test

The DNDC model is adopted to simulate the daily flux rates of CH_4 and N_2O from a winter wheat – single rice rotation system in this study. The DNDC model is available online at http://www.dndc.sr.unh.edu/, and consists of six sub-models for simulating soil climate, plant growth, decomposition, nitrification, denitrification and fermentation. Briefly, the soil climate sub-model calculates hourly and daily soil temperature and moisture fluxes in one dimension, the plant growth sub-model simulates plant biomass accumulation and partitioning, the decomposition sub-model simulates soil organic matter decay, N mineralization, CO_2 and dissolved organic carbon production, the nitrification and denitrification sub-models track the sequential biochemical reaction from ammonium to nitrate production and consumption, net NO and N_2O production and N_2 production, the fermentation sub-model simulates CH_4 production, consumption, transport and net flux. Numerous studies suggested that the DNDC model generally produced good performances for modeling SOC dynamics from the paddy cropping system across China [31,39,46]. Nonetheless, due to the lack of the long-term monitoring at the experimental site, the model validation and sensitivity test for SOC change cannot be conducted in this study.

To better understand the effects of both environmental factors and management practices on GHG emissions, a sensitivity test was conducted to isolate the most sensitive factors. A baseline scenario was chosen based on the local climatic and soil conditions and typical management for a winter wheat – single rice rotation system (Table 1). The sensitivity test was conducted by varying a single input parameter in a predefined range while keeping all other input parameters constant as those in the baseline scenario (Table 1). The DNDC model was run with each of the predefined scenarios for one rotation to produce annual emissions of CH_4 and N_2O and thereafter to calculate the total GWP of these gases on a 100-year time horizon.

Table 2. Measured and simulated data of cumulative emissions of CH_4 and N_2O from a winter wheat – single rice rotation system from November 13, 2009 to October 10, 2010.

	CH_4 (kg C ha^{-1})				N_2O (kg N ha^{-1})			
	Observed		Simulated		Observed		Simulated	
Treatment [a]	Wheat	Rice	Wheat	Rice	Wheat	Rice	Wheat	Rice
CK	0.78(3.05) [b]	83.9(29.9)	−0.28	5.3	0.18(0.07)	0.08(0.03)	0.22	0.01
N	1.41(2.02)	66.2(34.5)	−0.28	93.6	3.19(0.63)	0.22(0.18)	4.43	1.75
NS	1.67(1.87)	156.4(22.3)	−0.28	172.4	2.79(0.88)	0.18(0.04)	2.94	0.69

[a]CK, without both N fertilization and straw incorporation; N, with N fertilization; NS, with both N fertilization and straw incorporation;
[b]Data in the parenthesis indicate the standard deviation of three replicated experiments.

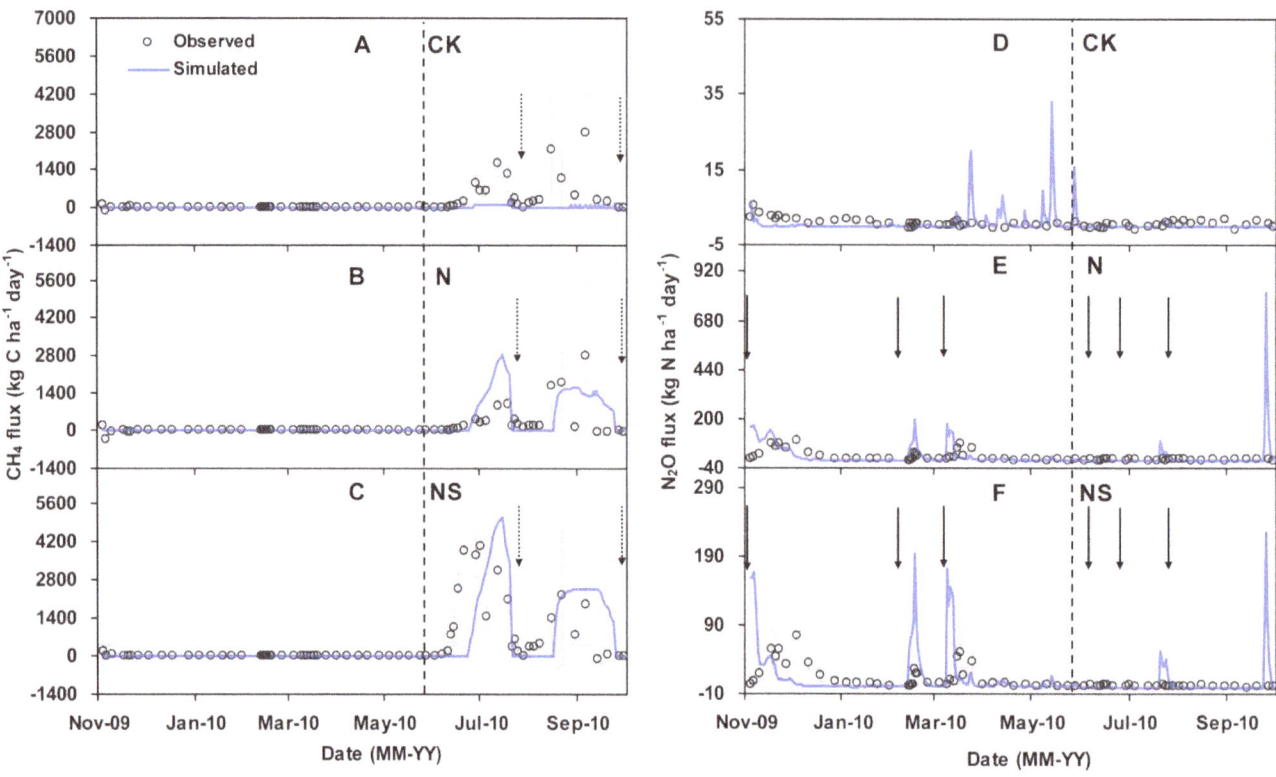

Figure 1. Dynamics variations of field observed (in dot) and simulated (in line) emissions of CH$_4$ (A, B, C) and N$_2$O (D, E, F) for the treatments of CK, N, and NS, respectively. CK, without both N fertilization and straw incorporation, N, with N fertilization, NS, with both N fertilization and straw incorporation. The vertical line in each panel divided the whole rotation into wheat (left) and rice (right) seasons. The vertical bars indicate the standard deviation of three replicates for each treatment. The dotted and solid arrows represent midseason or final drainage and N fertilization, respectively.

Design of Alternative Management Practices

Large uncertainties commonly existed in the evaluation of soil C and N processes when evaluating short-term anthropogenic perturbations. To accurately identify the consequences of major management practices, a predictive process-based model was employed to evaluate the long-term impacts for a 45-year period. Six alternative management practice scenarios were designed (Tables 1, 3). As compared to the baseline scenario, only the targeted parameters were changed by (1) reducing the annual inorganic N fertilizer rate to 300 kg N ha^{-1} (N300), (2) increasing the annual inorganic N fertilizer rate to 700 kg N ha^{-1} (N700), (3) reducing the amount of straw return by 50% (S1.5), (4) doubling the amount of straw return to 6 t ha^{-1} (S6), (5) replacing half of the annual inorganic N fertilizer (250 kg N ha^{-1}) with an equivalent N amount of bean cake (C:N = 6.8:1) incorporated as basal manure fertilizer (125 kg N ha^{-1}) for each crop season (OM250), and (6) changing the conventional tillage practice to no-tillage (No-tillage). The climate data used in these simulations were the present data from our field measurements during the corresponding wheat and rice seasons. Soil properties were obtained from the experimental measurements as those in the baseline scenario.

Analysis of Net GWP and GHGI

To quantitatively identify the impacts of alternative scenarios on net GWP over the 45-year period for this system, the IPCC factors [47] were adopted for calculating the combined GWPs on a time horizon of 100-year. The equation used was as follows:

Net GWP (kg CO$_2$-equiv. ha^{-1} yr^{-1}) = kg CH$_4$ ha^{-1} yr^{-1} ×25+ kg N$_2$O ha^{-1} yr^{-1} ×298– SOCSR ×44/12.

In addition, to associate the net GWP with crop production, greenhouse gas intensity (GHGI) was introduced and calculated by the following equation [38,48,49]:

GHGI (kg CO$_2$-equiv. kg^{-1} yield C) = Net GWP/grain yield.

Statistical Analysis

A linear regression analysis was performed to determine the variance between the cumulative emissions of CH$_4$ and N$_2$O from the three treatments, and then to reflect "the goodness of fit" of applying the DNDC model to test against our field measurement data. This statistical analysis was carried out using SigmaPlot 12.0 (Systat, San Jose, CA, USA).

Results and Discussion

Modeling Validation

Based on the baseline scenario as listed in Table 1 and also the same as the field managements, we used the DNDC model to simulate the daily flux rates of CH$_4$ and N$_2$O from a winter wheat – single rice rotation system. The simulated results were then compared with the field measurement data in daily flux dynamics (Figure 1) and in cumulative emissions (Figure 2).

During the winter wheat season, the rates of CH$_4$ were usually negligible in all the treatments, occasionally acted as a small source of atmospheric CH$_4$, and the simulated CH$_4$ emissions were in good agreement with the observed data (Figure 1A–C). The simulated seasonal CH$_4$ emissions fell within the statistical

Table 3. Impacts of alternative scenarios on averaged annual GHG emissions, soil organic carbon sequestration rate (SOCSR), N loss, grain yield, net GWP, and GHGI over the 45-year simulation.

Scenario [a]	CH_4 (kg C ha^{-1} yr^{-1})	N_2O (kg N ha^{-1} yr^{-1})	SOCSR (t C ha^{-1} yr^{-1})	N loss (kg N ha^{-1} yr^{-1})	Grain yield (kg C ha^{-1} yr^{-1})	Net GWP (kg CO_2-equiv. ha^{-1} yr^{-1})	GHGI (kg CO_2-equiv. kg^{-1} yield C)
Baseline	115.1	3.6	0.10	61.8	5682	5148	0.91
N300	114.0	2.2	0.10	30.9	5666	4439	0.78
N700	115.1	5.1	0.10	96.5	5683	5853	1.03
S1.5	113.9	3.6	0.06	60.9	5684	5291	0.93
S6	143.7	3.6	0.19	64.9	5689	5773	1.01
OM250	149.3	2.2	0.22	69.2	5714	5203	0.91
No-tillage	117.0	3.8	0.18	62.3	5686	5023	0.88

[a]The baseline scenario see Table 1; N300, total N fertilizer rate of 300 kg N ha^{-1} yr^{-1}; N700, total N fertilizer rate of 700 kg N ha^{-1} yr^{-1}; S1.5, straw return rate of 1.5 t ha^{-1} yr^{-1}; S6, straw return rate of 6 t ha^{-1} yr^{-1}; OM250, replacing half of the annual inorganic N fertilizer (250 kg N ha^{-1}) with an equivalent N amount of bean cake (C:N = 6.8:1) used as manure and incorporated as basal fertilizer for each season; No-tillage, zero-tillage.

standard deviation ranges of the observed data for all the treatments (Table 2). The simulated CH_4 emission rates were similar among treatments, suggesting that N fertilization did not affect CH_4 emissions during the winter wheat season (Table 2). During the rice season, the simulated and observed fluxes of CH_4 showed similar seasonal patterns in both the N and NS treatments (Figure 1B, C).

The simulated cumulative CH_4 rates of 93.6 kg C ha^{-1} and 172.4 kg C ha^{-1} for N and NS treatments, respectively, were close to the observed results of 66.2±34.5 kg C ha^{-1} and 156.4±22.3 kg C ha^{-1}, respectively (Table 2). In agreement with previous studies from rice paddies in Asia [15,24,25,34], CH_4 emissions were well simulated by the DNDC model with substrate input from straw return and crop growth (Figure 1B,C, 2B,C, Table 2). For the control treatment the DNDC model failed to simulate the peaks of CH_4 emission and resulted in a huge difference between the cumulative emissions in the simulation and measurement (Figure 1A, 2A, Table 2).

During the winter wheat season, the numbers of N_2O emission peaks were the same between the simulations and the measurements, the simulated peaks generally occurred earlier and greater than the observed ones when receiving N fertilizer (Figure 1D–F). Small discrepancies, with a mean value of 22%, existed between the observed and simulated seasonal rates of N_2O emission from the winter wheat season (Table 2). The simulated results of N_2O emissions coincide with previous studies from upland soils [29]. However, the differences between the observed and simulated cumulative emissions of N_2O during the rice season were big, ranging from −1.53 kg N ha^{-1} to 0.07 kg N ha^{-1} (Table 2). A previous study by Cai et al. [24] reported such discrepancies of up to eight-fold between the observed and simulated values for the rice field treated with urea of 300 kg N ha^{-1} at the same region. Great relative deviations, as high as -238 to 29%, were also reported by Babu et al. [25] in a study of several rice paddies in India. As compared to the automatic measurement, the manual chamber method might have missed the episodic N_2O emission peaks, particularly after midseason drainage and final drainage (Figure 1E, F) [2,24,44].

The regression slopes around 1 between the observed and simulated cumulative emissions of CH_4 and N_2O demonstrated the good performance of the DNDC model while their determination coefficients showed the covariance between the observed and simulated cumulative emissions (Figure 2). Thus, except for

the control treatment, the regression slopes were close to 1 (0.87–1.34) and the determination coefficients between the observed and simulated annual total emissions of CH_4 and N_2O were high ($r^2 = 0.84$–0.96) in the present study (Figure 2A–F), indicating that the applicability of the DNDC model is conservatively feasible for this site-specific rotation system of winter wheat – single rice in southern China.

Sensitivity Test

With one year rotation of winter wheat – single rice, sensitivity tests were conducted to identify the most sensitive factors that affect total GWP of CH_4 and N_2O by varying one single factor as listed in Table 1. An increase or reduction in CH_4 and N_2O emissions was thus converted into CO_2-equivalent in terms of their GWPs on 100-year time horizon as shown in Figure 3. The dominant source of the total GWP in this study was CH_4 emissions during the rice season and N_2O emissions during the wheat season, and their corresponding contributions were also shown in Figure 3.

Among all the selected environmental factors, soil texture and initial SOC content were the most influential factors associated with the total GWP. The GWP of this rotation system substantially decreased with increased clay fraction, the increased clay fraction had negative effect on CH_4 emissions although significantly promoted N_2O emissions (Figure 3A). The total GWP increased substantially in response to elevated SOC content due to the stimulatory effect on CH_4 emissions (Figure 3B). −20% low precipitation significantly increased the GWP mainly through the N_2O emissions during the wheat season while +20% precipitation had no obvious effect (Figure 3C). N_2O emissions during the wheat season increased in response to reduced annual mean temperature and remained similar in response to increased temperature, CH_4 emissions during the rice season increased with increasing temperature (Figure 3D). Due to the fact that the ratio of N_2O to N_2 decreases with increasing soil pH, soil pH was associated with lower N_2O emissions and with an insignificant impact on CH_4 emissions (Figure 3E). The above simulated results were in good agreement with previous field or model studies [29,34,35].

Among all selected management alternatives, N fertilization, straw return, and manure amendment were the most influential factors affecting the GWP of this rotation system (Figure 3F, G, H). For example, the GWPs rose from 6652 to 8191 kg CO_2-equiv.

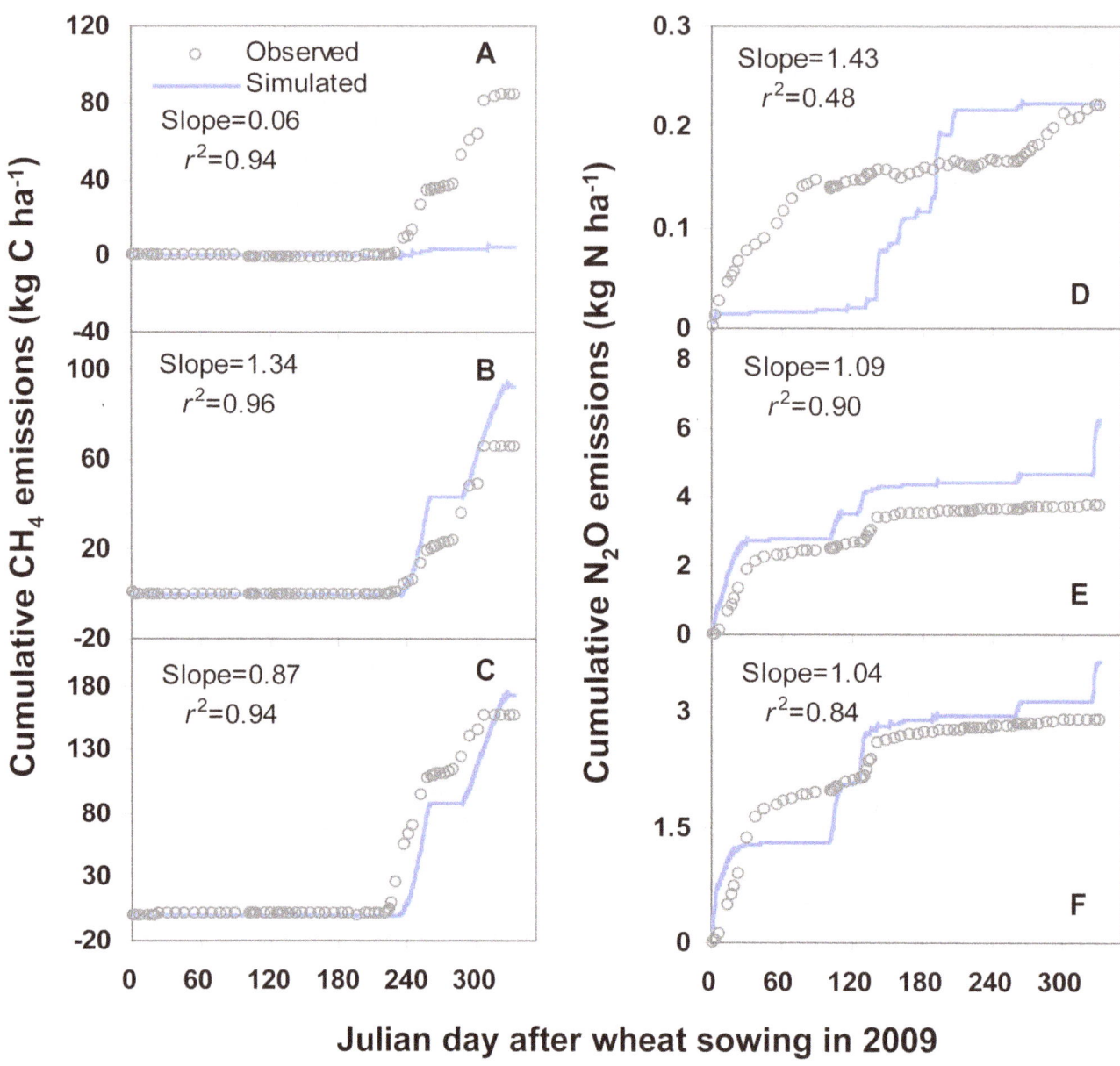

Figure 2. Cumulative emissions of CH₄ (A, B, C) and N₂O (D, E, F) for the treatments of CK, N, and NS, respectively. CK, without both N fertilization and straw incorporation, N, with N fertilization, NS, with both N fertilization and straw incorporation. The slopes and determination coefficients were calculated by the linear regression of observed and simulated cumulative emissions of CH₄ and N₂O.

ha^{-1} yr^{-1} with increasing the rates of N fertilizer from 300 to 700 kg N ha^{-1} yr^{-1}. And the increase in GWPs was mainly caused by increased N_2O emissions during the wheat season (Figure 3F). The increased GWPs were resulted from the stimulated CH_4 emissions when straw return and manure amendment increased (Figure 3G, H). As for the changes in tillage and the split ratio of N fertilization, there was no significant difference between the baseline scenario and the other alternative practices (Figure 3I, J).

Modeling the Long-term Effects of Alternative Management Practices on C and N Cycles and GHG Emissions

The long-term impacts of these practices on C and N cycles and GHG emissions were emphasized by repeatedly running the

DNDC model for 45-year period (Figure 4A, B). Compared with the baseline scenario, high straw return and manure amendment scenarios significantly increased CH_4 emissions by 25% and 30%, respectively, whereas other scenarios had negligible effects on CH_4 emissions (Table 3). Over the 45-year time course, CH_4 emissions under all management practices gradually decreased, which was probably due to the enhanced SOC stock that increased the capacity of soil to oxidize CH_4 [50]. The increasing trend of CH_4 emissions was in accordance with the short-term effect of organic matter incorporation [2,13,14,38]. No change in CH_4 emissions was observed due to varying inorganic N fertilizer rate (Table 3). Nitrogen fertilizer generally had statistically insignificant effect on CH_4 emissions (Figure 4B) [51], although some previous studies reported decreased or increased CH_4 emissions [2,38].

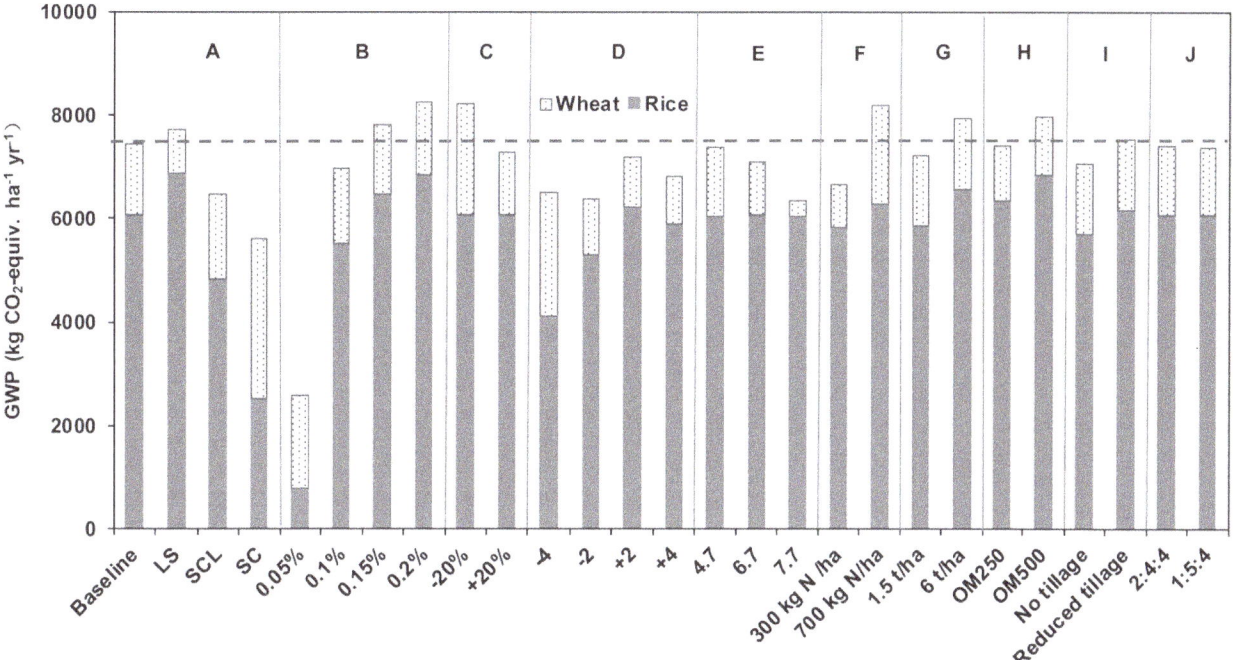

Figure 3. Sensitivity tests of GWP of CH₄ and N₂O emissions to environmental factors and alternative management practices. Starting from the baseline management conditions, change in (A) soil texture, (B) initial SOC content, (C) total precipitation, (D) annual mean temperature, (E) soil pH, (F) N fertilizer input, (G) straw return, (H) manure amendment, (I) tillage, and (J) split application ratio altered the GWP for the rotation system. Abbreviations for soil texture are as follows: LS, loamy sand; SCL, sandy clay loam; SC, sandy clay.

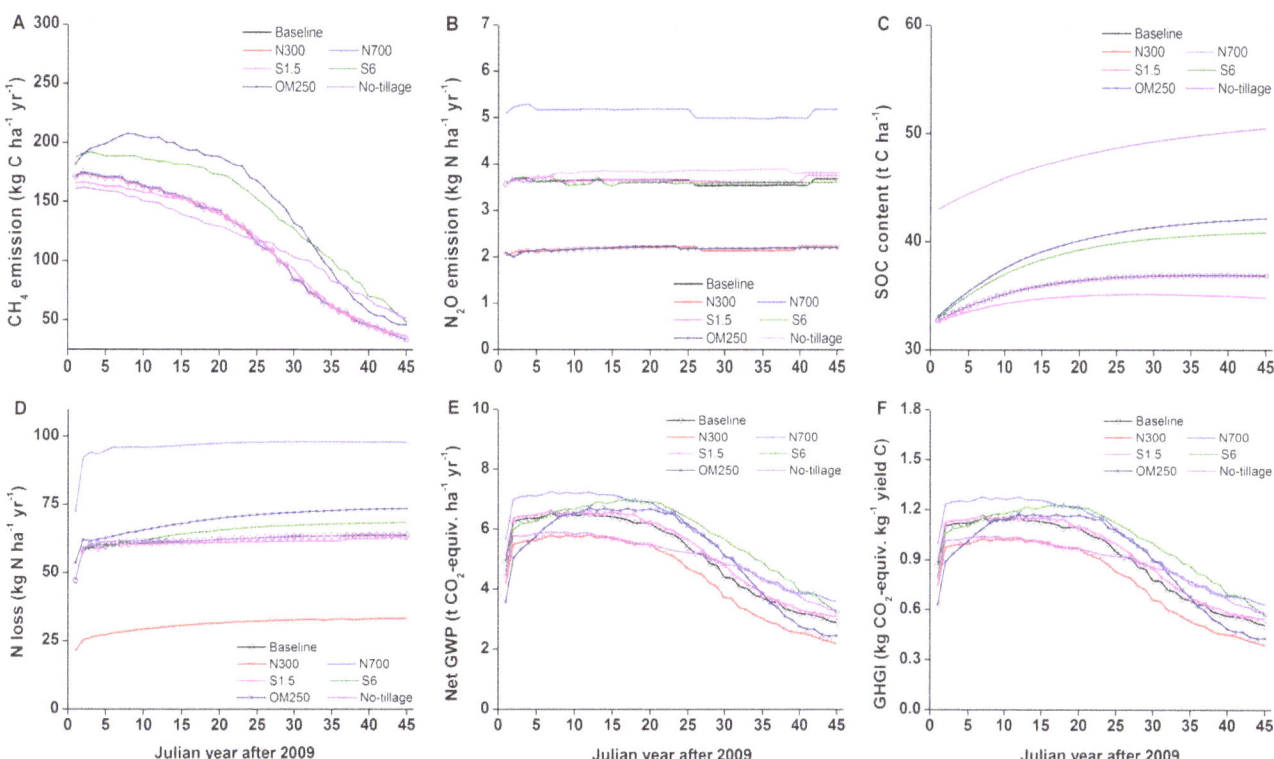

Figure 4. Long-term (45-year) impacts of alternative management scenarios on (A) CH₄, (B) N₂O, (C) SOC content, (D) N loss, (E) net GWP, and (F) GHGI. See Table 1 for the baseline scenario, N300 and N700, total N fertilizer rate of 300 and 700 kg N ha⁻¹ yr⁻¹, respectively, S1.5 and S6, straw return rate of 1.5 and 6.0 t ha⁻¹ yr⁻¹, respectively, OM250, replacing half of the annual inorganic N fertilizer (250 kg N ha⁻¹) with an equivalent N amount of bean cake (C:N = 6.8:1) used as manure and incorporated as basal fertilizer for each season, No-tillage, zero-tillage.

Among the alternative management scenarios, N_2O emissions had similar dynamics curves with small increasing trend over the 45-year time course, the increasing trend was more obvious under no-tillage scenario. N_2O emissions were significantly different under different levels of inorganic N fertilizer scenarios (Figure 4B). A significant linear relationship between total N input (inorganic plus organic N) and N_2O emissions was found for the six alternative scenarios (Table 3, $r^2 = 0.69$, $P = 0.02$), which was in good agreement with numerous previous studies [2,4,44]. Increasing straw return did not reduce N_2O emissions in this long-term simulation (Table 3), which was contrary to our previous short-term field measurement [44]. From the long term simulation, manure amendment was beneficial for reducing N_2O emissions as a result of reducing inorganic N fertilization rate (Table 3), although organic manure return would increase N_2O emissions in the long-term double rice cropping system [38]. No-tillage did not reduce N_2O emissions from soils, which was in support of previous findings [17,37,50,52].

Among the alternative management scenarios, SOC content (0–30 cm) gradually increased over the 45-year time course (Figure 4C), indicating that the soil has not reached its maximum capacity for C sequestration. The increasing trend of SOC supported previous model study that organic matter amendment on SOC sequestration would last across the 45-year period for paddy fields in the Yangtze Delta [41]. There may be great potential for C sequestration in this region due to low SOC density, and thus for mitigating the increasing atmospheric CO_2. Among the six management alternatives, the simulated SOC sequestration rates ranged between 0.06 and 0.22 t C ha^{-1} yr^{-1} (Table 3), which were comparable to previous field and modeling studies [39,40]. The amounts of SOC stock did not vary with inorganic N fertilizer rates, being around 0.10 t C ha^{-1} yr^{-1} of SOCSR (Figure 4C, Table 3). Scenarios with high straw return, manure amendment, and no-tillage produced greater SOCSR around 0.20 t C ha^{-1} yr^{-1} than other scenarios (Table 3). A positive correlation between soil C sequestration and the amount of incorporated C was found in this study ($r^2 = 0.76$, $P = 0.01$), suggesting that large amounts of crop residue inputs are necessary for enhancing SOC levels, especially in paddy soils [20,39–42,53]. Enhanced SOC helped to not only mitigate climate change but also enhance soil fertility and hence sustain crop productivity [20,40], and thus there was closely relationship between SOC and crop yield (Table 3). Thus straw return and manure amendment are of vital importance. No-tillage practices significantly elevated SOC content (Table 3, Figure 3C), suggesting that no-tillage as alternative practice should be deserved more attention as proposed by previous studies [49,50,53].

Among the management alternatives, N loss was simulated over the 45-year time course (Figure 4D). The elevated N loss was due to not only the increased N input but organic manure amendment [29] (Table 3). In terms of the GWPs of CH_4 and N_2O, the scenario of reduced inorganic N fertilizer was advocated for

attenuating global warming under this rotation system over the long-term period.

Modeling the Long-term effects of Alternative Management Practices on Grain Yield, Net GWP and GHGI

Due to the already-high N rate in the baseline scenario, increased N input level did not further increase grain yield under current managements without improving N use efficiency (Table 3). Moreover, it is not surprising that annual N fertilization reduced by 40% of the baseline value did not reduce crop yield in this study, which was confirmed by the previous study that 36% reduction in N fertilizer for the rice-wheat rotation in the same region did not reduce crop yield [54]. The similar trend for crop yield and SOC also reported by Wang et al. [39] through running the DNDC model at the same site for rice-wheat cropping system with similar alternative practices.

The temporal variations of net GWP and GHGI were different among these alternative scenarios across the 45-year time course (Figure 4E, F), indicating that different strategies should be employed at different time scales [29,32]. For example, during the first 25-year period, the scenarios of reduced inorganic N fertilizer and no-tillage obviously reduced net GWP, whereas during the following 20-year period, the reducing effect became complex for the no-tillage scenario (Figure 4E, F).

Scenarios of reduced inorganic N fertilizer and no-tillage reduced both the net GWP and GHGI, while the remaining scenarios tended to increase the net GWP and GHGI as compared with the baseline scenario (Table 3).

Overall, among all the alternative scenarios the scenarios of reduced inorganic N fertilizer and no-tillage could therefore contribute to mitigate global warming potential while sustain crop production, particularly for reduced inorganic N scenario with obviously the least N loss.

Conclusions

The applicability of the DNDC model and the long-term assessments on various management alternatives were tested for the rice-wheat rotation system in this study. The validation, sensitivity tests, and long-term prediction provided a sound basis for comprehensive understanding of the alternative management practices on soil C and N cycles involved in global warming. Therefore, reduced inorganic N fertilizer scenario followed by no-tillage scenario would be advocated for mitigating global warming without decreasing crop yield.

Author Contributions

Conceived and designed the experiments: ZQX JYW. Performed the experiments: XLZ YLL XJP PLL ZZC TQH. Analyzed the data: JYW XLZ. Contributed reagents/materials/analysis tools: PLL XLZ TQH. Wrote the paper: JYW ZQX.

References

1. USEPA (2006) Global anthropogenic non-CO_2 greenhouse gas emissions: 1990–2020 (June 2006 Revised), available at: http://www.epa.gov/climatechange/economics/downloads/GlobalAnthroEmissionsReport.pdf, Office of Atmospheric Programs, USEPA, Washington, DC.
2. Cai Z, Xing G, Yan X, Xu H, Tsuruta H, et al. (1997) Methane and nitrous oxide emissions from rice paddy fields as affected by nitrogen fertilisers and water management. Plant Soil 196 (1): 7–14.
3. Zheng X, Han S, Huang Y, Wang Y, Wang M (2004) Re-quantifying the emission factors based on field measurements and estimating the direct N_2O emission from Chinese croplands. Glob Biogeochem Cycles 18 (1): 1–19.
4. Zou J, Huang Y, Qin Y, Liu S, Shen Q, et al. (2009) Changes in fertilizer-induced direct N_2O emissions from paddy fields during rice-growing season in China between 1950s and 1990s. Glob Chang Biol 15 (1): 229–242.
5. Frolking S, Qiu J, Boles S, Xiao XM, Liu JY, et al. (2002) Combining remote sensing and ground census data to develop new maps of the distribution of rice agriculture in China. Glob Biogeochem Cycles 16 (4). Doi:10.1029/2001gb001425.
6. Xu Z, Lu Y (1992) Ecological environment of paddy soil. In: Li QK, editor. Paddy Soils of China. Science Press, Beijing, China. 108–125.
7. Huang Y, Zhang W, Zheng X, Li J, Yu Y (2004) Modeling methane emission from rice paddies with various agricultural practices. J Geophys Res 109 (D8): D08113.

8. Yan X, Akiyama H, Yagi K, Akimoto H (2009) Global estimations of the inventory and mitigation potential of methane emissions from rice cultivation conducted using the 2006 Intergovernmental Panel on Climate Change Guidelines. Glob Biogeochem Cycles 23 (2): GB2002.

9. Liu S, Qin Y, Zou J, Liu Q (2010) Effects of water regime during rice-growing season on annual direct N_2O emission in a paddy rice-winter wheat rotation system in southeast China. Sci Total environ 408 (4): 906–913.

10. Smith P, Martino D, Cai Z, Gwary H, Janzen H, et al. (2007) Agriculture. In: Metz B, Davidson OR, Bosch PR, et al (eds) Climate Change 2007: Mitigation. Contribution of Working Group III to the Fourth Assessment Report of the Intergovernmental Panel on Climate Change. Cambridge University Press, Cambridge, 497–540.

11. Yan X, Ohara T, Akimoto H (2003) Development of region-specific emission factors and estimation of methane emission from rice fields in the East, Southeast and South Asian countries. Glob Chang Biol 9 (2): 237–254.

12. Cai Z, Tsuruta H, Gao M, Xu H, Wei C (2003) Options for mitigating methane emission from a permanently flooded rice field. Glob Chang Biol 9 (1): 37–45.

13. Yan X, Yagi K, Akiyama H, Akimoto H (2005) Statistical analysis of the major variables controlling methane emission from rice fields. Glob Chang Biol 11 (7): 1131–1141.

14. Wang J, Zhang X, Xiong Z, Khalil MAK, Zhao X, et al. (2012) Methane emissions from a rice agroecosystem in South China: Effects of water regime, straw incorporation and nitrogen fertilizer. Nutr Cycl Agroecosyst 93: 103–112.

15. Li C, Frolking S, Xiao XM, Moore B III, Boles S, et al. (2005) Modeling impacts of farming management alternatives on CO_2, CH_4, and N_2O emissions: A case study for water management of rice agriculture of China. Glob Biogeochem Cycles 19 (3). Doi:10.1029/2004gb002341.

16. Smith P, Goulding KW, Smith KA, Powlson DS, Smith JU, et al. (2001) Enhancing the carbon sink in European agricultural soils: including trace gas fluxes in estimates of carbon mitigation potential. Nutr Cycl Agroecosyst 60 (1): 237–252.

17. Six J, Ogle SM, Breidt FJ, Conant R, Mosier AR, et al. (2004) The potential to mitigate global warming with no-tillage management is only realized when practised in the long term. Glob Chang Biol 10 (2): 155–160.

18. Robertson GP, Field C, Raupach M (2004) Abatement of nitrous oxide, methane, and the other non-CO_2 greenhouse gases: the need for a systems approach. The global carbon cycle: integrating humans, climate and the natural world. 493–506.

19. Fornara D, Steinbeiss S, McNamara N, Gleixner G, Oakley S, et al. (2011) Increases in soil organic carbon sequestration can reduce the global warming potential of long-term liming to permanent grassland. Glob Chang Biol 17(5): 1925–1934.

20. Lal R (2004) Soil carbon sequestration to mitigate climate change. Geoderma 123 (1–2): 1–22.

21. Li C, Frolking S, Frolking TA (1992) A model of nitrous oxide evolution from soil driven by rainfall events: 1. Model structure and sensitivity. J Geophy Res 97 (D9): 9759–9776.

22. Li C, Frolking S, Harriss R (1994) Modeling carbon biogeochemistry in agricultural soils. Glob Biogeochem Cycles 8 (3): 237–254.

23. Li C, Narayanan V, Harriss RC (1996) Model estimates of nitrous oxide emissions from agricultural lands in the United States. Glob Biogeochem Cycles 10 (2): 297–306.

24. Cai Z, Sawamoto T, Li C, Kang G, Boonjawat J, et al. (2003) Field validation of the DNDC model for greenhouse gas emissions in East Asian cropping systems. Glob Biogeochem Cycles 17(4): 1107.

25. Babu YJ, Li C, Frolking S, Nayak DR, Adhya TK (2006) Field validation of DNDC model for methane and nitrous oxide emissions from rice-based production systems of India. Nutr Cycl Agroecosyst 74 (2): 157–174.

26. Beheydt D, Boeckx P, Sleutel S, Li C, Vancleemput O (2007) Validation of DNDC for 22 long-term N_2O field emission measurements. Atmos Environ 41 (29): 6196–6211.

27. Abdalla M, Wattenbach M, Smith P, Ambus P, Jones M, et al. (2009) Application of the DNDC model to predict emissions of N_2O from Irish agriculture. Geoderma 151 (3–4): 327–337.

28. Li D, Lanigan G, Humphreys J (2011) Measured and simulated nitrous oxide emissions from ryegrass- and ryegrass/white clover-based grasslands in a moist temperate climate. PLoS ONE 6(10):e26176. Doi:10.1371/journal.pone.0026176.

29. Li H, Qiu J, Wang L, Tang H, Li C, et al. (2010) Modelling impacts of alternative farming management practices on greenhouse gas emissions from a winter wheat-maize rotation system in China. Agric Ecosyst Environ 135 (1–2): 24–33.

30. Kang X, Hao Y, Li C, Cui X, Wang J, et al. (2011) Modeling impacts of climate change on carbon dynamics in a steppe ecosystem in Inner Mongolia, China. J Soils Sediments 11: 562–576.

31. Pathak H, Li C, Wassmann R (2005) Greenhouse gas emissions from Indian rice fields : calibration and upscaling using the DNDC model. Climat Res: 113–123.

32. Li C, Salas W, DeAngelo B, Rose S (2006) Assessing alternatives for mitigating net greenhouse gas emissions and increasing yields from rice production in China over the next twenty years. J Environ Qual 35 (4): 1554–1565.

33. Zhang L, Yu D, Shi X, Weindorf DC, Zhao L, et al. (2009) Simulation of global warming potential (GWP) from rice fields in the Tai-Lake region, China by coupling 1: 50,000 soil database with DNDC model. Atmos Environ 43 (17): 2737–2746.

34. Zhang Y, Wang Y, Su S, Li C (2011) Quantifying methane emissions from rice paddies in Northeast China by integrating remote sensing mapping with a biogeochemical model. Biogeosciences 8 (5): 1225–1235.

35. Li C, Mosier A, Wassmann R, Cai Z, Zheng X, et al. (2004) Modeling greenhouse gas emissions from rice-based production systems: Sensitivity and upscaling. Glob Biogeochem Cycles 18 (1): 1–19.

36. Fumoto T, Kobayashi K, Li C, Yagi K, Hasegawa T (2008) Revising a process-based biogeochemistry model (DNDC) to simulate methane emission from rice paddy fields under various residue management and fertilizer regimes. Glob Chang Biol 14 (2): 382–402.

37. Ludwig B, Bergstermann A, Priesack E, Flessa H (2011) Modelling of crop yields and N_2O emissions from silty arable soils with differing tillage in two long-term experiments. Soil Till Res 112 (2): 114–121.

38. Shang Q, Yang X, Gao C, Wu P, Liu J, et al. (2011) Net annual global warming potential and greenhouse gas intensity in Chinese double rice-cropping systems: a 3-year field measurement in long-term fertilizer experiments. Glob Chang Biol 17 (6): 2196–2210.

39. Wang L, Qiu J, Tang H, Li H, Li C, et al. (2008) Modelling soil organic carbon dynamics in the major agricultural regions of China. Geoderma 147 (1–2): 47–55.

40. Pan G, Li L, Wu L, Zhang X (2004) Storage and sequestration potential of topsoil organic carbon in China's paddy soils. Glob Chang Biol 10: 79–92.

41. Rui W, Zhang W (2010) Effect size and duration of recommended management practices on carbon sequestration in paddy field in Yangtze Delta Plain of China: A meta-analysis. Agric Ecosyst Environ 135(3): 199–205.

42. Zhang W, Xu M, Wang X, Huang Q, Nie J, et al. (2012) Effects of organic amendments on soil carbon sequestration in paddy fields of subtropical China. J Soils Sediments 12: 457–470.

43. RGCST (Research Group on Chinese Soil Taxonomy (Institute of Soil Science, Chinese Academy of Sciences), Cooperative Research Group on Chinese Soil Taxonomy) (2001) Chinese Soil Taxonomy. Science Press, Beijing, New York. 1–203.

44. Wang J, Jia J, Xiong Z, Khalil MAK, Xing G (2011) Water regime–nitrogen fertilizer–straw incorporation interaction: Field study on nitrous oxide emissions from a rice agroecosystem in Nanjing, China. Agric Ecosyst Environ 141: 437–446.

45. Wang J, Pan X, Liu Y, Zhang X, Xiong Z (2012) Effects of biochar amendment in two soils on greenhouse gas emissions and crop production. Plant Soil. Doi:10.1007/s11104-012-12503.

46. Li C, Zhuang Y, Frolking S, Galloway J, Harriss R, et al. (2003) Modeling soil organic carbon change in croplands of China. Ecol Appl 13 (2): 327–336.

47. Intergovernmental Panel on Climate Change (IPCC) (2007) Changes in atmospheric constituents and in radiative forcing. In: Solomon S, Qin D, Manning M, et al (editors) Climate Change 2007: The Physical Science Basis, Contribution of Working Group I to the FourthAssessment Report of the Intergovernmental Panel on Climate Change. Cambridge, United Kingdom and New York, NY, USA, Cambridge University Press.

48. Jia J, Ma Y, Xiong Z (2012) Net ecosystem carbon budget, net global warming potential and greenhouse gas intensity in intensive vegetable ecosystems in China. Agric Ecosyst Environ 150(15): 27–37.

49. Mosier A, Halvorson A, Reule C, Liu X (2006) Net global warming potential and greenhouse gas intensity in irrigated cropping systems in northeastern Colorado. J Environ Qual 35 (4): 1584–1598.

50. Six J, Feller C, Denef K, Ogle SM, de Moraes Sa JC, et al. (2002) Soil organic matter, biota and aggregation in temperate and tropical soils - Effects of no-tillage. Agronomie 22: 755–775.

51. Cai Z, Shan Y, Xu H (2007) Effects of nitrogen fertilizer on CH_4 emissions from rice paddies. Soil Sci Plant Nutr 53: 353–361.

52. Rochette P (2008) No-till only increases N_2O emissions in poorly-aerated soils. Soil Till Res 101: 97–100.

53. Tang H, Qiu J, Van Ranst E, Li C (2006) Estimations of soil organic carbon storage in cropland of China based on DNDC model. Geoderma 134: 200–206.

54. Ju X, Xing G, Chen X, Zhang S, Zhang L, et al. (2009) Reducing environmental risk by improving N management in intensive Chinese agricultural system. Proc Nat Acad Sci 106: 3041–3046.

9

Effects of Different Regeneration Scenarios and Fertilizer Treatments on Soil Microbial Ecology in Reclaimed Opencast Mining Areas on the Loess Plateau, China

Junjian Li[1,2], Yuanming Zheng[2], Junxia Yan[1], Hongjian Li[1], Xiang Wang[3], Jizheng He[2]*, Guangwei Ding[4]

1 Institute of Loess Plateau, Shanxi University, Taiyuan Shanxi, China, 2 State Key Laboratory of Urban and Regional Ecology, Research Centre for Eco-environmental Sciences, Chinese Academy of Sciences, Beijing, China, 3 Biology Institute of Shanxi, Taiyuan Shanxi, China, 4 Chemistry Department, Northern State University, Aberdeen, South Dakota, United States of America

Abstract

The soil microbial community in reclaimed mining areas is fundamental to vegetative establishment. However, how this community responds to different regeneration scenarios and fertilizer treatments is poorly understood. This research evaluated plant and soil microbial communities from different regeneration scenarios and different fertilizer treatments. Regeneration scenarios significantly influenced soil bacterial, archaeal, and fungal rDNA abundance. The ratios of fungi to bacteria or archaea were increased with fertilizer application. The diversity of both plants and microbes was lowest in *Lotus corniculatus* grasslands. Regeneration scenario, fertilizer treatment, and their interaction influenced soil microbial richness, diversity and evenness indices. Labile carbon pool 2 was a significant factor affected plant and microbe communities in July, suggesting that plants and microbes may be competing for nutrients. The higher ratios of positive to negative association were found in soil bacteria and total microbe than in archaea and fungi. Stronger clustering of microbial communities from the same regeneration scenario indicated that the vegetative composition of regeneration site may have a greater influence on soil microbial communities than fertilizer treatment.

Editor: Vishal Shah, Dowling College, United States of America

Funding: This work was financially supported by the Natural Science Foundation of China (41271530) and the National Science and Technology Program (2012BAC10B04 and 2008BAD95B04). The funders had no role in study design, data collection and analysis, decision to publish, or preparation of the manuscript.

Competing Interests: The authors have declared that no competing interests exist.

* E-mail: jzhe@rcees.ac.cn

Introduction

The considerable growth of the mining industry in China over the last few decades has generated a vast amount of solid waste, which occupies a huge area of land [1]. In 2006, regulations were initiated to reclaim abandoned coal mining areas for agriculture and forestry in Shanxi Province [1,2]. It would be useful to have indicators to assess the effectiveness of reclamation treatments. Traditionally, the criteria for judging the effectiveness of such treatments largely involve vegetation coverage and diversity, soil erosion and physicochemical characteristics [3]. In contrast, soil microbial ecology is not well understood, and is important for establishment vegetation, soil formation and transformation of nutrients, especially during the initial stages of reclamation [4,5].

The majority on studies of the effects of mine reclamation on microorganisms have focused on specific fungal groups (particularly mycorrhizae) [6,7], and microbial biomass and activity [5,8–10]. Studies have demonstrated that mycorrhizae can promote plant growth [11], improve soil structure [12] and maintain plant biodiversity and ecosystem stability [13]. Microbial activity has been assessed by measuring enzyme activity or metabolic quotient [14,15]. Soil microbial biomass and activity have been found to be lower in reclaimed mining areas compared to undisturbed sites [3,14]; and many investigations support the hypothesis that soil microbial biomass and activity increase as reclamation progresses [3,9,10]. However, the increasing trend was found in only recently reclaimed areas [15]. In addition, Mummey et al. [4] found that the total bacterial and fungal biomasses showed opposite trends as reclamation progressed.

Recently, genetic profiling methods have produced more information on soil microbial ecology compared to the cultivation of isolated microbes [16,17]. The genetic characteristics of the soil microbial community in forest, grassland, and farmland were reported in literature [18–21]. But, few reports focus on soil microbial communities in reclaimed mining areas [5,15]. Bacterial RISA (ribosomal RNA intergenic spacer analysis) fingerprinting has been used to demonstrate the effects of raw parent material properties on the microbial indices of reclaimed mine spoils [5]. Both bacterial and fungal DGGE (denaturing gradient gel electrophoresis) profiles have clearly shown differences between reclaimed and natural sites, but could not discriminate between time since treatment or treatment types [15]. T-RFLP (terminal restriction fragment length polymorphism) analysis is a highly reproducible genetic profiling method that has also been proven to be very useful for describing differences and changes in soil microbial community structures [22–24].

Bacteria, archaea, and eukaryota form the three main domains of the phylogenic tree of life [25]. The diversity and ecological significance of archaea has received rather less attention compared to bacteria and fungi [26]. In this study, we investigated bacterial, archaeal and fungal communities in the initial development stage of post-mining rehabilitation in east central Lvliang Mountains, China. We investigated the effects of vegetation type and fertilizer treatment on microbial diversity and composition. We hypothesized that both the vegetation and application of fertilizer would produce significant effects on soil microbial community structure. Fertilizer was only applied once just prior to planting in our study, therefore, we further predict that vegetation type will have a larger impact on microbial properties than fertilizer treatment.

Materials and Methods

2.1. Study Sites

Research was conducted at Antaibao Mine in Northwest Plateau Loess (37°09.4′E; 111°31.1′N). The climate is terrestrial temperate, and the area experiences monsoons. Annual average precipitation is 480 to 510 mm, with rainfall occurring mainly in the summer. The annual average air temperature is about 10.1°C, with the lowest average minimum temperature in January (−5.6°C) and the highest average maximum temperature in July (23.7°C). The frost-free season ranges in length from 180 to 200 days. Undisturbed soil in the area is classified as Cumulic Anthrosol (WBR).

Ecological reconstruction was initiated on the abandoned land in 2009, and there were four regeneration scenarios including *Lotus corniculatus* (CO), *Medicago sativa* (SA) grasslands, *Pinus tabulaeformis* plantation (TA), and *Salix matsudana* -*Sabina chinensis* mixed forest (MF). **The grass seed was sow in rows (row space, 20 cm)**, and the seeding rates of *L. corniculatus* and *M. sativa* were 5 and

10 kg·hm^{-2}, respectively. 3-year-old *P. tabulaeformis* and *S. chinensis*, and 5-year-old *S.matsudana* seedling were planted 2 m apart. There were four fertilizer treatments in total: CK = no fertilizer added, IN = inorganic fertilizer (750 kg/hm^2; N: P$_2$O$_5$ = 18:12) added, OR = organic fertilizer (45 m^3/hm^2; N: 1.7%, Organic matter: 24.1%) added, and IO = combination of inorganic and organic fertilizer (375 kg/hm^2 inorganic fertilizer +22.5 m^3/hm^2 organic fertilizer) to soils.

We only investigated plant species in the field; we did not sample plant species. There were no endangered or protected species involed in this study. The location is Antaibao Mine Company that is state-owned enterprise. No specific permits were required for the described field studies.

2.2. Study Plots Survey and Soil Sampling

To study the characteristics of the various plant communities under different restoration types, quadrats were set up in the study areas. Quadrats of 20 m×20 m, and 1 m×1 m were established in forest and grassland communities, respectively. There were 3 replications for each of the 16 treatments, resulting in a total of 48 quadrates. Sampling occurred on two occasions, April and July 2011. The cover, height, diameter at breast height (DBH), individual number for each tree species, and the cover and height for herbs were recorded in each quadrat [1].

From each site, 6 soil cores (7.5 cm diameter×10 cm depth) were randomly collected in the middle of rows and mixed from the profile of each plot and bulked [5]. Subsamples for microbial

Table 1. Primers, probes and PCR conditions used for real-time PCR and T-RFLP.

Target group	Primer and probe	Sequence (5′–3′)	Reaction system	Thermal profile	Reference
Real-time PCR					
Bacteria	Primer Bact1369FB	CGGTGAATACGTTCYCGG	25 μl:12.5 μl *Premix Ex Taq*™, 1 μl BSA, 0.5 μl each primer and probe, 2 μl template, 8 μl H$_2$O	10 s at 95°C for initial denaturation; 35 cycles of 15 s at 95°C, 1 min at 56°C	[55]
	Primer Prok1492R	GGWTACCTTGTTACGACTT			
	Probe TM1389F	CTTGTACACACCGCCCGTC			
Archaea	Primer Ar364aF	CGGGGYGCASCAGGCGCGAA	25 μl: 12.5 μl of SYBR® *Premix Ex Taq*™, 1 μl BSA, 0.5 μl each primer, 2 μl template, 8.5 μl H$_2$O	30 s at 94°C for initial denaturation; 40 cycles of 20 s at 94°C, 30 s at 59°C, and 30 s at 72°C	[56]
	Primer Ar934b	GTGCTCCCCCGCCAATTCCT			
Fungi	Primer NS1	GTAGTCATATGCTTGTCC	25 μl: 12.5 μl of SYBR® *Premix Ex Taq*™, 1 μl BSA, 0.5 μl each primer, 2 μl template, 8.5 μl H$_2$O	3 min at 95°C for initial denaturation; 40 cycles of 10 s at 95°C, 30 s at 55°C, and 1 min at 72°C	[57]
	Primer FUNG	CATTCCCCGTTACCCGTTG			
PCR for T-RFLP					
Bacteria	Primer 27F-FAM	GAGTTTGATCCTGGCTCAG	50 μl: 5 μl 10×PCR buffer (MgCl$_2$, 2 mM), 4 μl 2.5 mM dNTPs, 0.5 μl EX-Taq polymerase (5 U μl^{-1}), 1 μl each primer, 1 μl BSA, 4 μl template, 33.5 μl H$_2$O(Reaction system for bacteria, archaea and fungi)	5 min at 94°C; 35 cycles of 45 s at 94°C, 45 s at 54°C, 72°C for 90 s; 10 min at 72°C.	[58]
	Primer 1492R	ACGGCTACCTTGTTACGACT			
Archaea	Primer Ar364aF	CGGGGYGCASCAGGCGCGAA		35 cycles of 45 s at 94°C, 45 s at 58°C, 60 s at 72°C; 10 min at 72°C.	[56]
	Primer Ar934b-FAM	GTGCTCCCCCGCCAATTCCT			
Fungi	Primer NS1-FAM	GTAGTCATATGCTTGTCC		5 min at 94°C; 35 cycles of 30 s at 94°C, 30 s at 56°C, 60 s at 72°C; 10 min at 72°C.	[57]
	Primer FUNG	CATTCCCCGTTACCCGTTG			

Table 2. Soil pH, bulk density, organic carbon, nitrogen, labile carbon pool 1 and 2 (LC1 and LC2), and recalcitrant pool carbon (RC) in the reclaimed mining area in July.

	pH	Bulk density	Organic carbon	Nitrogen	LC1	LC2	RC
CO-CK	7.96±0.01 bA	1.20±0.04aB	2.61±0.13aB	0.33±0.01aA	1.15±0.19aB	0.36±0.25aa	1.10±0.09abAB
CO-IN	7.87±0.03 aA	1.23±0.04aA	2.49±0.23aAB	0.37±0.02aB	1.47±0.24aB	0.42±0.16aa	0.60±0.17aA
CO-IO	7.92±0.04 abA	1.25±0.02aaA	2.18±0.40aB	0.34±0.01aB	1.15±0.18aB	0.15±0.09aa	0.88±0.45aAB
CO-OR	7.89±0.02 aA	1.16±0.08aA	3.22±0.19bC	0.43±0.02bB	1.28±0.29aB	0.47±0.24aa	1.47±0.30bA
SA-CK	8.01±0.06 aA	1.01±0.05aA	2.29±0.28abB	0.35±0.01aA	0.58±0.19aA	0.44±0.09aA	1.28±0.31bB
SA-IN	7.93±0.04 aAB	1.14±0.11abA	2.14±1.06abAB	0.29±0.01aA	0.70±0.10aA	0.35±0.05aA	1.69±0.05cC
SA-IO	8.05±0.17 aA	1.32±0.09cA	1.29±0.03aA	0.25±0.01aA	0.45±0.06aA	0.31±0.10aAB	0.54±0.05aA
SA-OR	8.06±0.28 aA	1.26±0.05bcAB	2.61±0.23bAB	0.46±0.04bB	0.67±0.10aA	0.49±0.09aA	1.45±0.24bcA
TA-CK	7.95±0.01 bA	1.35±0.04aC	1.68±0.23aA	0.34±0.04aA	0.69±0.19aA	0.27±0.09aA	0.72±0.18aA
TA-IN	7.87±0.06 aA	1.27±0.02aA	1.68±0.13aA	0.36±0.03aB	0.50±0.19aA	0.54±0.16aA	0.64±0.20aA
TA-IO	7.88±0.01 aA	1.33±0.17aA	1.86±0.08abB	0.35±0.02aB	0.70±0.19aA	0.50±0.18aB	0.67±0.15aA
TA-OR	7.91±0.02 abA	1.34±0.06aB	2.21±0.27bA	0.36±0.04aA	0.77±0.11aA	0.48±0.09aA	0.95±0.39bA
MF-CK	8.24±0.11 bB	1.17±0.07aB	2.73±0.34aB	0.33±0.00aA	0.72±0.13aB	0.89±0.16aB	1.11±0.41aAB
MF-IN	7.97±0.03 aB	1.29±0.18aA	3.09±0.27abB	0.52±0.03bC	0.84±0.15aA	1.11±0.10aB	1.14±0.36aB
MF-IO	7.91±0.10 aA	1.26±0.06aA	3.56±0.15bC	0.36±0.02aB	1.07±0.11aB	1.15±0.23aC	1.34±0.23aB
MF-OR	8.03±0.04 aA	1.27±0.08aAB	3.10±0.42abB	0.31±0.02aA	0.94±0.11aA	0.95±0.10aB	1.21±0.44aA

Data are means ± standard deviations. Treatments with the same lower case letters are not significantly different from one another for the same regeneration scenario ($P>0.05$). Treatments with the same capital letters are not significantly different from one another for the same fertilizer treatment ($P>0.05$).
CO, SA, TA and MF respectively represent Lotus corniculatus, Medicago sativa, Pinus tabulaeformis and Salix matsudana–Sabina chinensis mixed forest.
CK, IN, IO and OR respectively represent no, inorganic, organic and a combination of inorganic and organic fertilizer added to soils.

analysis were stored at 4°C until DNA extraction (<2 weeks). Subsamples for other analyses were air dried. All soil samples were sieved to pass 2 mm.

2.3. Soil Chemical Analysis

Values (g cm^{-3}) for soil bulk density (BD) were obtained using the gravimetric method. Soil pH in dH$_2$O was measured in subsets of field-moist soils at a soil:solution ratio of 1:5 (g:ml) after 0.5–1 h. Soil organic carbon (SOC) was determined using the dichromate oxidation method. Total nitrogen (TN) was analyzed by the Kjeldahl method [27]. The labile and recalcitrant carbon pools were quantified by the two-step acid hydrolysis procedure with HS$_2$O$_4$ [28,29].

2.4. DNA Extraction

DNA was extracted from 0.5 g fresh soil samples using Ultra-clean TM soil DNA Isolation Kits (MoBio Laboratory, USA) following the manufacturer's protocol. The extracted DNA was eluted with **5** ml of solution S5 (MoBio Laboratories, cat. no. 12800-100) and stored at −20°C. The DNA extracts were 10-fold diluted and used as template with a final content of 1–10 ng in each reaction mixture to amplify soil bacterial, archaeal, and fungal rRNA genes.

2.5. PCR Assay for Real-time Quantification and T-RFLP

Real-time PCR was performed on an iCycler iQ 5 thermocycler (Bio-Rad). The probe TM189F and *Premix Ex Taq*^{TM} (Takara Biotechnology, Japan) were applied into bacterial 16S rRNA gene quantitative assay. The reaction mixture for quantifying archaeal and fungal rRNA gene included SYBR® *Premix Ex Taq*^{TM} with Green I (Takara Biotechnology, Japan). Primers were labeled at the 5′ end with the reporter dye FAM (6-carboxy-fluorescein) for

T-RFLP analysis. The detailed information on primer, probe, and PCR condition are listed in table 1.

2.6. T-RFLP Analysis

All PCR products were verified using 1% agarose gel electrophoresis and purified with Wizard® SV Gel and PCR Clean-Up System (Promega, USA). Purified products were digested in separate reactions with restriction endonucleases *Hha*I, *Hae*III and *Msp*I (Takara Biotechnology, Japan) and incubated at 37°C for 3 h in the manufacturer's recommended reaction buffer. Digestions were in a total volume of 25 µl, including 4 U of enzyme and about 500 ng of DNA. The digestion products were further purified, and a portion was mixed with deionized formamide and the internal standard GeneScan-ROX1000 (bacteria)/LIZ 500 (archaea and fungi) (Applied Biosystems). The mixtures were denatured for 3 min at 95°C, and the DNA fragments were size separated using a 3130xl Genetic Analyzer (Applied Biosystems).

2.7. Statistical Analysis

The effects of regeneration scenario and fertilizer treatment on plant and soil microbe characteristics were examined with separately one-way analyses of variance (ANOVA). A three-way ANOVA was applied to analyses the effects of season, regeneration scenario and fertilizer treatment on soil microbial communities. Plant species-pairs and microbial RF (Restricted fragment)-pairs association were analyzed by Spearman rank correlation. These statistical analyses were performed using SPSS 13.0 for Windows.

Plant species importance values (IV) for each quadrat were calculated using the following formulas [2]:

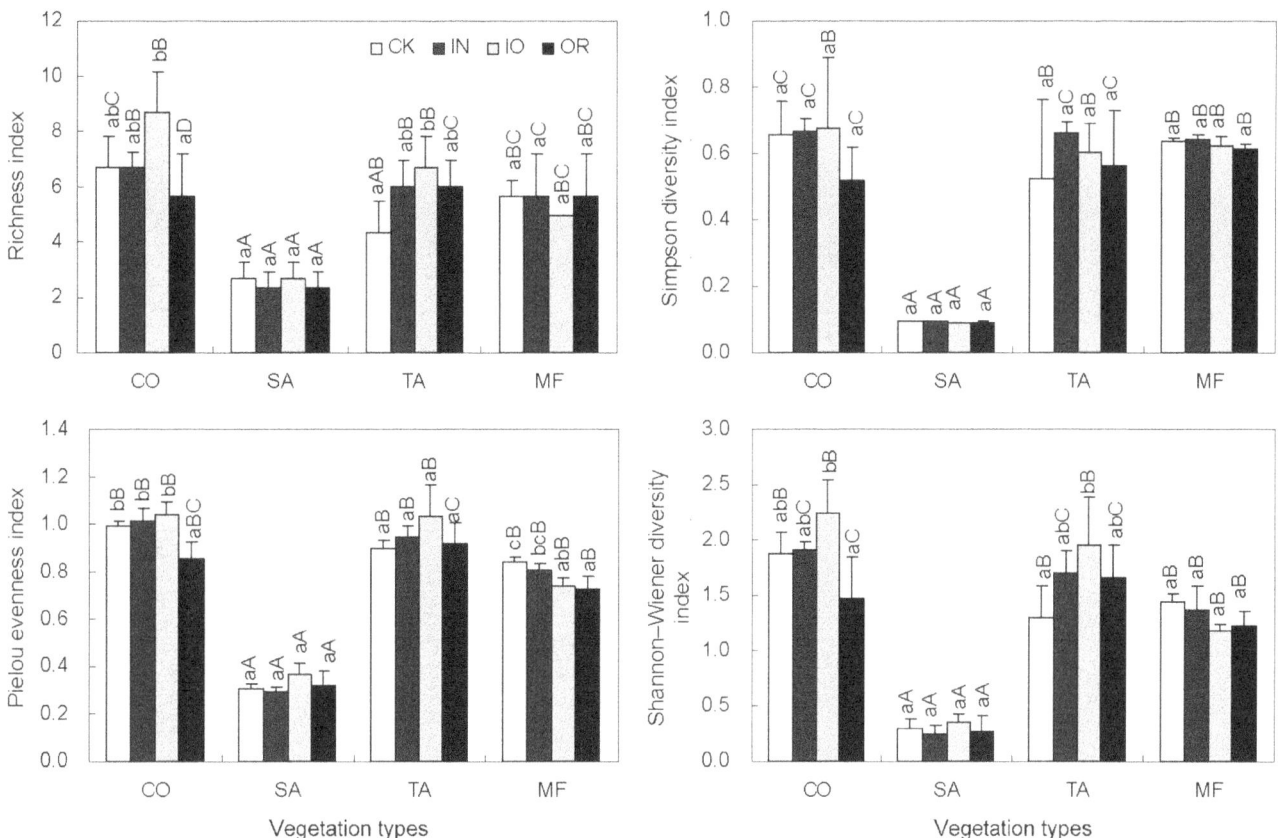

Figure 1. Plant species richness index, Shannon-Wiener diversity index, Simpson diversity index, and Pielou evenness index in the reclaimed mining area. Points show the means of three replicates, and vertical bars show standard deviations. Treatments with the same lower case letters are not significantly different from one another for the same regeneration scenario (P>0.05). Treatments with the same capital letters are not significantly different from one another for the same fertilizer treatment (P>0.05). CO, SA, TA, and MF represent *Lotus corniculatus*, *Medicago sativa*, *Pinus tabulaeformis*, and *Salix matsudana–Sabina chinensis* mixed forest. CK, IN, IO, and OR represent no fertilizer, inorganic, organic, and a combination of inorganic and organic fertilizer added to soils.

$$IV_{tree} = (\text{relative density} + \text{relative frequency} + \text{relative dominance}) \div 3 \quad (1)$$

$$IV_{shrub\ and\ herb} = (\text{relative coverage} + \text{relative height}) \div 2 \quad (2)$$

$$IV_{plant\ community} = (5 \times IV_{tree} + 2.5 \times IV_{shrub} + 2.5 \times IV_{grass}) \div 10 \quad (3)$$

To reveal their variation among different restoration types, plant species, and RF richness, diversity and evenness in each quadrat were determined. Four species diversity indices were employed:

(a) Species number as the richness index (S)

(b) Shannon–Wiener diversity index:

$$H' = - \sum P_i \ln P_i \quad (4)$$

(c) Simpson diversity index:

$$D = 1 - \sum P_i^2 \quad (5)$$

(d) Pielou evenness index:

$$E = H'/\ln S \quad (6)$$

S is the number of plant species. P_i is the relative importance value of plant species i. $Pi = IVi/IV$; IV_i and IV are the importance value of species i and all plant species in a quadrate, respectively [2]. For soil microbial community, S is the number of RF and P_i is the relative ratio of RF i [30].

A matrix of IVs for plant and matrices of RFs ratios for bacteria, archaea, fungi and total microbes were used as the basis of the community analysis. According to quantification PCR, the ratios of logarithm of the bacterial, archaeal and fungal rRNA gene copies approximate to 4:3:3. $Matrix_{(totalmicrobe)} = (4 \times ratio_{(RF\text{-}Bacteria)} + 3 \times ratio_{(RF\text{-}Archaea)} + 3 \times ratio_{(RFs\text{-}Fungi)}) \div 10$. A matrix of environmental values was also established, which were used to analyze the relationships between samples and environmental factors using canonical correspondence analysis (CCA). These calculations were

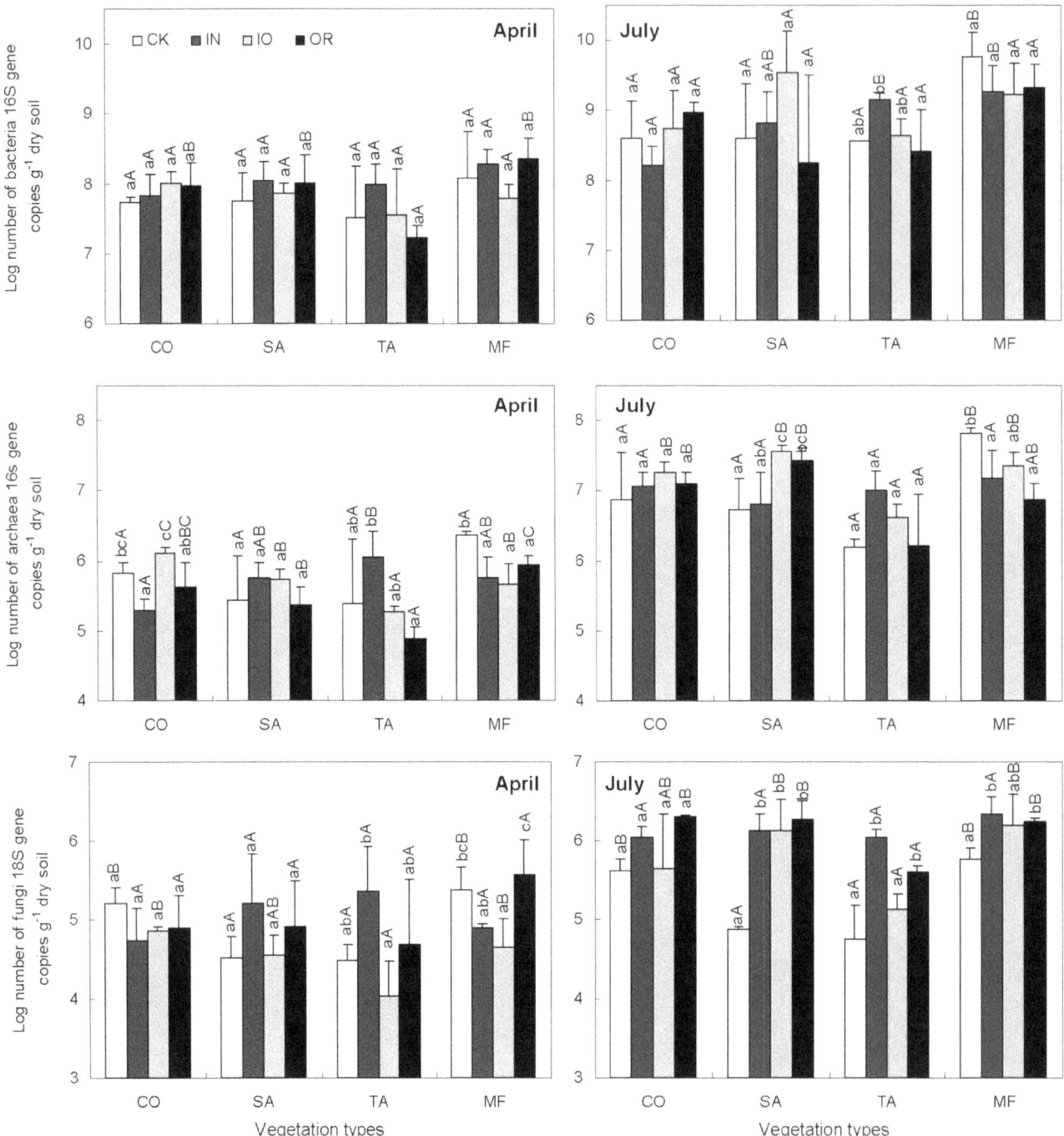

Figure 2. Abundance of soil bacteria, archaea, and fungi rRNA genes in the reclaimed mining area. Points show the means of three replicates, and vertical bars show standard deviations. Treatments with the same lower case letters are not significantly different from one another for the same regeneration scenario (P>0.05). Treatments with the same capital letters are not significantly different from one another for the same fertilizer treatment (P>0.05). CO, SA, TA, and MF represent *Lotus corniculatus, Medicago sativa, Pinus tabulaeformis,* and *Salix matsudana–Sabina chinensis* mixed forest. CK, IN, IO, and OR represent no fertilizer, inorganic, organic, and a combination of inorganic and organic fertilizer added to soils.

carried out, using the CANOCO 4.5 [31], and the plant and microbial data were logarithmic transformed for the analysis. The significance level (P<0.05) between species and environmental data was used the Monte Carlo permutation test (Number of permutation is 499), and the significant environmental factors were applied into final CCA.

Results

3.1 Soil Chemical Properties

Fertilizer treatments had significant effects (P<0.05) on soil pH in all vegetation types except SA grassland (Table 2). Soil pH in mixed forest was higher than in other regeneration scenarios under both CK and IN treatments. However, no significant

Table 3. Results of three-way ANOVA showing the effects of season, regeneration scenarios and fertilizer treatments for soil bacteria, archaea, fungi and total microbes RFs richness index (*S*), Shannon-Wiener diversity index (*H'*), Simpson diversity index (*D*) and Pielou evenness index (*E*).

	Bacteria				Archaea				Fungi				Total microbes			
	S	*H'*	*D*	*E*	*S*	*H'*	*D*	*E*	*S*	*H'*	*D*	*E*	*S*	*H'*	*D*	*E*
Season	ns	ns	**	ns	**	**	**	**	**	*	ns	**	**	**	**	**
Scenarios	**	**	**	**	**	**	**	**	**	**	ns	ns	ns	ns	ns	ns
Fertilizer	ns	*	**	*	**	**	**	ns	*	ns	ns	ns	**	*	ns	ns
Season × Scenarios	ns	ns	ns	*	*	**	**	**	**	**	ns	*	*	ns	ns	ns
Season × Fertilizer	ns	ns	**	*	*	**	**	**	ns	ns	ns	*	ns	*	*	**
Scenarios × Fertilizer	ns	ns	*	ns	ns	*	*	**	*	*	**	*	*	ns	ns	ns
Season × Scenarios × Fertilizer	ns	ns	ns	ns	ns	**	**	**	**	ns	*	*	*	ns	ns	ns

**Effect is significant at the 0.01 level;
*Effect is significant at the 0.05 level. ns, effect is not significant.

differences (P>0.05) were found between scenarios treated with IO or OR fertilizer. Fertilizer treatments significantly (P<0.05) increased soil organic carbon in SA grassland and less significant effects of fertilizer treatments on soil organic carbon were demonstrated in other regeneration scenarios. In most cases, there were significantly (P<0.05) higher levels of soil organic carbon in CO and MF sites with the same fertilizer treatment. OR fertilizer significantly improved soil nitrogen content in the two grasslands and only inorganic fertilizer affected soil nitrogen in the

mixed forests, but there were no significant differences among fertilizer treatments in the TA plantation. No significant differences in soil nitrogen were found between regeneration scenarios that did not have any fertilizer treatment. Under IN and IO fertilizer treatments, soil nitrogen from SA grassland was significantly (P<0.05) lower than that of the other vegetation types. Fertilizer did not produce significant effects (P<0.05) on labile carbon pools 1 and 2 (LC1 and LC2). Under the same fertilizer treatment, a significantly higher LC1 was found in CO grassland, and a significantly higher LC2 was found in MF. Different regeneration scenarios and fertilizer treatments both significantly (P<0.05) influenced soil recalcitrant carbon levels.

3.2 Plant Community Composition and Diversity

Plant community composition and the coverage of vegetation were investigated (Table S1). The vegetation coverage was lower in CO grassland than the other regeneration scenarios. IO fertilizer treatments significantly (P<0.05) improved the plant species richness and Shannon-Wiener diversity in CO and TA sites, but no significant effects were found in other regeneration scenarios (Figure 1). There were no significant (P>0.05) differences in the Simpson diversity index among fertilizer treatments. With the exception of SA grassland, fertilizer significantly (P<0.05) influenced the Pielou evenness index, although the effect was not consistent. Plant diversity indices from SA grassland were significantly lower than other sites treated with the same fertilizer.

3.3 Soil Bacterial, Archaeal, and Fungal Abundance

The seasonal effects on the abundance of soil microbial rRNA genes were significant (Figure 2). Fertilizer treatments produced no significant effects on soil bacterial 16S rRNA gene copy number, with the exception of the TA plantation in July. Either, there were no pronounced effects of fertilizer treatments on soil archaeal and fungal rRNA gene abundance in most cases. Different regeneration scenarios had a more obvious influence on the microbial rRNA gene abundance than fertilizer treatments, especially in archaea. In addition, the differences were more visible between regeneration scenarios in July than in April. In July, the log ratios of bacteria: fungi and archaea: fungi under SA and MF were lower with fertilizer treatments (Figure S1), which indicated that fertilizers show more positive effects on the growth of fungi than bacteria and archaea.

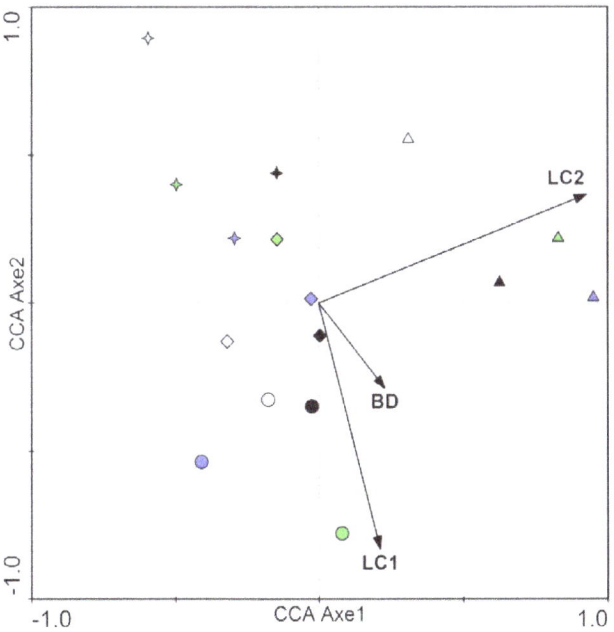

Figure 3. CCA ordination biplot of 16 quadrats and environmental factors for plant communities in the reclaimed mining area. Arrows indicate the direction and magnitude of measurable variables associated with plant community structures. Circle, Star, Diamond and Up-triangle symbol types respectively represent *Lotus corniculatus, Medicago sativa, Pinus tabulaeformis* and *Salix matsudana–Sabina chinensis* mixed forest. White, green, blue and black respectively represent no, inorganic, organic and a combination of inorganic and organic fertilizer added to soils. BD, LC1, and LC2 are bulk density, soil labile pool carbon 1 and 2, respectively.

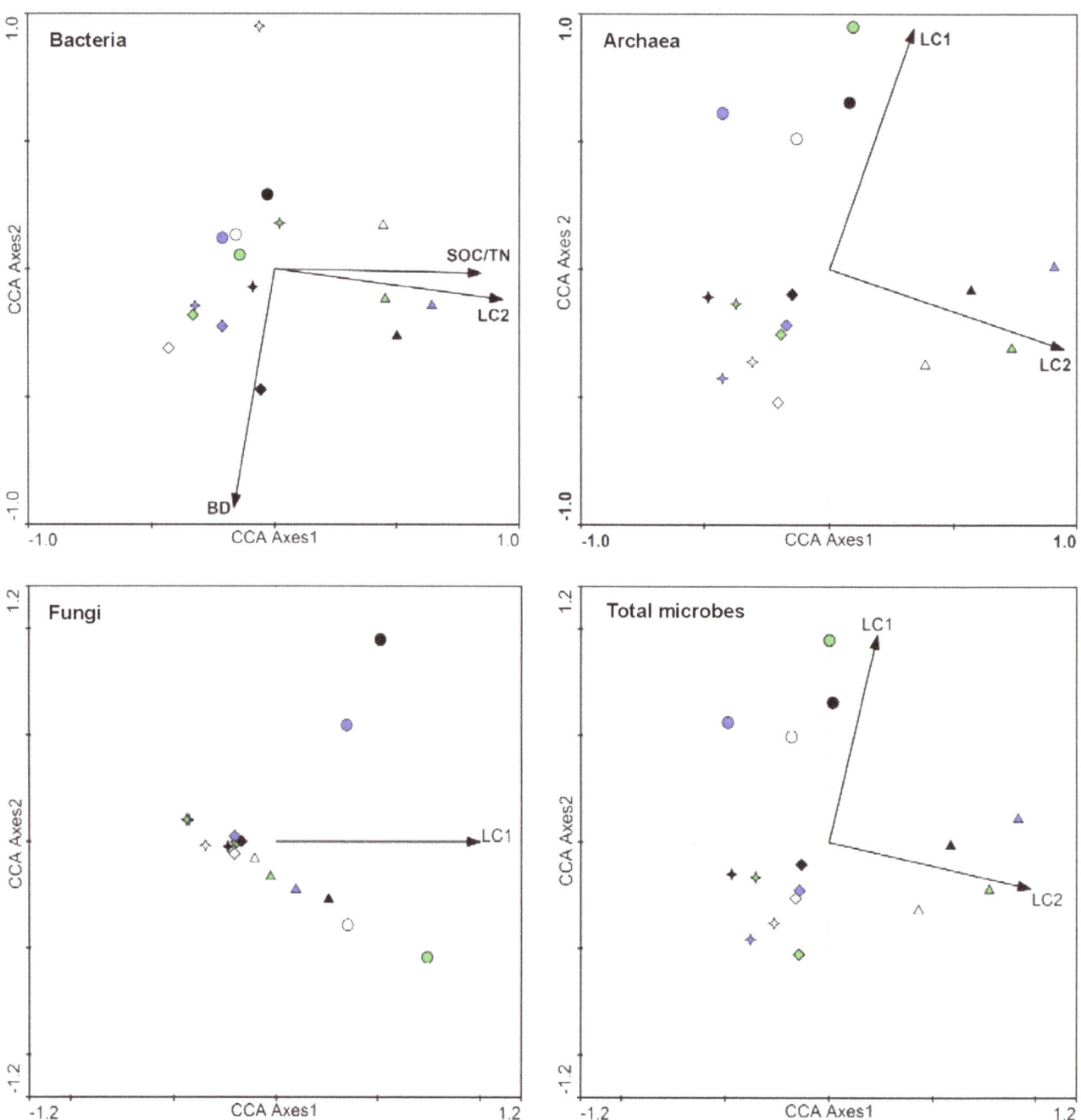

Figure 4. CCA ordination biplot of 16 quadrats and environmental factors for soil bacterial, archaeal, fungal, and total microbial communities in the reclaimed mining area in July. Arrows indicate the direction and magnitude of measurable variables associated with soil microbial communities structures. Circle, Star, Diamond and Up-triangle symbol types respectively represent *Lotus corniculatus*, *Medicago sativa*, *Pinus tabulaeformis* and *Salix matsudana–Sabina chinensis* mixed forest. White, green, blue and black respectively represent no, inorganic, organic and a combination of inorganic and organic fertilizer added to soils. BD, SOC/TN, LC1, and LC2 are the soil bulk density, ratio of soil organic carbon to total nitrogen, labile carbon pool 1 and 2, respectively.

3.4 Soil Microbial T-RFLP Profiles and Diversity

T-RFLP profiles of soil bacterial, archaeal, and fungal rRNA genes from study sites were produced using the endonuclease enzymes *Hha*I, *Hae*III and *Msp*I. The effects of sampling date, regeneration scenarios, and fertilizer treatments on microbial diversity indices produced by *Hha*I are listed in Table 3.

When soil bacteria 16S rRNA genes were restricted using *Hha*I, the 77 bp RF had the highest relative abundance and there were

RFs unique to each regeneration scenario. In T-RFLP profiles of soil archaeal 16S rRNA restricted using *Hha*I, the 320 bp RF had the highest relative abundance at all sites. Fragments of 136 and 141 bp were found in regeneration areas where trees grew, but not in grasslands; and the fragments lower than 88 bp were only observed in fertilizer treatments. Both regeneration scenario and fertilizer treatment demonstrated an influence on T-RFLP profiles of soil fungi 18S rRNA restricted using *Hha*I enzyme (Figure S2).

Soil bacterial and total microbial diversity indices inferred from fragments restricted using *Hha*I and *Msp*I were more liable to be influenced by sampling date, regeneration scenarios, fertilizer treatments and their interaction, than those restricted using *Hae*III(Table 3 and Table S2). Soil archaeal diversity indices from RFs produced using *Hha*I appeared to be more sensitive to these factors. Most of the diversity indices for fungal fragments restricted using *Msp*I were not significantly influenced by vegetation type or fertilizer treatment. Sampling date was the most significant factor affecting microbial diversity, and followed by regeneration scenario, then fertilizer treatment. The Pielou evenness index (E) and richness index (S) were the most and least susceptible for these factors, respectively.

3.5 Spearman Rank Correlation Test for Inter-species Correlation

Plant species-pair and microbial inter-RF ratios of positive and negative association of the Spearman rank correlation test are listed in Table S3. Plant species-pair ratios at the CO and TA sites were higher than in the other two treatments. Fertilizer treatments improved both plant and microbe ratios. Soil microbe inter-RFs ratios differed with regeneration scenarios. Inter-RF ratios of total soil microbes were the highest, and decreased in order from bacteria to fungi to archaea.

3.6 Effect of Environmental Factors on Plant and Soil Microbial Community Composition and Species Distribution

The results of CCA showed that the eigenvalues of axes 1 and 2 was respectively 62.3% and 44.0%, and the first two axes explained 64.7% of species–environment relations (i.e. Plant-environmental data). Liable carbon pool and bulk density were significantly linked to the plant community variability ($P<0.05$), and positively correlated with the first axes (Figure 3).

The relationships between environmental factors and soil microbial communities, according to T-RFLP profiles restricted using *Hha*I, were shown by the CCA method (Figure S3, Figure 4). In April, the significant environmental factors that affected bacterial, archaeal, fungal, and total microbial communities were pH, ratio of soil organic carbon to nitrogen, bulk density and LC2 (Figure S3). Soil pH was positively correlated with the first axes in bacterial and total microbial communities; however, negative effects were shown in fugal communities. Bulk density was significant environmental factors for archaeal and fungal communities, and positively correlated with the second axes. In July, LC1 was positively correlated with the first two axes, but the contrary correlations were demonstrated between LC2 and the first two axes (Figure 4). Treatments from the same regeneration scenario tended to cluster together more than those from the same fertilization treatment. Similar results were found when CCA was applied to test relationships between environmental factors and microbial RFs restricted using *Hae*III and *Msp*I enzymes (results not shown).

Discussion

The growth of *L. corniculatus* demands high fertility soil, and we observed that fertilizer significantly improved plant coverage in CO grasslands. *M. sativa* has been widely planted in reclamation sites in the Loess Plateau due to its strong resistance to drought and barren soil [32]. In our study, *M. sativa* grew well, the coverage being over 90% with the available space almost used; other species, therefore, could not easily invade and plant richness, diversity and evenness was significantly lower in these grasslands

(Figure 1). Possibly, due to the lower coverage in CO grasslands, plant diversity was higher as invasive species was able to establish easily. Inter-plant spacing provided gaps for shrub growth so plant diversity was also higher in TA and MF plantations.

A higher growth rate of vegetation results in more soil nutrients being consumed. Meanwhile, litter and dead roots are not degraded during the initial reclamation period. We observed higher coverage and above-ground biomass in SA than CO grasslands, and levels of soil organic carbon, total nitrogen, and labile carbon pool were lower in SA than CO grasslands (Table 2). Declining trends have been recorded in soil organic carbon, nitrogen, and microbial biomass in fast-growing poplar and *Eucalyptus* plantations [33–35].

In July, the different regeneration scenarios in our study demonstrated more obvious influences on soil microbial rDNA copy numbers (Figure 2). In addition, soil microbial RF richness and diversity were lowest in the SA grasslands where the richness and diversity of above-ground vegetation were also the lowest compared to other regenertion scenarios (Figure1). These results imply that these microbial communities are closely related to plant colonization. Previous studies have also supported the hypothesis that vegetation produces significant effects on the structuring of microbial communities during the pioneer stage of ecosystem development [36,37]. The relationship between microbial reproduction and plant colonization involves competition for nutrients primarily during short-term rehabilitation programs [38]. However, nutritive symbiosis was chiefly for long-term rehabilitation potentially between regeneration scenarios and soil microbial communities via the chemical composition, quantities and botanical forms of different plant residues [39,40].

In this study, fertilizer treatments were beneficial for vegetation establishment, particularly in the CO grasslands. Organic fertilizer produced significant ($P<0.05$) increases in soil organic and recalcitrant carbon levels and nitrogen content (Table 2). Fertilizers provide different substrates for soil microbial communities [41–43]. There were no significant differences between the three domains of soil microbial rDNA copy numbers (Figure 2). These results differ from previous studies where significant increases in soil microbial biomass due to fertilizer have been reported during the reclamation of barren soils or agricultural fields [44–46]. The difference may be explained by the fact that only a single application of fertilizer was in our study, whereas long-term application of fertilizers was used in other studies.

Soil pH was significantly ($P<0.05$) decreased as a result of fertilizer treatments, especially in the mixed forest (Table 2). However, the application rate of fertilizer used here did not produce significant detrimental effects on plant or microbial growth. Significant plant growth in part may have contributed to a decrease in pH via litter inputs, organic exudation, and proton extrusion [29,47]. The much lower pH from sites treated with fertilizer could also be explained by the influence of fertilizer on microbial community structure, e.g., lower pH levels may improve the competitive advantage of soil fungi that are more suited to weak acid conditions compared to bacteria and archaea. Lower ratios of bacteria and archaea to fungi were found, which was consistent with other studies [48]. In addition, soil pH influenced soil bacterial and fungal composition across samples [37,49] and was a significant environmental factor in determining soil bacterial and total microbial community composition in April (Figure S3).

Using a Monte Carlo permutation test in CCA, we showed that LC2 was a significant environmental factor affecting both plant and microbial communities in July. However, labile carbon was not the only significant environmental factor that affected soil microbial communities in April (Figure S3). The test also supported the view that there may be competition for nutrition between plants and microbes to meet their growth demands in infertile soil. In reclaimed coal mine soils, the development of microbial communities was shown to be stimulated by the presence of an easily available carbon source [9]. Previous studies have also shown that soluble carbon is a dominant influence on microbial communities in other conditions [50–52].

In our study, soil microbial diversity was more affected by the vegetative composition of the regeneration sites than by fertilizer treatments (Table 3). We also found microbial communities stronger clustering from similar regeneration scenarios than form similar fertilizer treatments (Figure 4, S3). These results suggest that regeneration scenarios have greater impacts on the microbial communities than fertilizer treatments. This is probably related to root exudates. In reclaimed post-mining sites near Sokolov (Czech Republic) and near Cottbus, Šourková et al. [14] reported that vegetation type played a more important role in the soil microbial community than substrate, which was mainly dependent on litter quality. We hypothesized that the degradation of litter would not be the primary substrate for soil microbial communities in 3-year-old reclaimed sites.

Higher ratios of positive to negative association from the Spearman rank correlation test indicate that biological community is more steady [53]. The low ratios listed in table S3 suggested that plant and soil microbe were still in the initial succession period. The higher ratios of bacteria suggested that bacteria would promote the succession of plant, archaea, and fungi, and play an important role in maintaining ecological stability. Our results are similar to Susyan et al. [54]. They reported that soil fungi clearly dominated the microbial communities later in the succession on abandoned arable soils, which suggested that the soil bacterial community prevailed during the initial period.

Conclusions

The results partly confirmed our hypothesis that soil microbial communities were significantly influenced by regeneration scenarios, however, fertilizer treatments produced less significant influence on soil microbial communities. Regeneration scenarios produced a significant effect on soil microbial rDNA copy numbers and microbial communities. Season showed pronounced effects on soil microbial growth and composition. There may be nutrient competition between vegetation and microbes for their growth during initial rehabilitation period, but mutual benefits would be demonstrated between vegetation and soil microbe with succession progressed in reclaimed mining areas of Shanxi, China.

Supporting Information

Figure S1 Ratios of soil bacterial, archaeal and fungal log numbers of rRNA gene copy number in the reclaimed mining area. Points show the means of three replicates, and vertical bars show standard deviations. Lower case letters indicate that the means are not significantly different among fertilizer treatments for the same regeneration scenario ($P<0.05$). Capital letters indicate that the means are not significantly different among regeneration scenarios for the same fertilizer treatment ($P<0.05$). CO, SA, TA and MF respectively represent *Lotus corniculatus, Medicago sativa, Pinus tabulaeformis* and *Salix matsudana–Sabina chinensis* mixed forest. CK, IN, IO and OR respectively represent no, inorganic, organic and a combination of inorganic and organic fertilizer added to soils.

Figure S2 Relative fluorescence of soil bacteria, archaea and fungi populations measured by T-RFLP electropherogram target on rRNA gene sequences digested using *Hha*I restriction enzymes in the reclaimed mining area. CO, SA, TA and MF respectively represent *Lotus corniculatus, Medicago sativa, Pinus tabulaeformis* and *Salix matsudana–Sabina chinensis* mixed forest. CK, IN, IO and OR respectively represent no, inorganic, organic and a combination of inorganic and organic fertilizer added to soils.

Figure S3 CCA ordination biplot of 16 quadrats and environmental factors for soil bacterial, archaeal, fungal, and total microbial communities in the reclaimed mining area in April. Arrows indicate the direction and magnitude of measurable variables associated with soil microbial communities structures. Circle, Star, Diamond and Up-triangle symbol types respectively represent *Lotus corniculatus, Medicago sativa, Pinus tabulaeformis* and *Salix matsudana–Sabina chinensis* mixed forest. White, green, blue and black respectively represent no, inorganic, organic and a combination of inorganic and organic fertilizer added to soils. BD, SOC/TN, and LC2 are the soil bulk density, ratio of soil organic carbon to total nitrogen, and labile carbon pool 2, respectively.

Table S1 Plant community composition and coverage in reclaimed mining area. CO, SA, TA and MF are respectively *Lotus corniculatus, Medicago sativa, Pinus tabulaeformis* and *Salix matsudana -Sabina chinensis* mixed forest. CK, IN, IO and OR are respectively no, inorganic, organic and combination of inorganic and organic fertilizer added to soils

Table S2 Results of three-way ANOVA showing the effects of season, regeneration scenarios and fertilizer treatments for Soil bacteria, archaea, fungi and total microbe RFs richness index (S), Shanon-Wiener diversity index (H'), Simpson diversity index (D) and Pielou evenness index (E). ** Effect is significant at the 0.01 level; * Effect is significant at the 0.05 level. ns Effect is not significant.

Table S3 Ratios of positive and negative association from Spearman rank correlation test of the inter-species correlation of plant and inter-RFs correlation of soil bacteria, archaea, fungi and total microbes in the reclaimed mining area. CO, SA, TA and MF represent *Lotus corniculatus, Medicago sativa, Pinus tabulaeformis* and *Salix matsudana–Sabina chinensis* mixed forest. CK, IN, IO and OR represent no, inorganic, organic and a combination of inorganic and organic fertilizer added to soils.

Acknowledgments

We are grateful to Professor J. Hughes for constructive comments and linguistic corrections. We thank Dr. L. Jiao for providing help in vegetation investigation.

Author Contributions

Conceived and designed the experiments: JL HL JH. Performed the experiments: JL JY. Analyzed the data: JL YZ. Contributed reagents/materials/analysis tools: XW YZ JH. Wrote the paper: JL JH GD.

References

1. Zhang J (2005) Succession analysis of plant communities in abandoned croplands in the Eastern Loess Plateau of China. J Arid Environ 63: 458–474.
2. Zhang J, Dong Y (2010) Factors affecting species diversity of plant communities and the restoration process in the loess area of China. Ecol Eng 36: 345–350.
3. Mummey DL, Stahl PD, Buyer JS (2002) Soil microbiological properties 20 years after surface mine reclamation spatial analysis of reclaimed and undisturbed sites. Soil Biol Biochem 34: 1717–1725.
4. Mummey DL, Stahl PD, Buyer JS (2002) Microbial biomass as an indicator of ecosystem recovery following surface mine reclamation. Appl Soil Ecol 21: 251–259.
5. Machulla G, Bruns MA, Scow KM (2005) Microbial properties of mine spoil materials in the initial stages of soil development. Soil Sci Soc Am J 69: 1069–1077.
6. Chen B, Zhu Y, Duan J, Xiao X, Smith SE (2007) Effects of the arbuscular mycorrhizal fungus Glomus mosseae on growth and metal uptake by four plant species in copper mine tailings. Environ Pollut 147 : 374–380.
7. Wang L, Zhang W, Guo G, Qian K, Huang X (2009) Selection experiments for the optimum combination of AMF-plant-substrate for the restoration of coal mines. Mining Sci Tech 19: 479–482.
8. Long J, Huang C, Teng Y, Yao HY (2003) Preliminary study on soil microbes and soil biochemical activities in mining wasteland. Acta Ecol Sin 23: 496–503.
9. Rumpel C, Kögel-Knabner I (2004) Microbial use of lignite compared to recent plant litter as substrates in reclaimed coal mine soils. Soil Biol Biochem 36: 67–75.
10. Baldrian P, Trög J, Frouz J, Šnajdr J, Valášková V, et?al. (2008) Enzyme activities and microbial biomass in topsoil layer during spontaneous succession in spoil heaps after brown coal mining. Soil Biol Biochem 40: 2107–2116.
11. Chen B, Roos P, Zhu Y, Jakobsen I (2008) Arbuscular mycorrhizas contribute to phytostabilization of uranium in uranium mining tailings. J Environ Radioactiv 99: 801–810.
12. Rillig MC, Steinberg PD (2002) Glomalin production by an arbuscular mycorrhizal fungus a mechanism of habitat modification? Soil Biol Biochem 34: 1371–1374.
13. Koide RT, Dickie IA (2002) Effects of mycorrhizal fungi on plant populations. Plant Soil 244 307–317.
14. Šourková M, Frouz J, Fettweis U, Bens O, Hüttl RF, et?al. (2005) Soil development and properties of microbial biomass succession in reclaimed post mining sites near Sokolov (Czech Republic) and near Cottbus (Germany). Geoderma 129: 73–80.
15. Dimitriu PA, Prescott CE, Quideau SA, Grayston SJ (2010) Impact of reclamation of surface-mined boreal forest soils on microbial community composition and function. Soil Biol Biochem 42: 2289–2297.
16. Griffiths RI, Whiteley AS, O'Donnell AG, Bailey MJ (2000) Rapid Method for Coextraction of DNA and RNA from Natural Environments for Analysis of Ribosomal DNA- and rRNA-Based Microbial Community Composition. Appl Environ Microbiol 66: 5488–5491.
17. Nemergut DR, Anderson SSP, Cleveland CC, Martin AP, Miller AE, et?al. (2007) Microbial community succession in an unvegetated recently deglaciated soil. Microbiol Ecol 53: 110–122.
18. Lukow T, Dunçeld PF, Liesack W (2000) Use of the T-RFLP technique to assess spatial and temporal changes in the bacterial community structure within an agricultural soil planted with transgenic and non-transgenic potato plants. FEMS Microbiol Ecol 32: 241–247.
19. Johnson MJ, Lee KY, Scow KM (2003) DNA fingerprinting reveals links among agricultural crops soil properties and the composition of soil microbial communities. Geoderma 114: 279–303.
20. Frey B, Stemmer M, Widmer F, Luster J, Sperisen C (2006) Microbial activity and community structure of a soil after heavy metal contamination in a model forest ecosystem. Soil Biol Biochem 38: 1745–1756.
21. Singh BK, Munro S, Potts JM, Millard P (2007) Influence of grass species and soil type on rhizosphere microbial community structure in grassland soils. Appl Soil Ecol 36: 147–165.
22. Leybo AI, Netrusov A, Conrad R (2006) Effect of hydrogen concentration on the community structure of hydrogenotrophic methanogens studied by T-RFLP analysis of 16S rRNA gene amplicons. Microbiol 75 683–688.
23. Bennett LT, Kasel S, Tibbits J (2008) Non-parametric multivariate comparisons of soil fungal composition Sensitivity to thresholds and indications of structural redundancy in T-RFLP data. Soil Biol Biochem 40: 1601–1611.
24. McMahon SK, Wallenstein MD, Schimel JP (2011) A cross-seasonal comparison of active and total bacterial community composition in Arctic tundra soil using bromodeoxyuridine labeling. Soil Biol Biochem 43: 287–295.
25. Woese CR, Fox GE (1977) Phylogenetic structure of the prokaryotic domain the primary kingdoms. PNAS 74: 5088–5090.
26. Woese CR, Kandler O, Wheelis ML (1990) Towards a natural system of organisms Proposal for the domains Archaea Bacteria and Eucarya. PNAS 87: 4576–4579.
27. Lu R (2000) Soil Agrochemical Analysis Method. Beijing, China Agricultural Science Technology Press.
28. Rovira P, Vallejo VR (2002) Labile and recalcitrant pools of carbon and nitrogen in organic matter decomposing at different depths in soil an acid hydrolysis approach. Geoderma 107: 109–141.
29. Li J, Li H, Zhou X, Zhao X, Yan J (2011) Labile and recalcitrant organic matter and microbial communities in soil after conversion of abandoned lands in the Loess Plateau China. Soil Sci 176: 313–325.
30. Ge Y, He J, Zhu Y, Zhang J, Xu Z, et?al. (2008) Differences in soil bacterial diversity driven by contemporary disturbances or historical contingencies? ISME 2: 254–264.
31. ter Braak CJF, Šmilauer P (2002) CANOCO Reference Manual and CanoDraw for Windows User's Guide Software for Canonical Community Ordination (version 4 5). Microcomputer Power (Ithaca NY USA) 500 p.
32. Li FM, Xu JZ, Sun GJ (2003) Restoration of degraded ecosystems and development of water-harvesting ecological agriculture in the semi-arid Loess Plateau of China. Acta Ecol Sin 23: 1901–1909.
33. Behera N, Sahani U (2003) Soil microbial biomass and activity in response to Eucalyptus plantation and natural regeneration on tropical soil. Forest Ecol Manage 174: 1–11.
34. Ashagrie Y, Zech W, Guggenberger G (2005) Transformation of a Podocarpus falcatus dominated natural forest into a monoculture Eucalyptus globulus plantation at Munesa Ethiopia soil organic C N and S dynamics in primary particle and aggregate-size fractions. Agr Ecosyst Environ 106: 89–98.
35. Zhang X, Li J, Shi F (2008) Organic carbon and nitrogen contents and microbial biomass in soils under rapid-growth Polar plantation. J Ecol Rural Environ 24: 32–35.
36. Tscherko D, Hammesfahr U, Zeltner G, Kandeler E, Böcker R (2005) Plant succession and rhizosphere microbial communities in a recently deglaciated alpine terrain. Basic Appl Ecol 6: 367–383.
37. Knelman JE, Legg TM, O'Neill SP, Washenberger CL, González A, et?al. (2012) Bacterial community structure and function change in association with colonizer plants during early primary succession in a glacier forefield. Soil Biol Biochem 46: 172–180.
38. Lipson DA, Monson RK (1998) Plant-microbe competition for soil amino acids in the alpine tundra effects of freeze-thaw and dry-rewet events. Oecologia 113: 406–414.
39. Fang M, Motavalli PP, Kremer RJ, Nelson KA (2007) Assessing changes in soil microbial communities and carbon mineralization in Bt and non-Bt corn residue-amended soils. Appl Soil Ecol 37: 150–160.
40. Shi F, Li J, Wang S (2008) Soil organic carbon nitrogen and microbial properties in contrasting forest ecosystems of Northeast China under different regeneration scenarios. Acta Agr Scan B Soil Plant Sci 58: 1–10.
41. Marschne P, Kandeler E, Marschner B (2003) Structure and function of the soil microbial community in a long-term fertilizer experiment. Soil Biol Biochem 35: 453–461.
42. Zhang Q, Wang G, Yao H (2007) Phospholipid fatty acid patterns of microbial communities in paddy soil under different fertilizer treatments. J Environ Sci 19: 55–59.
43. Li J, Zhao B, Li X, Jiang R, Bing SH (2008) Effects of long-term combined application of organic and mineral fertilizers on microbial biomass soil enzyme activities and soil fertility. Agr Sci China 7: 336–343.
44. Cai X (2002) Effect of different methods of application on degenerated soil I middle part of Tibet. J Soil Water Conserv 16: 12–17.
45. Li J, Zhao B, Li X, Jiang R, Bing SH (2008) Changes of soil microbial properties affected by different long-term fertilization regimes. J Plant Ecol 32: 891–899.
46. Liang B, Yang X, He X, Zhou J (2011) Effects of 17-year fertilization on soil microbial biomass C and N and soluble organic C and N in Loessial Soil during maize growth. Biol Fertil Soils 47: 121–128.
47. Menyailo OV, Hungate BA, Zech W (2002) The effect of single tree species on soil microbial activities related to C and N cycling in the Siberian artificial afforestation experiment. Plant Soil 242: 183–196.
48. Eskelinen A, Stark S, Männistö M (2009) Links between plant community composition soil organic matter quality and microbial communities in contrasting tundra habitats. Oecologia 161: 113–123.
49. Lauber CL, Hamady M, Knight R, Fierer N (2009) Pyrosequencing-based assessment of soil pH as a predictor of soil bacterial community structure at the continental scale. Appl Environ Microbiol 75 : 5111–5120.
50. Hu SJ, Van Bruggen AHC, Grünwald NJ (1999) Dynamics of bacterial populations in relation to carbon availability in a residueamended soil. Appl Soil Ecol 13: 21–30.
51. Edwards IP, Bürgmann H, Miniaci C, Zeyer J (2006) Variation in microbial community composition and culturability in the rhizosphere of Leucanthemopsis alpina (L.) Heywood and adjacent bare soil along an alpine chronosequence. Microb Ecol 52: 679–692.
52. McCrackin ML, Harms TK, Grimm NB, Hall SJ, Kaye JP (2008) Responses of soil microorganisms to resource availability in urban desert soils. Biogeochem 87: 143–165.
53. Zhou XY, Wang BS, Li MG, Zan QJ (2000) An analysis of interspecific associations in secondary succession forest communities in Heishiding Nature Reserve, Guangdong Province. Acta Phytoecologica Sinica 24: 332–339.
54. Susyan EA, Wirth S, Ananyeva ND, Stolnikova EV (2011) Forest succession on abandoned arable soils in European Russia - impacts on microbial biomass fungal-bacterial ratio and basal CO_2 respiration activity. Eur J Soil Biol 47: 169–174.

55. Suzuki MT, Taylor LT, Delong EF (2000) Quantitative analysis of small-subunit rRNA genes in mixed microbial populations via 50-nuclease assays. Appl Environ Microbiol 66 : 4605–4614.

56. Schellenberger S, Kolb S, Drake HL (2010) Metabolic responses of novel cellulolytic and saccharolytic agricultural soil bacteria to oxygen. Environ Microbiol 12: 845–861.

57. May LA, Smiley B, Schmidt MG (2001) Comparative denaturing gradient gel electrophoresis analysis of fungal communities associated with whole plant corn silage. Can J Microbiol 47: 829–841.

58. Baker GC, Smith JJ, Cowan DA (2003) Review and re-analysis of domain-specific 16S primers J Microbiol Meth 55: 541–555.

10

Soil Fertilization Leads to a Decline in Between-Samples Variability of Microbial Community δ^{13}C Profiles in a Grassland Fertilization Experiment

Stavros D. Veresoglou[1,4]*, Barry Thornton[2], George Menexes[3], Andreas P. Mamolos[1], Demetrios S. Veresoglou[1]

1 Laboratory of Ecology and Environmental Protection, Faculty of Agriculture, Aristotle University of Thessaloniki, Thessaloniki, Greece, 2 Environmental and Biochemical Sciences Group, The James Hutton Institute, Craigiebuckler, Aberdeen, United Kingdom, 3 Laboratory of Agronomy, School of Agriculture, Aristotle University of Thessaloniki, Thessaloniki, Greece, 4 Plant Ecology, Freie Universität Berlin, Berlin, Germany

Abstract

Gas chromatography combustion isotope ratio mass spectrometry (GC-C-IRMS) was used to measure the ^{13}C/^{12}C ratios of PLFAs at natural abundance levels from a temperate grassland nitrogen (N) and phosphorus (P) factorial fertilization experiment in northern Greece. In each plot two rhizosphere samples were derived centred around individual *Agrostis capillaris* and *Prunella vulgaris* plants. It was hypothesized that the isotopic signal of microbes that preferentially feed on recalcitrant litter such as fungi would be modified by fertilization more strongly than that of opportunistic microbes using labile C. Microbial community δ^{13}C was affected by both P and N fertilization regime and plant species identity. However, we have been unable to detect significant nutrient effects on individual groups of microbes when analyzed separately in contrast to our original hypothesis. Intra-treatment variability, as evaluated from Hartley's F_{max} tests in the five first PCA components axes as well as the size of the convex hulls in PCA scoreplots and Mahalanobis distances, was considerably higher in the non-fertilized controls. Moreover, a significant relationship was established between the change in PLFA abundances and their respective changes in δ^{13}C for the aggregate of samples and those simultaneously fertilized with N and P. We conclude that use of compound specific isotope analysis in the absence of labelling represents a valuable and overlooked tool in obtaining an insight of microbial community functioning.

Editor: A. Mark Ibekwe, U. S. Salinity Lab, United States of America

Funding: The project was partially funded through a PhD studentship to SDV by the Chloros Trust. The funders had no role in study design, data collection and analysis, decision to publish, or preparation of the manuscript. No additional external funding was received for this study.

Competing Interests: The authors have declared that no competing interests exist.

* E-mail: seby31@zeroone.net

Introduction

Carbon (C) isotopes represent a diagnostic tool for soil ecosystem functioning [1]. Carbon occurs in nature in the form of three isotopes, ^{12}C, ^{13}C, ^{14}C at abundances 98.89%, 1.1% and 0.01% respectively [2]. In isotope studies at natural abundance levels the ^{13}C content of a sample is usually measured through the δ^{13}C value defined as: δ^{13}C (‰) = $1000\times(R_{sample} - R_{standard})/R_{standard}$‰, where R_{sample} and $R_{standard}$ are the ^{13}C/^{12}C abundance ratios of the sample and a reference standard (Vienna-Pee Dee Belemnite) [3]. Compound specific, including phospholipid fatty acids (PLFA), isotope analysis through gas chromatography combustion isotope ratio mass spectrometry (GC-C-IRMS) can provide information of carbon pathways at the molecular scale [4].

The vast majority of soil microbes are heterotrophes and rely on C assimilation to meet their energy needs. Carbon limitation of microbial growth, that is commonly reported in the soil environment [5], results in most microbes aggregating in carbon "hot-spots" such as the rhizosphere [6]. However, assimilation speed of carbon by the microbial community may vary considerably for different carbon compounds [7]. Moreover, for some carbon compounds the range of potential degraders appears to be narrower than others. For example lignin is mainly degraded by white rot fungi; whilst glycine may be assimilated by a narrower range of microbial groups than other exudates [8,9].

In the short term organic residues in soil exhibit a δ^{13}C fractioning as more recalcitrant litter compounds such as waxes and lignin are known to be more depleted (possess more negative δ^{13}C) in ^{13}C [10,11,12]. Wedin et al. [13] were able to demonstrate, in an incubation study, that the proportion of lignin fraction of grass litter in organic matter increased with time and that it was consistently more depleted in ^{13}C than the "bulk" of organic matter. However, in contrast to their original hypothesis, the overall isotopic signature of organic matter, δ^{13}C, did not change towards that of lignin. This is in agreement with an increasing amount of literature that has shown an increase in ^{13}C content (more positive δ^{13}C signal) of soil organic matter when organic matter ages [11,14]. It is now believed that in the longer term, decomposability of soil compounds is determined by accessibility of the substrate to the microbial community and the rate of decomposition is determined by abiotic factors [15,16]. Selective use of C compounds from microbial groups, along with kinetic fractionation, the isotopic discrimination for carbon during microbial consumption [17], are perceived to be the two mechanisms that shape the isotopic composition of microbially respired CO_2 that may differ considerably from that of the organic matter [18]. Thereby, isotopic $\delta^{13}C$ signal of bulk tissue of microbial groups, that is believed to be depleted at a scale of 0.7–2.8‰ relative to respired CO_2 [19], may reflect that of the compounds been assimilated. Microbes that feed on older more recalcitrant litter could, consequently, be expected to possess less depleted isotopic $\delta^{13}C$ bulk tissue. Phospholipids, although further depleted in ^{13}C relative to bulk tissue at a scale of 6–8‰ [20], because of the rapid post-death degradation in the soil environment [21] may represent an ideal target to study microbial group $\delta^{13}C$ signals.

GC-C-IRMS analysis of soil samples has, mainly, been applied following ^{13}C feeding of the microbial community [22], $^{13}CO_2$ labelling of plants [23] or a shift from C_4 to C_3 vegetation [24]. While the specific approach has the potential to provide invaluable information on the microbial groups that utilize the labelled compounds, it may be of more limited applicability in long term studies because spiked $\delta^{13}C$ signal fade with time and high noise can result from natural variability in $\delta^{13}C$ of microbial cell membranes. An alternative way to conduct analysis of long-term dynamics of litter could be through GC-C-IRMS analysis of PLFAs at natural abundance levels. A challenge in this case remains the interpretation of natural isotopic signals of individual signatures. To date, natural abundance GC-C-IRMS analysis of PLFAs in soil ecology has been very restricted; the few studies which have adopted this specific approach have made limited attempts to draw inferences out of them [25,26]. By contrast, natural abundance GC-C-IRMS is routinely applied in marine studies [27].

It is well established that fertilization of soil may evoke considerable shifts with regards to terrestrial C cycling. The traditional view that fertilization may increase decomposition rates [28] has been challenged by Berg and Meentemeyer [29] who argued that at higher N levels (low C:N ratio) more litter remains at the stage when decomposition virtually ceases. In agreement with this view, the microbial N mining hypothesis [30] states that following N addition, rates of decomposition decline as microbes use the readily available N that is applied through fertilization instead of decomposing organic compounds, a hypothesis supported by results from Craine et al. [31]. By contrast, Galantini and Rossell [32] were able to demonstrate that N and phosphorus (P) long term fertilization resulted in an increase of the ratio of labile to recalcitrant organic fractions. While a considerable amount of literature has addressed the impact of fertilization practices in the short-term [33,34], there is increasing awareness that ecosystem feedback to pulsed nutrient addition on the longer term may differ [35,36,37]. The effects of fertilization on C-cycling are of high ecological relevance as they could implicate with the C sequestration ability of soils. Post industrial increase in N deposition had been predicted, through increased primary productivity, to result in increases in C sequestration [38] but empirical results often reveal that more complicated mechanisms may operate [39].

We launched a study of exploratory nature on the potential use of natural abundance GC-C-IRMS as a tool to get an insight of microbial community responses to medium-term (*i.e* 3–10 years) fertilization in a C_3 grassland in northern Greece. Absence of C_4 plants, which produce plant litter with distinct ^{13}C signal, restricted ^{13}C variability in plant litter and exudates and enabled us to attribute $\delta^{13}C$ signal of individual microbial groups mainly to isotopic discrimination during photosynthesis, litter decomposition, fractionation and discrimination during C assimilation. We

anticipated that despite the natural gradient of $\delta^{13}C$ that occurs amongst C compounds, the $\delta^{13}C$ isotope signal would reflect the persistence time in the soil and that the recalcitrant fraction of soil organic matter (older in age) would be characterized by less depleted ^{13}C. We, thus, expected that microbial group $\delta^{13}C$ largely reflected the isotopic signal of the compounds consumed. We primarily wanted to assess the extent of differences in the natural abundance $\delta^{13}C$ signal of individual microbial groups as affected by the fertilization practices. As stated earlier, fertilization results in longer residence time for the recalcitrant fraction of litter [29,30,31], a litter fraction with a $\delta^{13}C$ signal dependent on decomposition history [11,14]. We hypothesized that the isotopic signal of microbial groups that specialize on the assimilation of recalcitrant litter such as fungi would have been affected more strongly than the respective $\delta^{13}C$ of more opportunistic microbial groups such as bacteria that use labile C. We further hypothesized that the microbial groups that would have gained access, following fertilization, to less recalcitrant litter would have increased in relative abundance whereas those that utilised less readily available compounds would have declined. This could be reflected in a linear relationship between change in $\delta^{13}C$ ($\Delta^{13}C$ - a capital Δ is used for $\delta^{13}C$ differences as opposed to isotopic signal where the use of a small δ has been adopted) of individual PLFA signatures (since recalcitrant compounds may be expected to be less depleted in $\delta^{13}C$) and relative change in PLFA signature abundance. Further to the ecological significance (to the best of our knowledge the specific question has not been addressed in the past) of recovering a negative relationship between the two variables tested this would additionally confirm the sensitivity of the GC-C-IRMS approach adopted in the manuscript in the absence of labelling as a means of studying the physiology of the microbial community in soils.

Results

Kolmogorov-Smirnov tests did not detect violations of normality in PLFA $\delta^{13}C$ signals at a scale that would prohibit PCA. Normality of the microbial group datasets that were used for analysis of variance was also confirmed. The analysis of variance did not reveal significant effects of either the two manipulations (N and P fertilization) or the plant species (rhizosphere) effect in the microbial groups examined with the exception of fungi where a significant plant species effect was detected ($F_{1,28} = 10.52$, $P = 0.003$– after Bonferroni $P = 0.021$) (Fig. 1). However, a trend of P fertilization practice resulting in an increase of $\delta^{13}C$ value of gram negative bacteria existed ($F_{1,28} = 5.996$, $P = 0.02$– after Bonferroni $P = 0.14$) as well as a trend for an interactive effect of species and P fertilisation on AM fungal $\delta^{13}C$ value ($F_{1,28} = 6.532$, $P = 0.016$– after Bonferroni $P = 0.11$). The broken stick model revealed significance of the five first PCs that explained a cumulative 72.36% of total variance. PCA distance scoreplots revealed overlap of the imposed treatments (Fig. 2). According to MANOVA analysis, however, the effects of P ($P = 0.001$– significant for the 1st, 3rd and 5th principal components) N ($P = 0.011$– significant for the 2nd component) and plant ($P = 0.001$– significant for the 5th principal component) manipulations on $\delta^{13}C$ values were rendered significant as well as the interactive effect of plant and P ($P = 0.04$– the effect was not significant for any individual component considered) (Table 1).

We, subsequently, tested violation of the homogeneity of variance criterion in the PCA scores grouped according to N and P manipulations. For all five principal components the F_{max} test was significant ($P < 0.05$, Table 2). Non-fertilized samples consistently were those where the highest variability with respect to

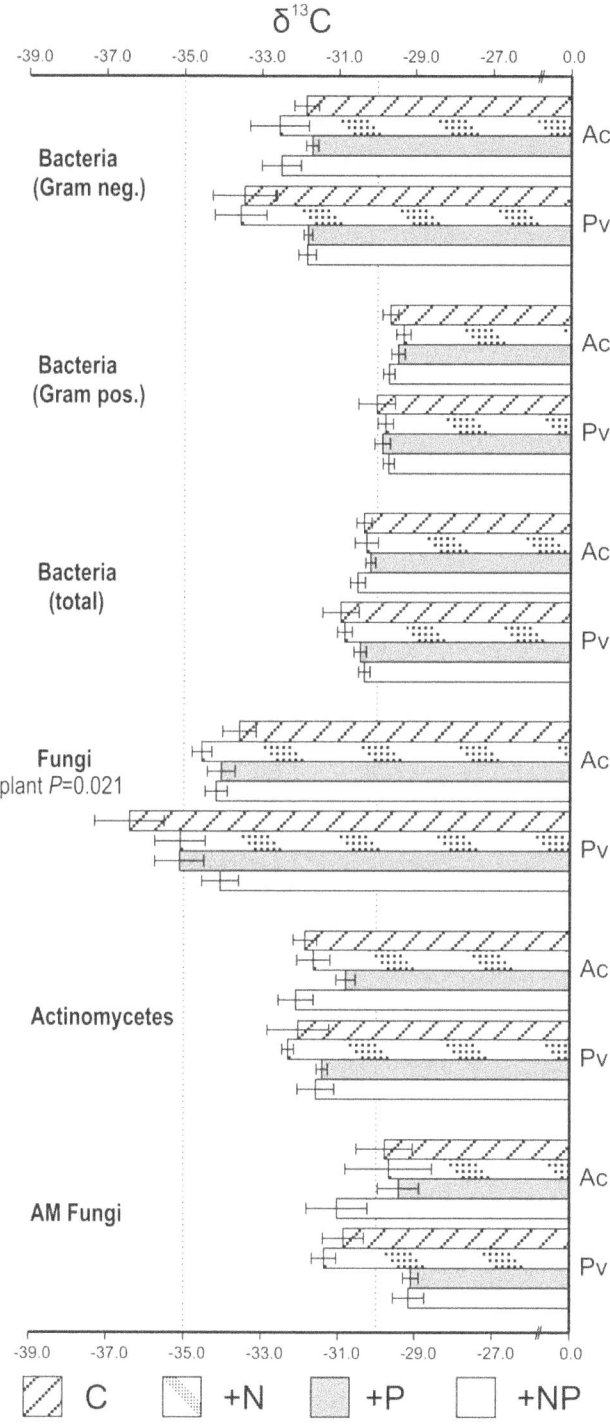

Figure 1. Isotopic signal of key microbial groups. $\delta^{13}C$ value (±S.E.) of main microbial groups in the fertilization experiment as affected by N and P fertilization practices and plant species identity. The only significant effect was that of plant species on fungi. The two dotted lines have been drawn for reference purposes for $\delta^{13}C$ signals equal to −30 and −35, respectively. Abbreviations stand for: C-control +N-N fertilized samples +P-P fertilized samples +NP = NP fertilized samples, Ac-*Agrostis capillaris*, Pv-*Prunella vulgaris*.

PLFA $\delta^{13}C$ signal was recorded. Convex hulls that were obtained from combination of the five first principal components verified that non-fertilized samples exhibit the highest variability with

Table 1. Pillai's Trace statistics following a MANOVA analysis of microbial community $\delta^{13}C$ PCA scores.

Effect	Pillai's Trace	F	P	PCs
Block	0.906	1.580	0.071	
Plant	0.538	5.585	**0.001**	5
N	0.444	3.840	**0.011**	2
P	0.564	6.214	**0.001**	1,3,5
Plant * N	0.293	1.988	0.117	
Plant * P	0.365	2.760	**0.042**	–
N * P	0.317	2.232	0.084	
Plant * N * P	0.199	1.194	0.342	

The broken stick model retrieved five significant PCA components. The column PCs highlights the principal components where significance of the factor has been detected.

respect to $\delta^{13}C$ signals (Table 3). Finally Mahalanobis distances of individual samples are presented in Table S1. The Kruskal-Wallis test, that explored the impact of the four fertilization treatments on the ranking of the Mahalanobis distances, was significant ($P = 0.039$). We therefore concluded that a significantly higher variance existed in unfertilized plots.

Correlation of the weighted difference in concentration with the respective non-fertilized control and ΔC signalling was significant for the NP treatment ($n = 42$, Spearman $\rho = -0.477$, $P < 0.01$) as well as for the entire dataset ($n = 126$, Spearman $\rho = -0.177$, $P < 0.05$) (Fig. 3).

Discussion

In C_3 grasslands a major source of fractionation of the plant community in $\delta^{13}C$ is the water use efficiency. Isotopic discrimination of ^{13}C greater than 22‰ may result during photosynthesis in C_3 plants [40,41]. Whilst the majority of this is resultant from metabolic discrimination, it is also in part due to stomatal aperture changes. However, in the long term plants show an intermediate ^{13}C fractioning response towards the typical moisture regime of their environment. The upland grassland where the experiment was located was typical Mediterranean semi-arid grassland where thaw of ice in April and drought during summer months result in a short plant growing season. Preliminary experimentation had revealed heterogeneity in water availability between different blocks (Fig. S1) and this could have accounted for the absence of detectable Euclidean space differences in $\delta^{13}C$ amongst fertilization treatments. On the other hand, persistence of this short growth season signified that temporal complementarity in plant growth was maintained to a minimum [42] and that this could constrain across-years variability with regards to plant litter $\delta^{13}C$ signal. Ehleringer et al. [11] and Bowling et al. [14] were able to detect a difference in $\delta^{13}C$ between litter and bulk organic matter deep in soil of approximately 2‰. Fertilization effects may have consequently fallen short of the natural soil C recalcitrance variation on $\delta^{13}C$.

MANOVA analysis revealed that P and N fertilization regimes and plant species identity had a significant effect on microbial $\delta^{13}C$ values (Fig. 2). N fertilization, through altering the water use efficiency (either positively or negatively) of plant species, has been repetitively demonstrated to be a key regulator of $\delta^{13}C$ signal of the plant community and, consequently, plant litter [43]. In Hobbie and Colpaert [43] N additions through deteriorating

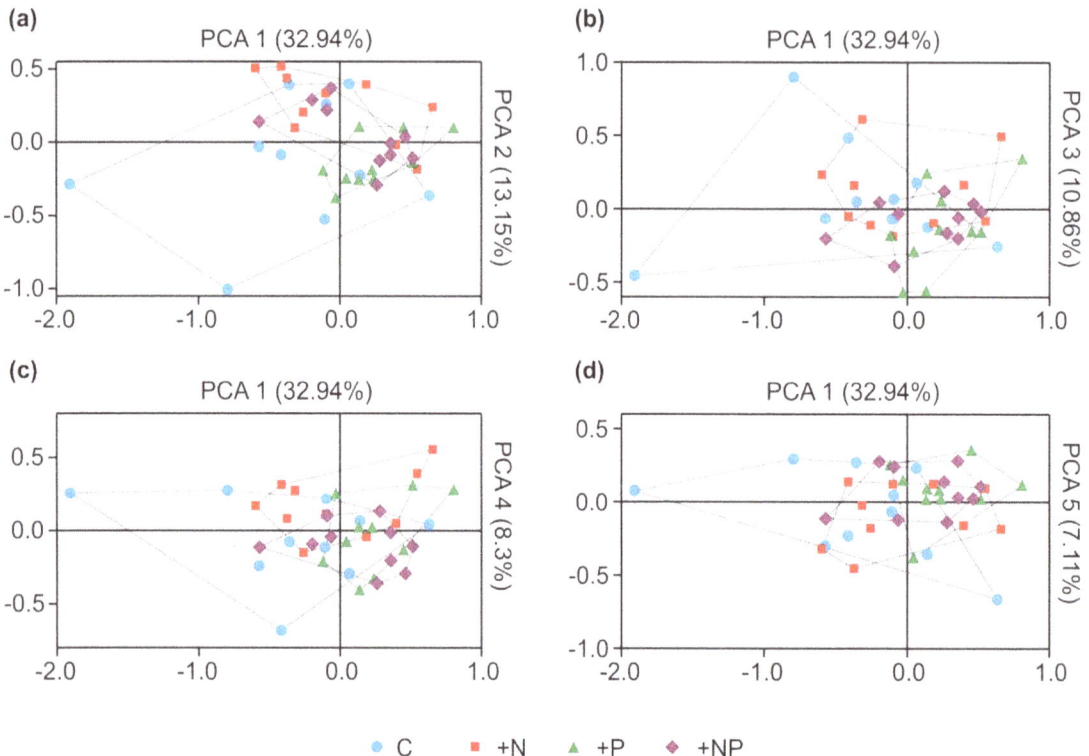

Figure 2. Ordination diagrams of the isotopic signal of PLFA signatures. Principal component analysis distance scoreplots and respective convex hulls of PCA scores grouped according to N and P fertilization regime. In all plots the x axis illustrates PCA scores of the first component whereas the y axis illustrates PCA scores of the (a) second; (b) third; (c) fourth and (d) fifth component, respectively. The five axes explained a cumulative 72.36% of variance. Abbreviations stand for: C-control +N-N fertilized samples +P-P fertilized samples +NP = NP fertilized samples.

water use efficiency resulted in plants grown in N-fertilized plots possessing 1–1.5‰ less depleted (more positive) $\delta^{13}C$. The relatively low significance of N fertilization practice on PLFA $\delta^{13}C$ was, thus, a surprising outcome of the analysis. However, in the specific sampling area P fertilization had a more pronounced impact than N in shaping microbial community clustering patterns [44].

Because of the exploratory nature of the study a conservative approach in interpreting data was adopted. Hence, the influence of different treatments was evaluated following a classical Bonferroni correction. In an earlier analysis of PLFA concentrations [44] no significant responses of the main microbial groups, as a direct effect of either fertilization or plant species identity, was observed. Therefore differences in $\delta^{13}C$ could not be attributed to

microbial community shifts and comparisons were facilitated. The specific approach revealed that the only microbial group whose $\delta^{13}C$ was significantly ($P<0.05$) affected by manipulations were fungi (AM fungi were separately analyzed and excluded from the fungal group as they assimilate plant derived labile carbon and, thus, exhibit a distinct $\delta^{13}C$ signal – no ectomycorrhizal plants were present in the grassland – Fig. 1). In specific, *P. vulgaris* appeared to support a fungal population with more depleted ^{13}C phospholipids than *A. capillaris* did. Soil fungi are a heterogeneous group of decomposers some of which (e.g. white rot fungi) exhibit high specificity on the carbon compounds they consume. Therefore, significance of the plant identity effect could reflect qualitative differences in litter quality produced that could have altered the relative composition within the fungal group.

Table 2. Variances of scores in the five first PCA component axes of the PCA grouped according to fertilization manipulation and results of the Hartley F_{max} test for homogeneity of variances.

	Variance control	Variance N-fertilized	Variance P-fertilized	Variance NP-fertilized	F value[a]	P[a]
PCA Comp. 1	0.46	0.20	0.08	0.12	5.93	0.007
PCA Comp. 2	0.18	0.05	0.03	0.04	5.83	0.007
PCA Comp. 3	0.15	0.07	0.07	0.02	6.33	0.006
PCA Comp. 4	0.09	0.04	0.06	0.02	3.56	0.036
PCA Comp. 5	0.10	0.04	0.04	0.03	3.69	0.033

[a]Values obtained from an F_{max} comparison between the non fertilized control and the less variable fertilization treatment.

Table 3. Convex hull sizes of samples grouped according to fertilization regime.

PCA Components	Convex hull size control	Convex hull size N-fertilized	Convex hull size P-fertilized	Convex hull size NP-fertilized
1,2	1.95	0.47	0.23	0.35
1,3	1.62	0.76	0.41	0.32
1,4	1.21	0.47	0.38	0.32
1,5	1.06	0.51	0.37	0.33
2,3	0.80	0.34	0.21	0.18
2,4	0.72	0.30	0.27	0.19
2,5	0.69	0.26	0.19	0.22
3,4	0.70	0.31	0.40	0.09
3,5	0.71	0.26	0.34	0.14
4,5	0.54	0.27	0.28	0.14

Alternatively, fungal PLFA signatures have been demonstrated to more readily respond to $^{13}CO_2$ labelling which signifies dependence on newly released photoassimilates [45]. Differences in exudation patterns and/or extent of mycorrhizal colonization between the two plants could justify the recorded differences in fungal phospholipid content in ^{13}C.

Lower inter-sample variance in $\delta^{13}C$ of microbial groups was recorded in the fertilized plots. The authors attribute this to microbes having narrowed the range of carbon compounds they assimilated. The N mining hypothesis predicts a decline in the degrading efficiency of the recalcitrant fraction of plant litter following N fertilization. Similarly, evidence suggests that the increase in microbial activity that is known to follow a fertilization event is transient and in the longer term microbial biomass declines [37]. The decline is more intense for fungi that are responsible for degrading the recalcitrant fraction litter [37,46]. Apparently, following chronic fertilization, the microbial community targets labile C and this makes microbial nutrition more predictable and constrains inter-sample variation in $\delta^{13}C$. With regards to the hypothesis that the microbial groups that increased in abundance following fertilization would have utilised less recalcitrant pools of carbon and would be characterized by a

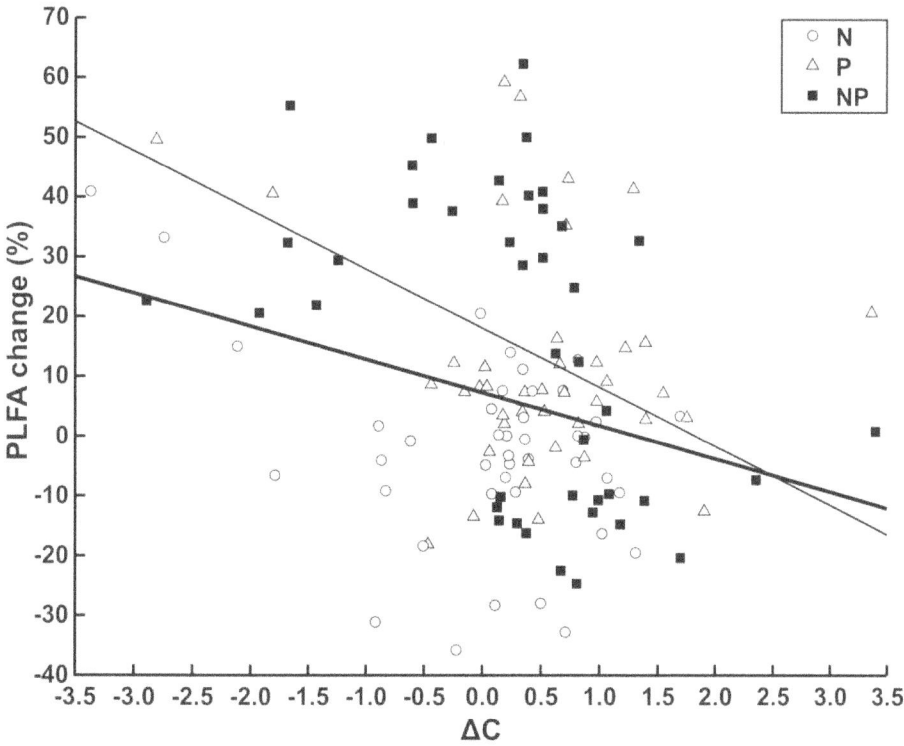

Figure 3. Modification of PLFA abundance vs isotopic signal diagram. Scatter plot of the relationship between percent change of individual PLFA concentrations and $\Delta^{13}C$ in the fertilized treatments compared to the non-fertilized controls. Open circles: N fertilized samples, triangles: P fertilized and squares: NP fertilized samples. The thin continuous line is the best fit least-squares line for the NP fertilized samples. The thick continuous line is the least-squares trendline for all three fertilization treatments when compared to the control. Note, non-significant trendlines for N and P fertilized samples were not drawn.

decline in $\delta^{13}C$ (although in Veresoglou et al. [44] the differences that had been detected were not significant) we were able to notice a linear relationship in agreement to the proposed hypothesis. The PLFA signatures that increased in abundance had actually been able to feed in more depleted in ^{13}C litter.

As stated in the introduction the specific study represents one of the very few attempts to utilize GC-C-IRMS in the absence of ^{13}C labelling in studies addressing soil microbial communities. The authors were able to use this data to report inferences on the inter-sample variability of $\delta^{13}C$ microbial profiles and the linear relationship between relative shifts in individual signature concentration and $\Delta^{13}C$ of signal. The main advantage of the approach has been the fact that it is fast to conduct and does not require the extensive optimization procedure that accompanies a ^{13}C labelling experiment. By contrast, the main drawback of the method may be its limited discrimination power as a maximum isotopic difference of 4‰ was recorded between microbial groups in the case between fungi and arbuscular mycorrhizal fungi. Such a small discrimination power may lead to a masking of significant effects because of "noise" generated by environmental and analytical variability and further complicate interpretation of data. Additionally, we had to confront the lack of background literature to compare our results to but the specific issue could be resolved following an increase in the number of studies in the near future that utilize this specific approach.

In conclusion, GC-C-IRMS in the absence of ^{13}C labelling may represent an interesting tool in evaluating responses of microbial groups to environmental manipulations. The technique suffers from low discrimination power due to low natural fractioning of C compounds but may provide a vehicle to address issues such as natural variability in isotopic composition of microbes between samples and overall patterns of PLFA concentration and $\Delta^{13}C$ responses to environmental factors.

Materials and Methods

Field Site and Experimental Design

The experimental area was located at a high elevation site 160 km west of Thessaloniki in Northern Greece (40o48′ N; 21o23′ E; 1,340 m a.s.l.) in a mesotrophic grassland with a plant community that resembles the MG5 plant community *Cynosurus cristatus-Centaurea nigra* grassland with a *Galium verum* sub-community [47]. Starting in spring 2004, two levels of N (0 and 15 g N m^{-2} $year^{-1}$ applied as NH_4NO_3) and two levels of P (0 and 10 g m^{-2} $year^{-1}$ applied as superphosphate) were applied annually on experimental plots (1.5×1.5 m) resulting in the formation of a factorial N and P fertilization experiment with eight blocks. Some details of the experimental site and analytical procedures have been described in Veresoglou et al. [44]; this previous manuscript described the microbial community shifts that resulted from fertilization practices. No specific permits were required for the described field study. The location was not privately-owned and was not designated as a protected area. Under Greek law no further permission was required to sample the area for scientific purposes. No endangered or protected species were involved in the field sampling.

Soil Sampling and Analysis

On the 10th of July 2008 a harvest was carried out on the abovementioned site. Five blocks were randomly selected and in each of the twenty plots (5 blocks × 4 treatments) two soil cores (2 cm internal diameter, 8 cm depth) were retrieved, one centered on an *Agrostis capillaris* stand and one around a *Prunella vulgaris* individual. The cores were immediately frozen in liquid N_2 and

were subsequently freeze-dried, then following removal of visible plant material were ball milled for PLFA extraction. PLFA extraction was implemented according to White et al., [48] and analysis followed in a compound specific gas chromatograph-isotope ratio mass spectrometer GC-C-IRMS (GC Trace Ultra with combustion column attached via a GC Combustion III to a Delta V Advantage isotope ratio mass spectrometer Thermo Finnigan, Bremen, Germany). Samples (2 μl) were injected in splitless mode onto a J&W Scientific HP-5 column, 50 m length, id 0.2 mm with a film thickness of 0.33 μm (Agilent Technologies Inc, Santa Clara, USA); otherwise running conditions were as described by Paterson et al. [9]. The C isotope ratios were calculated with respect to a CO_2 reference gas injected with every sample and traceable to International Atomic Energy Agency reference material NBS 19 TS-Limestone. Repeated analysis, over a two month period, of the $\delta^{13}C$ value of a 19:0 FAME internal standard gave a standard deviation of 1.11 ‰ ($n = 18$).

Data Analysis

Standard nomenclature was used for PLFAs [49] as follows: the number before the colon indicates the number of carbon atoms in the phospholipids and the number after the colon gives the number of double bonds and their location (ω). Prefixes Me, cy, i and a indicate presence of methyl- and cyclopropyl- groups and iso- or antesio- branching, respectively. The PLFA signatures 16:1ω7, 17:0cy, 19:0cy, 19:1w6, 15:0i, 15:0a, 16:0i, 17:0i, 17:0a 10-Me-17:0, 10-Me-18:0 were attributed to bacteria [23]. The fatty acid 18:2ω6,9 was used as a marker for fungi other than arbuscular mycorrhizal (AM) fungi [50] and the signatures 10-Me-17:0 and 10-Me-18:0 as markers for actinomycetes [51]. Signature 16:1ω5c was used as a biomarker of AM fungi [52].

Data were subjected to Kolmogorov-Smirnov tests for normality prior to analyses. To compute mean $\delta^{13}C$ of microbial groups that included more than one PLFA signature, a weighted mean was obtained with individual PLFA signature PLFA concentrations. Analyses of variance were conducted based on three-way ANOVA models with factors P fertilization, N fertilization and plant identity. Hypothesis testing significance level was corrected with a classical Bonferroni procedure. Principal component analysis (PCA) was conducted in the standardized matrices of PLFA $\delta^{13}C$ signals. The analysis was implemented with the package "vegan" in R 2.12.0 [53]. The number of significant components was retrieved following the broken-stick model [54]. PCA scores of the significant components were subjected to a MANOVA analysis in SPSS v. 15.0 (SPSS, Inc. IL: Chicago) with factors P fertilization, N fertilization and plant identity to test significance of experimental treatments. Presence of differences in variance amongst the fertilization treatments were investigated through (i) Hartley's Fmax tests separately for the scores in each principal component [55], (ii) calculation of the size of the convex hulls of combinations of the significant components as performed in earlier studies [56,57] and (iii) Mahalanobis distances [58] separately calculated for each sample that were analysed based on a Kruskal-Wallis test. Convex hull sizes were calculated with the function convexhull.xy in the package "spatstat" in R.

Finally, we tested significance of the correlation between the change in abundance of PLFAs and the change in their isotopic signal. We used as controls the PLFA $\delta^{13}C$ data from the non-fertilized plots. For the three fertilization treatments, calculated separately for the two plants, the weighted difference in concentration with the respective non-fertilized control was calculated and the difference in signal, $\Delta^{13}C$ approximated as $\Delta^{13}C \approx \delta^{13}C$ fertilized - $\delta^{13}C$ control (21 signatures × 2 plants × 3 fertilization treatments = 126 points). Due to deviation from normality of both metrics Spearman's rank test (Spearman's rho or ρ test) was used.

Supporting Information

Figure S1 Heterogeneity in Moisture Content. Means ($\pm$S.E) of moisture content of the samples obtained in the harvest of the 10th of July 2008 grouped according to the blocking factor.

Table S1 Mahalanobis-distances-based matrix. Mahalanobis distances for the centroids of the samples following a principal component analysis. Analysis of the effect of the four fertilization treatments on the ranking of the Mahalanobis

distances based on a Kruskal-Wallis test revealed significance of the effect of treatment.

Acknowledgments

The authors would like to thank Mrs Gillian Martin and Mrs Maureen Procee for technical support and the reviewers of the manuscript for useful suggestions.

Author Contributions

Conceived and designed the experiments: SDV DSV. Performed the experiments: SDV APM DSV. Analyzed the data: SDV BT GM. Contributed reagents/materials/analysis tools: SDV BT DSV. Wrote the paper: SDV BT.

References

1. Staddon PL (2004) Carbon isotopes in functional soil ecology. Trends Ecol Evol 19: 148–154.
2. Weast RC (1978) CRC Handbook of Chemistry and Physics, 58th ed. Boca Raton (FL): CRC Press Inc. 2348 p.
3. Lajtha K, Marshall JD (1994) Sources of variation in the stable isotopic composition of plants. In: Lajtha K, Marshall JD, editors. Stable isotopes in ecology and environmental science. London: Blackwell Scientific Publications, pp. 1–21.
4. Zhang CL (2002) Stable carbon isotopes of lipid biomarkers: analysis of metabolites and metabolic fates of environmental microorganisms. Curr Opin Biotechnol 13: 25–30.
5. Demoling F, Figueroa D, Bååth E (2007) Comparison of factors limiting bacterial growth in different soils. Soil Biol Biochem 39: 2485–2495.
6. Nannipieri P, Ascher J, Ceccherini MT, Landi L, Pietramellara G, et al. (2003) Microbial diversity and soil functions. Eur J Soil Sci 54: 655–670.
7. Entry JA (2000) Influence of nitrogen on cellulose and lignin mineralization in blackwater and redwater forested wetland soils. Biol Fertil Soils 31: 436–440.
8. Leonowicz A, Matuszewska A, Luterek J, Ziegenhagen D, Wojtaś-Wasilewska M, et al. (1999) Biodegradation of lignin by white rot fungi. Fungal Genet Biol 27: 175–185.
9. Paterson E, Gebbic T, Abel C, Sim A, Telfer G (2007) Rhizodeposition shapes rhizosphere microbial community structure in organic soil. New Phytol 173: 600–610.
10. Gleixner G, Danier HJ, Werner RA, Schmidt HL (1993) Correlation between the 13C content and secondary plant-products in different cell compartments and that in decomposing basidiomycetes. Plant Physiol 102: 1287–1290.
11. Ehleringer JR, Buchmann N, Flanagan LB (2000) Carbon isotope ratios in belowground carbon cycle processes. Ecol Appl 10: 412–422.
12. Tiunov AV (2007) Stable isotopes of carbon and nitrogen in soil ecological studies. Biol Bull 34: 395–407.
13. Wedin DA, Tieszen LL, Drwby B, Pastor J (1995) Carbon isotope dynamics during grass decomposition and soil organic matter formation. Ecology 76: 1283–1392.
14. Bowling DR, McDowell NG, Bond BJ, Law BE, Ehleringer JR (2002) 13C content of ecosystem respiration is linked to precipitation and vapor pressure deficit. Oecologia 131: 113–124.
15. Kemmitt SJ, Lanyon CV, Waite IS, Wen Q, Addiscott TM, et al. (2008) Mineralization of native soil organic matter is not regulated by the size, activity or composition of the soil microbial biomass-a new perspective. Soil Biol Biochem 40: 61–73.
16. Schmidt MWI, Torn MS, Abiven S, Dittmar T, Guggenberger G, et al. (2011) Persistence of soil organic matter as an ecosystem property. Nature 478: 49–56.
17. Blair N, Leu A, Munoz E, Olsen J, Kwong E, et al. (1985) Carbon isotopic fractionation in heterotrophic microbial metabolism. Appl Environ Microbiol 50: 996–1001.
18. Ngao J, Cotruto MF (2011) Carbon isotope discrimination during litter decomposition can be explained by selective use of substrate with differing δ^{13}C. Biogeosci Discuss 8: 51–82.
19. Werth M, Kuzyakov Y (2010) 13C fractionation at the root microorganisms soil interface: A review and outlook for partitioning studies. Soil Biol Biochem 42: 1372–1384.
20. Ruess L, Chamberlain PM (2010) The fat that matters: Soil food web analysis using fatty acids and their carbon stable isotope signature. Soil Biol Biochem 42: 1898–1910.
21. Palojärvi A (2006) Phospholipid fatty acid (PLFA) analyses. In: Bloem J, Hopkins DW, Benedetti A, editors. Microbiological Methods for Assessing Soil Quality. Reading (UK): CABI Publishing, 204–211.
22. Waldrop MP, Firestone MK (2004) Microbial community utilization of recalcitrant and simple carbon compounds: impact of oak-woodland plant communities. Oecologia 138: 275–284.
23. Paterson E, Osler G, Dawson LA, Gebbing T, Sim A, et al. (2008) Labile and recalcitrant plant fractions are utilized by distinct microbial communities in soil:
24. independent of the presence of roots and mycorrhizal fungi. Soil Biol Biochem 40: 1103–1113.
24. Ruess L, Schütz K, Haubert D, Häggblom MM, Kandeler E, et al. (2005) Applications of lipid analysis to understand trophic interactions in soil. Ecology 86: 2075–2082.
25. Billings SA, Ziegler SE (2008) Altered patterns of soil carbon substrate usage and heterotrophic respiration in a pine forest with elevated CO_2 and N fertilization. Glob Chang Biol 14: 1–12.
26. Cusack DF, Silver WL, Torn MS, Burton SD, Firestone MK (2011) Changes in microbial community characteristics with nitrogen additions and effects on soil organic matter in two tropical forests. Ecology 92: 621–32.
27. Boschker HTS, Middelburg JJ (2002) Stable isotopes and biomarkers in microbial ecology. FEMS Microbiol Ecol 40: 85–95.
28. Garten CT, Cooper LW, Post WM, Hanson PJ (2000) Climate controls on forest soil isotope ratios in the Southern Appalachian Mountains. Ecology 81: 1108–1119.
29. Berg B, Meentemeyer V (2002) Litter quality in a north European transect versus carbon storage potential. Plant Soil 242: 83–92.
30. Moorhead DL, Sinsabaugh RL (2006) A theoretical model of litter decay and microbial interaction. Ecol Monogr 76: 151–174.
31. Craine JM, Morrow C, Fierer N (2007) Microbial nitrogen limitation increases decomposition. Ecology 88: 2105–2113.
32. Galantini J, Rosell R (2006) Long-term fertilization effects on soil organic matter quality and dynamics under different production systems in semiarid Pampean soils. Soil Tillage Res 87: 72–79.
33. Bardgett RD, Mawdsley JL, Edwards S, Hobbs PJ, Rodwell JS, et al. (1999) Plant species and nitrogen effects on soil biological properties of temperate upland grasslands. Funct Ecol 13: 650–660.
34. Bradley K, Drijjber RA, Knops J (2006) Increased N availability in grassland soils modifies their microbial communities and decreases the abundance of arbuscular mycorrhizal fungi. Soil Biol Biochem 38: 1583–1595.
35. Aber JD, McDowell W, Nadelhoffer K, Magill A, Berntson G, et al. (1998) Nitrogen saturation in temperate forest ecosystems: Hypotheses revisited. Biosci 48: 921–933.
36. Emmett BA (1999) The impact of nitrogen on forest soils and feedbacks on tree growth. Water Air Soil Pollut 116: 65–74.
37. Wallenstein MD, McNulty S, Fernandez IJ, Boggs J, Schlesinger WH (2006) Nitrogen fertilization decreases forest soil fungal and bacterial biomass in three long-term experiments. For Ecol Manage 222: 459–468.
38. Lloyd J (1999) The CO_2 dependence of photosynthesis, plant growth responces to elevated CO_2 concentrations and their interactions with soil nutrient status. II. Temperature and boreal forest productivity and the combined effects of increasing CO_2 concentrations and increased nitrogen deposition. Func Ecol 13: 439–459.
39. Mack MC, Schuur EAG, Bret-Harte MS, Shaver GR, Chapin FS III (2004) Ecosystem carbon storage in arctic tundra reduced by long-term nutrient fertilization. Nature 431: 440–443.
40. Evans JR, Sharkey TD, Berry JA, Farquhar GD (1986) Carbon isotope discrimination measured concurrently with gas-exchange to investigate CO_2 diffusion in leaves of higher-plants. Aust J Plant Physiol 13: 281–292.
41. Schnyder H (1992) Long-term steady-state labeling of wheat plants by use of natural $^{13}CO_2/^{12}CO_2$ mixtures in an open, rapidly turned-over system. Planta 187: 128–135.
42. Mamolos AP, Vasilikos CV, Veresoglou DS (2011) Temporal patterns of growth and nutrient accumulation of plant species in a Mediterranean mountainous grassland. Ecol Res 26: 583–593.
43. Hobbie EA, Colpaert JV (2004) Nitrogen availability and mycorrhizal colonization influence water use efficiency and carbon isotope patterns in *Pinus sylvestris*. New Phytol 164: 515–525.
44. Veresoglou SD, Mamolos AP, Thornton B, Voulgari OK, Sen R, et al. (2011) Medium-term fertilization of grassland plant communities masks plant species-linked effects on soil microbial community structure. Plant Soil 344: 187–196.

45. Denef K, Roobroeck D, Wadu MCWM, Lootens P, Boeckx P (2009) Microbial community composition and rhizodeposit-carbon assimilation in differently managed temperate grassland soils. Soil Biol Biochem 41: 144–153.

46. Frey SD, Knorr M, Parrent JL, Simpson RT (2004) Chronic nitrogen enrichment affects the structure and function of soil microbial community in temperate hardwood and pine forests. For Ecol Manage 196: 159–171.

47. Rodwell JS, Pigott CD, Ratcliffe DA, Maloch AJC, Birks HLB, et al. (2000) British Plant Communities. Cambridge: Cambridge University Press, 528 p.

48. White DC, Davis WM, Nickels JS, King JD, Bobbie RJ (1979) Determination of the sedimentary microbial biomass by extractable lipid phosphate. Oecologia 40: 51–62.

49. Frostegård A, Tunlid A, Bååth E (1993) Shifts in the structure of soil microbial communities in limed forests as revealed by phospholipid fatty acid analysis. Soil Biol Biochem 25: 723–730.

50. Frostegård A, Bååth E (1996) The use of phospholipid fatty acid analysis to estimate bacterial and fungal biomass in soil. Biol Fertil Soils 22: 59–65.

51. Kroppenstedt RM (1992) The genus *Nocardiopsis*. In: Balows A, Trüper HG, Dworkin M, Harder W, Schleifer KH, editors. The prokaryotes: a handbook on the biology of bacteria: ecophysiology, isolation, identification, applications, second ed., vol II. Berlin: Springer-Verlag, pp. 1139–1156.

52. Olsson PA, Bååth E, Jakobsen I, Söderström B (1995) The use of phospholipids and neutral fatty acids to estimate biomass of arbuscular mycorrhizal fungi in soil. Mycol Res 99: 623–629.

53. R Development Core Team (2008) R: A Language and Environment for Statistical Computing. Vienna: R Foundation for Statistical Computing. 2673 p.

54. Jackson DA (1993) Stoping rules in principal components analysis: a comparison of heuristical and statistical approaches. Ecology 74: 2204–2214.

55. Hartley HO (1950) The use of range in analysis of variance. Biometrica 37: 271–280.

56. Islam K, McBratney A, Singh B (2005) Rapid estimation of soil variability from the convex hull biplot area of topsoil ultra-violet, visible and near-infrared diffuse reflectance spectra. Geoderma 128: 249–257.

57. Gubsch M, Buchmann N, Schmid B, Schulze ED, Lipowsky A, et al. (2011) Differential effects of plant diversity on functional trait variation of grass species. Ann Bot 107: 157–169.

58. Penny K (1996) Appropriate critical values when testing for a single multivariate outlier using the Mahalanobis distance. Appl Stat 45: 73–81.

11

Stable Isotope Biogeochemistry of Seabird Guano Fertilization: Results from Growth Chamber Studies with Maize (*Zea Mays*)

Paul Szpak[1]*, Fred J. Longstaffe[2], Jean-François Millaire[1], Christine D. White[1]

1 Department of Anthropology, The University of Western Ontario, London, Ontario, Canada, 2 Department of Earth Sciences, The University of Western Ontario, London, Ontario, Canada

Abstract

Background: Stable isotope analysis is being utilized with increasing regularity to examine a wide range of issues (diet, habitat use, migration) in ecology, geology, archaeology, and related disciplines. A crucial component to these studies is a thorough understanding of the range and causes of baseline isotopic variation, which is relatively poorly understood for nitrogen ($\delta^{15}N$). Animal excrement is known to impact plant $\delta^{15}N$ values, but the effects of seabird guano have not been systematically studied from an agricultural or horticultural standpoint.

Methodology/Principal Findings: This paper presents isotopic ($\delta^{13}C$ and $\delta^{15}N$) and vital data for maize (*Zea mays*) fertilized with Peruvian seabird guano under controlled conditions. The level of ^{15}N enrichment in fertilized plants is very large, with $\delta^{15}N$ values ranging between 25.5 and 44.7‰ depending on the tissue and amount of fertilizer applied; comparatively, control plant $\delta^{15}N$ values ranged between -0.3 and 5.7‰. Intraplant and temporal variability in $\delta^{15}N$ values were large, particularly for the guano-fertilized plants, which can be attributed to changes in the availability of guano-derived N over time, and the reliance of stored vs. absorbed N. Plant $\delta^{13}C$ values were not significantly impacted by guano fertilization. High concentrations of seabird guano inhibited maize germination and maize growth. Moreover, high levels of seabird guano greatly impacted the N metabolism of the plants, resulting in significantly higher tissue N content, particularly in the stalk.

Conclusions/Significance: The results presented in this study demonstrate the very large impact of seabird guano on maize $\delta^{15}N$ values. The use of seabird guano as a fertilizer can thus be traced using stable isotope analysis in food chemistry applications (certification of organic inputs). Furthermore, the fertilization of maize with seabird guano creates an isotopic signature very similar to a high-trophic level marine resource, which must be considered when interpreting isotopic data from archaeological material.

Editor: John P. Hart, New York State Museum, United States of America

Funding: The authors hank the Natural Sciences and Engineering Research Council of Canada, Canada Research Chairs Program and the Canada Foundation for Innovation for funding in support of this research. This is Laboratory for Stable Isotope Science Contribution #279. The funders had no role in study design, data collection and analysis, decision to publish, or preparation of the manuscript.

Competing Interests: The authors have declared that no competing interests exist.

* E-mail: pszpak@uwo.ca

Introduction

Seabird excrement (guano) was arguably the most economically significant organic fertilizer in the world prior to the twentieth century. The guano was mined from small, nearshore islands off the arid western coast of South America in the Peru-Humboldt upwelling region. The guano islands of Peru and Chile are typically composed of rocky cliffs essentially devoid of vascular plants, with a relatively small number of fauna (ants, spiders, scorpions, lizards) that are supported by allochthonous inputs from the guano birds (guano, carcasses, feathers, eggshells) [1]. Because the region receives virtually no precipitation, the guano accumulates in sedimentary layers. The once thick deposits of seabird guano (>50 m in some cases) were mined extensively during the guano boom of the 1800 s, and today the islands rarely have more than several years worth of droppings accumulated [1]. The trade in guano peaked during the middle of the nineteenth century, with

20 million tons being exported to Europe and North America between 1848 and 1875 [2]. The popularity of guano with European and North American farmers waned in the latter part of the nineteenth century for a number of reasons, including: increasing guano prices, irregular availability, unsuitability for particular crops (especially turnips), a dwindling supply, and the development of the chemical fertilizer industry [3]. In recent years, however, there has been a resurgence in its popularity (particularly in horticulture) as worldwide demand for organically grown produce has increased [4,5]. The importance of guano as a fertilizer prior to the nineteenth century is less well known, but is mentioned by Spanish chroniclers and in colonial administrative documents [6,7]. On this basis, some have suggested that it may have been of some importance in prehispanic agriculture [8,9].

From an ecological perspective, the importance of ornithogenic nitrogen to marine and terrestrial ecosystems has long been recognized [10–12]. A number of studies conducted in tropical, temperate, subpolar, and polar regions have shown that seabird guano alters the concentration of soil nutrients (particularly NH_4^+, NO_3^-, PO_4^{3-}, K^+, Mg^{2+}), plant tissue nutrients (N, P, K), and plant productivity [13–21]. Seabird guano may also affect the diversity of plant species present, though results from such studies are inconsistent [21]. Numerous factors other than the presence of guano may also affect the chemistry, physiology, and ecology of plants growing within or near seabird colonies. In field studies it is often difficult, or impossible, to rule out the effects of these factors, which include: physical disturbance caused by birds such as plant clipping or trampling [22,23], deposition of seabird carcasses, feathers and eggshells [24–26], and avian-aided seed dispersal [27].

Particularly large ^{15}N enrichments in soils, plants, and animals (5–40‰) have been recorded in and around seabird nesting sites, allowing for the relative contribution of avian-derived nutrients to be assessed (Table 1). Despite this large body of literature, there have been no investigations that examine the biogeochemical effects of seabird guano on the western coast South America, with

the majority of studies focusing on Oceania, Japan, California, and Antarctica [21]. Furthermore, no studies have addressed the isotopic biogeochemistry of seabird guano from an agricultural or horticultural standpoint. The purpose of this study, therefore, is to assess the isotopic and vital effects of Peruvian seabird guano fertilization on maize (*Zea mays*) under controlled conditions. In particular we examine the extent of the enrichment in plant ^{15}N resulting from guano fertilization.

Plants are capable of utilizing several different soil N sources, both organic (amino acids) and inorganic (NH_4^+, NO_3^-, N_2). From a biogeochemical perspective, the uptake, assimilation, and allocation/reallocation of N compounds are all significant. Uptake of NO_3^- in plant root cells occurs through at least three different NO_3^- transport systems [28]. Once inside the root, NO_3^- can be assimilated into organic N, or translocated to the shoot for assimilation by nitrate reductase (NR), nitrite reductase (NiR), and glutamine synthetase (GS) [29]. Little or no fractionation of ^{15}N is reported to be associated with the uptake of NO_3^- [30–32];

Table 1. Summary of studies examining the effects of seabird guano on the isotopic composition ($\delta^{15}N$) of plants and soils.

Location	Bird Species	Guano $\delta^{15}N$ (‰, AIR)	Plant $\delta^{15}N$ (‰, AIR)		Plant $\delta^{15}N$ (‰, AIR)		Reference
			Bird	Non-bird	Bird	Non-bird	
California	Mixed	–	36.4±2.7	5.3±0.9	33.6±2.0	7.2±1.2	[13]
California	Mixed	–	24.3 to 24.5	6.8 to 7.8	–	–	[173]
Falkland Islands and Antarctica	Mixed	10.9±1.9	−1.8 to 15.8	−7.9 to 7.6	14.0±0.3	0.0 to 9.7	[174]
Antarctica	Snow petrel	–	−3.1 to 25.8	−17.6 to −0.5	13.1 to 25.9	−13.4 to −1.0	[175]
Australia	Gentoo penguin	14.6	7.2 to 18.8	−9.8 to −0.7	–	–	[176]
North Africa	Various gulls	–	9.8 to 17.4	–	10.5 to 13.4	16.8 to 20.8	[51]
New Zealand	Westland petrel	–	–	–	14.1±0.3	–	[177]
New Zealand	Mixed	–	−3.9 to 9.1	–	–	–	[178]
New Zealand	Sooty shearwater	7.7	14.2±3.1	−6.1±1.7	–	–	[179]
Japan	Great cormorant	–	16.4 to 16.9	−2.5±0.6	10.6 to 16.0	0.4±0.3	[180]
Japan	Great cormorant	13.2±1.3	10.0 to 14.7	−2.3 to 6.8	–	–	[181]
Sweden	Great cormorant	–	13.6±1.7	1.7	–	–	[182]
New Zealand	Mixed	–	4.6 to 6.7	14.4 to 15.9	10.2±1.0	16.2±0.3	[183]
Fiji	Mixed	39.1 to 50.1[a]	13.6 to 36.7	−1.3 to 0.8	15.1 to 31.6[a]	−4.1 to −1.3[b]	[184]
Fiji	Mixed	14.9 to 23.3[b]	–	–	13.5 to 33.0[b]	–	[184]
Japan	Black-tailed gull	10.2 to 10.5	3.9 to 14.6	−4.1 to −2.1	10.1 to 43.3[a]	−4.3 to −2.9[b]	[185]
Japan	Black-tailed gull	–	–	–	−0.2 to 33.7[b]	–	[185]
Japan	Black-tailed gull	9.1 to 12.8	–	–	18.5 to 44.1[a]	–	[53]
Japan	Black-tailed gull	–	–	–	−4.1 to 42.2[b]	–	[53]
Japan and Antarctica	Penguin and gull	8.0 to 9.4	13.6 to 38.1	–	–	–	[186]
Antarctica	Penguin	7.4	–	–	32.1	–	[75]
Japan	Mixed	–	–	–	9.1 to 37.9	−4.6 to 8.6	[54]
New Zealand	Rockhopper penguin	7.0±0.4	–	–	23.8±3.3	−0.5±0.2	[187]
Australia	Mixed	9.9	9.5±2.2	7.0±2.6	–	–	[55]
California	Mixed	–	27.2 to 27.3	8.3 to 9.5	28.3±5.4	–	[188]
Pribilof Islands	Mixed	12.5	22.0	11.3	–	–	[162]
California	Mixed	–	–	–	35.6±2.6	7.5±0.3	[56]
Pacific (Palmyra Atoll)	Mixed	13.9	14.0±1.4	9.3±0.9	16.2±0.3	11.0±0.7	[189]
Antarctica	Penguin	20.9±4.2	–	–	10.4±3.1	–	[190]

[a]NH_4^+.
[b]NO_3^-.

fractionation of ^{15}N does not appear to vary with respect to source $[NO_3^-]$ [33–35]. Some variability in fractionation is associated with NR activity, and it has been difficult in some cases to differentiate between isotopic fractionation associated with N uptake and assimilation, respectively [36]. Ledgard et al. [37] report the fractionation for the entire process to be $-15‰$, while a range of 0 to $-19‰$ is reported by Robinson [38].

NH_4^+ is taken up by plants via high or low affinity transporters depending on extracellular $[NH_4^+]$ [39]. NH_4^+ is assimilated into organic N only in the roots via GS and most estimated $\Delta^{15}N$ values for NH_4^+ uptake and assimilation fall between -5 and $-20‰$ [38,40]. Unlike NO_3^-, however, there are substantial differences in $\Delta^{15}N$ with source $[NH_4^+]$. For example, in two different rice cultivars, Yoneyama et al. [33] found $\Delta^{15}N$ for NH_4^+ uptake to be -6.1 to $-12‰$ at low source $[NH_4^+]$, and -13.4 to $-28.9‰$ at high source $[NH_4^+]$.

Materials and Methods

Materials

All plants were grown in a walk-in growth chamber at the Biotron Centre for Experimental Climate Change Research at the University of Western Ontario. The substrate utilized for all treatments was Pro-mix® for containers (75–85% sphagnum moss, 15–25% perlite and limestone). Peruvian seabird guano (Guano Company International, Cleveland, Ohio, United States) was obtained from an organic gardening outlet. The nitrogen content of the guano was reported to be 10% and determined to be $11.2\pm0.2\%$ based on five analyses of dried, powdered guano as described for plant samples below. The 'Early Sunglow' maize cultivar was used (*Zea mays* cv. Early Sunglow, Lot E1, 2010, Ferry Morse, Fulton, Kentucky, United States) for all experiments because it is a relatively small variety of maize that accommodated physical restrictions on plant height imposed by the growth chamber.

Growth Chamber Conditions

Growth chamber temperature was $25/18°C$ (day/night), with a photoperiod of 13 h provided by 185 W fluorescent bulbs. Relative humidity was set at 80% for the first four daylight hours, and 60% for the remainder of the day. These conditions were monitored electronically, and did not deviate from these parameters for the duration of the experiment.

Maize Germination Experiment

Guano (well-mixed with soil) was applied to 1.2 L plastic containers (1.0 L of soil) in the following amounts: 0 g, 1.0 g, 2.5 g, 5.0 g, 7.5 g, 10.0 g and 15.0 g. Six replicates of each treatment were prepared. One hour after addition of the guano, maize seeds were planted ~2.5 cm below the surface in the containers. Emergence and growth of the plants were recorded every 2–3 days for 35 days.

Maize Fertilization Experiment

Fifteen maize seeds were planted ~2.5 cm below the surface in 1.2 L plastic containers (1.0 L of soil). At this time, guano was mixed with soil in free-draining (perforated at the base) 18.9 L plastic buckets containing 16 L of soil in the following amounts: 0 (C0), 80 g (G1, 5 g guano/L), 160 g (G2, 10 g guano/L). Five replicates of each treatment were prepared. Maize is typically fertilized prior to planting, and sometimes again approximately three weeks after emergence, although this second application is uncommon [41]. To avoid complications associated with additional fertilizer applications, only one fertilizer application was

employed. After germination (7 days after sowing) maize plants were moved into the 18.9 L plastic buckets. Plants were watered every 2–3 days and the height and general growth of the plants was monitored. Distal leaf samples (~3 cm×6 cm) were taken at 30 and 75 days after planting (d). Plants at 30 d were characterized by only vegetative growth, while plants sampled at 75 d had begun reproductive growth (tassels fully emerged, silks beginning to appear). Anthers were sampled at 75 d. At completion of the experiment (115 d), the following tissues were sampled: leaves, grains, roots, and stalks. All buckets were relocated randomly within the growth chamber five times (30, 45, 60, 75, 100 d) during the course of the experiments to account for any micro-variations in light, temperature or humidity, although such changes were not expected.

Stable Isotope Analysis

All plant materials were stored at $-25°C$ following sampling until needed for analysis. Samples were then dried at $90°C$ under normal atmosphere for 72 hours, ground using a Wig-L-Bug (Crescent, Lyons, Illinois, United States) and the resulting powders stored at room temperature in sealed glass vials. Isotopic compositions ($\delta^{13}C$ and $\delta^{15}N$ values determined separately) and relative percentages of carbon and nitrogen were determined using a Delta V isotope ratio mass spectrometer (Thermo Scientific, Bremen, Germany) coupled to an elemental analyzer (Costech Analytical Technologies, Valencia, California, United States). For the analysis of $\delta^{15}N$, excess CO_2 was removed using a Carbo-Sorb trap (Elemental Microanalysis, Okehampton, Devon, United Kingdom). Sample reproducibility was $\pm0.09‰$ for $\delta^{13}C$ and $\pm0.90\%$ for %C (6 replicates), and $\pm0.12‰$ for $\delta^{15}N$ and $\pm0.10\%$ for %N (24 replicates). A $\delta^{15}N$ value of $20.31\pm0.18‰$ was obtained for 37 analyses of IAEA-N2, which compared well with its accepted value of $20.30‰$. A $\delta^{13}C$ value of $-29.87\pm0.29‰$ was obtained for 11 analyses of NBS-22, which compared well with its accepted value of $-30.00‰$.

Statistical Analyses

Comparisons between treatments and between organs were completed using one-way analysis of variance (ANOVA). Levene's test was used to assess homogeneity of variance; if variance was homoscedastic, a *post hoc* Tukey's honestly significant difference (HSD) test was applied and if variance was not homoscedastic, a *post hoc* Dunnett's T3 test was applied. All statistical analyses were conducted at a significance level of 5% ($p<0.05$). All statistical analyses were performed in SPSS 16 for Windows.

Results and Discussion

Maize Germination and Seedling Establishment

All unfertilized plants germinated and commenced normal growth (Figure 1). There was a clear trend towards the inhibition of germination and seedling emergence with increasing rate of guano applied (Figure 1). It is apparent that the presence of seabird guano in the soil has the potential to inhibit germination and that this effect is concentration dependent. Ishida [42] found lower germination rates in oak and pine trees within, compared to outside of, cormorant colonies but did not offer a detailed explanation for this pattern. Mulder and Keall [43] also found that seabird guano negatively affected seed germination and seedling survival. Germination inhibition with increasing concentrations of guano probably results from a number of factors, including reduced soil pH and the presence of a high concentration of soluble salts, both of which are characteristic of ornithogenic soils [20]. Very high concentrations of NO_3^- and especially NH_4^+ are

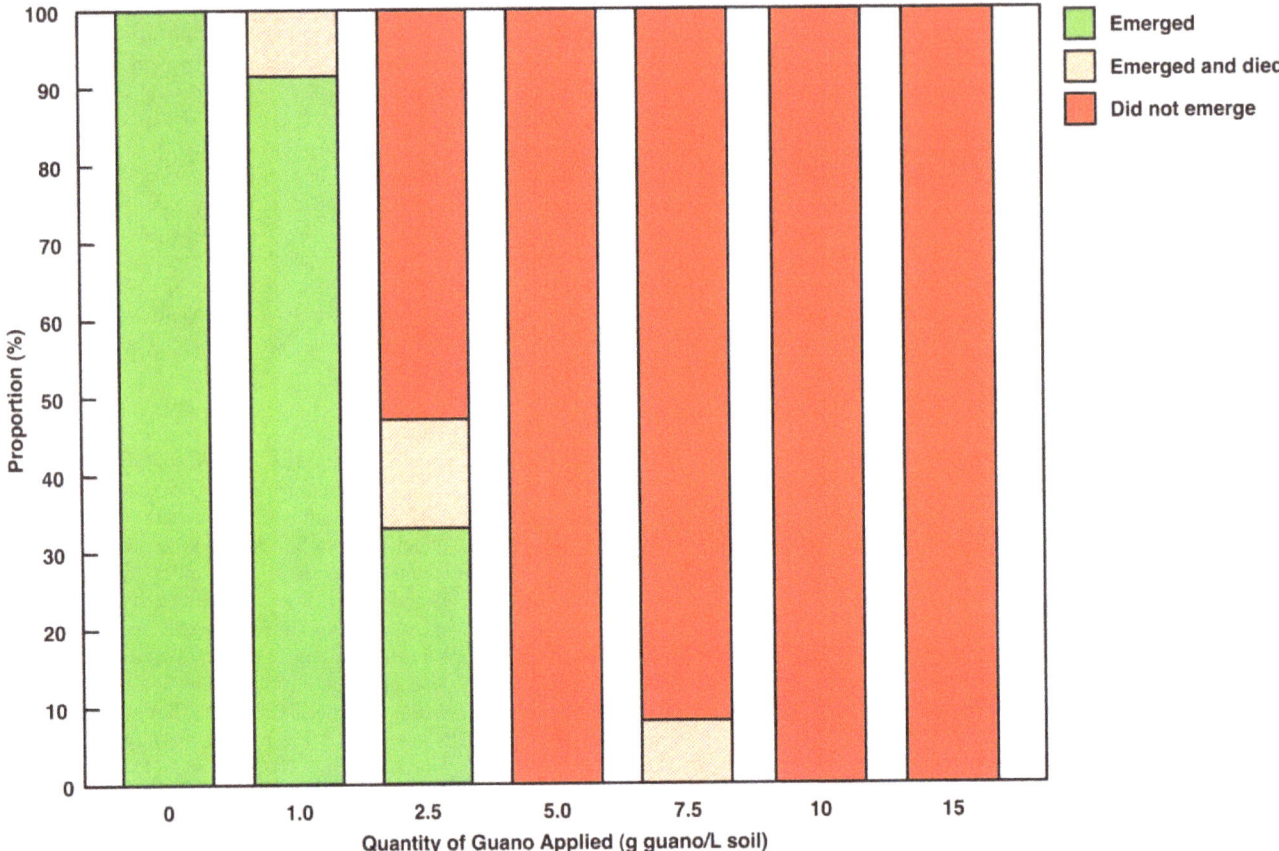

Figure 1. Relative percentages of seedlings that germinated and emerged with differing amounts of seabird guano applied.

also characteristic of ornithogenic soils and these characteristics can inhibit maize germination [44], with the early stages of growth being the most detrimental for plants under NH_4^+ stress [45,46].

Vital Effects of Guano Fertilization

Plant growth was strongly inhibited in the heavy guano treatment (G2). Maximum plant heights were significantly lower in G2 compared to C0 ($p = 0.02$) and G1 ($p = 0.008$) (Figure 2). While the G1 plants did not attain greater maximum heights than the C0 plants ($p = 0.83$), they yielded significantly more grain ($p = 0.004$). The G2 plants yielded less grain than the G1 plants ($p = 0.03$) and more grain than the C0 plants, although this difference was not statistically significant ($p = 0.42$) (Figure 2).

In this study, we observed a positive influence of guano on maize growth at moderate concentrations (G1), but a negative influence at high concentrations (G2). A number of studies have found that plant abundance and/or species richness tends to be lower within seabird colonies, but is often higher in areas in relatively close proximity to the colonies [14,22,47,48].

Very high levels of soil P can have deleterious effects on plant growth [49]. Ornithogenic soils may contain fifty times more available phosphorous than normal, but the P salts in bird excrement tend to be immobile in soil because of their limited solubility, making them generally unavailable for uptake by plants [4,50,51]. It is thus unlikely that the reduced growth observed in the G2 plants is the result of P toxicity. The most likely cause for the reduced growth of the G2 plants is NH_4^+ toxicity.

Very high $[NH_4^+]$ is a ubiquitous trait of ornithogenic soils [52–56]. High soil NH_4^+ can negatively impact plants in several ways:

(1) soil acidification, particularly of the rhizosphere [57], 'scorching' of root hairs [46]; (2) accumulation of free NH_4^+ in plant tissues, which has the capacity to uncouple plastid energy gradients [46]; (3) assimilation of NH_4^+ in the roots and associated translocation of carbon skeletons from the shoot, which is metabolically expensive and places 'carbon stress' on roots [58]; (4) suppression of the expression of certain proteins (aquaporins), which can have detrimental effects on the uptake of water [59]; and (5) the influx and efflux of NH_4^+ through root cells, which is associated with a very high metabolic cost when source $[NH_4^+]$ is high [60].

Both the G1 and G2 plants exhibited significantly reduced growth compared to the control plants for the first 45 days of the experiment (Figure 2; $p = 0.01$), but this trend did not continue as the G1 plants produced the greatest yields, and had similar maximum heights to the control plants. This is likely the result of initially very high soil $[NH_4^+]$, which negatively impacted the growth of the fertilized plants, followed by increased soil NO_3^- resulting from nitrification of guano-derived NH_4^+. When plants largely supplied with NH_4^+ as an N source are supplemented with NO_3^-, NH_4^+ uptake is suppressed and plants are able to resume normal growth [58]. The fact that the G2 plants still produced grain even though they were characterized by reduced heights and less above–ground biomass than either the control or G1 plants suggests that there was some acclimatization of these plants to the high $[NH_4^+]$, and/or nitrification was substantially delayed and $[NH_4^+]$ remained high in the soil for a much longer period of time. Schortemeyer et al. [46] observed a similar result in maize plants grown with NH_4^+ as the sole N source.

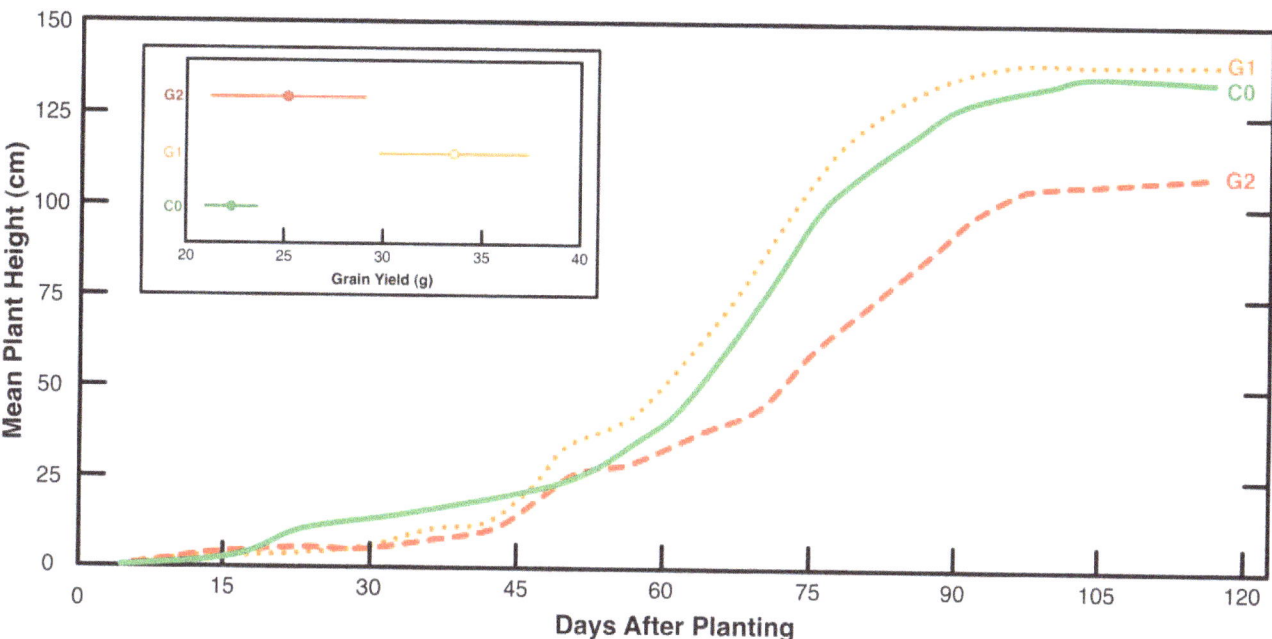

Figure 2. Maximum heights of maize plants throughout experiment. Harvest occurred at 115 d. Inset: grain yield for each experiment.

The effects of guano on plants are difficult to generalize. There is considerable variability at the community level and also within a community in accordance with plant physiology (nutrient demands, salt tolerance) at the species level [47,52]. Even within maize there are differences in NH_4^+ tolerance, with some varieties being able to survive higher concentrations than others [46]. Therefore, it cannot be assumed that the results of this study are directly applicable to all maize varieties.

Nitrogen Isotope Composition of Seabird Guano

Most inorganic N fertilizers have $\delta^{15}N$ values close to 0‰, with organic fertilizers generally having highly variable but positive $\delta^{15}N$ values (Table 2). The $\delta^{15}N$ value of the seabird guano used in this experiment was 26.7±0.6‰ (5 analyses), which is much higher than any other organic fertilizer analyzed to date. This is the product of avian nitrogen metabolism and excretion, which is quite different than in mammals, combined with the high trophic position of the guano-producing birds. Guano contains 9–21% nitrogen, which is composed primarily of uric acid (~80%), with smaller amounts of protein (~10%), ammonia (~7%), and nitrate (~0.5%) [23,61–67]. In addition, guano contains ~4% phosphorous (~50% of which is PO_4^{3-}) and 2% potassium [62,67,68].

A simplified pathway for guano nitrogen, with associated nitrogen-isotope fractionation factors, is shown in Figure 3. The principal producers of guano on the western coast of South America are the Peruvian booby (*Sula variegata*), brown pelican (*Pelecanus occidentalis thagus*), and guanay cormorant (*Phalacrocorax bougainvilli*) [1,12]. These birds, and similar species, feed at high trophic levels, and typically have tissue $\delta^{15}N$ values in the range of 17 to 20‰ [69–71], suggesting a $\delta^{15}N_{diet}$ of 14 to 18‰ assuming a diet–tissue fractionation of 3–4‰ for $\delta^{15}N$ [72]. Thus, the high trophic level of the birds only partially explains the very high $\delta^{15}N_{bulk\ guano}$ of 26.7‰.

After deposition in the soil, the uric acid in guano is rapidly mineralized to NH_4^+, and this process occurs much more rapidly in the presence of water [68,73,74]. Based on results presented by Mizutani and Wada [65], uric acid quickly decomposed (75% in

ten days) in soil, but the $\delta^{15}N$ value of the remaining uric acid was unchanged. A very large isotopic fractionation (−40 to −60‰) occurs during NH_3 volatilization, leaving the remaining soil NH_4^+ highly enriched in ^{15}N [38,75]. Ammonia volatilization is largely responsible for the high $\delta^{15}N$ values in ornithogenic soils and in some cases, seabird guano (Table 1). The relatively high $\delta^{15}N$ value of the guano utilized in this study suggests that some of the NH_4^+ in the guano had been subject to volatilization prior to deposition in the soil during the experiment; similar observations have been made concerning other avian manures [76].

^{15}N Enrichment in Guano Fertilized Plants

Plant isotopic compositions are summarized in Table 3; raw data are presented in Table S1. Plant organs of fertilized plants (G1, G2) sampled at 115 d were significantly enriched in ^{15}N compared to control plants in every case (Tables 3, 4; Figure 4). Also, the $\delta^{15}N$ values of plant tissues were significantly higher for heavily fertilized (G2) versus more lightly fertilized (G1) plants (Tables 3, 4). The difference in mean $\delta^{15}N$ values between the G1 and G2 plant organs was fairly consistent: 6.2‰ for stalks and roots, 6.4‰ for leaves (at 115 d), 7.6‰ for grain, and 7.8‰ for anthers.

A growing body of literature has emerged in recent years demonstrating that organic fertilizers, specifically those derived from animal waste, can cause large ^{15}N enrichments of plant tissues (Table 5). The $\delta^{15}N$ values reported here for plants grown in guano-fertilized soils are significantly higher than any published $\delta^{15}N$ values for plants grown on other organic fertilizers to date (Table 5), but comparable to $\delta^{15}N$ values for plants growing in ornithogenic soils (Table 2). The higher $\delta^{15}N$ values in the G1 and G2 compared to the C0 plants is the result of the uptake of ^{15}N-enriched guano-derived nitrogen. Moreover, the significantly higher tissue $\delta^{15}N$ values in the G2 compared to G1 plants reflects, at least in part, the greater availability of guano-derived nitrogen throughout the course of the experiment. This does not imply that guano-derived N was absent in the G1 treatment towards the end of the experiment, but it is possible that N

Table 2. $\delta^{15}N$ values of organic and inorganic fertilizers.

Type	Fertilizer	Fertilizer $\delta^{15}N$ (‰, AIR)	Reference
Organic	Blood	6.0 ± 1.3	[191]
	Bonemeal	4.9 ± 0.3	[191]
	Cattle manure	5.0 ± 0.8	[192]
	Cattle manure	2.9 ± 0.5	[193]
	Cattle manure	4.5	[133]
	Cattle manure	3.1 ± 0.2	[85]
	Chicken manure	6.2 ± 1.9	[191]
	Fishmeal	7.1 ± 3.6	[191]
	Hoof and horn	6.4 ± 0.2	[191]
	Livestock manure	8.8 ± 4.4	[191]
	Livestock manure	8.7 ± 0.2	[132]
	Pig manure	13.9	[102]
	Pig manure	16.9	[194]
	Pig manure	11.3	[133]
	Pig manure	6.5	[133]
	Pig manure	16.4	[195]
	Poultry manure	8.6 ± 0.3	[132]
	Poultry manure	2.7	[133]
	Seabird guano	26.7 ± 0.6	This study
	Seaweed	2.5 ± 1.5	[191]
	Various composts	17.4 ± 1.2	[196]
Inorganic	$(NH_4)_2H_2PO_4$	-0.6 ± 0.4	[191]
	$(NH_4)_2SO_4$	1.7 ± 3.4	[191]
	$(NH_4)_2SO_4$	-1.6	[133]
	$(NH_4)_2SO_4$	-2.6	[197]
	KNO_3	-1.2 ± 0.3	[191]
	NH_4NO_3	-1.3	[128]
	NH_4NO_3	-0.6 ± 1.7	[191]
	NH_4NO_3	-1.7	[133]
	Urea	-2.4 ± 2.1	[191]
	Urea	-1.7	[133]
	Urea	-0.7	[195]

immobilization had overtaken N mineralization, reducing the amount of guano-derived N available to the plants.

Elemental Concentration in Plant Parts

There were significant differences in N content between fertilized and control plants, with fertilized plants tending to have significantly higher N (Tables 3, 4). There were no significant differences in C content between control and fertilized plants for all organs, with the exception of the stalks, which had significantly lower %C in the fertilized plants compared to the control, and in G2 compared to G1 plants.

In general, the differences in C and N content between fertilized and unfertilized plants can be attributed to the accumulation of proteins, particularly those related to the GS-GOGAT pathway, that assimilate NH_4^+ and amino acids. Free amino acids tend to accumulate unabated in plant tissues with increasing supply of N irrespective of source, although different amino acids may accumulate at different rates depending on plant species and N

source [45,77–80]. Moreover, many studies have noted an increase in proteins, such as GS, in plant tissue in accordance with increasing NH_4^+ supply [80,81]. Thus, the relatively high N content of the organs of fertilized plants likely reflects the accumulation of these N compounds.

The two amino acids that dominate the free amino acid pool when plants are supplied with excess N are glutamine and arginine [77,82]. Arginine, which has a very low C:N ratio (6:4), has been implicated as an important product for the accumulation of excess N, possibly as a buffering mechanism against NH_4^+ toxicity [45,83,84]. Again, the accumulation of high levels of arginine in NH_4^+-fed plants fits with the pattern observed in the G1, and particularly the G2 plants. The very high levels of N and low levels of C in the stalks of the fertilized plants (compared to the control) suggests that the stalk was the most important accumulator for metabolites produced from excess N.

A notable exception to the pattern of increased N with fertilization is the grain, for which there was no significant difference in N content between treatments (Table 4). Our results suggest that at different levels of N supply and plant N content, there was no preferential allocation of accumulated N to the grain, and N that was absorbed post-silking was probably not allocated to the grain. A similar pattern was observed by Ma and Dwyer [85], although it is important to bear in mind the variability among maize hybrids in N metabolism during grain filling [86].

As plants progress through various stages of growth, their uptake, metabolism and partitioning of N may change dramatically. In maize, a significant portion (45–65%) of the grain N is obtained from endogenous N reallocated primarily from the stalk and leaves, while the remaining grain N is obtained from uptake of exogenous soil N [87–90]. Leaf N content at 75 d and 115 d varied as a function of the amount of guano applied (ie. C<G1≤G2), although this was not the case for leaves sampled at 30 d, where there was no clear relationship between quantity of fertilizer applied and leaf N content (Figure 5a). This likely reflects both a reliance on stored seed N early in growth, and the short period of growth prior to transplanting (7 d) during which no fertilizer N was available.

We observed decreases in leaf N content over time, with leaf N content decreasing by 77.9% (C0), 46.9% (G1) and 47.1% (G2) between 30 and 115 d (Figure 5a). The maintenance of very high levels of N in G1 and G2 relative to C0 plants suggests the accumulation of plant N as a result of excess source N [91].

Based on the results of this study, seabird guano fertilization has the potential to significantly alter the C and N economy of maize plants. Specifically, fertilization results in increased N and decreased C:N ratio in plant tissues, which likely arises because of increased accumulation of N–rich metabolites such as arginine, glutamine, and proteins related to NH_4^+ metabolism.

Intraplant Variation in $\delta^{15}N$

Intraplant variability in nitrogen isotopic composition for all treatments was large, with maximum differences between mean organ $\delta^{15}N$ being 3.0‰ for C0, 12.9‰ for G1 and 11.4‰ for G2 (Figure 4). We found significant differences in the $\delta^{15}N$ values between maize plant organs for both control ($F_{4,20} = 7.41$, $p<0.001$) and fertilized ($F_{4,20} = 18.60$, $p<0.001$ for G1; $F_{4,20} = 28.73$, $p<0.001$ for G2) treatments (Figure 4). In all treatments, the grain possessed the lowest $\delta^{15}N$ value, while anthers had the highest $\delta^{15}N$ values in the control treatment and the second-highest $\delta^{15}N$ values in the fertilized treatments, following stalks (Figure 4).

Significant variability in $\delta^{15}N$ within plants has been recorded in several studies [30,92–98]. Evans [99] suggests that, in general,

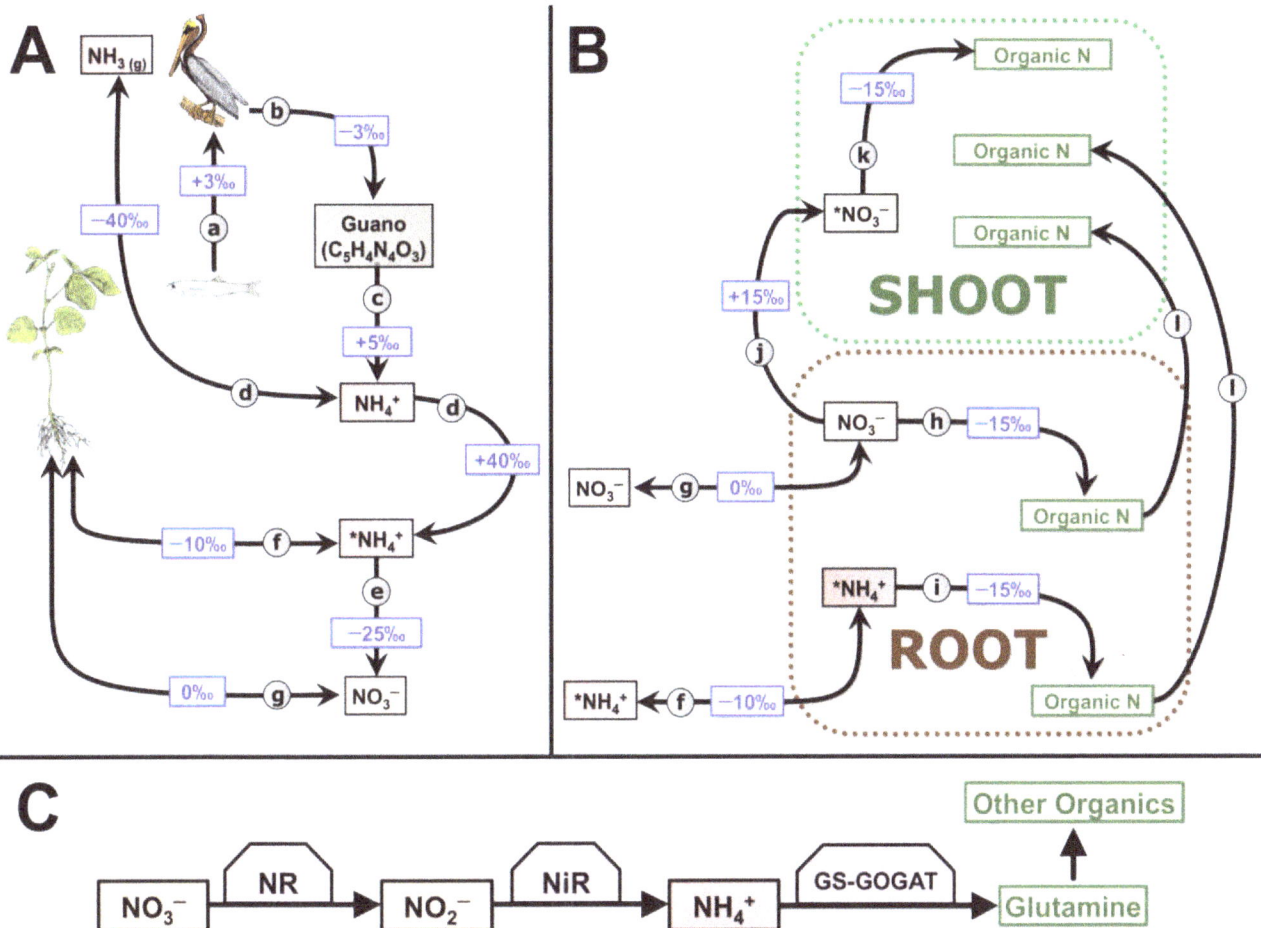

Figure 3. Simplified schematic of fractionation factors associated with decomposition and uptake of seabird guano N. A) Simplified pathway for guano-derived nitrogen. (**a**) Incorporation of dietary N into consumer tissue N. Tissue–diet fractionation for birds has been calculated to be ~3‰ for most tissues [160,161]. (**b**) Excretion of dietary N as uric acid. Wainright et al. [162] found bulk guano to be depleted of ^{15}N by 2.5‰ relative to seabird blood. Moreover, Mizutani et al. [75,163] and Bird et al. [164] found δ^{15}N of uric acid to be very similar to bulk guano δ^{15}N. (**c**) Conversion of uric acid to NH_4^+, according to the experiment performed by Mizutani et al. [163]. (**d**) Ammonia volatilization. Many studies have found this process to be associated with a large equilibrium fractionation that concentrates ^{15}N in the remaining substrate (*NH_4^+ in the diagram) [54,75,165]. (**e**) Nitrification. The fractionation factor for the entire process of nitrification in the soil ($NH_4^+{\rightarrow}NO_2^-{\rightarrow}NO_3^-$) is estimated to be between -12 and -35‰ [38,166,167]. (**f**) Uptake of NH_4^+ is associated with a nitrogen isotope fractionation ranging from -6 to -30‰ and appears to depend on the concentration of the source NH_4^+ [33,168]. (**g**) Uptake of NO_3^- by the plant does not appear to be associated with any fractionation [33,169,170]. Both NO_3^- and NH_4^+ may be effluxed from the plant, passively and in some cases actively [171]. **B**) (**h**) NO_3^- assimilation into organic N occurs in the root by the NR-NiR (nitrate reductase-nitrite reductase) and GS-GOGAT (glutamine synthetase–glutamine:oxoglutarate aminotransferase) pathways (see Figure 4C). The reduction of NO_3^- to NH_4^+ is associated with a fractionation factor of -15‰ [37,172]. (**i**) NH_4^+ assimilation occurs in the root via the GS-GOGAT pathway and is associated with a fractionation factor of -10 to -15‰ [40,94]. (**j, k**) NO_3^- may also be mobilized to the shoot for assimilation. In this case, this NO_3^- pool has already been exposed to NO_3^- assimilation in the root and is enriched in ^{15}N [95]. Therefore, organic N formed from NO_3^- in the shoot (*NO_3^-) will have a higher δ^{15}N value than organic N formed from NO_3^- in the root. (**l**) Organics may be moved between the root and shoot. **C**) Simplified schematic for the assimilation of N by plants. For a more detailed description see Miller and Cramer [171]. All fractionation factors are approximate values representing medians of ranges, which may be large (see text for discussion).

plants with NO_3^- as the primary N source are characterized by significant intraplant variability, while this is not true for plants with NH_4^+ as their primary N source. This general pattern results largely from the fact that NH_4^+ is assimilated into organic N only in the root, while NO_3^- assimilation occurs both in roots and shoots (Figure 3B) [57,100,101]. Therefore, organics derived from NH_4^+ are assimilated from the same N pool in the roots, while NO_3^- that has been translocated to the shoot prior to assimilation has already undergone some fractionation (in the roots) and is thus enriched in ^{15}N [30,95,99].

The δ^{15}N values of the roots were intermediate compared to other above-ground tissues, which does not fit with the scenario described above for NO_3^- fed plants in which shoot tissues have

higher δ^{15}N values than roots. In the C0 and G1 plants, the roots did not differ significantly from stalks, grains, or leaves in terms of δ^{15}N (Table 6). In the G2 plants, root δ^{15}N was significantly lower relative to the stalk, but significantly higher than the leaf or grain (Table 6). The lack of a consistent pattern of root vs. shoot δ^{15}N observed in this study likely reflects complex N metabolism, with relative reliance on NH_4^+ and NO_3^-, as well as guano-derived N changing over time.

The relatively low grain δ^{15}N values observed in this study are indicative of the reallocation of stored N. Choi et al. [102] also observed that grain tended to be depleted of ^{15}N compared to stalks and leaves. This can be attributed to a kinetic isotope effect associated with catabolism and remobilization of stored plant N,

Table 3. Isotopic and elemental compositions of plant samples (mean±1σ).

Treatment	Tissue	Sampling Date	δ¹⁵N (‰, AIR)	δ¹³C (‰, VPDB)	%N	%C
C0	Leaf	30 d	−0.3±3.6	–	5.1±0.7	–
	Leaf	75 d	5.5±1.4	–	2.7±0.6	–
	Leaf	115 d	3.9±1.0	−15.7±0.8	1.1±0.3	39.5±2.3
	Grain	115 d	2.7±0.7	−12.9±0.1	2.4±0.3	42.5±1.4
	Anther	75 d	5.7±0.4	−14.0±0.4	2.4±0.4	47.0±2.3
	Root	115 d	3.6±0.9	−15.0±0.5	0.8±0.2	44.7±1.1
	Stalk	115 d	3.2±1.4	−15.2±0.2	1.0±0.5	47.0±1.8
G1	Leaf	30 d	−5.0±10.0	–	4.9±0.9	–
	Leaf	75 d	32.4±2.2	–	4.1±0.3	–
	Leaf	115 d	26.8±2.0	−15.1±0.5	2.6±0.7	40.6±4.7
	Grain	115 d	25.5±1.6	−14.1±0.8	2.5±0.2	45.8±2.8
	Anther	75 d	34.2±3.4	−13.5±0.4	3.2±0.1	48.4±2.1
	Root	115 d	33.1±4.1	−15.2±0.3	1.4±0.7	44.2±4.2
	Stalk	115 d	38.4±1.9	−15.4±0.5	2.9±0.8	36.4±2.6
G2	Leaf	30 d	6.0±4.3	–	5.7±0.4	–
	Leaf	75 d	38.2±0.9	–	4.8±0.2	–
	Leaf	115 d	33.3±2.7	−15.4±1.0	3.2±0.6	42.6±3.9
	Grain	115 d	33.1±2.8	−13.3±0.3	2.6±0.2	44.6±1.7
	Anther	75 d	41.8±2.6	−13.5±0.4	3.3±0.5	45.2±4.4
	Root	115 d	40.1±2.6	−14.5±0.6	2.1±0.7	41.4±1.8
	Stalk	115 d	44.7±0.8	−14.7±0.6	3.4±0.2	29.7±0.8

which discriminates against ¹⁵N [103]. The high δ¹⁵N values of stalks suggest that this organ is an important source of accumulated N that is remobilized during grain filling. This supports the findings of Ta [104], who found that maize stalks functioned as a significant temporary storage reservoir for N-compounds. It is surprising that the leaves at 115 d are not characterized by higher δ¹⁵N values in comparison to the grain, as they are thought to be a significant contributor to grain N [105,106]; this is discussed in more detail below. The importance of stalk, compared to leaf, N during grain filling may be specific to

Table 4. Results of ANOVA for differences in isotopic and elemental tissue compositions between treatments.

Tissue	Treatment	G1 δ¹⁵N (‰, AIR)	δ¹³C (‰, VPDB)	%N	%C	G2 δ¹⁵N (‰, AIR)	δ¹³C (‰, VPDB)	%N	%C
Leaf 30 d	C0	0.707	–	0.889	–	0.096	–	0.347	–
	G1	–	–	–	–	0.171	–	0.176	–
Leaf 75 d	C0	**<0.001**	–	**0.008**	–	**<0.001**	–	**0.002**	–
	G1	–	–	–	–	**<0.001**	–	**0.002**	–
Leaf 115 d	C0	**<0.001**	0.509	**0.003**	0.884	**<0.001**	0.857	**<0.001**	0.414
	G1	–	–	–	–	**0.001**	0.819	0.357	0.686
Stalk	C0	**<0.001**	0.640	**<0.001**	**<0.001**	**<0.001**	0.249	**<0.001**	**<0.001**
	G1	–	–	–	–	**<0.001**	0.056	0.415	**<0.001**
Grain	C0	**<0.001**	0.066	0.760	0.092	**<0.001**	0.150	0.463	0.348
	G1	–	–	–	–	**<0.001**	0.221	0.869	0.632
Anther	C0	**<0.001**	0.118	**0.010**	0.746	**<0.001**	0.135	**0.006**	0.669
	G1	–	–	–	–	**0.017**	0.997	0.940	0.280
Root	C0	**<0.001**	0.746	0.227	0.958	**<0.001**	0.249	**<0.001**	0.163
	G1	–	–	–	–	**<0.001**	0.076	0.415	0.250

Values in boldface are statistically significant (p<0.05).

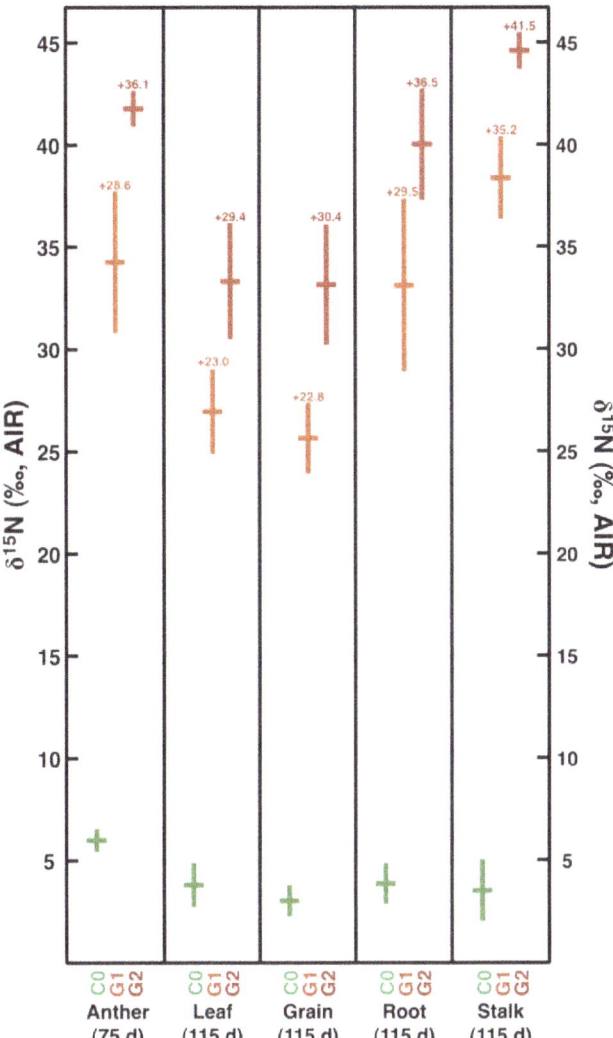

Figure 4. Mean nitrogen isotope composition of maize organs; horizontal bars represent means, vertical bars represent standard deviations. Values above G1 and G2 maize represent differences in nitrogen isotopic composition relative to C0 maize.

this variety of maize. Further study of the nitrogen metabolism of different maize hybrids is needed to clarify this issue.

Temporal Variation in Plant $\delta^{15}N$ Values

There was significant variability in maize leaves over the course of the experiment (Figure 5B). Maize leaves sampled at 115 d had lower $\delta^{15}N$ values than those sampled at 75 d for all treatments; these differences were statistically significant for the fertilized groups, but not for the control group (Table 4). For all treatments, leaf $\delta^{15}N$ values were significantly lower at 30 d compared to 75 d (Table 4).

Several studies have attempted to document changes in plant $\delta^{15}N$ values over time and/or arising from natural leaf senescence. Kolb and Evans [97] and Garten [107] found no significant differences in the $\delta^{15}N$ values of living and abscised leaves, which suggested a lack of ^{15}N discrimination with N remobilization. Conversely, several other studies have found older or senescent plant leaves to be characterized by higher $\delta^{15}N$ values, which has been attributed to a kinetic isotopic fractionation associated with N

catabolism and reallocation [108–110]. We observed no significant difference between leaf $\delta^{15}N$ at 75 d and 115 d for the control group, suggesting that under normal circumstances, there is no significant fractionation associated with N remobilization from leaves for this variety of maize. That there was a concurrent decrease in N content and $\delta^{15}N$ for leaves between 75 and 115 d in the fertilized plants is counterintuitive, as the reallocation of leaf N to the grain should result in a ^{15}N-enriched leaf. As was previously suggested for the stalk, we suspect that a significant portion of the leaf N pool consisted of accumulated N in the form of free amino acids (especially arginine and glutamine) as a result of high N supply and, in particular, high source [NH$_4^+$]. The reason that older or senescent plant parts are characterized by higher $\delta^{15}N$ values is because the metabolic processes involved (e.g. deamination, transamination) are associated with large kinetic fractionations that concentrate the remaining substrate in ^{15}N [111]. Therefore, if the majority of the decrease in leaf N between 75 and 115 d is the result of the transfer of organic N products (amino acids) to another part in the plant (e.g. the stalk), which is not associated with any known ^{15}N fractionation [112], this would help to explain why the leaves are not relatively enriched in ^{15}N at 115 compared to 75 d.

Leaf $\delta^{15}N$ values were more variable at 30 d than at either 75 or 115 d (Figure 5c). This is likely a result of variable reliance on stored and absorbed N sources. Kolb and Evans [97] found that young leaves (*Quercus* and *Encelia*) had an isotopic composition ($\delta^{15}N$) that reflected both stored and absorbed N, while mature leaf $\delta^{15}N$ values reflected primarily absorbed N. Very low leaf $\delta^{15}N$ values (−12.4, −12.4, −10.2‰) were observed at 30 d for three of the guano-fertilized maize plants. These compositions probably arise from physiological responses to high soil [NH$_4^+$]. At high extracellular [NH$_4^+$], influx of NH$_4^+$ occurs only via a low-affinity transport system, with high-affinity transport system proteins being down-regulated; this process occurs in concert with the active efflux of NH$_4^+$ from the roots [34]. Yoneyama et al. [33] suggest that when NH$_4^+$ assimilation is slow (because extracellular [NH$_4^+$] is high), NH$_4^+$-N isotopic fractionation is larger, with relatively more ^{15}N-enriched NH$_4^+$ being effluxed from the cell. Ariz et al. [34] found plants that were most sensitive to NH$_4^+$ toxicity also had the lowest tissue $\delta^{15}N$ values. The fact that not all plants in the present study were characterized by low leaf $\delta^{15}N$ values is difficult to explain, but may be the result of heterogeneous distribution of the guano throughout the soil or genotypic variability in resilience to NH$_4^+$ toxicity.

Temporal patterns in plant $\delta^{15}N$ values are complicated and are determined by a number of factors. We suspect that significant changes in the N source occurred over time as a result of soil nitrification, and there were also significant changes in [source N] over time. This complication, however, is a reality of working with animal fertilizers, rather than hydroponic solutions, and must be taken into account when interpreting data from field settings.

Guano Fertilization and Plant Carbon Isotopic Composition

We observed no difference in plant $\delta^{13}C$ values resulting from guano fertilization for any of the organs analyzed (Tables 3, 4). In earlier studies, variable plant N sources have been associated with small, but significant variations in $\delta^{13}C$ values [113]. It is thought that this association arises because different N sources (and different N source concentrations) may alter plant water-use efficiency and thus change the carbon isotope composition of plant tissues [114].

Previous studies have found plant $\delta^{13}C$ values to be distinct in organic vs. inorganic fertilization regiments, an outcome ascribed

Table 5. Summary of studies examining the influence of organic fertilization on plant $\delta^{15}N$ values.

Fertilizer	Fertilizer $\delta^{15}N$ (‰, AIR)	Plant	Plant $\delta^{15}N$ (‰, AIR)	$\Delta^{15}N_{fertilized-control}$	Reference
Pig manure	13.9	Maize	7.7	+1.1	[102]
Various composts	17.4±1.2	Maize	17.7	+13.5	[196]
Various composts	17.4±1.2	Nightshade	13.4	+10.7	[196]
Various composts	17.4±1.2	Pepper	14.5	+9.8	[196]
Various composts	17.4±1.2	Mustard	16.3	+12.7	[196]
Various composts	17.4±1.2	Melon	13.3	+10.1	[196]
Various composts	17.4±1.2	Lettuce	13.5	+9.4	[196]
Various composts	17.4±1.2	Spinach	9.5	+3.9	[196]
Various composts	17.4±1.2	Beefsteak plant	19.9	+15.4	[196]
Various composts	17.4±1.2	Sesame	17.8	+12.1	[196]
Pig manure	16.9	Chrysanthemum	10.3	+3.5	[194]
Pig manure	16.9	Cabbage	13.3	+5.6	[194]
Sheep manure	–	Sweet pepper	10.0	–	[198]
Chicken manure	–	Sweet pepper	10.2	–	[198]
Horse manure	–	Sweet pepper	9.8	–	[198]
Livestock manure	8.7±0.2	Orange (pulp)	9.0	–	[132]
Poultry manure	8.6±0.3	Orange (pulp)	8.5	–	[132]
Livestock manure	8.7±0.2	Orange (juice)	8.5	–	[132]
Poultry manure	8.6±0.3	Orange (juice)	7.9	–	[132]
Pig manure	16.4	Chinese cabbage	12.5	+11.0	[195]
Mixed (Cattle+poultry manure)	16.7	Tomato	13.5	+10.2	[199]
Mixed (Cattle+poultry manure)	9.9	Tomato	7.9	+4.6	[199]

to higher rates of soil microbiological activity [115,116]. Specifically, Georgi et al. [116] suggest that CO_2 released during decomposition is depleted of ^{13}C. Because control and fertilized plants were grown in the same growth chamber, there would be no differences in the $\delta^{13}C$ of CO_2 utilized by either group of plants, although this may not be true for an agricultural field fertilized with guano. In general, the influence on nitrogenous fertilizers (both organic and inorganic) on plant $\delta^{13}C$ is unclear. Experimental results have been conflicting, with studies finding $\delta^{13}C$ values to increase [117–122], decrease [123], or be unaffected [120,124] in response to N fertilization. The relationship between N fertilizer application and plant $\delta^{13}C$ is likely mediated by several factors and warrants further study. We likely did not detect any difference in plant $\delta^{13}C$ values resulting from fertilization because the magnitude of difference would be quite small [113] and our sample size was also quite small ($n = 5$ per treatment).

Implications for Food Chemistry

Seabird guano is becoming increasingly popular as an organic alternative among farmers in the United States and Europe [5]. Moreover, as the demand for organically grown produce soars worldwide [125], there is an increased incentive for farmers in areas in close proximity to guano deposits (e.g. Peru, Ecuador, Chile, and Namibia) to use this fertilizer and market their produce as organic [5]. In recent years, there has been a surge in isotopic research directed at demonstrating isotopic distinctions between conventional and organically grown produce [126–136]. The reason that this technique may sometimes be effective is primarily that inorganic fertilizers tend to have $\delta^{15}N$ values close to 0‰, while organic fertilizers tend to have higher $\delta^{15}N$ values, although

there is great variability (Table 2). Based on the results of this study, the application of seabird guano in an organic fertilization regime would result in a very large ^{15}N enrichment of all plant tissues in comparison to unfertilized plants, or to plants treated with chemical fertilizers. The magnitude of this difference is much greater than what has been observed for other organic fertilizers (Table 5), and thus isotopic data would be useful in verifying use of seabird guano. Moreover, the very high $\delta^{15}N$ value of the guano itself suggests that its presence in mixed organic fertilizers should also be detectable via isotope ratio mass spectrometry.

Implications for Archaeology

Stable isotope analysis ($\delta^{13}C$ and $\delta^{15}N$ in particular) plays an increasingly important role in the reconstruction of prehistoric diet. Dietary reconstruction requires a thorough understanding of the sources of isotopic variation in the foods that were consumed [137]. Recently, the notion that animal manure may have influenced the $\delta^{15}N$ values of plants grown in prehistoric Europe has been proposed [138–140] and integrated into regional paleodietary studies. In the Andean region, several fertilizers are thought to have been of some importance in prehispanic agriculture including llama dung [141] and seabird guano [9,142,143]. Based on the large settlements that developed on the coast of Peru (e.g. Moche, Chimú) and the relative infertility of local soils, Nordt et al. [8] have suggested that the application of some kind of nitrogenous fertilizer, possibly seabird guano, would have been necessary to maintain agricultural productivity in at least some parts of the region. Direct evidence for fertilization, however, is very difficult to come by. One of the primary goals of this study was to determine whether or not the enrichment in ^{15}N

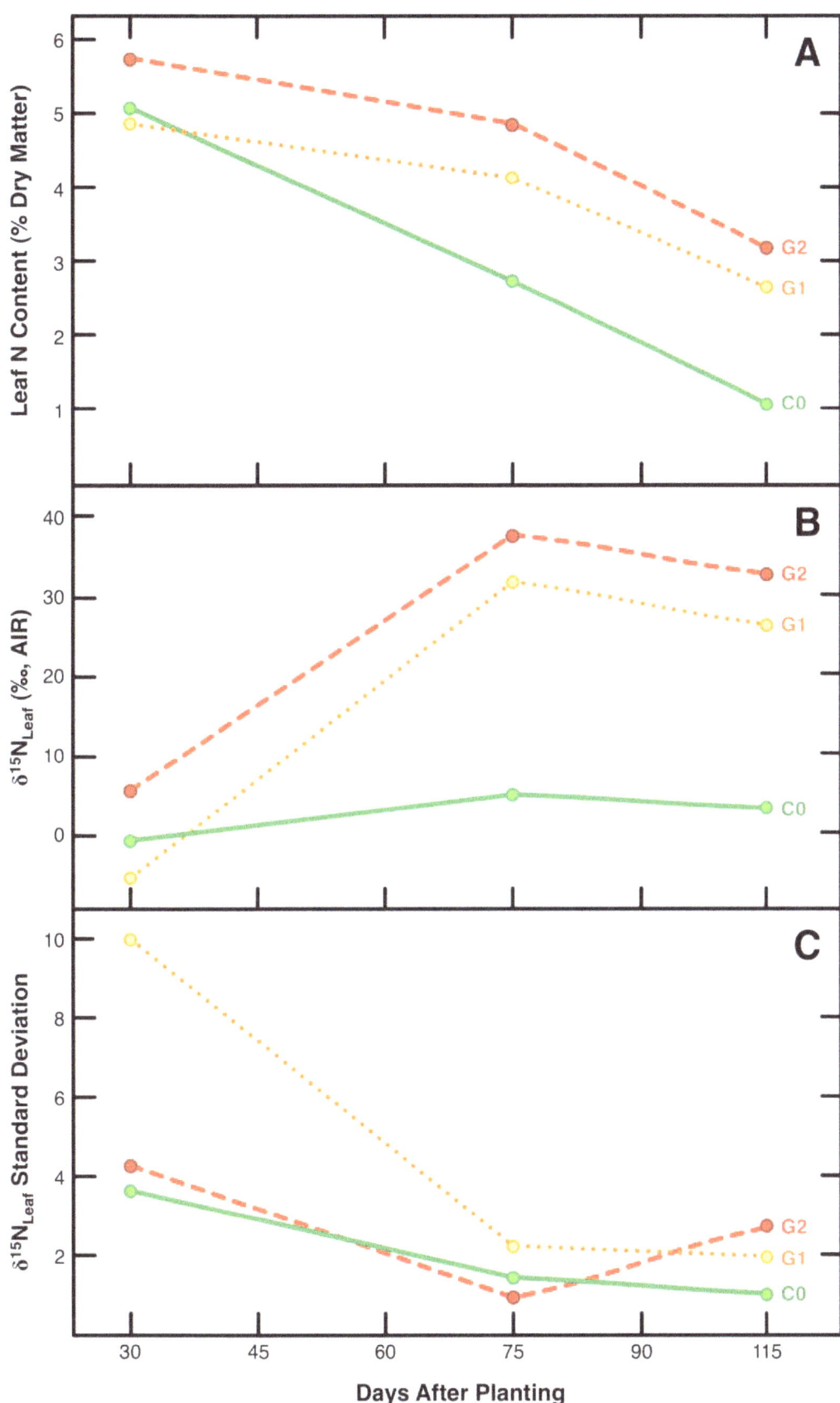

Figure 5. Temporal patterns in isotopic and element composition. (**A**) Leaf N content, (**B**) Leaf $\delta^{15}N$, and (**C**) standard deviation for Leaf $\delta^{15}N$.

Table 6. Results of ANOVA for differences in nitrogen isotopic composition between plant parts.

Treatment	Tissue	Leaf	Anther	Root	Stalk
C0	Grain	0.319	**<0.001**	0.626	0.908
	Leaf	–	**0.041**	0.981	0.803
	Anther	–	–	**0.013**	**0.004**
	Root	–	–	–	0.980
G1	Grain	0.915	**0.017**	0.077	**<0.001**
	Leaf	–	**0.035**	0.152	**<0.001**
	Anther	–	–	0.999	0.309
	Root	–	–	–	0.252
G2	Grain	0.999	**<0.001**	**<0.001**	**<0.001**
	Leaf	–	**<0.001**	**<0.001**	**<0.001**
	Anther	–	–	0.709	0.259
	Root	–	–	–	**0.022**

Values in boldface are statistically significant ($p<0.05$).

resulting from guano fertilization would be sufficient to detect this agricultural practice in the isotopic composition of a human or animal consuming the fertilized plant. Based on the results of this study and others that have examined the biogeochemistry of seabird-associated sites (summarized in Table 2), the application of seabird guano to agricultural fields would have caused a significant increase in the $\delta^{15}N$ value of plants and of animals consuming these plants. In archaeological bone collagen from western South America, high $\delta^{15}N$ values are usually accompanied by high $\delta^{13}C$ values. This pattern applies to both humans [144–146] and domestic animals [147], and has generally been attributed to the consumption of high trophic-level marine resources (e.g. predatory fish, marine mammals). Conversely, this pattern may also be caused by the consumption of maize (a C_4 plant) fertilized with seabird guano, which appears (isotopically) very much like a high-trophic level marine organism. As such, it is important to be mindful of the possibility of guano-fertilization when interpreting diet, not just on the coast, but in the interior highland region as well. According to ethnohistoric documents, guano was moved great distances and prized by groups living in the highlands as an essential component in maize agriculture [9].

The Andes were certainly not the only region in which seabird guano was used extensively as a fertilizer. Millions of tonnes of guano were exported to Europe and North America during the nineteenth century and Peruvian seabird guano was the most highly prized fertilizer at that time [148–150]. Isotopic analysis is being employed with increased frequency within the context of historical archaeology [151–159], a period during which the possible influence of seabird guano must also be considered.

Supporting Information

Table S1 Raw isotopic and elemental data for all samples analyzed.

Acknowledgments

Steve Bartlett (Biotron), Kim Law and Li Huang (LSIS) provided technical assistance. Sharon Buck assisted with sample preparation. This is Laboratory for Stable Isotope Science Contribution #279.

Author Contributions

Conceived and designed the experiments: PS FJL JFM CDW. Performed the experiments: PS. Analyzed the data: PS FJL CDW. Wrote the paper: PS.

References

1. Duffy DC (1994) The guano islands of Peru: the once and future management of a renewable resource. In: Nettleship DN, Burger J, Gochfeld M, eds. Seabirds on Islands, Threats, Case Studies and Action Plans. Cambridge: BirdLife Conservation Series, No. 1. pp 68–76.
2. Murphy RC (1981) The guano and the anchoveta fishery. In: Glantz MH, Thompson JD, eds. Resource Management and Environmental Uncertainty: Lessons from Coastal Upwelling Fisheries. New York: Wiley. pp 81–106.
3. Mathew WM (1970) Peru and the British Guano Market, 1840–1870. The Economic History Review 23: 112–128.
4. Zapata F, Arrillaga JL (2002) Agronomic evaluation of guano sources by means of isotope techniques. In: Sikora F, ed. Assessment of soil phosphorus status and management of phosphatic fertilisers to optimise crop production. Vienna: International Atomic Energy Agency. pp 83–89.
5. Romero S (2008) Peru guards its guano as demand soars again. New York Times. May 30 ed. New York.
6. Garcilaso de la Vega I (1966) Royal Commentaries of the Incas and General History of Peru. Livermore HL, translator. Austin: University of Texas Press. 1530 p.
7. Cieza de León Pd (1964) The Travels of Pedro de Cieza de León: A.D. 1532–50, contained in the first part of his Chronicle of Peru. Markham CR, translator. New York: Franklin. 438 p.
8. Nordt L, Hayashida F, Hallmark T, Crawford C (2004) Late prehistoric soil fertility, irrigation management, and agricultural production in northwest coastal Peru. Geoarchaeology 19: 21–46.
9. Julien CJ (1985) Guano and resource control in sixteenth-century Arequipa. In: Masuda S, Shimada I, Morris C, eds. Andean Ecology and Civilization: An Interdisciplinary Perspective on Andean Ecological Complementarity. Tokyo: University of Tokyo Press. pp 185–231.
10. Powell GVN, Fourqurean JW, Kenworthy WJ, Zieman JC (1991) Bird colonies cause seagrass enrichment in a subtropical estuary: Observational and experimental evidence. Estuarine, Coastal and Shelf Science 32: 567–579.
11. Ishizuka K (1966) Ecology of the ornithocoprophilous plant communities on breeding places of the black-tailed gull, *Larus crassirostris*, along the coast of Japan: I. Vegetation analysis. Ecological Review 16: 229–244.
12. Hutchinson GE (1950) Survey of Existing Knowledge of Biogeochemistry: 3. The Biogeochemistry of Vertebrate Excretion. Bulletin of the American Museum of Natural History 96: 1–554.
13. Anderson WB, Polis GA (1999) Nutrient fluxes from water to land: seabirds affect plant nutrient status on Gulf of California islands. Oecologia 118: 324–332.
14. Ryan PG, Watkins BP (1989) The influence of physical factors and ornithogenic products on plant and arthropod abundance at an Inland Nunatak group in Antarctica. Polar Biology 10: 151–160.
15. Burger AE, Lindeboom HJ, Williams AJ (1978) The mineral and energy contributions of guano of selected species of birds to the Marion Island terrestrial ecosystem. South African Journal of Antarctic Research 8: 59–70.
16. Engelskjon T (1986) Botany of two Antarctic mountain ranges: Gjelsvikfjella and Mühlig-Hofmannfjella, Dronning Maud Land. Polar Research 4: 205–224.
17. Leentvaar P (1967) Observations in guanotrophic environments. Hydrobiologia 29: 441–489.
18. McColl JG, Burger J (1976) Chemical Inputs by a Colony of Franklin's Gulls Nesting in Cattails. American Midland Naturalist 96: 270–280.
19. Smith VR (1978) Animal-plant-soil nutrient relationships on Marion Island (Subantarctic). Oecologia 32: 239–253.
20. Speir TW, Cowling JC (1984) Ornithogenic soils of the Cape Bird Adelie penguin rookeries, Antarctica: 1. Chemical Properties. Polar Biology 2: 199–205.
21. Ellis JC (2005) Marine Birds on Land: A Review of Plant Biomass, Species Richness, and Community Composition in Seabird Colonies. Plant Ecology 181: 227–241.
22. Vidal E, Jouventin P, Frenot Y (2003) Contribution of alien and indigenous species to plant-community assemblages near penguin rookeries at Crozet archipelago. Polar Biology 26: 432–437.
23. Lindeboom HJ (1984) The Nitrogen Pathway in a Penguin Rookery. Ecology 65: 269–277.
24. Williams AJ, Berruti A (1978) Mineral and energy contributions of feathers moulted by penguins, gulls and cormorants to the Marion Island terrestrial ecosystem. South African Journal of Antarctic Research 8: 71–74.

25. Williams AJ, Burger AE, Berruti A (1978) Mineral and energy contributions of carcasses of selected species of seabirds to the Marion Island terrestrial ecosystem. South African Journal of Antarctic Research 8: 53–58.

26. Siegfried WR, Williams AJ, Burger AE, Berruti A (1978) Mineral and energy contributions of eggs of selected species of seabirds to the Marion Island terrestrial ecosystem. South African Journal of Antarctic Research 8: 75–87.

27. Gillham ME (1956) Ecology of the Pembrokeshire Islands: V. Manuring by the Colonial Seabirds and Mammals, with a Note on Seed Distribution by Gulls. Journal of Ecology 44: 429–454.

28. Crawford NM, Glass ADM (1998) Molecular and physiological aspects of nitrate uptake in plants. Trends in Plant Science 3: 389–395.

29. Andrews M (1986) The partitioning of nitrate assimilation between root and shoot of higher plants. Plant, Cell & Environment 9: 511–519.

30. Yoneyama T, Kaneko A (1989) Variations in the Natural Abundance of 15N in Nitrogenous Fractions of Komatsuna Plants Supplied with Nitrate. Plant and Cell Physiology 30: 957–962.

31. Mariotti A, Mariotti F, Amargar N, Pizelle G, Ngambi JM, et al. (1980) Fractionnements isotopiques de l'azote lors des processus d'absorption des nitrates et de fixation de l'azote atmosphérique par les plantes. Physiologie Végétale 18: 163–181.

32. Yoneyama T, Ito O, Engelaar WMHG (2003) Uptake, metabolism and distribution of nitrogen in crop plants traced by enriched and natural ^{15}N: Progress over the last 30 years. Phytochemistry Reviews 2: 121–132.

33. Yoneyama T, Matsumaru T, Usui K, Engelaar WMHG (2001) Discrimination of nitrogen isotopes during absorption of ammonium and nitrate at different nitrogen concentrations by rice (Oryza sativa L.) plants. Plant, Cell & Environment 24: 133–139.

34. Ariz I, Cruz C, Moran J, González-Moro M, García-Olaverri C, et al. (2011) Depletion of the heaviest stable N isotope is associated with NH_4^+/NH_3 toxicity in NH_4^+-fed plants. BMC Plant Biology 11: 83.

35. Flores-Delgadillo L, Fedick SL, Solleiro-Rebolledo E, Palacios-Mayorga S, Ortega-Larrocea P, et al. (2011) A sustainable system of a traditional precision agriculture in a Maya homegarden: Soil quality aspects. Soil and Tillage Research 113: 112–120.

36. Pritchard ES, Guy RD (2005) Nitrogen isotope discrimination in white spruce fed with low concentrations of ammonium and nitrate. Trees - Structure and Function 19: 89–98.

37. Ledgard SF, Woo KC, Bergersen FJ (1985) Isotopic fractionation during reduction of nitrate and nitrite by extracts of spinach leaves. Australian Journal of Plant Physiology 12: 631–640.

38. Robinson D (2001) δ^{15}N as an integrator of the nitrogen cycle. Trends in Ecology & Evolution 16: 153–162.

39. Glass ADM, Brito DT, Kaiser BN, Kronzucker HJ, Kumar A, et al. (2001) Nitrogen transport in plants, with an emphasis on the regulation of fluxes to match plant demand. Journal of Plant Nutrition and Soil Science 164: 199–207.

40. Yoneyama T, Kamachi K, Yamaya T, Mae T (1993) Fractionation of Nitrogen Isotopes by Glutamine Synthetase Isolated from Spinach Leaves. Plant and Cell Physiology 34: 489–491.

41. Subedi KD, Ma BL (2005) Nitrogen Uptake and Partitioning in Stay-Green and Leafy Maize Hybrids. Crop Science 45: 740–747.

42. Ishida A (1997) Seed germination and seedling survival in a colony of the common cormorant, Phalacrocorax carbo. Ecological Research 12: 249–256.

43. Mulder C, Keall S (2001) Burrowing seabirds and reptiles: impacts on seeds, seedlings and soils in an island forest in New Zealand. Oecologia 127: 350–360.

44. Bremner JM, Krogmeier MJ (1989) Evidence that the adverse effect of urea fertilizer on seed germination in soil is due to ammonia formed through hydrolysis of urea by soil urease. Proceedings of the National Academy of Sciences 86: 8185–8188.

45. Roosta HR, Schjoerring JK (2007) Effects of Ammonium Toxicity on Nitrogen Metabolism and Elemental Profile of Cucumber Plants. Journal of Plant Nutrition 30: 1933–1951.

46. Schortemeyer M, Stamp P, Feil B (1997) Ammonium Tolerance and Carbohydrate Status in Maize Cultivars. Annals of Botany 79: 25–30.

47. Wootton JT (1991) Direct and indirect effects of nutrients on intertidal community structure: variable consequences of seabird guano. Journal of Experimental Marine Biology and Ecology 151: 139–153.

48. Weseloh DV, Brown RT (1971) Plant Distribution within a Heron Rookery. American Midland Naturalist 86: 57–64.

49. Clarkson DT, Scattergood CB (1982) Growth and Phosphate Transport in Barley and Tomato Plants During the Development of, and Recovery from, Phosphate-stress. Journal of Experimental Botany 33: 865–875.

50. Ligeza S, Smal H (2003) Accumulation of nutrients in soils affected by perennial colonies of piscivorous birds with reference to biogeochemical cycles of elements. Chemosphere 52: 595–602.

51. García LV, Marañón T, Ojeda F, Clemente L, Redondo R (2002) Seagull influence on soil properties, chenopod shrub distribution, and leaf nutrient status in semi-arid Mediterranean islands. Oikos 98: 75–86.

52. Young HS, McCauley DJ, Dirzo R (2011) Differential responses to guano fertilization among tropical tree species with varying functional traits. American Journal of Botany 98: 207–214.

53. Mizota C (2009) Temporal variations in the concentration and isotopic signature of ammonium- and nitrate-nitrogen in soils under a breeding colony of Black-tailed Gulls (Larus crassirostris) on Kabushima Island, northeastern Japan. Applied Geochemistry 24: 328–332.

54. Mizutani H, Hasegawa H, Wada E (1986) High nitrogen isotope ratio for soils of seabird rookeries. Biogeochemistry 2: 221–247.

55. Schmidt S, Dennison WC, Moss GJ, Stewart GR (2004) Nitrogen ecophysiology of Heron Island, a subtropical coral cay of the Great Barrier Reef, Australia. Functional Plant Biology 31: 517–528.

56. Wait DA, Aubrey DP, Anderson WB (2005) Seabird guano influences on desert islands: soil chemistry and herbaceous species richness and productivity. Journal of Arid Environments 60: 681–695.

57. Raven JA, Smith FA (1976) Nitrogen assimilation and transport in vascular land plants in relation to intracellular pH regulation. New Phytologist 76: 415–431.

58. Deignan MT, Lewis OAM (1988) The inhibition of ammonium uptake by nitrate in wheat. New Phytologist 110: 1–3.

59. Guo S, Kaldenhoff R, Uehlein N, Sattelmacher B, Brueck H (2007) Relationship between water and nitrogen uptake in nitrate- and ammonium-supplied Phaseolus vulgaris L. plants. Journal of Plant Nutrition and Soil Science 170: 73–80.

60. Britto DT, Siddiqi MY, Glass ADM, Kronzucker HJ (2001) Futile transmembrane NH_4^+ cycling: A cellular hypothesis to explain ammonium toxicity in plants. Proceedings of the National Academy of Sciences 98: 4255–4258.

61. Gillham ME (1960) Destruction of indigenous heath vegetation in Victorian sea-bird colonies. Australian Journal of Botany 8: 277–317.

62. Hartz TK, Johnstone PR (2006) Nitrogen availability from high-nitrogen-containing organic fertilizers. HortTechnology 16: 39–42.

63. Mizutani H, Kabaya Y, Wada E (1991) Linear correlation between latitude and soil ^{15}N enrichment at seabird rookeries. Naturwissenschaften 78: 34–36.

64. McNabb EMA, McNabb RA, Prather ID, Conner RN, Adkisson CS (1980) Nitrogen Excretion by Turkey Vultures. The Condor 82: 219–223.

65. Mizutani H, Wada E (1985) High-performance liquid chromatographic determination of uric acid in soil. Journal of Chromatography A 331: 359–369.

66. Gaskell M, Smith R (2007) Nitrogen sources for organic vegetable crops. HortTechnology 17: 431–441.

67. Staunton Smith J, Johnson CR (1995) Nutrient inputs from seabirds and humans on a populated coral cay. Marine Ecology Progress Series 124: 189–200.

68. Hadas A, Rosenberg R (1992) Guano as a nitrogen source for fertigation in organic farming. Nutrient Cycling in Agroecosystems 31: 209–214.

69. Forero MG, Bortolotti GR, Hobson KA, Donazar JA, Bertellotti M, et al. (2004) High trophic overlap within the seabird community of Argentinean Patagonia: a multiscale approach. Journal of Animal Ecology 73: 789–801.

70. Schoeninger MJ, DeNiro MJ (1984) Nitrogen and carbon isotopic composition of bone collagen from marine and terrestrial animals. Geochimica et Cosmochimica Acta 48: 625–639.

71. Szpak P, Orchard TJ, Gröcke DR (2009) A Late Holocene vertebrate food web from southern Haida Gwaii (Queen Charlotte Islands, British Columbia). Journal of Archaeological Science 36: 2734–2741.

72. Szpak P, Orchard TJ, McKechnie I, Gröcke DR (2012) Historical ecology of late Holocene sea otters (Enhydra lutris) from northern British Columbia: isotopic and zooarchaeological perspectives. Journal of Archaeological Science 39: 1553–1571.

73. Loder TC, Ganning B, Love JA (1996) Ammonia nitrogen dynamics in coastal rockpools affected by gull guano. Journal of Experimental Marine Biology and Ecology 196: 113–129.

74. Kirchmann H (1991) Carbon and nitrogen mineralization of fresh, aerobic and anaerobic animal manures during incubation with soil. Swedish Journal of Agricultural Research 21: 165–173.

75. Mizutani H, Kabaya Y, Wada E (1985) Ammonia volatilization and high ^{15}N/^{14}N ratio in a penguin rookery in Antarctica. Geochemical Journal 19: 323–327.

76. Burger M, Venterea RT (2008) Nitrogen Immobilization and Mineralization Kinetics of Cattle, Hog, and Turkey Manure Applied to Soil. Soil Science Society of America Journal 72: 1570–1579.

77. Okano K, Chutani K, Matsuo K (1997) Suitable level of nitrogen fertilizer for tea (Camellia sinensis L.) plants in relation to growth, photosynthesis, nitrogen uptake and accumulation of free amino acids. Japanese Journal of Crop Science 66: 279–287.

78. Näsholm T, Ericsson A (1990) Seasonal changes in amino acids, protein and total nitrogen in needles of fertilized Scots pine trees. Tree Physiology 6: 267–281.

79. Warren CR, Adams MA (2000) Capillary electrophoresis for the determination of major amino acids and sugars in foliage: application to the nitrogen nutrition of sclerophyllous species. Journal of Experimental Botany 51: 1147–1157.

80. Ruan J, Gerendás J, Härdter R, Sattelmacher B (2007) Effect of root zone pH and form and concentration of nitrogen on accumulation of quality-related components in green tea. Journal of the Science of Food and Agriculture 87: 1505–1516.

81. Garnica M, Houdusse F, Zamarreño AM, Garcia-Mina JM (2010) Nitrate modifies the assimilation pattern of ammonium and urea in wheat seedlings. Journal of the Science of Food and Agriculture 90: 357–369.

82. Ruan J, Haerdter R, Gerendás J (2010) Impact of nitrogen supply on carbon/nitrogen allocation: a case study on amino acids and catechins in green tea [Camellia sinensis (L.) O. Kuntze] plants. Plant Biology 12: 724–734.

83. Smolders AJP, den Hartog C, van Gestel CBL, Roelofs JGM (1996) The effects of ammonium on growth, accumulation of free amino acids and nutritional status of young phosphorus deficient Stratiotes aloides plants. Aquatic Botany 53: 85–96.

84. Potel F, Valadier M-H, Ferrario-Méry S, Grandjean O, Morin H, et al. (2009) Assimilation of excess ammonium into amino acids and nitrogen translocation in Arabidopsis thaliana– roles of glutamate synthases and carbamoylphosphate synthetase in leaves. FEBS Journal 276: 4061–4076.

85. Ma BL, Dwyer LM (1998) Nitrogen uptake and use of two contrasting maize hybrids differing in leaf senescence. Plant and Soil 199: 283–291.

86. Rajcan I, Tollenaar M (1999) Source: sink ratio and leaf senescence in maize: II. Nitrogen metabolism during grain filling. Field Crops Research 60: 255–265.

87. Gallais A, Coque M, Quilléré I, Prioul J-L, Hirel B (2006) Modelling postsilking nitrogen fluxes in maize (Zea mays) using ^{15}N-labelling field experiments. New Phytologist 172: 696–707.

88. Gallais A, Coque M (2005) Genetic variation and selection for nitrogen use efficiency in maize: A synthesis. Maydica 50: 531–547.

89. Tsai CY, Huber DM, Warren HL (1980) A Proposed Role of Zein and Glutelin as N Sinks in Maize. Plant Physiology 66: 330–333.

90. Below FE, Christensen LE, Reed AJ, Hageman RH (1981) Availability of Reduced N and Carbohydrates for Ear Development of Maize. Plant Physiology 68: 1186–1190.

91. Binford GD, Blackmer AM, El-Hout NM (1990) Tissue test for excess nitrogen during corn production. Agronomy Journal 82: 124–129.

92. Dijkstra P, Williamson C, Menyailo O, Doucett R, Koch G, et al. (2003) Nitrogen stable isotope composition of leaves and roots of plants growing in a forest and a meadow. Isotopes in Environmental and Health Studies 39: 29–39.

93. Hobbie EA, Macko SA, Williams M (2000) Correlations between foliar δ^{15}N and nitrogen concentrations may indicate plant-mycorrhizal interactions. Oecologia 122: 273–283.

94. Yoneyama T, Omata T, Nakata S, Yazaki J (1991) Fractionation of Nitrogen Isotopes during the Uptake and Assimilation of Ammonia by Plants. Plant and Cell Physiology 32: 1211–1217.

95. Evans RD, Bloom AJ, Sukrapanna SS, Ehleringer JR (1996) Nitrogen isotope composition of tomato (Lycopersicon esculentum Mill. cv. T-5) grown under ammonium or nitrate nutrition. Plant, Cell & Environment 19: 1317–1323.

96. Högberg P, Högberg MN, Quist ME, Ekblad ALF, Näsholm T (1999) Nitrogen isotope fractionation during nitrogen uptake by ectomycorrhizal and non-mycorrhizal Pinus sylvestris. New Phytologist 142: 569–576.

97. Kolb KJ, Evans RD (2002) Implications of leaf nitrogen recycling on the nitrogen isotope composition of deciduous plant tissues. New Phytologist 156: 57–64.

98. Högberg P, Högbom L, Schinkel H, Högberg M, Johannisson C, et al. (1996) ^{15}N abundance of surface soils, roots and mycorrhizas in profiles of European forest soils. Oecologia 108: 207–214.

99. Evans RD (2001) Physiological mechanisms influencing plant nitrogen isotope composition. Trends in Plant Science 6: 121–126.

100. Lewis OAM, Leidi EO, Lips SH (1989) Effect of nitrogen source on growth response to salinity stress in maize and wheat. New Phytologist 111: 155–160.

101. Murphy AT, Lewis OAM (1987) Effect of nitrogen feeding source on the supply of nitrogen from root to shoot and the site of nitrogen assimilation in maize (Zea mays L. CV. R201). New Phytologist 107: 327–333.

102. Choi W-J, Lee S-M, Ro H-M, Kim K-C, Yoo S-H (2002) Natural ^{15}N abundances of maize and soil amended with urea and composted pig manure. Plant and Soil 245: 223–232.

103. Tcherkez G (2011) Natural ^{15}N/^{14}N isotope composition in C₃ leaves: are enzymatic isotope effects informative for predicting the ^{15}N-abundance in key metabolites?. Functional Plant Biology 38: 1–12.

104. Ta CT (1991) Nitrogen Metabolism in the Stalk Tissue of Maize. Plant Physiology 97: 1375–1380.

105. Beauchamp EG, Kannenberg LW, Hunter RB (1976) Nitrogen Accumulation and Translocation in Corn Genotypes Following Silking. Agronomy Journal 68: 418–422.

106. Donnison IS, Gay AP, Thomas H, Edwards KJ, Edwards D, et al. (2007) Modification of nitrogen remobilization, grain fill and leaf senescence in maize (Zea mays) by transposon insertional mutagenesis in a protease gene. New Phytologist 173: 481–494.

107. Garten CT (1993) Variation in Foliar ^{15}N Abundance and the Availability of Soil Nitrogen on Walker Branch Watershed. Ecology 74: 2098–2113.

108. Gebauer G, Giesemann A, Schulze E, Jäger H (1994) Isotope ratios and concentrations of sulfur and nitrogen in needles and soils of Picea abies stands as influenced by atmospheric deposition of sulfur and nitrogen compounds. Plant and Soil 164: 267–281.

109. Näsholm T (1994) Removal of nitrogen during needle senescence in Scots pine (Pinus sylvestris L.). Oecologia 99: 290–296.

110. Choi W-J, Chang SX, Ro H-M (2005) Seasonal Changes of Shoot Nitrogen Concentrations and ^{15}N/^{14}N Ratios in Common Reed in a Constructed Wetland. Communications in Soil Science and Plant Analysis 36: 2719–2731.

111. Högberg P (1997) Tansley Review No. 95 ^{15}N natural abundance in soil-plant systems. New Phytologist 137: 179–203.

112. Robinson D, Handley LL, Scrimgeour CM (1998) A theory for ^{15}N/^{14}N fractionation in nitrate-grown vascular plants. Planta 205: 397–406.

113. Yin Z-H, Raven JA (1998) Influences of different nitrogen sources on nitrogen- and water-use efficiency, and carbon isotope discrimination in C₃ Triticum aestivum L. and C₄ Zea mays L. plants. Planta 205: 574–580.

114. Raven JA, Wollenweber B, Handley LL (1992) A comparison of ammonium and nitrate as nitrogen sources for photolithotrophs. New Phytologist 121: 19–32.

115. Camin F, Perini M, Bontempo L, Fabroni S, Faedi W, et al. (2011) Potential isotopic and chemical markers for characterising organic fruits. Food Chemistry 125: 1072–1082.

116. Georgi M, Voerkelius S, Rossmann A, Graßmann J, Schnitzler W (2005) Multielement Isotope Ratios of Vegetables from Integrated and Organic Production. Plant and Soil 275: 93–100.

117. Serret M, Ortiz-Monasterio I, Pardo A, Araus J (2008) The effects of urea fertilisation and genotype on yield, nitrogen use efficiency, δ^{15}N and δ^{13}C in wheat. Annals of Applied Biology 153: 243–257.

118. Kondo M, Pablico P, Aragones D, Agbisit R (2004) Genotypic variations in carbon isotope discrimination, transpiration efficiency, and biomass production in rice as affected by soil water conditions and N. Plant and Soil 267: 165–177.

119. Cabrera-Bosquet L, Molero G, Bort J, Nogués S, Araus JL (2007) The combined effect of constant water deficit and nitrogen supply on WUE, NUE and Δ^{13}C in durum wheat potted plants. Annals of Applied Biology 151: 277–289.

120. Jenkinson D, Coleman K, Harkness D (1995) The influence of fertilizer nitrogen and season on the carbon-13 abundance of wheat straw. Plant and Soil 171: 365–367.

121. Zhao LJ, Xiao HL, Liu XH (2007) Relationships Between Carbon Isotope Discrimination and Yield of Spring Wheat Under Different Water and Nitrogen Levels. Journal of Plant Nutrition 30: 947–963.

122. Iqbal MM, Akhter J, Mohammad W, Shah SM, Nawaz H, et al. (2005) Effect of tillage and fertilizer levels on wheat yield, nitrogen uptake and their correlation with carbon isotope discrimination under rainfed conditions in north-west Pakistan. Soil and Tillage Research 80: 47–57.

123. Shangguan ZP, Shao MA, Dyckmans J (2000) Nitrogen nutrition and water stress effects on leaf photosynthetic gas exchange and water use efficiency in winter wheat. Environmental and Experimental Botany 44: 141–149.

124. Clay DE, Engel RE, Long DS, Liu Z (2001) Nitrogen and Water Stress Interact to Influence Carbon-13 Discrimination in Wheat. Soil Science Society of America Journal 65: 1823–1828.

125. Winter CK, Davis SF (2006) Organic Foods. Journal of Food Science 71: R117–R124.

126. Schmidt H-l, Roβmann A, Voerkelius S, Schnitzler WH, Georgi M, et al. (2005) Isotope characteristics of vegetables and wheat from conventional and organic production. Isotopes in Environmental and Health Studies 41: 223–228.

127. Šturm M, Lojen S (2011) Nitrogen isotopic signature of vegetables from the Slovenian market and its suitability as an indicator of organic production. Isotopes in Environmental and Health Studies 47: 214–220.

128. Bateman AS, Kelly SD, Jickells TD (2005) Nitrogen Isotope Relationships between Crops and Fertilizer: Implications for Using Nitrogen Isotope Analysis as an Indicator of Agricultural Regime. Journal of Agricultural and Food Chemistry 53: 5760–5765.

129. Bateman AS, Kelly SD, Woolfe M (2007) Nitrogen Isotope Composition of Organically and Conventionally Grown Crops. Journal of Agricultural and Food Chemistry 55: 2664–2670.

130. Flores P, Fenoll J, Hellín P (2007) The Feasibility of Using δ^{15}N and δ^{13}C Values for Discriminating between Conventionally and Organically Fertilized Pepper (Capsicum annuum L.). Journal of Agricultural and Food Chemistry 55: 5740–5745.

131. Rapisarda P, Calabretta ML, Romano G, Intrigliolo F (2005) Nitrogen Metabolism Components as a Tool To Discriminate between Organic and Conventional Citrus Fruits. Journal of Agricultural and Food Chemistry 53: 2664–2669.

132. Rapisarda P, Camin F, Fabroni S, Perini M, Torrisi B, et al. (2010) Influence of Different Organic Fertilizers on Quality Parameters and the δ^{15}N, δ^{13}C, δ^{2}H, δ^{34}S, and δ^{18}O Values of Orange Fruit (Citrus sinensis L. Osbeck). Journal of Agricultural and Food Chemistry 58: 3502–3506.

133. Rogers KM (2008) Nitrogen Isotopes as a Screening Tool To Determine the Growing Regimen of Some Organic and Nonorganic Supermarket Produce from New Zealand. Journal of Agricultural and Food Chemistry 56: 4078–4083.

134. Camin F, Moschella A, Miselli F, Parisi B, Versini G, et al. (2007) Evaluation of markers for the traceability of potato tubers grown in an organic versus conventional regime. Journal of the Science of Food and Agriculture 87: 1330–1336.

135. Šturm M, Kacjan-Maršić N, Lojen S (2011) Can δ^{15}N in lettuce tissues reveal the use of synthetic nitrogen fertiliser in organic production? Journal of the Science of Food and Agriculture 91: 262–267.

136. Flores P, Murray PJ, Hellín P, Fenoll J (2011) Influence of N doses and form on ^{15}N natural abundance of pepper plants: considerations for using δ^{15}N values

as indicator of N source. Journal of the Science of Food and Agriculture 91: 2255–2258.

137. Schwarcz HP (1991) Some theoretical aspects of isotope paleodiet studies. Journal of Archaeological Science 18: 261–275.

138. Bogaard A, Heaton THE, Poulton P, Merbach I (2007) The impact of manuring on nitrogen isotope ratios in cereals: archaeological implications for reconstruction of diet and crop management practices. Journal of Archaeological Science 34: 335–343.

139. Commisso RG, Nelson DE (2007) Patterns of plant δ^{15}N values on a Greenland Norse farm. Journal of Archaeological Science 34: 440–450.

140. Fraser RA, Bogaard A, Heaton T, Charles M, Jones G, et al. (2011) Manuring and stable nitrogen isotope ratios in cereals and pulses: towards a new archaeobotanical approach to the inference of land use and dietary practices. Journal of Archaeological Science 38: 2790–2804.

141. Chepstow-Lusty AJ (2011) Agro-pastoral and social change in the Cuzco heartland of Peru: a brief history using environmental proxies. Antiquity 85: 570–582.

142. Kubler G (1948) Towards Absolute Time: Guano Archaeology. Memoirs of the Society for American Archaeology 4: 29–50.

143. Netherly PJ (1977) Local Level Lords on the North Coast of Peru [Unpublished Ph.D. Dissertation]. Ithaca: Cornell University. 366 p.

144. Slovak NM, Paytan A (2011) Fisherfolk and farmers: Carbon and nitrogen isotope evidence from Middle Horizon Ancón, Peru. International Journal of Osteoarchaeology 21: 253–267.

145. Knudson KJ, Aufderheide AE, Buikstra JE (2007) Seasonality and paleodiet in the Chiribaya polity of southern Peru. Journal of Archaeological Science 34: 451–462.

146. Tomczak PD (2003) Prehistoric diet and socioeconomic relationships within the Osmore Valley of southern Peru. Journal of Anthropological Archaeology 22: 262–278.

147. DeNiro MJ (1988) Marine food sources for prehistoric coastal Peruvian camelids: isotopic evidence and implications. In: Wing ES, Wheeler JC, eds. Economic Prehistory of the Central Andes. Oxford: British Archaeological Reports International Series 427. pp 119–128.

148. Cushman GT The Lords of Guano: Science and the Management of Peru's Marine Environment, 1800–1973 [Unpublished Ph.D. Dissertation]: University of Texas, Austin. 762 p.

149. Cordle C (2007) The Guano Voyages. Rural History 18: 119–133.

150. Simmons D (2006) Waste Not, Want Not: Excrement and Economy in Nineteenth-Century France. Representations 96: 73–98.

151. Cox G, Sealy J (1997) Investigating Identity and Life Histories: Isotopic Analysis and Historical Documentation of Slave Skeletons Found on the Cape Town Foreshore, South Africa. International Journal of Historical Archaeology 1: 207–224.

152. Katzenberg MA (1995) Nitrogen isotope evidence for weaning age in a nineteenth century Canadian skeletal sample. In: Grauer AL, ed. Bodies of Evidence: Reconstructing History through Skeletal Analysis. Cambridge: Wiley-Liss. pp 221–235.

153. Cox G, Sealy J, Schrire C, Morris A (2001) Stable Carbon and Nitrogen Isotopic Analyses of the Underclass at the Colonial Cape of Good Hope in the Eighteenth and Nineteenth Centuries. World Archaeology 33: 73–97.

154. Klippel WE (2001) Sugar Monoculture, Bovid Skeletal Part Frequencies, and Stable Carbon Isotopes: Interpreting Enslaved African Diet at Brimstone Hill, St Kitts, West Indies. Journal of Archaeological Science 28: 1191–1198.

155. Valentin F, Bocherens H, Gratuze B, Sand C (2006) Dietary patterns during the late prehistoric/historic period in Cikobia island (Fiji): insights from stable isotopes and dental pathologies. Journal of Archaeological Science 33: 1396–1410.

156. Katzenberg M, Saunders S, Abonyi S (2000) Bone Chemistry, Food and History: A Case Study from 19th Century Upper Canada. In: Ambrose SH, Katzenberg MA, eds. Biogeochemical Approaches to Paleodietary Analysis. New York: Kluwer Academic. pp 1–22.

157. Roy DM, Hall R, Mix AC, Bonnichsen R (2005) Using stable isotope analysis to obtain dietary profiles from old hair: A case study from Plains Indians. American Journal of Physical Anthropology 128: 444–452.

158. Schroeder H, O'Connell TC, Evans JA, Shuler KA, Hedges REM (2009) Trans-Atlantic slavery: Isotopic evidence for forced migration to Barbados. American Journal of Physical Anthropology 139: 547–557.

159. Sealy J, Armstrong R, Schrire C (1995) Beyond lifetime averages: tracing life histories through isotopic analysis of different calcified tissues from archaeological human skeletons. Antiquity 69: 290–300.

160. Hobson KA, Clark RG (1992) Assessing avian diets using stable isotopes II: factors influencing diet-tissue fractionation. The Condor 94: 189–197.

161. Hobson KA (1995) Reconstructing Avian Diets Using Stable-Carbon and Nitrogen Isotope Analysis of Egg Components: Patterns of Isotopic Fractionation and Turnover. The Condor 97: 752–762.

162. Wainright SC, Haney JC, Kerr C, Golovkin AN, Flint MV (1998) Utilization of nitrogen derived from seabird guano by terrestrial and marine plants at St. Paul, Pribilof Islands, Bering Sea, Alaska. Marine Biology 131: 63–71.

163. Mizutani H, Kabaya Y, Wada E (1985) High-performance liquid chromatographic isolation of uric acid from soil for isotopic determination. Journal of Chromatography A 331: 371–381.

164. Bird MI, Tait E, Wurster CM, Furness RW (2008) Stable carbon and nitrogen isotope analysis of avian uric acid. Rapid Communications in Mass Spectrometry 22: 3393–3400.

165. Kirshenbaum I, Smith JS, Crowell T, Graff J, McKee R (1947) Separation of the Nitrogen Isotopes by the Exchange Reaction between Ammonia and Solutions of Ammonium Nitrate. Journal of Chemical Physics 15: 440–446.

166. Shearer G, Kohl DH (1986) N_2-Fixation in Field Settings: Estimations Based on Natural ^{15}N Abundance. Australian Journal of Plant Physiology 13: 699–756.

167. Feigin A, Shearer G, Kohl DH, Commoner B (1974) The Amount and Nitrogen-15 Content of Nitrate in Soil Profiles from two Central Illinois Fields in a Corn-Soybean Rotation. Soil Science Society of America Journal 38: 465–471.

168. Hoch MP, Fogel ML, Kirchman DL (1992) Isotope Fractionation Associated with Ammonium Uptake by a Marine Bacterium. Limnology and Oceanography 37: 1447–1459.

169. Yoneyama T, Fujiwara H, Wilson JW (1998) Variations in fractionation of carbon and nitrogen isotopes in higher plants: N-metabolism and partitioning in phloem and xylem. In: Griffiths H, ed. Stables Isotopes, Integration of Biological, Ecological and Geochemical Processes. Oxford: BIOS Scientific Publishers. pp 99–109.

170. Mariotti A, Mariotti F, Champigny M-L, Amarger N, Moyse A (1982) Nitrogen Isotope Fractionation Associated with Nitrate Reductase Activity and Uptake of NO_3^- by Pearl Millet. Plant Physiology 69: 880–884.

171. Miller A, Cramer M (2005) Root Nitrogen Acquisition and Assimilation. Plant and Soil 274: 1–36.

172. Tcherkez G, Farquhar GD (2006) Isotopic fractionation by plant nitrate reductase, twenty years later. Functional Plant Biology 33: 531–537.

173. Barrett K, Anderson WB, Wait DA, Grismer LL, Polis GA, et al. (2005) Marine subsidies alter the diet and abundance of insular and coastal lizard populations. Oikos 109: 145–153.

174. Bokhorst S, Huiskes A, Convey P, Aerts R (2007) External nutrient inputs into terrestrial ecosystems of the Falkland Islands and the Maritime Antarctic region. Polar Biology 30: 1315–1321.

175. Cocks MP, Balfour DA, Stock WD (1998) On the uptake of ornithogenic products by plants in the inland mountains of Dronning Maud Land, Antarctica, using stable isotopes. Polar Biology 20: 107–111.

176. Erskine PD, Bergstrom DM, Schmidt S, Stewart GR, Tweedie CE, et al. (1998) Subantarctic Macquarie Island – a model ecosystem for studying animal-derived nitrogen sources using ^{15}N natural abundance. Oecologia 117: 187–193.

177. Harding JS, Hawke DJ, Holdaway RN, Winterbourn MJ (2004) Incorporation of marine-derived nutrients from petrel breeding colonies into stream food webs. Freshwater Biology 49: 576–586.

178. Hawke DJ, Holdaway RN (2005) Avian assimilation and dispersal of carbon and nitrogen brought ashore by breeding Westland petrels (*Procellaria westlandica*): a stable isotope study. Journal of Zoology 266: 419–426.

179. Hawke DJ, Newman J (2007) Carbon-13 and nitrogen-15 enrichment in coastal forest foliage from nutrient-poor and seabird-enriched sites in southern New Zealand. New Zealand Journal of Botany 45: 309–315.

180. Hobara S, Koba K, Osono T, Tokuchi N, Ishida A, et al. (2005) Nitrogen and phosphorus enrichment and balance in forests colonized by cormorants: Implications of the influence of soil adsorption. Plant and Soil 268: 89–101.

181. Kameda K, Koba K, Hobara S, Osono T, Terai M (2006) Pattern of natural ^{15}N abundance in lakeside forest ecosystem affected by cormorant-derived nitrogen. Hydrobiologia 567: 69–86.

182. Kolb G, Jerling L, Hambäck P (2010) The Impact of Cormorants on Plant-Arthropod Food Webs on Their Nesting Islands. Ecosystems 13: 353–366.

183. Markwell TJ, Daugherty CH (2002) Invertebrate and lizard abundance is greater on seabird-inhabited islands than on seabird-free islands in the Marlborough Sounds, New Zealand. Ecoscience 9: 293–299.

184. Mizota C, Naikatin A (2007) Nitrogen isotope composition of inorganic soil nitrogen and associated vegetation under a sea bird colony on the Hatana islands, Rotuma Group, Fiji. Geochemical Journal 41: 297–301.

185. Mizota C (2009) Nitrogen isotopic patterns of vegetation as affected by breeding activity of Black-tailed Gull (*Larus crassiostris*): A coupled analysis of feces, inorganic soil nitrogen and flora. Applied Geochemistry 24: 2027–2033.

186. Mizutani H, Wada E (1988) Nitrogen and Carbon Isotope Ratios in Seabird Rookeries and their Ecological Implications. Ecology 69: 340–349.

187. Mizutani H, Kabaya Y, Moors PJ, Speir TW, Lyon GL (1991) Nitrogen Isotope Ratios Identify Deserted Seabird Colonies. The Auk 108: 960–964.

188. Stapp P, Polis GA, Sanchez Pinero F (1999) Stable isotopes reveal strong marine and El Nino effects on island food webs. Nature 401: 467–469.

189. Young HS, McCauley DJ, Dunbar RB, Dirzo R (2010) Plants cause ecosystem nutrient depletion via the interruption of bird-derived spatial subsidies. Proceedings of the National Academy of Sciences 107: 2072–2077.

190. Zhu R, Liu Y, Ma E, Sun J, Xu H, et al. (2009) Nutrient compositions and potential greenhouse gas production in penguin guano, ornithogenic soils and seal colony soils in coastal Antarctica. Antarctic Science 21: 427–438.

191. Bateman AS, Kelly SD (2007) Fertilizer nitrogen isotope signatures. Isotopes in Environmental and Health Studies 43: 237–247.

192. Dijkstra P, Menyailo OV, Doucett RR, Hart SC, Schwartz E, et al. (2006) C and N availability affects the 15N natural abundance of the soil microbial

biomass across a cattle manure gradient. European Journal of Soil Science 57: 468–475.

193. Kerley SJ, Jarvis SC (1996) Preliminary studies of the impact of excreted N on cycling and uptake of N in pasture systems using natural abundance stable isotopic discrimination. Plant and Soil 178: 287–294.

194. Lim S-S, Choi W-J, Kwak J-H, Jung J-W, Chang S, et al. (2007) Nitrogen and carbon isotope responses of Chinese cabbage and chrysanthemum to the application of liquid pig manure. Plant and Soil 295: 67–77.

195. Yun S-I, Ro H-M, Choi W-J, Chang SX (2006) Interactive effects of N fertilizer source and timing of fertilization leave specific N isotopic signatures in Chinese cabbage and soil. Soil Biology and Biochemistry 38: 1682–1689.

196. Choi W-J, Ro H-M, Hobbie EA (2003) Patterns of natural ^{15}N in soils and plants from chemically and organically fertilized uplands. Soil Biology and Biochemistry 35: 1493–1500.

197. Choi W-J, Ro H-M (2003) Differences in isotopic fractionation of nitrogen in water-saturated and unsaturated soils. Soil Biology and Biochemistry 35: 483–486.

198. del Amor FM, Navarro J, Aparicio PM (2008) Isotopic Discrimination as a Tool for Organic Farming Certification in Sweet Pepper. Journal of Environmental Quality 37: 182–185.

199. Nakano A, Uehara Y (2007) Effects of different kinds of fertilizer and application methods on δ^{15}N values of tomato. Japan Agricultural Research Quarterly 41: 219–226.

12

In-Situ Effects of Simulated Overfishing and Eutrophication on Benthic Coral Reef Algae Growth, Succession, and Composition in the Central Red Sea

Christian Jessen[1]*, Cornelia Roder[2], Javier Felipe Villa Lizcano[2], Christian R. Voolstra[2], Christian Wild[1,3]

1 Coral Reef Ecology Group (CORE), Leibniz Center for Tropical Marine Ecology (ZMT), Bremen, Germany, **2** Red Sea Research Center, King Abdullah University of Science and Technology (KAUST), Thuwal, Saudi Arabia, **3** Faculty of Biology and Chemistry, University of Bremen, Bremen, Germany

Abstract

Overfishing and land-derived eutrophication are major local threats to coral reefs and may affect benthic communities, moving them from coral dominated reefs to algal dominated ones. The Central Red Sea is a highly under-investigated area, where healthy coral reefs are contending against intense coastal development. This *in-situ* study investigated both the independent and combined effects of manipulated inorganic nutrient enrichment (simulation of eutrophication) and herbivore exclosure (simulation of overfishing) on benthic algae development. Light-exposed and shaded terracotta tiles were positioned at an offshore patch reef close to Thuwal, Saudi Arabia and sampled over a period of 4 months. Findings revealed that nutrient enrichment alone affected neither algal dry mass nor algae-derived C or N production. In contrast, herbivore exclusion significantly increased algal dry mass up to 300-fold, and in conjunction with nutrient enrichment, this total increased to 500-fold. Though the increase in dry mass led to a 7 and 8-fold increase in organic C and N content, respectively, the algal C/N ratio (18 ± 1) was significantly lowered in the combined treatment relative to controls (26 ± 2). Furthermore, exclusion of herbivores significantly increased the relative abundance of filamentous algae on the light-exposed tiles and reduced crustose coralline algae and non-coralline red crusts on the shaded tiles. The combination of the herbivore exclusion and nutrient enrichment treatments pronounced these effects. The results of our study suggest that herbivore reduction, particularly when coupled with nutrient enrichment, favors non-calcifying, filamentous algae growth with high biomass production, which thoroughly outcompetes the encrusting (calcifying) algae that dominates in undisturbed conditions. These results suggest that the healthy reefs of the Central Red Sea may experience rapid shifts in benthic community composition with ensuing effects for biogeochemical cycles if anthropogenic impacts, particularly overfishing, are not controlled.

Editor: Christopher Fulton, The Australian National University, Australia

Funding: This study was jointly funded by the Leibniz Center for Tropical Marine Ecology (ZMT) and the King Abdullah University of Science and Technology (KAUST). The funders had no role in study design, data collection and analysis, decision to publish, or preparation of the manuscript.

Competing Interests: The authors have declared that no competing interests exist.

* E-mail: christian.jessen@zmt-bremen.de

Introduction

Both global stressors, such as emerging climate change resulting in ocean warming and acidification, and local factors are critically threatening coral reefs. Two of the most significant local stressors are eutrophication and overfishing [1].

Eutrophication stems from the over-enrichment of nutrients in water bodies. Sources of eutrophication in coastal marine environments are often anthropogenic in nature and include agriculture runoff, human sewage, urban waste, industrial effluent, and fossil fuel combustion [2]. Scleractinian corals, the primary reef ecosystem engineers [3], are mostly negatively impacted by eutrophication. The effects of eutrophication vary from reducing growth [4,5] and calcification rates, [6,7] to impairing reproduction [4,8], lowering bleaching resistance [9], and advancing coral disease [10]. Algae is also affected by increased nutrient levels. Among those affected can be crustose coralline algae (CCA) [11–13], an important settlement substrates for corals [14], as well as turf and macroalgae [15–18].

Overfishing is the second local stressor simulated in this study. It has caused more than 90% worldwide decline of predators [19], and this lack of predators in an ecosystem has dramatic cascading effects. For example, in kelp forests, sea urchin populations exploded and led to immense deforestation following the removal of apex predators by fishing [19,20]. In coral reefs, protection from overfishing can mitigate starfish outbreaks [21] and healthy herbivorous fish communities support higher resilience since they limit growth and establishment of algal communities [22]. Herbivore grazing in coral reefs helps maintain low algal turf growths, reduces the number and duration of coral-algal interactions, and increases space for coral settling by promoting encrusting coralline algae growth over macroalgae [23].

The pressures of eutrophication, overfishing and a combination thereof can cause benthic algae proliferation [24]. Once macroalgae are well established in a reef, herbivorous fish recruitment can be impeded by their natural avoidance of reef patches with high densities of macroalgae [25]. Macroalgae also compete for space with encrusting coralline algae, resulting in diminished coral larvae recruitment [26,27], and the frequency and intensity of interactions between corals and algae can also increase [28]. As a consequence, excessive algal growth can lead to a reduction in coral recruitment [29] and can directly impact corals via allelochemicals [30–33] or decrease O_2 availability in the direct vicinity [34–36]. In addition to a reduction in habitat complexity [37], the change in benthic community composition towards algal dominance also leads to an increase in algae-derived dissolved organic carbon (DOC) [38,39]. Higher concentrations of DOC are known to stimulate microbial growth and metabolism [38–40] which in turn can negatively affect corals, presumably by unbalancing the coral-associated microbial community whose growth concomitantly generates hypoxic reef conditions [34–36,41,42].

Benthic algae can be useful bioindicators due to their fast growth and turnover rates [43,44]. The predictions of the Relative Dominance Model (RDM) by Littler and Littler [45], state that a high cover of CCA over turf and frondose macroalgae is generally found in reef environments with elevated nutrient levels and an intact herbivorous community. Higher relative abundances of turf algae may indicate low nutrient and low grazing levels, while abundant frondose macroalgae represent the worst scenario, a combination of high nutrient and low herbivory levels. Until today, only limited support exists for this model. Though numerous studies compared the individual and combined effects of herbivory and nutrient availability on benthic algal community composition [11–13,24,46–51], many of these studies were of limited duration. While the RDM is still under debate [13,17,24,46,49], no comparative studies exist for the Red Sea, and the individual effects of nutrient enrichment and herbivory exclusion have received little attention in this area (bottom-up: [52,53]; top-down: [54–56]). Meanwhile, emerging coastal development together with overfishing and land-derived nutrient runoff are threatening many healthy Red Sea coral reefs, particularly around the fast developing and wealthy Jeddah region [1,57].

The study presented was designed to answer the following questions: (1) What influence, if any, do increased nutrient availability (bottom-up factor) and herbivore exclusion (top-down factor) have on benthic algae development, in terms of dry mass, organic carbon (C) and nitrogen (N) production, O2 consumption, and community composition? (2) Which factor, bottom-up or top-down, demonstrates a larger effect in this context? (3) Does the availability of light compound the benthic algae development? To answer these questions, we conducted an in-situ experiment in an offshore reef in the Central Red Sea over 4 months, simulating the individual and combined effects of eutrophication and overfishing.

Materials and Methods

Ethics Statement

The study site of Al Fahal reef does not fall under any legislative protection or special designation as a marine/environmental protected area. No special permit is required for the inshore coastal, reef, and intertidal areas around Thuwal. The Saudi Coast Guard Authority under the auspices of KAUST University issued sailing permits to the site, which included sample (algae) collection.

Study Site

The study was carried out from June to September 2011 over a period of 16 wks at the patch reef Al Fahal about 13 km off the Saudi Arabian coast in the Central Red Sea (N22.18.333, E38.57.768; Figure S1). Al Fahal is located >80 km to urban areas (next large city is Jeddah, >3 Mio inhabitants), with only a small village (Thuwal) located on shore. Neither are river deltas located in this region nor is any land of the surrounding region

allocated for agriculture. This reef was chosen in particular, due to its relatively large distance from shore and minimal impacts from land-based nutrient import and large-scale fishing.

Benthic Cover

Benthic reef community composition was assessed using the linear point intercept (LPI) method [58]. Benthic coverage was classified every 0.5 m along a 70 m transect that ran along the investigated reef site into the following categories: hard coral, soft coral, coral rubble (<20 cm), rock (bare substrate and rubble >20 cm), CCA, macroalgae (erected non-filamentous algae, e.g. Padina, Halimeda, Turbinaria, Ulva), filamentous algae (>2 mm), and other.

Cage Setups

Sixteen polyvinyl chloride (PVC) frames (50×75 cm) were deployed in the reef at 5–6 m water depths along a 70 m transect with 2–5 m distance in between. Each frame was equipped with 12 terracotta tiles, each with 100 cm^2 surface area. Prior to the start of the experiment, the tiles were autoclaved to remove any interfering compounds that could have accumulated during tile production and transported to the study site in a sealed plastic bag to avoid contamination. Tiles were installed pairwise on top of each other with the unglazed sides facing outside, resulting in an upper (light-exposed) and lower (shaded) tile. To avoid excessive sedimentation, tiles were installed at an angle of 45 degrees approximately 10 cm above the reef substrate using stainless steel screws, nuts, and washers. Four different treatments were applied to the frames (each with a replication of n = 4): (1) control (only the equipped frame), (2) fertilizer tubes (see nutrient enrichment section), (3) cage (hemispherical zinc galvanized cages with a mesh size of 4 cm and a diameter of 100 cm), and (4) a combination of cage and fertilizer tubes. The cages served to exclude larger herbivores; smaller fish (e.g. small parrotfish, wrasses, and surgeonfish) were still able to gain access to the tiles. High numbers of mobile grazing invertebrates (e.g. crustaceans, polychaetes, or gastropods) were not observed in any of the cages. Cage controls were not used, since studies showed that similar cages even with a lower mesh size did not affect water movement, light availability, and sedimentation rates [48,59,60].

Nutrient enrichment was simulated by deploying 4 fertilizer tubes around the frame, consisting of perforated PVC tubes filled with Osmocote fertilizer (Scotts; 15% total nitrogen as nitrate & ammonium, 9% phosphate as phosphoric pentoxide, and 12% potassium oxide) embedded in 3% agarose. Fertilizer dry mass was 580 g per frame. Fertilizer was deployed once without replenishments, but regular monitoring of inorganic nutrient concentrations assured continuous enrichment levels (actual values will be presented in the results section).

On each of the 5 sampling events one pair of tiles (light-exposed and shaded) was collected per frame, after 1, 2, 4, 8, and 16 wk(s) using SCUBA. All tiles were pre-scored and first divided in half (each 50 cm^2; an area which had been chosen from the asymptote of species-area curves by Hixon and Brostoff [61]) and then wrapped separately in ziplock bags. They were brought on board within 30 min where half of them were immediately flash frozen in liquid nitrogen for subsequent microbial analyses (results reported elsewhere), while the other half was handled as described below.

Incubations

O_2 consumption rates were measured after a modified method by Wild et al. [36]. Tiles were stored without air bubble inclusions in 1 L airtight incubation glass jars, that were kept in 4 large (70 L volume), opaque polyethylene (PE) containers filled with reef water to keep samples at constant ambient temperatures during incubations (monitored with Onset HOBO pendant temperature loggers in each container). Incubations were run in closed and dark containers. Temperature differences between *in-situ* temperatures (measured at PVC frames) and incubation jars ranged from 0.5 to 1.6°C. Net O_2 consumption rates were calculated for each incubation jar by dividing the difference between initial and end O_2 concentrations by the incubation duration (1.5–1.7 h) and corrected by subtracting mean O_2 consumption rate of 4 seawater controls without tiles. During incubations, the boxes were carefully moved by hand every 5 minutes on one side to mix the water inside the jars. O_2 measurements were carried out using a Hach O_2 probe (Hach HQ40d) that was placed a few cm above each tile in the incubation jars. All samples were stored on ice until further processing.

Response Variables on the Tiles

Light-exposed and shaded tiles were rinsed with fresh water to remove salt, attached sediment, and mobile invertebrates, resulting in light-exposed tiles that were almost exclusively covered with algal material with very rare invertebrate cover. Tiles were then photographed with a digital camera, before algal cover was carefully removed by using spatula and scalpel (only light-exposed tiles). The removed algae cover was dried in an oven at 37°C to constant weight, and dry mass (non-decalcified) was measured with a precision balance (Mettler Toledo XS205, accuracy: 0.01 mg). Until further processing, samples were kept dry at 37°C.

To quantify the proportional coverage of functional groups on the light-exposed and shaded tiles, 100 points were randomly overlaid on the digital picture of each tile using the software Coral Point Count with Excel extensions (CPCe) 4.1 [62]. Applied categories were: open space (non biotic cover or bare terracotta surface), filamentous algae (≥2 mm), crustose coralline algae (CCA), green crusts (non-coralline light green crusts), red crusts (non-coralline red crusts, e.g. *Peyssonnelia* spp.), brownish crusts (non-coralline dark-green and brownish crusts, e.g. filamentous algae <2 mm), cyanobacteria (whitish & mucilaginous), red macroalgae (fleshy upright red algae), and invertebrates (sessile forms).

For the elemental analyses of algae tissue, samples were homogenized using mortar and pestle and subsequently either acidified (organic C) or directly measured (N) with a EuroVector elemental analyzer (EURO EA 3000). Carbon and nitrogen contents were derived from calculation using elemental standards (apple leaf standard; Hekatech: HE34010100; analytical precision ≤0.1% (N) and ≤0.6% (C) of the standard value). Isotopic analysis of δ15N signatures of dried algal material relative to atmospheric nitrogen was run with an isotope ratio mass spectrometer (Finnigan Corp., San Jose, CA).

One of the 4 cage barriers deployed in the combined treatment seemed to have been breached by large herbivores, as evidenced by tile appearance and cage warping; the data (i.e. algal dry mass, organic C, N, O_2 consumption, and functional group assemblages) from said replicate were removed from the subsequent analysis after application of Grubb's outlier tests.

Water Parameters

Directly before sampling of the tiles, samples of ~5 L seawater (in total n = 80; 40 enriched and 40 non-enriched) were collected with large ziplock bags directly from above each frame. From this stock, 1000 mL were filtered on untreated Whatman-GF/F filters (Chlorophyll *a* (Chl *a*)) and 1000–2500 mL on pre-combusted and pre-weighted filters for particulate organic matter (POM). Due to laboratory mishap there were no samples for wk 1 for particulate

organic nitrogen (PON) and only 1 sample from 1 treatment for particulate organic carbon (POC). Elemental analyses of N and organic C of POM were performed using an EuroVector elemental analyzer (EURO EA 3000). The remaining filtrate was further used for nutrient (50 mL) and dissolved organic matter (DOM) measurements (40 mL). Analyses of dissolved inorganic nitrogen ($DIN = NH_4^+ + NO_3^- + NO_2^-$) and soluble reactive phosphorous ($SRP = PO_4^{3-}$) were performed using a continuous flow analyzer (FlowSys Alliance Instruments). Dissolved organic matter (DOM) measurements were carried out with the Teledyne Tekmar Apollo 9000 Combustion TOC/TN Analyzer. Chl *a* filters were stored at −20°C prior to acetone-extraction (90%) and measured fluorometrically according to the method described in Environmental Protection Agency (EPA) 445.0 [63].

Over the study period, temperature data were continuously measured (at 5 minutes intervals) at all PVC frames using HOBO pendant and Pro v2 loggers (Onset Computer Corporation, Pocasset, MA).

Herbivore Biomass

Visual surveys of herbivorous fish and sea urchins were carried out along the 70 m long transect of the frames in 5 m water depth with 4 replicates from June to July 2011. The fish surveys were conducted at noon between 11:15 am and 12:15 pm, 2.5 m left and 2.5 m right from the 70 m transect line, surveying a total area of 350 m². All herbivorous species ≥5 cm were counted, their size estimated, and grouped in one of 4 size classes (5–10 cm, 10–20 cm, 20–30 cm, and 30–40 cm). Species identification followed Randall [64], Debelius [65], and Lieske and Myers [66]. Classifying fish into herbivorous and non-herbivorous groups was based on Randall [64], Khalaf and Disi [67], Lieske and Myers [66], and own observations of grazing species (Table S1). Classification of herbivores took place according to their ability to remove algal material from the reef and not on their physiological ability to digest algal material [68]. Biomass of herbivorous fish was calculated on basis of the average length of the size class following length-weight ratios of the species or when not available of their family published by Green and Bellwood [69] and in FishBase [70].

No sea urchin species were observed during the 4 daytime surveys, so the sea urchin survey was conducted after the sun had fully set at 8 pm. The survey area was reduced to 1 m in width, resulting in a total surveyed area of 70 m². All sea urchins encountered along a 1 m polyethylene (PE) bar were counted and their test diameters were measured with a caliper to the nearest cm. Biomass was calculated on the basis of published length-weight relationships [71–73].

Statistical Data Analysis

Data from nutrient concentrations were analyzed using 2-sided t-tests. Water parameter data of Chl *a*, PON, POC, DON, DOC, as well as algal dry mass, organic C, N content, C_{org}/N ratio, δ15N signatures of exposed tile cover, and O_2 consumption rates (log transformation of values from light-exposed tiles) were analyzed using a 3-factorial ANOVA with backward stepwise deletion of variables, containing cage (present/absent), fertilizer (present/absent), time (5 sampling times), and their interactions as fixed factors. Functional algal group compositions were analyzed using a 3-factorial generalized linear model (GLM) with quasibinomial distribution and logit function. ANOVA and GLM analyses were carried out with the R statistical software version 2.15.2 [74]. To meet test assumptions of normal distribution and homoscedasticity, data of algal dry mass were log(x+1) transformed.

Results

Reef Background Parameters

Linear point intercept surveys revealed coral as dominating benthic feature (49%; with 32% hard coral and 17% soft coral), followed by rock (27%), coral rubble (13%), CCA (7%), filamentous algae (2%), and other (2%). Macroalgae were not observed.

During 4 transect surveys, 532 herbivorous fish were counted. Sixteen different species from 8 families with a total abundance of 0.4 ± 0.1 ind. m^{-2} (mean$\pm$SE) and biomass of 22.4 ± 8.0 g m^{-2} were found. Scaridae (8.9 g m^{-2}) and Acanthuridae (9.8 g m^{-2}) had the largest biomass (Table S1). During the sea urchin survey, 120 individuals of 4 species (*Echinometra mathaei*, *Echinothrix calamaris*, *Eucidaris metularia*, and *Heterocentrotus mammillatus*) were counted. Sea urchins exhibited a mean total abundance of 1.71 ind. m^{-2} and a biomass of 37.5 g m^{-2} (Table S2).

Experimental Background Parameters

The fertilizer and combined treatment led to an increase in DIN concentrations in the water column above the frames with significant differences for wk 1 and 4 in comparison to the non-enriched treatments. DIN concentrations changed over time with a peak after 4 wks (Figure 1A). In contrast, SRP concentrations remained rather constant, but enriched and non-enriched treatments significantly differed over all sampling times (Figure 1A).

Only Chl *a* (Figure 1B), but not POM (Figure 2A and 2B, Table 1) and DOM (Figure 2C and 2D, Table 1) concentrations in the water column directly above the setup were influenced by the treatments. Chl *a* values above the caged treatments were significantly higher than those of the non caged treatments (Figure 1B). Chl *a* together with PON and POC concentrations were significantly influenced by time. Chl *a* levels peaked after 4 wks and increased again after 16 wks following a drop at wk 8, while PON and POC concentrations declined and DON and DOC concentrations remained constant.

Effects on Tile Cover

Nutrient enrichment effects. Nutrient enrichment had no effect on algal dry mass, organic C, and N on the light-exposed tiles compared to controls when applied individually (Figure 3,

Table 2). This result is contrasted with the δ15N values, which were significantly decreased in the enriched treatments compared to controls (Figure S3; Table S3). Additionally, benthic cover was not significantly altered by nutrient enrichment except for decreasing cyanobacteria cover on the light-exposed tiles (Figure 4G, Table 3) and green crusts on the shaded tiles (Figure 4D, Table 4) compared to controls. Furthermore, O_2 respiration rates of the light-exposed and shaded tiles did not significantly differ between controls and nutrient addition (Figure 5).

Herbivore exclusion effects. In contrast, herbivore exclusion significantly increased algal dry mass, organic C, and N content and decreased the organic C/N ratio on the light-exposed tiles at all sampling times compared to the control treatment (Figure 3, Table 2). Furthermore, on the light-exposed tiles, filamentous algae grew exclusively in the caged treatments, while the cover of green (40% decrease compared to controls) and brownish crusts (50% decrease) and cyanobacteria (7% decrease) were significantly decreased (Figure 4B, 4D, 4F and 4G; Table 3). Shaded tiles revealed a very different picture; herbivore exclusion significantly enhanced cover of green crusts (20% increase compared to controls) (Figure 4D, Table 4) and invertebrates (7% increase), while red crusts (15% decrease) (Figure 4C, Table 4) and CCA (20% decrease) (Figure 4E, Table 4) were suppressed. Together with algal dry mass, O_2 consumption rates increased when herbivores were excluded on the light-exposed tiles (Pearson correlation, $r = 0.65$, $p<0.05$), but no treatment effect was detectable for the shaded tiles (Table 5).

Combined effects. The interaction of herbivore exclusion and nutrient enrichment was significant on the light-exposed tiles and further increased algal biomass in terms of algal dry mass, organic C, N, and O_2 consumption rates compared to the cage treatment (Figures 3 and 5). Filamentous algae cover was increased by a further 50%, compared to cage treatments (Figure 4B, Table 3) and cyanobacteria decreased a further 5%. (Figure 4G, Table 3). Red crusts on the shaded tiles had their percent cover further reduced by 9% in the combined treatments compared to the cage treatments (Figure 4E, Table 4).

Temporal changes. The temporal patterns in the development of algal biomass in terms of dry mass, organic C, N, and O_2 consumption rates on the light-exposed tiles were similar: while the

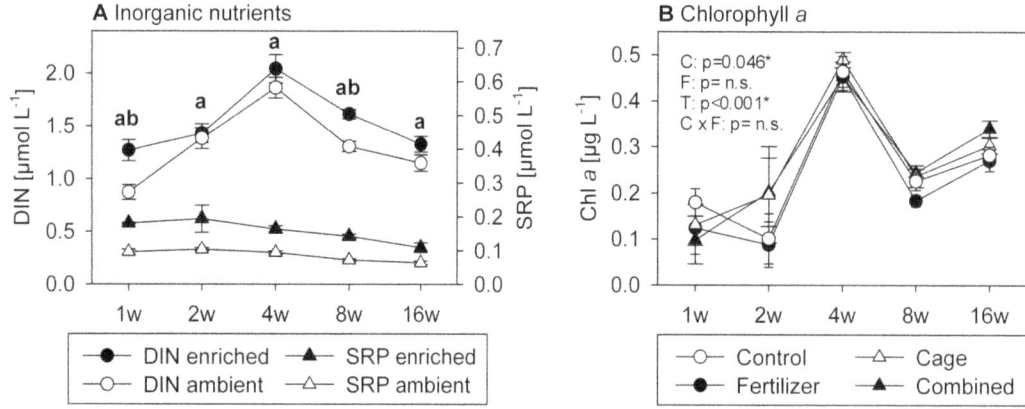

Figure 1. Inorganic nutrient (A) and Chlorophyll *a* (B) concentrations. A: Inorganic nutrient concentrations (µmol L^{-1}; means$\pm$SE) in the nutrient enrichment treatments (fertilizer & combined) and the non-enriched treatments (control & cage). Small letters (a for SRP; b for DIN) indicate statistical significant differences between enriched and non-enriched plots of $p<0.05$ (t-test). DIN: dissolved inorganic nitrogen; SRP: soluble reactive phosphate. B: Chlorophyll *a* concentrations (µg L^{-1}, means$\pm$SE) from water samples taken directly above the tile setups at all 5 sampling times. P-values were calculated from 3-factorial ANOVA and originate from analysis across the whole study period (see Table 1 for full results). P-values were tagged as n.s. (= not significant), when the model reduction step excluded the corresponding factor(s).

Table 1. Results of the 3-factorial ANOVA of the water parameters.

| | Chlorophyll *a* | | | PON | | | POC | | | DON | | | DOC | | |
	df	F	P	df	F	P	df	F	P	df	F	P	df	F	P
C	1	4.17	0.046*	–	–	–	1	2.56	0.116	–	–	–	–	–	–
F	–	–	–	–	–	–	1	0.20	0.657	1	2.42	0.124	–	–	–
T	4	43.97	0.000*	4	5.10	0.001*	4	6.56	0.000*	–	–	–	–	–	–
C×F	–	–	–	–	–	–	–	–	–	–	–	–	–	–	–
T×C	4	2.43	0.057	–	–	–	–	–	–	–	–	–	–	–	–
T×F	–	–	–	–	–	–	3	3.63	0.019*	–	–	–	–	–	–
T×C×F	–	–	–	–	–	–	–	–	–	–	–	–	–	–	–

Response variables (1st row) are chlorophyll *a*, particulate organic nitrogen (PON), particulate organic carbon (POC), dissolved organic nitrogen (DON), dissolved organic carbon (DOC). Independent factors (1st column) are Cage (C), Fertilizer (F), and Time (T). Significant results are indicated by asterisks. P-values of 0.000 symbolize values <0.001. Dashes represent factors that have been excluded by the model reduction.

non-caged treatments had no significant effects, the herbivore exclusion treatments exhibited a gradual increase of these data markers over the course of the first 4 wks of the study. Compared to the control, 300-fold, 7-fold, 8-fold, and 5-fold increases were observed in algal dry mass, organic C and N, and O_2 consumption rates, respectively (Figures 3 and 5). This peak at wk 4 was followed by a drop to lower values in wks 8 and 16. The algal dry mass in the cage treatment decreased rapidly down to wk 2 levels, unlike the combined treatment, where the peak after wk 4 was even higher (500 times in algal dry mass, 9 times in organic C, 11

times in N, and 6 times in O_2 consumption rates compared to the controls) and the decline was much less pronounced (Figures 3 and 5).

Discussion

Status of the Reef

High coral cover and lack of macroalgae at Al Fahal reef suggest a healthy reef [43] that ranks highly compared to Indo-Pacific reefs [75] and more closely to the pristine reefs from the northern Line Islands [76]. The rock and rubble proportion of the benthic

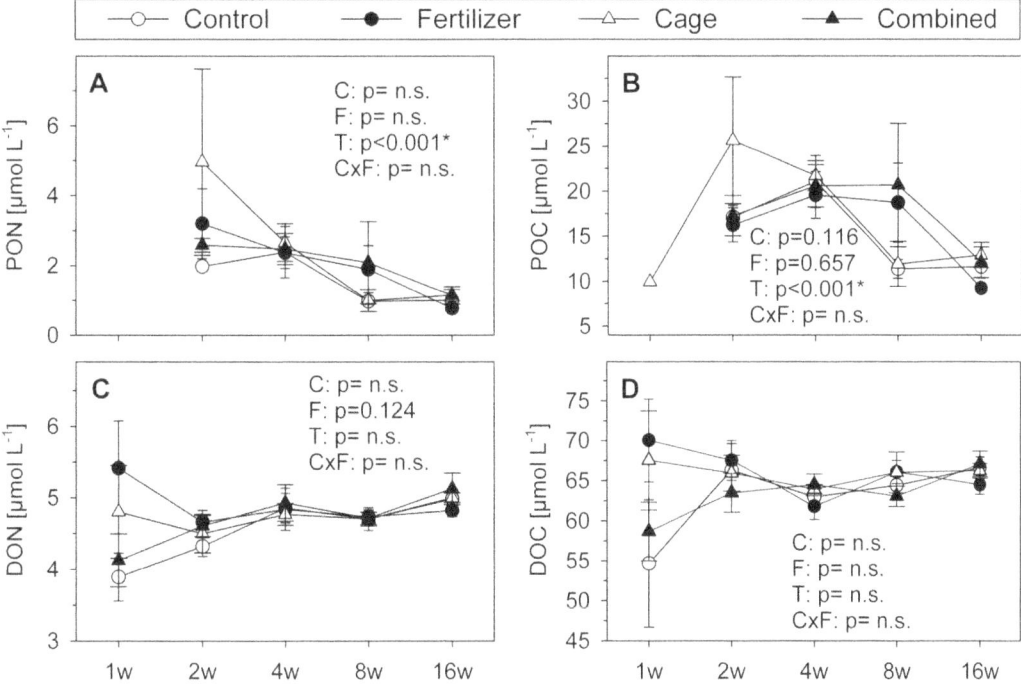

Figure 2. Concentrations of particulate and dissolved organic matter. Particulate (mg cm^{-2}, means±SE) and dissolved organic matter concentrations (μmol L^{-1}, means±SE) in water samples taken directly above the installations A: particulate organic nitrogen (PON), B: particulate organic carbon (POC), C: dissolved organic nitrogen (DON), and D: dissolved organic carbon (DOC). Shown are data of all treatments for all 5 sampling times. P-values were calculated from 3-factorial ANOVA and originate from analysis across the whole study period (see Table 1 for full results). Abbreviations: C = Cage, F = Fertilizer, T = Time. Missing values of 1wk for PON and POC resulted from insufficient algal dry mass for analysis. Shown P-values originate from analysis across the whole study period. P-values were tagged as n.s. (= not significant), when the model reduction step excluded the corresponding factor(s).

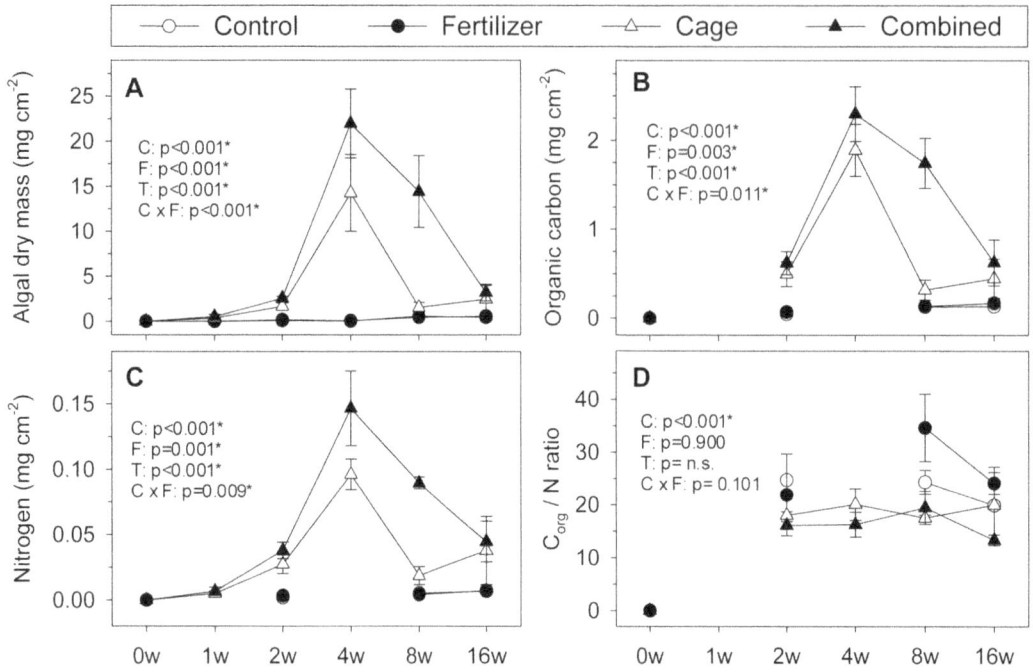

Figure 3. Development of algal dry mass (A), organic carbon (B), nitrogen content (C), and organic C/N ratio (D) on light-exposed tiles. Shown are means±SE of all treatments in mg cm^{-2} over the 5 sampling points after 1, 2, 4, 8 and 16 wk(s). P-values were calculated from a 3-factorial ANOVA and originate from analysis across the whole study period (see Table 1 for full results). P-values were tagged as n.s. (=not significant), when the model reduction step excluded the corresponding factor(s). Missing connections between data points are due to insufficient algal material for analysis. Abbreviations: C=Cage, F=Fertilizer, T=Time.

cover of the reef may have originated from a recent bleaching event in the region [77]. Our measurements of herbivorous fish biomass (22 g m^{-2}) were below the pristine reefs of Kingman (32 g m^{-2}) [76], the average Indo-Pacific values (29 g m^{-2}) [75], and data from recent studies in the Red Sea (63 g m^{-2} in 5 m water depth by Brokovich et al. [78] and 27 g m^{-2} by Khalil et al. [79]. However, other studies suggest that the measured biomass values of our study correspond to unfished reefs (e.g. [80,81]). This is supported by the sea urchin biomass at our study site (38 g m^{-2}), typical for unfished reefs [82,83]. Ambient concentrations of SRP ranged under the thresholds of increased macroalgae growth of 1.0 µmol L^{-1} for DIN and 0.1 µmol L^{-1} for SRP proposed by Bell [84] & Lapointe [15], though these values are under discussion [17,85] and many field studies have not found data supporting these thresholds [16,47,48,51,86–89]. In contrast, DIN ambient concentrations exceeded the threshold after the 1st wk. However, the low DOC and Chl a values (DOC: [41,90], Chl a: [84]) suggest that the reef is little impacted by eutrophication.

Effects of Treatments

Nutrient concentrations in the enriched treatments constantly exceeded ambient conditions and ranged above the suggested thresholds of Bell [84] and Lapointe [15], showing the successful enrichment. However, nutrient concentrations of the enriched treatments in this study are less enhanced than in similar experiments (e.g. [12,13]). We assume that the large water sampling volumes and the concomitant dilution of samples prevented the detection of higher nutrient levels in the enrichment treatments. This view is supported by the Chl a, POM, and DOM concentrations in the water column just above the treatments that were not significantly influenced by fertilizer addition or other treatments.

Algal Biomass. Nutrient enrichment altered algal biomass on the light-exposed tiles only in interaction with herbivore exclusion in terms of algal dry mass, organic C, and N. However, it is likely that a larger effect of nutrient enrichment was masked by compensatory feeding by herbivores [12]. In contrast to the nutrient treatment, herbivore exclusion had an immediate and direct influence on most measured algal parameters, which was further extended by the combined treatment.

C and N removal rates are strongly connected to algal wet and dry mass. However, C and N data analyses provide a more neutral method than other biomass measures because values are independent of algal species and their calcified structures, if any, and permit greater comparability between studies, albeit data available are scarce. Only one recent study from the Egyptian Red Sea [55] showed N removal rates and their maxima were similar to the results found here. The consistently lower organic C/N ratio in the caged treatments indicates that herbivore preferentially graze on N rich algae [91–93], which did not accumulate outside the cages. Furthermore, C/N ratio data suggest that extra N provided by the fertilizer was directly used for growth and not stored in the algal tissue as previously reported for depleted but not for enriched algal tissue [94]. The uptake of extra N from the fertilizer could therefore not be proven by the C/N ratio, but by the isotope analysis. The δ15N ratio of the fertilizer was close to 0, and the incorporation of the fertilizer therefore should reduce the δ15N ratio of the algal material. This reduction could be shown in the enriched frames over the non-enriched frames (Figure S3; Table S3).

If not controlled, algal biomass can increase to huge quantities, in our experiment up to 19 mg cm^{-2} wk^{-1}. This would be 190 t wk^{-1} if extrapolated to a reef of 1 km^2.

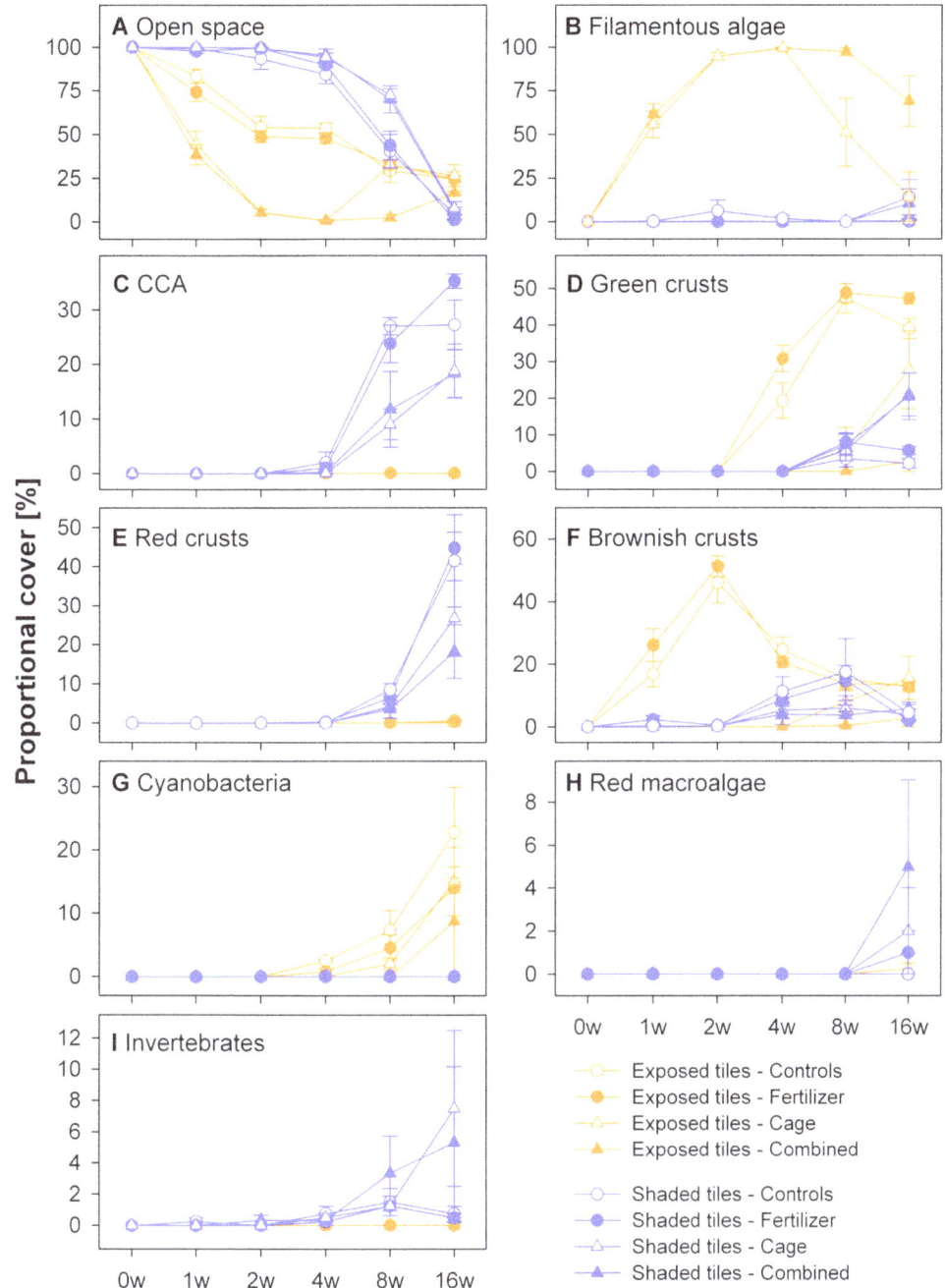

Figure 4. Percent cover of functional groups on light-exposed (orange) and shaded tiles (blue). Shown is the proportional cover (means±SE) over the study period of 4 months of functional groups in the 4 treatments: control, fertilizer, cage, and combined. See Tables 3 and 4 for statistical results.

Our findings from the Red Sea demonstrated that decreased herbivory has a stronger influence on algal biomass than increased nutrients, corresponding to the majority of comparative studies from reefs around the world that compared herbivory versus nutrient enrichment on algal growth (Australia: [86,87,95]; Caribbean: [12,47,48,50]; Hawaii: [13,60]; Guam: [11,51]. Yet, other studies collected evidence that nutrient enrichment can also have larger and delayed influence on algal development and the ability of algae to overgrow corals [13,96,97].

Our data clearly show that nutrient enrichment alone was not able to increase algal biomass, even when the proposed threshold concentrations of 1.0 µmol L^{-1} of DIN and 0.1 µmol L^{-1} of SRP [15,84] were exceeded for most of the study time. One may argue, that the ambient nutrient levels already saturated the nutrient needs of most algae and field and laboratory studies revealed maximum growth rates for some algae at DIN concentrations of about 0.5–0.8 µmol L^{-1} [98,99]. However, the interactive effects of nutrient enrichment and herbivore exclusion on biomass (algal dry mass, organic C, N), and community composition on the light-exposed tiles showed the potential of nutrient enrichment on algal growth and composition.

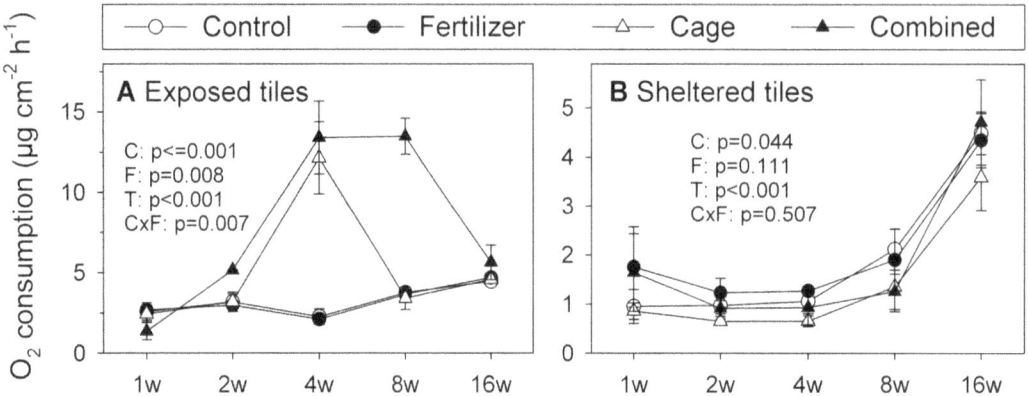

Figure 5. O₂ consumption rates. O$_2$ consumption rates of A: light-exposed tiles and B: shaded tiles in µg cm^{-2} h^{-1} (means±SE). P-values are calculated from 3-factorial ANOVA and originate from analysis across the whole study period (see Table 5 for full test results). Significant p-values ($p < 0.05$) are indicated by asterisks. Abbreviations: C = Cage, F = Fertilizer, T = Time.

Since microbial activity is enhanced by algal derived DOC [38,39], we expected DOC concentrations in the water column to rise with increasing algal biomass. Surprisingly, no correlation patterns between DOC and biomass were detectable, possible due to a dilution effect. Nevertheless, a parallel study [100], conducted under the same conditions, resulted in treatment specific responses of coral associated bacterial communities.

Algae community structure. Filamentous algae benefited directly from herbivore exclusion since they are a main feeding substratum for many herbivores [101–104]. Concordant with a study by McClanahan et al. [47], filamentous algae on the light-exposed tiles grew best under the combined treatment with herbivore exclusion and elevated nutrient concentrations. The rapid response of the algae and the clearly distinguishable differences between the caged and non-caged treatments, together with a low abundance outside the frames (CJ pers. obs.) strongly suggest filamentous algae to be an indicator for herbivore overfishing in the investigated area [43,44].

In contrast to a recent study by Jessen and Wild [55] in the Egyptian Red Sea, who found frondose brown algae within 4 wks after the start of a similar experiment, this algal group was not observed during the present study. Other studies from other oceans found frondose brown algae also within 4 months on their tiles [11–13,50,51,86], though some of the examined substrates

were likely affected by preconditioning. The absence of certain genera is likely due to a combination of seasonality and predation preferences [105,106].

Concordant with Jessen and Wild [55] from the Red Sea, but contrary to other studies [11–13,49], CCA cover was not found on the light-exposed tiles. Though Belliveau and Paul [11] and Smith et al. [13] preconditioned their tiles for 2 months, CCA appeared no later than 1 and 2 months respectively, indicating that settling and growth of CCA on the light-exposed tiles was inhibited in this study. The lack of CCA can be due to sediment trapping that can result in anoxic conditions coupled with decreased survivorship and recruitment of CCA [107–109]. The findings in the present study support this hypothesis: CCA grew on the shaded tiles where no filamentous algae dominated. The lower light conditions on the shaded tiles did not prevent CCA from growing, presumably due to their slow growing speed [110,111].

Littler and Littler [45] proposed the Relative Dominance Model (RDM) that predicts benthic community structure in response to anthropogenic threats of overfishing (grazer reduction), elevated nutrients, and a combination thereof. Although, the present study was conducted in a limited time frame of 4 months, the results for this time period can neither confirm that CCA dominated in the high nutrient, high grazing treatment (shown by [12,13] but not by [49]), nor the domination of frondose macroalgae under the

Table 2. Results of the 3-factorial ANOVA of the algal parameters.

	Algal dry mass			Algal C organic			Algal N			C$_{org}$/N		
	df	F	P	df	F	P	df	F	P	df	F	P
C	1	253.83	0.000*	1	30.49	0.000*	1	37.13	0.000*	1	13.38	0.000*
F	1	15.42	0.000*	1	11.87	0.000*	1	12.10	0.001*	1	0.02	0.900
T	4	26.63	0.000*	3	32.41	0.000*	4	33.24	0.000*	–	–	–
C×F	1	15.44	0.000*	1	9.50	0.000*	1	7.51	0.009*	1	2.83	0.101
T×C	4	28.71	0.000*	2	2.36	0.000*	2	0.97	0.387	–	–	–
T×F	4	2.75	0.037*	3	2.94	0.037*	4	2.81	0.039*	–	–	–
T×C×F	4	4.46	0.003*	2	4.08	0.003*	2	3.10	0.057	–	–	–

Response variables (1st row) are algal dry mass, algal organic C, algal N of the exposed tiles. Independent factors (1st column) are Cage (C), Fertilizer (F), and Time (T). Algal dry mass data were log (x+1) transformed to meet parametric assumptions. Significant results are indicated by asterisks. P-values of 0.000 symbolize values <0.001. Dashes represent factors that have been excluded by the model reduction.

Table 3. Results of 3-factorial generalized linear model (GLM; binomial distribution and logit function) of functional algal groups of light-exposed tiles.

	Open space			Filamentous algae			CCA		
	df	F	P	df	F	P	df	F	P
C	1	210.20	0.000*	1	693.74	0.000*	x	x	x
F	1	9.77	0.002*	1	30.99	0.000*	x	x	x
T	4	44.04	0.000*	4	31.55	0.000*	x	x	x
C×F	1	6.07	0.017*	–	–	–	x	x	x
T×C	4	27.34	0.000*	–	–	–	x	x	x
T×F	4	0.50	0.739	4	4.45	0.003*	x	x	x
T×C×F	4	4.83	0.002*	–	–	–	x	x	x

	Green crusts			Red crusts			Brownish crusts		
	df	F	P	df	F	P	df	F	P
C	1	192.46	0.000*	1	23.29	0.000*	1	250.90	0.000*
F	1	0.19	0.667	1	2.04	0.159	1	0.44	0.512
T	4	96.84	0.000*	4	14.44	0.000*	4	16.82	0.000*
C×F	1	30.13	0.000*	–	–	–	1	14.21	0.000*
T×C	4	7.97	0.000*	–	–	–	4	23.10	0.000*
T×F	4	0.49	0.746	–	–	–	4	1.04	0.394
T×C×F	4	0.12	0.973	–	–	–	4	0.11	0.978

	Cyanobacteria			Red algae			Invertebrates		
	df	F	P	df	F	P	df	F	P
C	1	13.08	0.000*	1	24.91	0.000*	x	x	x
F	1	8.93	0.004*	1	20.12	0.000*	x	x	x
T	4	51.75	0.000*	4	14.00	0.000*	x	x	x
C×F	1	0.17	0.677	–	–	–	x	x	x
T×C	4	1.73	0.157	–	–	–	x	x	x
T×F	4	0.11	0.980	–	–	–	x	x	x
T×C×F	4	0.43	0.789	–	–	–	x	x	x

Response variables are shown in the 1st row and in the first column the independent factors: Cage (C), Fertilizer (F), and Time (T). Significant results are indicated by asterisks. P-values of 0.000 symbolize values <0.001. Dashes represent factors that have been excluded by the model reduction and 'x' stands for insufficient data for analysis.

combined treatments (shown by [13,46], but not by [12,47,49]). However, following the model, filamentous algae predominated under low grazing levels (shown by [12,13,46]). Though, in contrast to the model, best conditions for filamentous algae in terms of biomass and cover were found in the combined treatment (shown by [47], but not by [13,49]).

Differences between light-exposed and shaded tiles. This is the first study that compared the individual and combined effects of manipulated herbivory exclusion and nutrient enrichment on the reef algae community composition on light-exposed versus light-shaded tiles in coral reefs. The open space data (Figure 4A, Tables 3 and 4) showed that the tile surface colonization occurred faster on the light-exposed tiles than on the shaded tiles. Higher light availability, easier access for grazers, and the putative higher supply of recruits from the water column on the light-exposed tiles may be responsible for this difference.

CCA and non-coralline red crusts were found almost exclusively and were predominant on the shaded tiles, which have been found

to either enhance [14,112–115] or impair coral recruitment [116–118]. The light-exposed tiles featured neither CCA nor invertebrate cover and only slight amounts of red crusts and red macroalgae. The lack of these algal groups on the light-exposed tiles could be due to out-competition by filamentous algae [119]. In contrast, (mucilaginous) cyanobacteria were the only group that grew exclusively on the light-exposed tiles and not on the shaded tiles (Figure 4G, Tables 3 and 4).

Our results corroborate the observation by Burkepile and Hay [12] that studies from deeper reefs (6–18 m, except [60]) showed minimal effects of nutrient enrichment on overall algal abundance and moderate effects on community structures. They supposed that these differences may resulted from high light conditions in shallow areas allowing macrophytes to take full advantage of nutrient enrichment and enable them to grow rapidly. However, if it is assumed that the lower light conditions on the shaded tiles simulate reduced water depths, the lower influence of nutrient

Table 4. Results of 3-factorial generalized linear model (GLM; binomial distribution and logit function) of functional algal groups of light shaded tiles.

	Open space			Filamentous algae			CCA		
	df	F	P	df	F	P	df	F	P
C	1	39.64	0.000*	1	2.19	0.144	1	60.94	0.000*
F	1	0.88	0.354	1	2.90	0.094	1	0.01	0.909
T	4	225.73	0.000*	4	11.39	0.000*	4	150.47	0.000*
C×F	1	0.01	0.919	1	26.21	0.000*	1	2.60	0.113
T×C	4	0.39	0.816	4	0.35	0.844	4	2.90	0.030*
T×F	4	1.05	0.391	4	0.45	0.768	4	1.23	0.309
T×C×F	4	0.67	0.618	4	0.16	0.999	–	–	–

	Green crusts			Red crusts			Brownish crusts		
	df	F	P	df	F	P	df	F	P
C	1	40.47	0.000*	1	38.78	0.000*	1	9.53	0.003*
F	1	4.38	0.041*	1	2.84	0.098	1	0.43	0.512
T	4	66.75	0.000*	4	132.64	0.000*	4	15.76	0.000*
C×F	1	2.13	0.150	1	8.20	0.006*	1	0.07	0.797
T×C	4	3.18	0.020*	4	0.02	0.999	4	2.03	0.103
T×F	4	0.17	0.952	4	0.28	0.889	4	0.55	0.702
T×C×F	4	0.13	0.971	4	0.01	0.999	4	0.43	0.787

	Cyanobacteria			Red algae			Invertebrates		
	df	F	P	df	F	P	df	F	P
C	x	x	x	1	18.62	0.000*	1	14.16	0.000*
F	x	x	x	1	7.65	0.007*	1	0.13	0.721
T	x	x	x	4	21.06	0.000*	1	12.94	0.000*
C×F	x	x	x	1	2.32	0.133	2	0.71	0.403
T×C	x	x	x	–	–	–	2	2.22	0.078
T×F	x	x	x	–	–	–	2	0.87	0.488
T×C×F	x	x	x	–	–	–	2	0.09	0.984

Response variables are shown in the 1st row and in the first column the independent factors: Cage (C), Fertilizer (F), and Time (T). Significant results are indicated by asterisks. P-values of 0.000 symbolize values <0.001. Dashes represent factors that have been excluded by the model reduction and 'x' stands for insufficient data for analysis.

enrichment there suggests an important role of water depth and light availability on the effect of nutrient enrichment [120,121].

Seasonality. It remains unclear whether the algal community was still in the succession process or already at a final stage. In contrast to other successional studies that compared the effects of herbivore exclusion and nutrient enrichment (e.g. [12,13]), filamentous algae on the light-exposed tiles declined after wk 4. Temperature is an important controlling factor for algae [122,123] and the Central Red Sea is subject to strong seasonal temperature fluctuations [124]. However, ambient condition data from temperature loggers in this study (Figure S2) did not reveal correlating patterns of temperature and biomass, nor did CTD data of several parameters (turbidity, O₂ saturation and Chl *a* along the transect) (data not shown). DIN concentrations in ambient and enriched treatments that peaked after wk 4 and declined afterwards may be an important factor.

Consequences & conclusions. Cascading negative effects have been reported when reef ecosystems were continuously

exposed to overfishing of herbivores and increased nutrient concentrations. Algae can gain dominance over corals [22], resulting in less settling substrate for coral spat [26,27], decreased herbivore grazing rates [25], and changes in C and N fluxes [36,125]. Predicted climate change effects of ocean warming and acidification may further exacerbate these processes [126,127].

The study underlines the importance of herbivory for the Red Sea, especially in the light of the relatively low herbivore biomass compared to other Indo-Pacific reefs and the high algal growth potential when herbivory was impeded. Surprisingly, macroalgal (here particularly filamentous algae) growth rates in the first 4 wks of this study greatly exceeded average patterns of the Indo-Pacific and even those of the Caribbean [75]. However, after 4 wks, coverage declined and resembled the average Caribbean cover (at 8 wk) and the lower Indo-Pacific values (at 16 wk). Our data suggest that the surveyed reef is not resistant against herbivore overfishing or a combination together with increased nutrient concentrations that has been simulated in this study. However, the

Table 5. Results of the 3-factorial ANOVA of O_2 consumption of exposed and shaded tiles.

	Exposed Tiles			Shaded Tiles		
	df	F	P	df	F	P
C	1	35.15	0.000*	1	3.70	0.056
F	1	3.87	0.054	1	1.92	0.170
T	4	18.27	0.000*	4	39.84	0.000*
C×F	1	3.68	0.060	–	–	–
T×C	4	10.10	0.000*	–	–	–
T×F	4	1.68	0.006*	–	–	–
T×C×F	4	5.79	0.000*	–	–	–

Response variables are shown in the 1st row. In the 1st column are the independent factors: Cage (C), Fertilizer (F), and Time (T). Significant results are indicated by asterisks. P-values of 0.000 symbolize values <0.001 and dashes represent factors that have been excluded by the model reduction. n.s.=not significant.

potential compensatory feeding and the present herbivore biomass suggest that the benthic community is resistant against enhanced nutrient concentrations even when exceeding proposed thresholds.

Supporting Information

Figure S1   Study site. Right panel shows position of the study area in the Red Sea. The circle on the left panel indicates the study site at the Northern tip of Al Fahal-reef, located about 13 km off the Saudi-Arabian coast.

Figure S2   Temperature development at Al Fahal reef. Daily average temperatures (± max/min) of the 16 experimental frames at 5 m water depths at Al Fahal reef over the study period from June to September 2011. Sampling times are indicated by vertical lines.

References

1. Burke LM, Reytar K, Spalding M, Perry A (2011) Reefs at Risk Revisited. Washington, DC: World Resources Institute.
2. Selman M, Greenhalgh S, Diaz R, Sugg Z (2008) Eutrophication and hypoxia in coastal areas: a global assessment of the state of knowledge. World Resources Institute. Washington, DC.
3. Wild C, Hoegh-Guldberg O, Naumann MS, Colombo-Pallotta MF, Ateweberhan M, et al. (2011) Climate change impedes scleractinian corals as primary reef ecosystem engineers. Mar Freshw Res 62: 205–215.
4. Koop K, Booth D, Broadbent A, Brodie J, Bucher D, et al. (2001) ENCORE: the effect of nutrient enrichment on coral reefs. Synthesis of results and conclusions. Mar Pollut Bull 42: 91–120.
5. Fabricius KE, Cséke S, Humphrey C, De'ath G (2013) Does trophic status enhance or reduce the thermal tolerance of scleractinian corals? A review, experiment and conceptual framework. PLOS ONE 8: e54399.
6. Ferrier-Pagès C, Gattuso JP, Dallot S, Jaubert J (2000) Effect of nutrient enrichment on growth and photosynthesis of the zooxanthellate coral Stylophora pistillata. Coral Reefs 19: 103–113.
7. Kinsey DW, Davies PJ (1979) Effects of elevated nitrogen and phosphorus on coral reef growth. Limnol Oceanogr 24: 935–940.
8. Loya Y, Lubinevsky H, Rosenfeld M, Kramarsky-Winter E (2004) Nutrient enrichment caused by in situ fish farms at Eilat, Red Sea is detrimental to coral reproduction. Mar Pollut Bull 49: 344–353.
9. Wiedenmann J, D'Angelo C, Smith EG, Hunt AN, Legiret FE, et al. (2012) Nutrient enrichment can increase the susceptibility of reef corals to bleaching. Nature Clim Change 3: 160–164.
10. Voss JD, Richardson LL (2006) Nutrient enrichment enhances black band disease progression in corals. Coral Reefs 25: 569–576.

Figure S3   δ15N isotopic signatures of homogenized cover of light-exposed tiles. δ15N values (mean±SE) are shown for each treatment over 5 sampling times. Missing values of wk 1 and wk 4 resulted from insufficient algal material for analysis. P-values are calculated from 3-factorial ANOVA and originate from analysis across the whole study period (see Table S3 for full test results).

Table S1   List of counted herbivorous fish. Listed are families, species names, abundance (normalized to ind. m^{-2}), and their biomass (normalized to g m^{-2}).

Table S2   List of counted sea urchins. Listed are species names, abundance (ind. m^{-2}), and their biomass (g m^{-2}).

Table S3   Results of the 3-factorial ANOVA of d15N isotopic signatures of cover from light exposed tiles. Significant results are indicated by asterisks. Abbreviations: C = Cage, F = Fertilizer, T = Time.

Acknowledgments

Many thanks for support goes to the CMOR team at KAUST, L. Smith, F. Mallon, E. Aravantinos, and the boat crew K. Al-Moullad, E. Al-Jahdali, G. Al-Jahdali. We also would like to acknowledge the field and laboratory assistance of T. Bayer, A. Roik, C. Arif, T. Brown, M. Kruse, C. Jhen, and V. Gregoire. We thank the KAUST ACL lab team N. Fayad and M. Masry, as well as ZMT laboratory team D. Dassbach and M. Birkicht for invaluable advice and support in the laboratory, A. Wolf and L. Yum for improving previous versions of the manuscript, and W. Wosniok for statistical advice. Furthermore, the authors want to thank the editor and 2 anonymous reviewers who provided valuable comments to improve the manuscript.

Author Contributions

Conceived and designed the experiments: CJ CR JVL CRV CW. Performed the experiments: CJ CR JVL. Analyzed the data: CJ. Wrote the paper: CJ CW.

11. Belliveau S, Paul V (2002) Effects of herbivory and nutrients on the early colonization of crustose coralline and fleshy algae. Mar Ecol Prog Ser 232: 105–114.
12. Burkepile DE, Hay ME (2009) Nutrient versus herbivore control of macroalgal community development and coral growth on a Caribbean reef. Mar Ecol Prog Ser 389: 71–84.
13. Smith JE, Hunter CL, Smith CM (2010) The effects of top-down versus bottom-up control on benthic coral reef community structure. Oecologia 163: 497–507.
14. Harrington L, Fabricius K, De'ath G, Negri A (2004) Recognition and selection of settlement substrata determine post-settlement survival in corals. Ecology 85: 3428–3437.
15. Lapointe B (1997) Nutrient thresholds for bottom-up control of macroalgal blooms on coral reefs in Jamaica and southeast Florida. Limnol Oceanogr 42: 1119–1131.
16. McClanahan TR, Cokos BA, Sala E (2002) Algal growth and species composition under experimental control of herbivory, phosphorus and coral abundance in Glovers Reef, Belize. Mar Pollut Bull 44: 441–451.
17. McClanahan TR, Sala E, Mumby PJ, Jones S (2004) Phosphorus and nitrogen enrichment do not enhance brown frondose "macroalgae". Mar Pollut Bull 48: 196–199.
18. Miller MW, Hay ME (1996) Coral-seaweed-grazer-nutrient interactions on temperate reefs. Ecol Monogr 66: 323–344.
19. Jackson JBC, Kirby MX, Berger WH, Bjorndal KA, Botsford LW, et al. (2001) Historical overfishing and the recent collapse of coastal ecosystems. Science 293: 629–637.
20. Steneck RS, Graham MH, Bourque BJ, Corbett D, Erlandson JM, et al. (2002) Kelp forest ecosystems: biodiversity, stability, resilience and future. Environ Conserv 29: 436–459.

21. Sweatman H (2008) No-take reserves protect coral reefs from predatory starfish. Curr Biol 18: R598–R599.
22. Hughes T, Rodrigues M, Bellwood D, Ceccarelli D, Hoegh-Guldberg O, et al. (2007) Phase shifts, herbivory, and the resilience of coral reefs to climate change. Curr Biol 17: 360–365.
23. Mumby PJ (2009) Herbivory versus corallivory: are parrotfish good or bad for Caribbean coral reefs? Coral Reefs 28: 683–690.
24. Burkepile DE, Hay ME (2006) Herbivore vs. nutrient control of marine primary producers: context-dependent effects. Ecology 87: 3128–3139.
25. Hoey AS, Bellwood DR (2011) Suppression of herbivory by macroalgal density: a critical feedback on coral reefs? Ecol Lett 14: 267–273.
26. Birrell CL, McCook LJ, Willis BL, Diaz-Pulido GA (2008) Effects of benthic algae on the replenishment of corals and the implications for the resilience of coral reefs. Oceanogr Mar Biol 46: 25–64.
27. Schaffelke B, Mellors J, Duke NC (2005) Water quality in the Great Barrier Reef region: responses of mangrove, seagrass and macroalgal communities. Mar Pollut Bull 51: 279–296.
28. Done T (1992) Phase shifts in coral reef communities and their ecological significance. Hydrobiologia 247: 121–132.
29. Arnold SN, Steneck RS, Mumby PJ (2010) Running the gauntlet: inhibitory effects of algal turfs on the process of coral recruitment. Mar Ecol Prog Ser 414: 91–105.
30. Rasher DB, Hay ME (2010) Chemically rich seaweeds poison corals when not controlled by herbivores. Proc Natl Acad Sci U S A 107: 9683–9688.
31. Rasher DB, Stout EP, Engel S, Kubanek J, Hay ME (2011) Macroalgal terpenes function as allelopathic agents against reef corals. Proc Natl Acad Sci U S A 108: 17726–17731.
32. Barott KL, Rohwer FL (2012) Unseen players shape benthic competition on coral reefs. Trends Microbiol 20: 621–628.
33. Paul VJ, Kuffner IB, Walters LJ, Ritson-Williams R, Beach KS, et al. (2011) Chemically-mediated interactions between macroalgae, Dictyota spp., and multiple life-history stages of the coral Porites astreoides. Mar Ecol Prog Ser 426: 161–170.
34. Barott K, Smith J, Dinsdale E, Hatay M, Sandin S, et al. (2009) Hyperspectral and physiological analyses of coral-algal interactions. PLOS ONE 4: e8043.
35. Smith JE, Shaw M, Edwards RA, Obura D, Pantos O, et al. (2006) Indirect effects of algae on coral: algae-mediated, microbe-induced coral mortality. Ecol Lett 9: 835–845.
36. Wild C, Niggl W, Naumann MS, Haas AF (2010) Organic matter release by Red Sea coral reef organisms: potential effects on microbial activity and in situ O2 availability. Mar Ecol Prog Ser 411: 61–71.
37. Mumby PJ, Steneck RS (2011) The resilience of coral reefs and its implications for reef management. In: Dubinsky Z, Stambler N, editors. Coral Reefs: An ecosystem in transition. Amsterdam Springer. 509–519.
38. Haas AF, Nelson CE, Kelly LW, Carlson CA, Rohwer F, et al. (2011) Effects of coral reef benthic primary producers on dissolved organic carbon and microbial activity. PLOS ONE 6: e27973.
39. Wild C, Haas A, Naumann M, Mayr C, El-Zibdah M (2009) Comparative investigation of organic matter release by corals and benthic reef algae - implications for pelagic and benthic microbial metabolism. Proc 11th Int Coral Reef Sym: 1319–1323.
40. Haas AF, Jantzen C, Naumann MS, Iglesias-Prieto R, Wild C (2010) Organic matter release by the dominant primary producers in a Caribbean reef lagoon: implication for in situ O2 availability. Mar Ecol Prog Ser 409: 27–39.
41. Kline DI, Kuntz NM, Breitbart M, Knowlton N, Rohwer F (2006) Role of elevated organic carbon levels and microbial activity in coral mortality. Mar Ecol Prog Ser 314: 119–125.
42. Kuntz NM, Kline DI, Sandin SA, Rohwer F (2005) Pathologies and mortality rates caused by organic carbon and nutrient stressors in three Caribbean coral species. Mar Ecol Prog Ser 294: 173–180.
43. Cooper TF, Gilmour JP, Fabricius KE (2009) Bioindicators of changes in water quality on coral reefs: review and recommendations for monitoring programmes. Coral Reefs 28: 589–606.
44. Littler MM, Littler DS (2007) Assessment of coral reefs using herbivory/nutrient assays and indicator groups of benthic primary producers: a critical synthesis, proposed protocols, and critique of management strategies. Aquat Conserv 17: 195–215.
45. Littler MM, Littler DS (1984) Models of tropical reef biogenesis: the contribution of algae. In: Round FE, Chapman DJ, editors. Progress in Phycological Research, Vol 3. Bristol: Biopress. 323–364.
46. Littler MM, Littler DS, Brooks BL (2006) Harmful algae on tropical coral reefs: bottom-up eutrophication and top-down herbivory. Harmful Algae 5: 565–585.
47. McClanahan TR, Sala E, Stickels P, Cokos B, Baker A, et al. (2003) Interaction between nutrients and herbivory in controlling algal communities and coral condition on Glover's Reef, Belize. Mar Ecol Prog Ser 261: 135–147.
48. Miller MW, Hay ME, Miller SL, Malone D, Sotka EE, et al. (1999) Effects of nutrients versus herbivores on reef algae: a new method for manipulating nutrients on coral reefs. Limnol Oceanogr 44: 1847–1861.
49. Rasher DB, Engel S, Bonito V, Fraser GJ, Montoya JP, et al. (2012) Effects of herbivory, nutrients, and reef protection on algal proliferation and coral growth on a tropical reef. Oecologia 169: 187–198.
50. Sotka EE, Hay ME (2009) Effects of herbivores, nutrient enrichment, and their interactions on macroalgal proliferation and coral growth. Coral Reefs 28: 555–568.
51. Thacker R, Ginsburg D, Paul V (2001) Effects of herbivore exclusion and nutrient enrichment on coral reef macroalgae and cyanobacteria. Coral Reefs 19: 318–329.
52. Fishelson L (1973) Ecology of coral reefs in the Gulf of Aqaba (Red Sea) influenced by pollution. Oecologia 12: 55–67.
53. Genin A, Lazar B, Brenner S (1995) Vertical mixing and coral death in the Red Sea following the eruption of Mount Pinatubo. Nature 377: 507–510.
54. Vine PJ (1974) Effects of algal grazing and aggressive behaviour of the fishes Pomacentrus lividus and Acanthurus sohal on Coral-Reef Ecology. Marine Biology 24: 131–136.
55. Jessen C, Wild C (2013) Herbivory effects on benthic algal composition and growth on a coral reef flat in the Egyptian Red Sea. Mar Ecol Prog Ser 476: 9–21.
56. Korzen L, Israel A, Abelson A (2011) Grazing effects of fish versus sea urchins on turf algae and coral recruits: possible implications for coral reef resilience and restoration. Journal of Marine Biology, vol 2011.
57. Wilkinson C (2008) Status of coral reefs of the world: 2008. Townsville: Australian Institute of Marine Science.
58. Nadon M, Stirling G (2006) Field and simulation analyses of visual methods for sampling coral cover. Coral Reefs 25: 177–185.
59. Burkepile DE, Hay ME (2007) Predator release of the gastropod Cyphoma gibbosum increases predation on gorgonian corals. Oecologia 154: 167–173.
60. Smith J, Smith C, Hunter C (2001) An experimental analysis of the effects of herbivory and nutrient enrichment on benthic community dynamics on a Hawaiian reef. Coral Reefs 19: 332–342.
61. Hixon MA, Brostoff WN (1996) Succession and herbivory: effects of differential fish grazing on Hawaiian coral-reef algae. Ecol Monogr 66: 67–90.
62. Kohler KE, Gill SM (2006) Coral Point Count with Excel extensions (CPCe): A Visual Basic program for the determination of coral and substrate coverage using random point count methodology. Comput Geosci 32: 1259–1269.
63. Arar EJ, Collins GB, United States Environmental Protection Agency (1997) Method 445.0: In vitro determination of chlorophyll a and pheophytin a in marine and freshwater algae by fluorescence: United States Environmental Protection Agency, Office of Research and Development, National Exposure Research Laboratory.
64. Randall JE (1983) Red Sea Reef Fishes. London: Immel Publishing.
65. Debelius H (2007) Riff-Führer Rotes Meer: Ägypten, Israel, Jordanien, Sudan, Saudi-Arabien, Jemen, Arabische Halbinsel. Stuttgart: Franckh-Kosmos-Verlag.
66. Lieske E, Myers R (2009) Korallenriff-Führer Rotes Meer: Rotes Meer bis Golf von Aden, Südoman. Stuttgart: Franckh-Kosmos-Verlag.
67. Khalaf MA, Disi A (1997) Fishes of the Gulf of Aqaba: Marine Science Station Aqaba, Jordan.
68. Choat J, Robbins W, Clements K (2004) The trophic status of herbivorous fishes on coral reefs. Marine Biology 145: 445–454.
69. Green AL, Bellwood DR (2009) Monitoring functional groups of herbivorous reef fishes as indicators of coral reef resilience - A practical guide for coral reef managers in the Asia Pacific Region. IUCN, Gland, Switzerland: IUCN working group on climate change and coral reefs.
70. Froese R, Pauly D (2012) FishBase. World Wide Web electronic publication. Available: www.fishbase.org, version (08/2012).
71. Dotan A (1990) Population structure of the echinoid Heterocentrotus mammillatus (L.) along the littoral zone of south-eastern Sinai. Coral Reefs 9: 75–80.
72. Ebert TA (1975) Growth and mortality of post-larval echinoids. Am Zool 15: 755–775.
73. Muthiga N, McClanahan T (1987) Population changes of a sea urchin (Echinometra mathaei) on an exploited fringing reef. Afr J Ecol 25: 1–8.
74. R Development Core Team (2012) R: A Language and Environment for Statistical Computing (Vienna: R Foundation for Statistical Computing). Available: http://www.r-project.org/.
75. Roff G, Mumby PJ (2012) Global disparity in the resilience of coral reefs. Trends Ecol Evol 27: 404–413.
76. Sandin SA, Smith JE, DeMartini EE, Dinsdale EA, Donner SD, et al. (2008) Baselines and degradation of coral reefs in the northern Line Islands. PLOS ONE 3: e1548.
77. Furby KA, Bouwmeester J, Berumen ML (2013) Susceptibility of central Red Sea corals during a major bleaching event. Coral Reefs 32: 505–513.
78. Brokovich E, Ayalon I, Einbinder S, Segev N, Shaked Y, et al. (2010) Grazing pressure on coral reefs decreases across a wide depth gradient in the Gulf of Aqaba, Red Sea. Mar Ecol Prog Ser 399: 69–80.
79. Khalil MT, Cochran JEM, Berumen ML (2013) The abundance of herbivorous fish on an inshore Red Sea reef following a mass coral bleaching event. Environ Biol Fishes DOI 10.1007/s10641-012-0103-5:.
80. McClanahan T, Muthiga N, Kamukuru A, Machano H, Kiambo R (1999) The effects of marine parks and fishing on coral reefs of northern Tanzania. Biol Conserv 89: 161–182.
81. Williams I, Polunin N (2001) Large-scale associations between macroalgal cover and grazer biomass on mid-depth reefs in the Caribbean. Coral Reefs 19: 358–366.

82. McClanahan T (1997) Primary succession of coral-reef algae: differing patterns on fished versus unfished reefs. J Exp Mar Biol Ecol 218: 77–102.

83. O'Leary JK, Potts DC, Braga JC, McClanahan TR (2012) Indirect consequences of fishing: reduction of coralline algae suppresses juvenile coral abundance. Coral Reefs 31: 547–559.

84. Bell P (1992) Eutrophication and coral reefs-some examples in the Great Barrier Reef lagoon. Water Res 26: 553–568.

85. Bell PRF, Lapointe BE, Elmetri I (2007) Reevaluation of ENCORE: support for the eutrophication threshold model for coral reefs. AMBIO 36: 416–424.

86. Diaz-Pulido G, McCook L (2003) Relative roles of herbivory and nutrients in the recruitment of coral-reef seaweeds. Ecology 84: 2026–2033.

87. Hatcher BG, Larkum AWD (1983) An experimental analysis of factors controlling the standing crop of the epilithic algal community on a coral reef. J Exp Mar Biol Ecol 69: 61–84.

88. Larkum AWD, Koop K (1997) ENCORE, algal productivity and possible paradigm shifts. Proc 8th Int Coral Reef Sym 1: 881–884.

89. Szmant AM (1997) Nutrient effects on coral reefs: a hypothesis on the importance of topographic and trophic complexity to reef nutrient dynamics. Proc 8th Int Coral Reef Sym 2: 1527–1532.

90. Dinsdale EA, Rohwer F (2011) Fish or germs? Microbial dynamics associated with changing trophic structures on coral reefs. In: Dubinsky Z, Stambler N, editors. Coral Reefs: An Ecosystem in Transition. 231–240.

91. Boyer KE, Fong P, Armitage AR, Cohen RA (2004) Elevated nutrient content of tropical macroalgae increases rates of herbivory in coral, seagrass, and mangrove habitats. Coral Reefs 23: 530–538.

92. Furman BT, Heck K (2008) Effects of nutrient enrichment and grazers on coral reefs: an experimental assessment. Mar Ecol Prog Ser 363: 89–101.

93. Goecker ME, Heck KL Jr, Valentine JF (2005) Effects of nitrogen concentrations in turtlegrass Thalassia testudinum on consumption by the bucktooth parrotfish Sparisoma radians. Mar Ecol Prog Ser 286: 239–248.

94. Fong P, Boyer KE, Kamer K, Boyle KA (2003) Influence of initial tissue nutrient status of tropical marine algae on response to nitrogen and phosphorus additions. Mar Ecol Prog Ser 262: 111–123.

95. Jompa J, McCook LJ (2002) The effects of nutrients and herbivory on competition between a hard coral (Porites cylindrica) and a brown alga (Lobophora variegata). Limnol Oceanogr 47: 527–534.

96. Littler MM, Littler DS, Brooks BL, Lapointe BE (2006) Nutrient manipulation methods for coral reef studies: A critical review and experimental field data. J Exp Mar Biol Ecol 336: 242–253.

97. Vermeij MJA, van Moorselaar I, Engelhard S, Hörnlein C, Vonk SM, et al. (2010) The effects of nutrient enrichment and herbivore abundance on the ability of turf algae to overgrow coral in the Caribbean. PLOS ONE 5: e14312.

98. DeBoer JA, Guigli HJ, Israel TL, D'Elia CF (1978) Nutritional studies of two red algae. I. Growth rate as a function of nitrogen source and concentration. J Phycol 14: 261–266.

99. Lapointe BT, Tenore K (1981) Experimental outdoor studies with Ulva fasciata Delile. I. Interaction of light and nitrogen on nutrient uptake, growth, and biochemical composition. J Exp Mar Biol Ecol 53: 135–152.

100. Jessen C, Lizcano JFV, Bayer T, Roder C, Aranda M, et al. (2013) In-situ Effects of Eutrophication and Overfishing on Physiology and Bacterial Diversity of the Red Sea Coral Acropora hemprichii. PLOS ONE 8: e62091.

101. Bonaldo RM, Bellwood DR (2008) Size-dependent variation in the functional role of the parrotfish Scarus rivulatus on the Great Barrier Reef, Australia. Mar Ecol Prog Ser 360: 237–244.

102. Ferreira CEL, Gonçalves JEA (2006) Community structure and diet of roving herbivorous reef fishes in the Abrolhos Archipelago, south-western Atlantic. J Fish Biol 69: 1533–1551.

103. Fox RJ, Bellwood DR (2007) Quantifying herbivory across a coral reef depth gradient. Mar Ecol Prog Ser 339: 49–59.

104. Wilson SK, Bellwood DR, Choat JH, Furnas MJ (2003) Detritus in the epilithic algal matrix and its use by coral reef fishes. Oceanogr Mar Biol 41: 279–310.

105. Benayahu Y, Loya Y (1977) Seasonal occurrence of benthic-algae communities and grazing regulation by sea urchins at the coral reefs of Eilat, Red Sea. Proc 3rd Int Coral Reef Sym 2: 383–389.

106. Lotze HK, Worm B, Sommer U (2000) Propagule banks, herbivory and nutrient supply control population development and dominance patterns in macroalgal blooms. Oikos 89: 46–58.

107. Fabricius K, De'ath G (2001) Environmental factors associated with the spatial distribution of crustose coralline algae on the Great Barrier Reef. Coral Reefs 19: 303–309.

108. Figueiredo MAO, Steneck RS (2002) Floristic and ecological studies of crustose coralline algae on Brazil's Abrolhos reefs. Proc 9th Int Coral Reef Sym 1: 493–498.

109. Steneck R (1997) Crustose corallines, other algal functional groups, herbivores and sediments: complex interactions along reef productivity gradients. Proc 8th Int Coral Reef Sym 1: 695–700.

110. Littler MM (1972) The crustose corallinaceae. Oceanogr Mar Biol 10: 103–120.

111. Littler MM, Littler DS (2011) Algae, coralline. In: Hopley D, editor. Encyclopedia of Modern Coral Reefs: Structure, Form and Process. Dordrecht: Springer. 20.

112. Tanner J (1995) Competition between scleractinian corals and macroalgae: An experimental investigation of coral growth, survival and reproduction. J Exp Mar Biol Ecol 190: 151–168.

113. Baird AH, Morse ANC (2004) Induction of metamorphosis in larvae of the brooding corals Acropora palifera and Stylophora pistillata. Mar Freshw Res 55: 469–472.

114. Heyward AJ, Negri AP (1999) Natural inducers for coral larval metamorphosis. Coral Reefs 18: 273–279.

115. Morse ANC, Iwao K, Baba M, Shimoike K, Hayashibara T, et al. (1996) An ancient chemosensory mechanism brings new life to coral reefs. Biol Bull 191: 149–154.

116. Diaz-Pulido G, Harii S, McCook L, Hoegh-Guldberg O (2010) The impact of benthic algae on the settlement of a reef-building coral. Coral Reefs 29: 203–208.

117. Golbuu Y, Richmond RH (2007) Substratum preferences in planula larvae of two species of scleractinian corals, Goniastrea retiformis and Stylaraea punctata. Marine Biology 152: 639–644.

118. Suzuki G, Hayashibara T (2011) Do epibenthic algae induce species-specific settlement of coral larvae? J Mar Biol Assoc U K 91: 677–683.

119. Carpenter RC (1990) Competition among marine macroalgae: a physiological perspective. J Phycol 26: 6–12.

120. Duhamel S, Björkman KM, Karl DM (2012) Light dependence of phosphorus uptake by microorganisms in the subtropical North and South Pacific Ocean. Aquat Microb Ecol 67: 225–238.

121. MacIsaac J, Dugdale R (1972) Interactions of light and inorganic nitrogen in controlling nitrogen uptake in the sea. Deep-Sea Res 19: 209–232.

122. Ferrari R, Gonzalez-Rivero M, Ortiz JC, Mumby PJ (2012) Interaction of herbivory and seasonality on the dynamics of Caribbean macroalgae. Coral Reefs 31: 683–692.

123. Ateweberhan M, Bruggemann JH, Breeman AM (2006) Effects of extreme seasonality on community structure and functional group dynamics of coral reef algae in the southern Red Sea (Eritrea). Coral Reefs 25: 391–406.

124. Davis KA, Lentz SJ, Pineda J, Farrar JT, Starczak VR, et al. (2011) Observations of the thermal environment on Red Sea platform reefs: a heat budget analysis. Coral Reefs 30: 26–36.

125. Davey M, Holmes G, Johnstone R (2008) High rates of nitrogen fixation (acetylene reduction) on coral skeletons following bleaching mortality. Coral Reefs 27: 227–236.

126. Anthony K, Maynard JA, Diaz-Pulido G, Mumby PJ, Marshall PA, et al. (2011) Ocean acidification and warming will lower coral reef resilience. Glob Change Biol 17: 1798–1808.

127. Diaz-Pulido G, Gouezo M, Tilbrook B, Dove S, Anthony K (2011) High CO2 enhances the competitive strength of seaweeds over corals. Ecol Lett 14: 156–162.

13

Dissecting Quantitative Trait Loci for Boron Efficiency across Multiple Environments in *Brassica napus*

Zunkang Zhao[1,2], Likun Wu[1,2], Fuzhao Nian[2], Guangda Ding[1,2], Taoxiong Shi[1,2], Didi Zhang[1,2], Lei Shi[1,2], Fangsen Xu[1,2]*, Jinling Meng[1]

1 National Key Laboratory of Crop Genetic Improvement, Huazhong Agricultural University, Wuhan, Hubei, China, 2 Microelement Research Centre, Huazhong Agricultural University, Wuhan, Hubei, China

Abstract

High yield is the most important goal in crop breeding, and boron (B) is an essential micronutrient for plants. However, B deficiency, leading to yield decreases, is an agricultural problem worldwide. *Brassica napus* is one of the most sensitive crops to B deficiency, and considerable genotypic variation exists among different cultivars in response to B deficiency. To dissect the genetic basis of tolerance to B deficiency in *B. napus*, we carried out QTL analysis for seed yield and yield-related traits under low and normal B conditions using the double haploid population (TNDH) by two-year and the BQDH population by three-year field trials. In total, 80 putative QTLs and 42 epistatic interactions for seed yield, plant height, branch number, pod number, seed number, seed weight and B efficiency coefficient (BEC) were identified under low and normal B conditions, singly explaining 4.15–23.16% and 0.53–14.38% of the phenotypic variation. An additive effect of putative QTLs was a more important controlling factor than the additive-additive effect of epistatic interactions. Four QTL-by-environment interactions and 7 interactions between epistatic interactions and the environment contributed to 1.27–4.95% and 1.17–3.68% of the phenotypic variation, respectively. The chromosome region on A2 of *SYLB-A2* for seed yield under low B condition and *BEC-A2* for BEC in the two populations was equivalent to the region of a reported major QTL, *BE1*. The *B. napus* homologous genes of Bra020592 and Bra020595 mapped to the A2 region and were speculated to be candidate genes for B efficiency. These findings reveal the complex genetic basis of B efficiency in *B. napus*. They provide a basis for the fine mapping and cloning of the B efficiency genes and for breeding B-efficient cultivars by marker-assisted selection (MAS).

Editor: Ivan Baxter, United States Department of Agriculture, Agricultural Research Service, United States of America

Funding: The work was supported by the National Natural Science Foundation of China [30971861]: http://www.nsfc.gov.cn/Portal0/default166.htm; and the Fundamental Research Funds for the Central Universities [2011PY150]. The funders had no role in study design, data collection and analysis, decision to publish, or preparation of the manuscript.

Competing Interests: The authors have declared that no competing interests exist.

* E-mail: fangsenxu@mail.hzau.edu.cn

Introduction

Boron (B) is an essential micronutrient for the growth and development of higher plants [1]. A key role of B in plants is to cross-link rhamnogalacturonan II (RG-II) in the cell wall to form a dimer (RG-II-B-RG-II), which is important for both the formation and the structural integrity of the cell wall [2,3]. It has also been reported that B seems to play important roles in many diverse processes in vascular plants, such as root elongation, sugar translocation, carbohydrate metabolism, nucleicacid synthesis, pollen tube growth and nitrogen fixation [4,5].

It is well known that plants absorb B from soil in the form of boric acid. It has been widely thought that passive diffusion is a major mechanism of B trans-membrane transport, with B translocation in plants being a passive process through the transpiration stream [6]. However, genotype differences in response to B deficiency, or in uptake and utilization of B among species, suggest that energy-dependent active transport and channel-mediated diffusion are involved in B transport [7]. Several transporters and channel proteins for B have been discovered in plants. The *AtBOR1* gene from *Arabidopsis* shows the characteristics of an efflux-type transporter for xylem loading

of B [8] and is mainly regulated by post-transcriptional mechanisms [9]. Four *BOR1* homolog have been validated in rice, and *OsBOR1* is required for efficient uptake of B and xylem loading under low B conditions [10]. Sun et al. (2011) cloned six *BOR1*-like homologs in *B. napus*: *BnBOR1;3a* and *BnBOR1;3c* showed ubiquitous expression in all of the investigated tissues, whereas the other four genes showed similar tissue-specific expression profiles. However, the expression of *BnBOR1;1c* and *BnBOR1;2a* was clearly induced by B deficiency [11]. Another type of B transporter, *ATR1* in yeast, is up-regulated at the transcriptional level by B [12]. Some members of the major intrinsic protein (MIP) family have been identified as boric acid channels in plants. *NIP5;1* and *NIP6;1* function as channel proteins for boric acid transport in *Arabidopsis* [13,14]. Kasajima et al (2010) first identified the transcription factor gene *WRKY6* as essential for root growth under B deficiency [15]. In general, plants growing in nutrient-deficient conditions up-regulate the expression of transporters and thereby increase transport of the deficient nutrient [4]. The induction of transporter expression in plants suggests that further enhancement of transporter genes would result in addition transport capacity and allow plants to perform better under conditions of limited nutrients. Several examples of enhanced B transport activity and better growth (more biomass or seed yield) due to the increased expression of transport genes in *Arabidopsis* have been reported. Over-expression of *AtBOR1* increased the seed yield of *Arabidopsis* under low B condition [16]. Enhanced expression of *AtNIP5;1* can significantly promote B uptake under low B stress and increase seed yield [17]. However, there are few reports on B uptake and transport-related genes and on the relationship between these genes and seed yield in *B. napus*.

B. napus (genome AACC, 2n = 38), plants are commonly used to derive food oil for humans and as a new type of bio-fuel [18]. After soybeans, *B. napus* is the second most important oilseed crop in the world. However, *B. napus* requires higher levels of B than do other species to maintain normal growth and shows high sensitivity to B deficiency [4]. B deficiency causes severe reduction in yield and can even cause a lack of seed setting, which is becoming an important limiting factor for *B. napus* growth in a large area of the world [19]. B deficiency can be alleviated by the application of B fertilizers. However the extreme use of B fertilizers could cause environmental problems. Moreover, B ore resources are limited. The various *B. napus* cultivars have genetic variations that affect their B efficiency. Thus, the development of cultivars with enhanced B-use efficiency would be a more efficient way of genetic improvement. Unveiling the genetic mechanisms involved in plant responses to B deficiency will be the first important step to solve this issue.

Most nutritional traits of crop plants are quantitative and have a complex genetic basis [20,21]. Quantitative trait locus (QTL) mapping has proved to be a powerful genetic approach for dissecting the genetic mechanism of complex traits [22,23]. In some plants, a number of QTLs for complex traits have been mapped [24,25]. QTLs associated with yield and yield-related traits in *B. napus* have been mapped, including plant height [26], yield and yield components [27,28], seed weight [29], and other complex traits [30]. Xu et al. (2001) first revealed that the B efficiency trait is a quantitative trait. A major QTL (*BE1*) and three minor QTLs for B efficiency were detected from an F_2 population derived from a cross between two cultivars in *B. napus* [21]. Furthermore, *BE1* was located in a narrow interval using an $F_{2:3}$ family population derived from a cross between the two cultivars described above [31]. A new locus, *BnBE2*, that controls B use efficiency was validated by bulked segregant analysis (BSA) in a back crossed population of *B. napus* [32]. However, it is still difficult to accurately predict potential candidate genes involved in B efficiency. Additionally, only a few B efficiency loci (rarely >10) have been found to be associated with seed yield or biomass; thus, the genetic architecture of B efficiency remains ambiguous.

B efficiency can be defined as the ability of a genotype to grow well and produce a high yield under B-deficient conditions. The objective of this study is to investigate natural variations for yield and yield-related traits under different B conditions and dissect the genetic basis of B efficiency in *B. napus* using two DH populations, TNDH and BQDH, in five environments. The TNDH population was derived from the F_1 progeny of a cross between Ningyou7 (B-efficient) and Tapidor (B-inefficient). Its genetic linkage map has become an international reference map for *B. napus* (http://brassica.bbsrc.ac.uk) [33,34,35]. The BQDH population was derived from the F1 progeny of a cross between QY10 (B-efficient) and Bakow (B-inefficient) [21]. In this study, QTLs associated with seven yield and yield-related traits were identified using the two DH populations under low and normal B field treatments in three growth seasons. Genes involved in B uptake and utility, yield and yield-related traits in *Arabidopsis* were mapped to the QTL interval. Some QTLs expressed during a low B supply may provide valuable information for improving the seed yield of *B. napus* in soils with low B levels via MAS.

Results

Construction of the genetic linkage map

A genetic map, named the BQDH map, was constructed based on the BQDH population and spanned 19 linkage groups corresponding to 19 chromosomes in *B. napus* named as A1-A10 and C1-C9 according to the new standardized nomenclature for *B. napus* linkage groups (http://www.brassica.info/resource/maps/lg-assignments.php). The BQDH genetic map comprises a total of 486 molecular markers, including 468 SSR, 9 GBM and 7 SRAP markers. This map covers a total length of 1873.9 cM with an average interval of 3.86 cM between adjacent markers. The largest distance between two adjacent markers is 41.98 cM on chromosome C3. The length of the 19 linkage groups varies from 23.31 to 194.71 cM (Table S1).

A comparative analysis between the BQDH and TNDH genetic map by common markers indicated that the BQDH genetic map was collinear with the TNDH genetic map. The two genetic maps were employed in the QTL analysis.

Phenotypic variation and genetic correlation analysis among traits

In the TNDH population, seed yield (SY) under low B (LB) and normal B (NB) conditions, as well as the B efficiency coefficient (BEC), were investigated in a two-year field trial (Table 1). Considerable variation was observed for SY and the BEC between the two parents and among the DH lines during the two years at both B conditions. The B-efficient parent Ningyou7 (NY7) showed

Table 1. Mean values and ranges of seed yield and BEC traits in the parental and DH lines of TN population.

| Traits | Year | Parents | | | | DH lines | | | | | |
| | | LB | | NB | | LB | | | NB | | |
		NY7	Tapitor	NY7	Tapitor	Mean	Range	CV(%)	Mean	Range	CV(%)
SY(g/plant)	2004	11.060	4.840	14.750	9.420	4.260	(0.760–9.931)	48.0	7.252	(0.745–13.356)	33.6
	2005	9.100	3.820	12.770	9.990	5.640	(0.912–10.443)	35.5	6.994	(0.881–12.882)	28.9
BEC	2004	0.750	0.514			0.607	(0.078–1.339)	49.5			
	2005	0.713	0.382			0.804	(0.210–1.299)	29.7			

Note: SY, seed yield; BEC, B efficiency coefficient; CV, coefficient of variation.

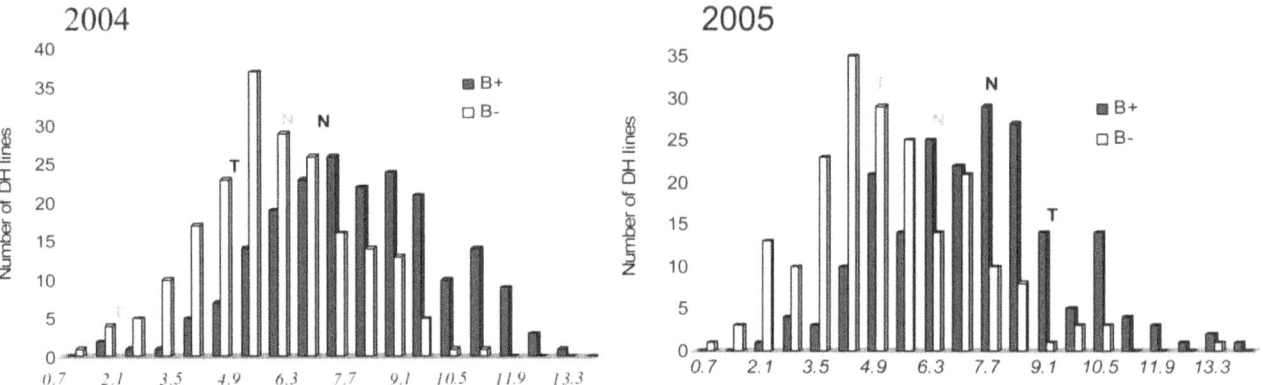

Figure 1. Frequency distribution of seed yield for the TNDH population and its parents at low B (B−) and normal B (B+) conditions in 2004 and 2005 (T: Tapidor; N: Ningyou7; low B grey; normal B black).

a higher SY and larger BEC than the B-inefficient parent Tapidor under the LB condition. In the two trials, SY displayed a continuous normal distribution in the TNDH population, and significant transgressive segregation was observed under both B conditions (Fig. 1).

In the BQDH population, a broad variation for seed yield and six yield-related traits was observed among the parents and DH lines under LB and NB conditions (Table 2). In general, the B-efficient parent Qingyou 10 (QY10) was significantly taller plant (PH 114.9–142.4 cm) and had a heavier seed weight (SW 3.55–5.24 g/1000), more seeds (SN 10.1–14.0), more pods (83.5–191.9), a better yield (5.135–10.962 g) and a larger BEC (0.627–0.950) than the B-inefficient parent Bakow under the LB condition. Compared to the NB condition, both parents and the DH lines showed smaller phenotypic values under the LB condition for all of the six measured traits, except for BN and SW. The BEC value was higher in QY10 than in Bakow in the three trials. All traits were highly variable, SY and the BEC were the most variable traits, with CVs of 37.1%–114.1%, and these traits showed more variability under the LB condition than under the NB condition. In the three field trials, all seven traits showed continuous phenotypic variation and significant transgressive segregation in both directions, implying that multiple genes were involved (Fig. 2). The ANOVA results suggest that genotype, B level, year and the interactions among these variable had significant effect on the six measured traits ($p<0.001$).

For all environments, significantly positive correlation was observed between the SY under the LB condition and the BEC in the two populations (Table S2). These results indicate that SY and the BEC may be controlled by some common genetic determinants for tolerance in a B-limited environment. In the BQDH population, significant positive correlations were observed between SY and all traits except BN and between the BEC and all traits except PH and BN under the LB condition. Under the NB condition, significant negative correlations were observed between the BEC and all traits, except SW and SN, and between SY and SW. Significant positive correlations were observed between SY and PH and between SN and PN.

Detection of putative QTLs in the TNDH and BQDH populations

Seed yield under LB and NB conditions and BEC were used to detect QTLs in the TNDH population. A total of 10 putative QTLs (6 for SY, 4 for BEC) during the two-year trials were detected in the 6 linkage groups, A2, A4, A6, A9, C3 and C4

(Table 3). The phenotypic variation explained (PVE) for the QTLs ranged from 5.7% to 18.5%. One of the four putative QTLs (SYLB-A6) for SY under the LB condition was detected across two years and explained 6.1–6.4% of the phenotypic variation of the favorable alleles from Tapidor. One of the four putative QTLs for the BEC (BEC-A2) was detected across two years for the favorable alleles from NY7. Additionally, two putative QTLs (SYLB-A2a and SYLB-A2b) for the SY under the LB condition and 2 putative QTLs (SYNB-A2 and SYNB-A9) for the SY under the NB condition showed positive effects, indicating that the favorable allele was from NY7.

A total of seventy putative QTLs under both B conditions were identified contributing 4.15%–23.16% of the phenotypic variation in the BQDH population. These QTLs were distributed among 15 linkage groups, which included 13 for SW, 15 for PH, 9 for SN, 10 for PN, 9 for BN, 9 for SY and 5 for the BEC (Table 4). Interestingly, the A and C genome shared an equal number of QTLs with 34 and 36 QTLs, respectively.

Thirty putative QTLs were detected in thirteen linkage groups under the LB condition (Table 4). One QTL (SWLB-A7b) was detected across three years, contributing 6.03–9.07% of the phenotypic variation. Five QTLs (SWLB-A7a, PHLB-A3a, PHLB-A3b, BNLB-C9 and SYLB-A7b) were detected across two years, contributing 6.01–11.93% of the phenotypic variation. Seven putative QTLs for seed yield in A2, A7, C3, C4 and C7 were identified. Among these QTLs, the QTL SYLB-A7, which was detected across two years with favorable alleles from QY10, co-localized with the QTL BEC-A7b. In general, fifteen QTLs had positive additive effects, showing that these alleles for higher phenotypic variation came from QY10. QTLs for different traits were clustered on A4, A7 and C4. Two robust QTLs for seed yield (SYLB-A7) and for seed weight (SWLB-A7a) co-localized on A7 with the favorable alleles contributed by QY10. Two QTL-by-environment interactions for PN and SN contributed 4.5% of the total phenotypic variation and were detected in A6 and C6 (Table 5).

Thirty-five putative QTLs were detected in eleven linkage groups under the NB condition (Table 4). Six QTLs were detected across three years and explained 5.10–23.26% of the phenotypic variation, and eleven QTLs were detected across two years and explained 4.75–16.98% of the phenotypic variation. Fifteen QTLs showed a positive effect, suggesting that these alleles for the advantageous phenotype were contributed by QY10. Two QTLs for pod number (PNNB-C9) and seed weight (SWNB-C9), as well as one QTL for branch number (BNNB-C9b), were clustered on C9.

Table 2. Mean values and ranges of seed yield and yield-related traits in the parental and DH lines of BQDH population.

Traits	Year	Parents				DH lines					
		LB		NB		LB			NB		
		QY10	Bakow	QY10	Bakow	Mean	Range	CV(%)	Mean	Range	CV(%)
PH(cm)	2009	142.4bcd	134.1de	151.9ab	134.4de	138.2	(82.0–179.4)	13.1	147.2	(90.3–186.5)	11.6
	2010	114.9f	98.3g	125.3ef	114.9f	81.7	(30.0–141.3)	23.4	123.5	(66.8–169.1)	14.8
	2011	140.0cd	135.0de	146.7abc	154.9a	141.1	(69.9–191.0)	13.1	149.7	(87.6–185.1)	11.5
BN	2009	7.2bc	7.1bcd	6.9bcd	6.8bcd	7.4	(3.3–11.2)	17.6	8.2	(5.8–14.5)	16.8
	2010			6.8bcd	6.5bcd				6.9	(2.4–11.8)	22.4
	2011	7.6ab	8.4a	6.0d	6.1cd	9.0	(3.9–14.0)	18.7	6.5	(3.4–10.2)	19.3
SN(/pod)	2009	14.0b	15.3b	17.3a	17.5a	13.3	(2.7–24.4)	37.0	14.3	(3.1–25.0)	33.1
	2010			11.1c	9.9c				12.2	(2.0–22.7)	39.6
	2011	10.1c	5.8d	17.4a	9.9c	7.04	(0.20–23.20)	66.2	13.52	(1.59–24.36)	40.1
PN(/plant)	2009	191.9c	130.3ef	225.5b	139.3cd	142.8	(22.4–380.0)	41.0	133.8	(22.4–264.9)	35.1
	2010	83.5g	40.3h	270.5a	129.1ef	45.5	(2.0–198.7)	75.4	135.0	(14.4–329.0)	41.4
	2011	102.8fg	51.3h	185.0c	148.2de	47.3	(3.4–154.3)	57.8	128.4	(21.3–236.8)	34.7
SW(g/1000)	2009	3.55de	3.02f	3.26ef	2.54g	3.035	(1.200–4.570)	17.6	3.111	(1.600–4.483)	15.4
	2010	4.84b	3.79de	4.20c	3.30ef	3.503	(1.613–5.053)	17.8	3.485	(1.917–5.163)	18.0
	2011	5.24a	3.72d	3.58de	3.06f	3.933	(1.641–5.633)	16.6	3.153	(1.830–5.100)	17.2
SY(g/plant)	2009	10.962a	4.582d	11.541a	9.426b	5.219	(0.230–13.107)	52.1	6.044	(0.505–13.175)	43.2
	2010	5.135d	0.321e	8.191c	5.311d	0.745	(0.006–4.717)	114.1	5.680	(0.407–12.210)	48.8
	2011	7.403c	0.712e	11.289a	5.068d	1.474	(0.013–7.618)	97.9	5.736	(0.256–12.598)	46.3
BEC	2009	0.950	0.486			0.854	(0.029–1.784)	37.1			
	2010	0.627	0.124			0.140	(0.001–0.666)	98.5			
	2011	0.656	0.140			0.265	(0.003–0.990)	79.8			

Note:
SY, seed yield; BEC, B efficiency coefficient; PH, plant height; BN, branch number per plant; SN, seed number; PN, pod number per plant; SW, seed weight; LB, low B condition; NB, normal B condition; CV, coefficient of variation. Different small letters indicate significant difference at the level of $P<0.05$.

One QTL-by-environment interaction for PN explained 1.27% of the total phenotypic variation and was detected on C6 (Table 5).

Five putative QTLs for the BEC were detected on the A2, A7, C3 and C8 linkage groups and explained 6.8–11.44% of the phenotypic variation (Table 4). No QTL for the BEC was detected across two or three years, but seven QTLs for the BEC overlapped with QTLs for other tested traits under the LB condition. For example, *BEC-C8* co-localized with *SWLB-C8* on C8, contributing 7.33% of the phenotypic variation. One QTL-by-environment interaction for the BEC on C3 explained 4.95% of the total phenotypic variation (Table 5). Taken together, these imply that environment has a considerable effect on the performance of traits.

Detection of epistatic interactions in the TNDH and BQDH populations

In total, ten epistatic interactions were detected for the SY under LB condition and the BEC in the TNDH population. These interactions contributed to 5.44%–14.38% of the phenotypic variation (Table 6). Eight of these interactions exhibited a negative effect, indicating that recombinant allele combinations could improve seed yield and the BEC. Two and three epistatic interactions were detected for the SY under the LB condition and the BEC in 2004 and contributed to 15.77% and 28.94% of phenotypic variation, respectively. Two and three epistatic interactions were detected for the SY under the LB condition

and the BEC in 2005 and explained 15.39% and 20.34% of the total phenotypic variation, respectively.

Thirty-two epistatic interactions involving 50 loci covering the whole genome, except A8, were detected for six measured traits under both B conditions, with the BEC contributing 0.53–14.26% of the phenotypic variation in the BQDH population (Table S3).

For SY, 10 epistatic interactions involving 19 loci were detected under the LB and NB conditions explaining 52.65% of the total phenotypic variation. One locus with an additive effect under the NB condition was involved in the epistatic interaction. For the BEC, 2 epistatic interactions involving 4 loci explained 7.1% of the total phenotypic variation with a positive effect. An interaction with the environment accounted for 6.23% of total phenotypic variation. For PN, one epistatic interaction involving a locus with an additive effect was detected explaining 1.62% of the phenotypic variation under the NB condition. For SN, a total of seven epistatic interactions involving 13 loci were detected under both conditions contributing 26.37% of the total phenotypic variation. Three loci with additive effects under the NB condition were involved in epistatic interactions. For BN, one epistatic interaction involving two loci with additive effect under the NB condition was found. Only one interaction with the environment contributed 1.74% of the phenotypic variation. For SW, six epistatic interactions involving eleven loci under the LB and the NB conditions were detected, and these interactions contributed 19.18% of the phenotypic variation. Six loci with additive effects were involved

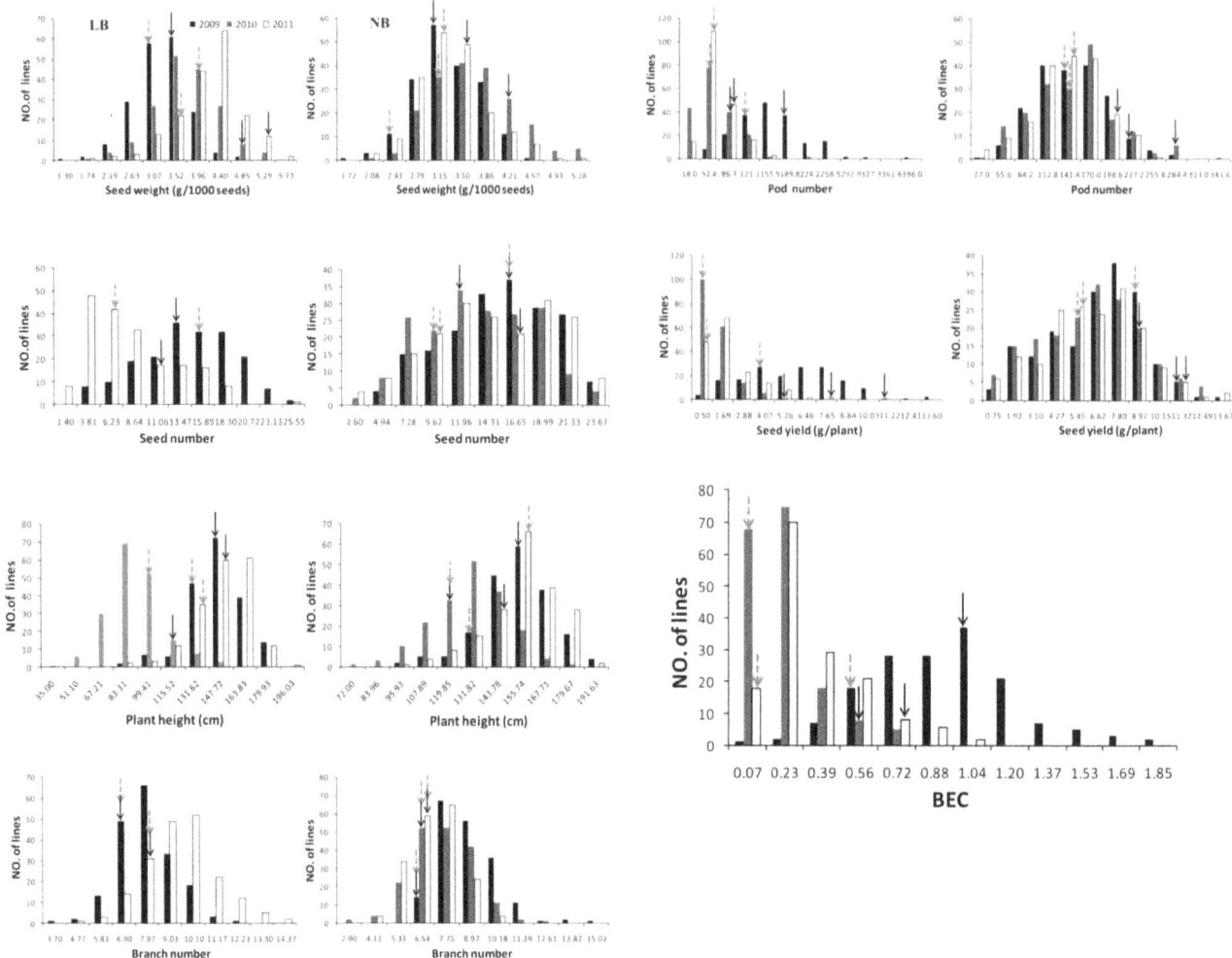

Figure 2. Frequency distributions of seed yield and yield related traits in the BQDH population under low B (left) and normal B (right) levels and BEC in the three-year field trials. Solid arrows indicate QY10, and dashed arrows indicate Bakow.

in epistatic interactions. For PH, five epistatic interactions involving nine loci explained 15.79% of the total phenotypic variation under the LB and NB conditions. Two loci with additive effects were involved. These results suggest that additive and additive-additive effects simultaneously control phenotypic variation under both B conditions. The additive effect was a major factor because the cumulative contribution from significant additive effects (4.15%–23.16%) was higher than the contribution from epistatic effects (0.53%–14.38%) for an individual trait.

Associations of QTL with functional genes by *in silico* mapping

Twenty synteny blocks (A genome: 13, C genome: 7) and 123 insertion fragment islands (A genome: 77, C genome: 46) were identified between *Arabidopsis* pseudochromosomes and BQDH genetic linkage groups by the *in silico* mapping approach (Table S1). In total, 159 orthologous genes in *Arabidopsis* were mapped to the confidence intervals of 70 putative QTLs, corresponding to 30 blocks. Among the 159 genes, 23 were involved in B metabolism, and 136 were involved in yield-related trait control (Table S4). The 23 genes involved in B metabolism may be candidate genes underlying the QTLs specifically expressed under the LB condition. For example, *BOR7*, a member of the anion transporter

family and homolog of *BOR1*, was located in the confidence intervals of *SYLB-A2a (b)* in the BQDH population and in the confidence intervals of *BEC-A2* and *SYLB-A2b* in the TNDH population (data not shown). Thus, *BOR7* is a potential candidate gene for B efficiency.

Discussion

High yield is the most important goal in modern crop production, and high yield during abiotic stress, such as nutrient deficiency, represents an important area of study. The nutrient efficiency of a genotype is defined as the ability to produce higher yields in soils with limited nutrient supplies [36,37]. We previously defined the BEC, which is the ratio of the yield under low B condition to that under normal B conditions, as B efficiency or the tolerance of a genotype to B deficiency [21]. Using the BEC as a screening index, B-efficient *B. napus* cultivars and *Arabidopsis* ecotypes were screened [38]. However, the BEC involving yield is a complex trait in plants regardless of the growth condition or environmental stress. In the present study, two DH populations were employed to identify QTLs for SY, SW, SN, PN, BN and PH under the LB and the NB conditions, and the BECs from two-year and three-year field trials were evaluated to determine the B efficiency of *B. napus*. Numerous QTLs for the seven traits were

Table 3. Putative QTLs for seed yield and BEC traits in the TNDH population under different B conditions.

Trait[a]	Chr.	QTL[b]	Add.[c]	2004			2005		
				PVE(%)[d]	CI[e]	Peak[f]	PVE(%)	CI	Peak
BEC	A2	BEC-A2	+	8.3	64–72	65.5	7.7	65–100	66.4
	A4	BEC-A4	−	8.4	103–120	110.5			
	A6	BEC-A6	−				10.2	15–26	18
	C3	BEC-C3	−	10.3	114–123	120			
QTLs detected under low boron (LB) condition									
SY	A2	SYLB-A2a	+				8.1	35–46	42
		SYLB-A2b	+	18.5	60–81	70			
	A6	SYLB-A6	−	6.4	13–30	23	6.1	13–29	23
	C4	SYLB-C4	−	5.7	75–80	77			
QTLs detected under normal boron (NB) condition									
SY	A2	SYNB-A2	+	7.3	33.8–38.1	36	7.4	32.7–45.7	40
	A9	SYNB-A9	+	8.1	80.1–85.2	83			

Note:
[a], short names of the traits: BEC, boron efficiency coefficient; SY, seed yield; SW, seed weight; PH, plant height; SN, seed number; PN, pot number; BN, branch number.
[b], Nomenclature for QTL: an trait abbreviation following a boron-level designator (LB, low B level; NB, normal B level), a hyphen (-), chromosome (A1-A10 or C1-C9) on which the QTL located and the serial letter (a, b, c...) in the same linkage group.
[c], Additive effect. Positive additive effects are associated with increased effects from B-efficient parent allele, and negative additive effects are associated with increased effects from B-inefficient parent allele.
[d], Percentage of phenotypic variation explained (PVE) by each identified QTL.
[e], The 2-LOD confidence interval (CI) of QTL, given in cM.
[f], The peak position is denoted by the number in parentheses.

detected for both B conditions. These QTLs, especially those expressed in the LB condition, are valuable for dissecting the mechanism of B efficiency.

Abundant variation during different B conditions

Phenotypic investigations showed abundant genetic variation in seven yield and yield-related traits between the two pairs of parents and between the two B conditions for the same genotype (Table 1, 2). In this study, the CV was calculated for both B conditions in the two populations, and a range from 11.5% (PHNB) to 114.1% (SYLB) throughout both B conditions was observed in the TNDH and BQDH populations. These traits were highly variable, especially the SY under the LB condition and the BEC in the two populations. In addition, there was considerable transgressive segregation for most of the traits under both B conditions. These results imply that yield and yield-related traits under the LB condition are highly variable and can be improved genetically.

Classical genetics assumed that the correlation among different traits is due to the tight linkage of genes affecting different traits or to genes with pleiotropic effects [39]. In this study, seed yield showed the highest correlation with the other measured traits (Table S2), and a high percentage (82%) of QTLs for SY co-localized with QTLs for other yield-related traits. On average, a QTL for SY involved 2 QTLs for yield-related traits under both B conditions, and a QTL for SY involved 1.4 QTLs for yield-related traits under the LB condition (Fig. S2). These results create a flexible approach for identifying the genetic basis of co-localized QTLs for SY under LB condition (i.e. B efficiency) by estimating which QTL of the yield-related trait(s) is the most probable locus for having linked genes or genes with pleiotropic effects on seed yield. Shi et al. (2009) facilitated the cloning of *qSY.A2-2*, a major QTL for seed yield, based on the hypothesis that *qFT.A2-4*

controlled flowering time, which was considered as an indicator QTL [40]. The successful cloning of QTL for yield in rice [24] was facilitated by the indicator QTL for biomass yield.

B efficiency QTLs and the genetic basis of B efficiency

In this study, the BEC was defined as the ability of a genotype to produce a high yield during B deficiency. Thus, the BEC was tightly associated with the SY under the LB condition. In the TNDH population, four putative QTLs for seed yield under the LB condition, *SYLB-A2a* and *SYLB-A2b* on A2, *SYLB-A6* on A6 and *SYLB-C4* on C4, and four putative QTLs for the BEC, *BEC-A2* on A2, *BEC-A4* on A4, *BEC-A6* on A6 and *BEC-C3* on C3, were identified. Among them, *BEC-A6*, *BEC-C3* and *SYLB-A2b* contributed 10.2%–18.5% of the phenotypic variation. In the BQDH population, seven putative QTLs for SY on A2, A7, C3, C4 and C7 under the LB condition, as well as seven putative QTLs for the BEC on A2, A7, C3 and C8, were identified. Among them, *BEC-C3*, *BEC-A7b* and *SYLB-A7* contributed 11.24%–14.25% of the phenotypic variation. These results suggest that B efficiency is regulated by a number of QTLs but is mainly controlled by the major QTLs, with higher than 10% of the PVE. Additionally, *SYLB-A6* and *BEC-A2* co-localized with *BEC-A6* and *SYLB-A2b*, respectively, in the TNDH population. *BEC-A2* and *BEC-C3* co-localized with *SYLB-A2* and *SYLB-C3*, respectively, in the BQDH population. This co-localization indicates that there is a close genetic associations between SYLB and the BEC and suggests that a single gene with pleiotropic effects or two linked genes can create tolerance to low B conditions.

A comparative analysis between the TNDH and BQDH populations and R2 of the *B. rapa* genome validated the loci for the BEC and SY under the LB condition at A2 (Fig. 3). Previous studies have reported some loci for B efficiency in *B. napus* and

Table 4. Putative QTLs for seed yield and yield-related traits in the BQDH population under different B conditions.

Trait[a]	Chrom.	QTL[b]	Add.[c]	2009			2010			2011		
				PVE(%)[d]	Cie	Peak[f]	PVE(%)	CI	Peak	PVE(%)	CI	Peak
BEC	A2	BEC-A2	+				6.80	71.2–76.4	72.81			
	A7	BEC-A7a	+							8.25	0–14.8	5.01
		BEC-A7b	+				11.24	19.4–38	27.41			
	C3	BEC-C3	−	11.44	68.9–75.3	71.91						
	C8	BEC-C8	+							7.33	11.9–23.4	16.41
QTLs detected under low boron (LB) condition												
SW	A1	SWLB-A1a	−				6.00	35.8–46.5	38.31			
		SWLB-A1b	+	6.49	68.8–78.1	73.11						
	A4	SWLB-A4	+	8.40	7.7–22.3	11.61						
	A7	SWLB-A7a	+				6.20	19.4–41	29.71	6.02	34.9–41.7	37.21
		SWLB-A7b	−	7.18	75.1–77.9	76.21	6.03	71.9–87.9	75.11	9.07	71.9–77.9	76.21
	C6	SWLB-C6	+	6.96	29.5–39.7	31.51						
	C8	SWLB-C8	+				7.05	11.9–20.2	13.61			
PH	A3	PHLB-A3a	−	11.89	19.8–23	21.41				6.09	22.5–25	23.01
		PHLB-A3b	−	10.51	31.3–34.1	32.31				11.93	29.9–35.9	34.31
	A7	PHLB-A7	+	5.25	58.4–73.7	68.71						
	C3	PHLB-C3a	−				8.80	0–23.4	12.01			
		PHLB-C3b	−							5.52	26.3–39.9	31.21
	C4	PHLB-C4	+	6.78	37.8–48.6	43.81						
SN	A4	SNLB-A4	−							8.26	7.9–15.9	11.61
	C6	SNLB-C6	−	20.05	5.3–17.8	9.31						
	C9	SNLB-C9	+							11.20	78.5–85.2	84.21
PN	A2	PNLB-A2	−				10.32	78–84.4	83.01			
	A6	PNLB-A6	+	8.32	6.5–33.8	26.81						
	A7	PNLB-A7a	+				8.21	6.9–29.5	21.41			
		PNLB-A7b	+							5.64	29.6–56.2	44.71
		PNLB-A7c	+				5.77	85.4–99.2	98.21			
	C4	PNLB-C4	−							6.01	76.3–83.2	81.31
	C5	PNLB-C5	−	6.55	41.2–64.5	54.31						
BN	A1	BNLB-A1	+	6.35	39–56.3	48.51						
	C9	BNLB-C9	−	6.01	39.6–53	41.81				7.09	39.6–48.9	41.81
SY	A2	SYLB-A2	+				9.11	71.6–83	76.4			
	A7	SYLB-A7	+				11.24	19.4–38	27.41	6.73	15.6–29.6	20.41
	C3	SYLB-C3	−	7.19	64.9–77.2	68.91						
	C4	SYLB-C4	−							7.85	76.3–104.2	84.21
	C7	SYLB-C7	+							8.09	22.7–53.9	44.11
QTLs detected under normal boron (NB) condition												
SW	A7	SWNB-A7	−	14.41	73.8–77.9	76.21	12.23	77.9–94.3	79.91	9.10	76.2–80.8	77.91
	A9	SWNB-A9	−	5.20	65–80.7	69.01						
	A10	SWNB-A10	+				9.11	19–41.5	32.21	7.17	18.2–41.4	29.91
	C6	SWNB-C6a	+				16.92	4.5–21.4	13.31			
		SWNB-C6b	+	12.24	26.8–35.4	30.51				10.97	29.5–37	30.61
	C9	SWNB-C9	+	5.43	11.1–37.8	32.31	5.72	18.2–36.9	31.41	8.43	22.7–32.3	30.41
PH	A3	PHNB-A3a	−	12.30	18.1–24.2	21.41	9.66	18.8–24.3	23.01	5.16	22.4–24.5	23.01
		PHNB-A3b	−	7.42	31.3–32.8	32.31	5.10	31.3–33.8	32.31	9.83	29.2–36.1	34.31
	A7	PHNB-A7	+	7.44	77.9–91.8	80.81						
	C3	PHNB-C3	−	5.12	25.8–45.5	35.21				5.45	26.3–45.8	36.21

Table 4. Cont.

Trait[a]	Chrom.	QTL[b]	Add.[c]	2009			2010			2011		
				PVE(%)[d]	Cie	Peak[f]	PVE(%)	CI	Peak	PVE(%)	CI	Peak
	C4	PHNB-C4	+	8.61	60–61.4	60.91						
	C5	PHNB-C5	−	7.10	8.1–38.6	17.41	4.75	8.2–38.6	24.61			
	C6	PHNB-C6	−							5.35	0–4.1	0.01
	C9	PHNB-C9a	−	4.15	8.1–37.8	33.91						
		PHNB-C9b	−				6.37	58.2–77.1	65.61			
SN	A2	SNNB-A2	−				6.25	74–85.1	83.01			
	A6	SNNB-A6	+				8.64	76–97.5	88.01			
	A7	SNNB-A7	+	7.95	80.4–98.2	90.81						
	C6	SNNB-C6a	−	23.26	5.3–17.2	9.31	16.55	8–24.5	17.31	15.14	6.9–22.2	15.31
		SNNB-C6b	−				13.22	30.6–37.4	31.51	8.93	30.6–37.6	32.51
	C9	SNNB-C9	+				6.21	70.3–78.2	73.01	5.06	66.9–78.2	73.21
PN	A6	PNNB-A6	+	8.29	74.5–99	91.01						
	C6	PNNB-C6	−	5.68	17.2–26.8	24.51	10.19	4.1–21.4	8.31			
	C9	PNNB-C9	−	12.67	30.5–33.9	32.31	8.87	13.6–39.6	31.41	6.70	9.5–40.4	33.31
BN	A6	BNNB-A6	+	6.21	61.4–74	61.71						
	A7	BNNB-A7	+				5.01	61.3–75.1	71.91			
	C3	BNNB-C3a	+	7.46	15.5–26.3	21.91						
		BNNB-C3b	+				5.25	76.5–92.8	80.41			
	C6	BNNB-C6	−				9.26	12.6–29.3	25.51			
	C9	BNNB-C9a	−	9.61	30.3–33.9	31.41	16.98	32.3–36.9	33.31			
		BNNB-C9b	−	10.71	37.8–47.5	41.41				10.97	39.6–43.5	41.81
SY	A1	SYNB-A1	−				12.07	37.4–44.5	40.31	7.18	28.7–49.6	39.31
	A7	SYNB-A7	+	7.95	19.4–41.4	26.41						
	C6	SYNB-C6a	−				13.92	5.3–24.5	15.31	8.63	4.1–21.6	8.31
		SYNB-C6b	−	10.18	22.2–30.6	25.51						

[a–f]See footnotes of Table 3 for explanations.

Arabidopsis. Xu et al. (2001) mapped a major B efficiency QTL (*BE1*) to LG9 (LG9 was equivalent to recognized A2 linkage group) using an F$_2$ population in *B. napus* [21]. Further, the *BE1* region mapped to an interval of 110.8–117.2 cM in length on *Arabidopsis* chromosome 1 using comparative mapping of two flanking RFLP markers (PB134-3~PA28) [41]. Interestingly, the *BE1*-aligned region in *Arabidopsis* overlapped with the *Arabidopsis* B

efficiency QTL *AtBE1-2* [38]. In this study, *BEC-A2* and *SYLB-A2b* in the TNDH population mapped to *AtBE1-2* in *Arabidopsis* by comparative mapping (Fig. 4). Therefore, we speculate that the genomic region on A2 acts as the major locus conferring B efficiency in *B. napus* and that this could be beneficial for fine mapping *BE1* in *B. napus*. Additionally, the QTLs *SYLB-A6* in

Table 5. QTL-by-environment interactions identified in the BQDH population.

Trait	Chr.	Interval	Position	A	AE1	AE2	PVE%(A)	PVE%(AE)
LBPN	A6	CNU400-CNU325a	28.8	9.0235***	10.2712**		2.89	2.41
NBPN	C6	BoGMS1497-06Au-4	15.3	−8.9574***	−6.4949*	3.38	1.27	
LBSN	C6	BoGMS1497-06Au-4	12.3	−1.3858***	0.7099*	7.01	2.09	
BEC	C3	BoGMS0576-CB10427	65.9	−0.0407***	−0.0716***	0.0395*	2.45	4.95

Note:
PNLB, pot number under low B condition; PNNB, pot number under normal B condition; SNLB, seed number under low B condition; BEC, B efficiency coefficient.
Significance:
*$P < 0.05$;
**$< P < 0.01$;
***$P < 0.001$.
PVE: phenotypic variation explained.

Table 6. Epistatic interactions for seed yield at low B condition (SYLB) and boron efficient coefficient (BEC) in the TN DH population of 2004 and 2005 field trials.

Trait	Chr-Int *i*	Markers	Chr-Int *j*	Markers	LOD	*Ai*	*R²(Ai)*	*Aj*	*R²(Aj)*	*AAij*	*R²(AAij)*
2004											
SYLB	A2-25	**S02M08-1-180/S08M15-170**	A9-49	CNU263/P13M10-265	4.72	−0.22	0.8	−0.08	0.1	−0.74	8.96
	A5-36	P10M6-190/sR9477	C8-16	CB10092/CB10028	4.46	−0.02	0.01	0.21	0.72	−0.64	6.81
BEC	A3-51	Na14G02/IGF0568c′	A9-19	pW235/CB10022	3.25	0.05	2.07	−0.01	0.04	0.08	5.28
	A4-14	E6HM40-160/CNU246	C7-4	sNRH63/Na10C01b	3.96	0.02	0.38	−0.03	0.97	−0.1	9.28
	C1-27	S08M15-85/IGF3141e	C3-19	SA27/sN2032	4.99	0.03	0.59	0.02	0.34	−0.13	14.38
2005											
SYLB	A4-10	IGF5193a/sN13034	A8-4	Na12B05a/CNU208	4.54	0.09	0.16	0.09	0.18	−0.62	8.2
	A9-37	pX150/S15M04-2-150	C1-14	E7HM40-590/S13M08-1-380	3.79	−0.15	0.5	0.14	0.41	0.58	7.19
BEC	A4-10	IGF5193a/sN13034	A10-16	AP1a/E5HM40-205	6.53	0.01	0.23	0	0.02	−0.08	8.33
	A5-28	IGF3165a/WG2E2	A7-18	sR7223/sNRA59	4.73	0	0.02	0	0	−0.07	6.57
	A9-48	pW123aH/CNU263	C2-14	SA12a/P5M5-1000	3.82	−0.02	0.77	0	0	−0.06	5.44

Note:
LOD score calculated by QTLmapper 2.0 at P≤0.005 level of probability.
A: The estimates of additive effect for testing point *i*; *j* and additive×additive epistasis *ij*.
R^2: Proportion of phenotypic variation explained by the marker genotypes at the locus *i* or *j* and between the two testing points *i* and *j*.
Markers with bold means that the locus was located in QTL interval.

TNDH and *SYLB-A7* in BQDH detected throughout the two seasons could be considered robust loci for fine mapping.

Several QTLs identified in other *B. napus* genetic populations were projected onto the BQDH genetic map based on the alignment analysis for common molecular markers between different genetic maps using the map projection function of BioMercator 2.1 software [42]. A total of 29 QTLs for six traits (SY, PH, BN, PN, SN and SW) [27,29,40,43] were projected onto the BQDH population (Fig. S1). Among them, 2 QTLs were co-localized with QTLs identified under the NB condition (QTLs for BN on C6), 18 QTLs were co-localized with QTLs identified under the LB condition (such as 3 QTLs for PN on A7 and 2 QTLs for SN on A4), and 9 QTLs were co-localized with QTLs identified under both B conditions (such as 2 QTLs for SY on A7 and 2 QTLs for SW on C6). These results imply that common genetic factors could exist for some traits in different genetic backgrounds and that environments and some genetic factors may be associated with B efficiency.

Epistatic interactions are considered an important genetic mechanism for regulating phenotypic variation [44,28]. In the present study, a number of epistatic interactions for yield and yield-related traits were identified. Interestingly, one locus (S02M08-1-180/S08M15-170) conferring the major QTL (*SYLB-A2b*) was associated with epistatic interactions, suggesting that epistatic interactions play important roles in controlling B efficiency in *B. napus*. Liu et al. (2009) identified 74 epistatic interactions regulating shoot mineral concentrations under different B conditions at the seedling stage in *B. napus* [45]. Previous studies successfully identified 45 differentially expressed proteins under B-limited condition at the seedling stage using the B-efficient cultivar QY10 [46]. The differentially expressed proteins covered eight metabolic pathways including antioxidant and detoxification, defense, signaling and regulation, carbohydrate and energy metabolism, amino acid and fatty acid metabolism and transport. These results suggest that a number of genes and interactions among associated genes could be involved in B efficiency and that they should not be neglected in MAS.

B-related genes

Based on the speculated genomic region on A2 in *B. napus*, five candidate genes located in the equivalent region on R2 in *B. rapa* were screened (Table 7). The five candidate genes function as small molecule transporters. The homologous genes in *Arabidopsis*, AT5G27350 and AT5G27360, function as a superfamily of monosaccharide transporters [47]. AT5G28470 functions as a major facilitating protein and is associated with pollen tube growth and development [48]. AtNIP5;1 is a B transport channel protein [13].

In order to identify whether the homologous genes in *B. napus* localized to the A2 region for B efficiency, twelve pairs of gene-based simple sequence repeat (GB-SSR) primers were developed for four of the five candidate genes. The 9 and 8 GB-SSR markers that correspond to 10 and 8 loci were polymorphic in the BQDH and TNDH populations, respectively (Table 8). Moreover, the 10 and 8 loci were integrated into the expected intervals of the BQDH and TNDH genetic maps, respectively. Further, the latest results of QTL mapping for the BQDH population using an improved BQDH genetic map suggest that the loci Bra020595-1, Bra020595-2 and Bra020592-1a are located in the SYLB-A2 and BEC-A2 intervals. These results indicate that the Bra020592 and Bra020595 genes in *B. napus* are candidate genes for B efficiency.

The comparative genome mapping analysis between *B. napus* and *Arabidopsis* revealed a common ancestor for the two species [49,50]. It would be beneficial to resolve the genetic mechanisms of the complex *B. napus* genome using genetic information from the model plant *Arabidopsis*. Based on *in silico* mapping, some important genes involved in different biological processes in *Arabidopsis* were mapped to the target QTL intervals in *B. napus*. These QTLs were for flowering time [33], shoot mineral concentrations [45], seed yield and yield-related traits [40,43], and P-efficiency traits [51]. In this study, a total of 44 genes involved in B uptake and transport [8,13,14,15,52] or genes induced by B-limited stress [46,53] were used for the *in silico* mapping of the BQDH population. Twenty-three homologous genes were mapped to QTL intervals (Table S4), including B

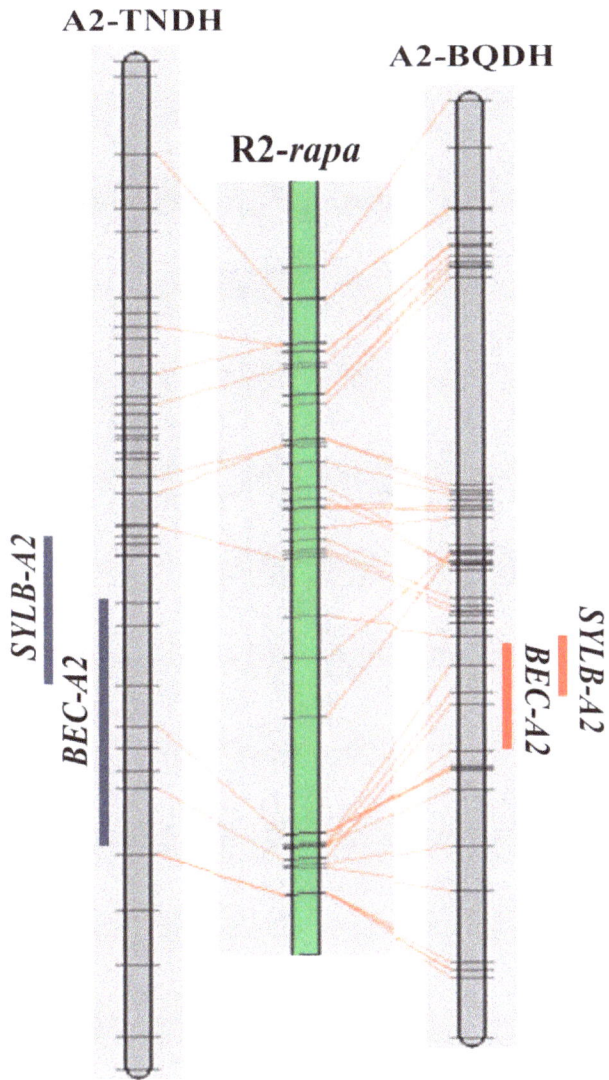

Figure 3. Co-location comparison of QTL interval for seed yield under low B (SYLB) and B efficiency coefficient (BEC) on A2 between the BQDH and the TNDH genetic maps by comparing with R2 of *B. rapa* genome. R2-rapa represented R2 chromosome in *B. rapa*.

transporter family genes, B channel genes, biosynthetic process genes, KDOP synthase-related genes, cell structure-related genes, transport-related genes, abiotic stress-related genes, other genes induced by B deficiency, and transcription factors expressed during low B levels. For example, *NIP2;1*, a member of the NOD26-like intrinsic protein family, was localized to the interval of *SYLB-C3* and *BEC-C3*, explaining 7.19% and 11.44% of the phenotypic variation, respectively. Long et al (2007) identified *BnFLC10* as a potential candidate gene that controls flowering time by *in silico* mapping [33]. This finding suggests that it is possible to accelerate the process of cloning genes by *in silico* mapping. However, further fine mapping and analysis of near isogenic lines or association mapping will be needed to confirm the involvement of potential candidate genes.

A total of 35 pairs of gene-based marker (GBM) primers were developed for *Arabidopsis* B transporter and channel genes, tobacco cell wall pectin glucuronosyltransferase gene and for genes induced

by B deficiency. Nine GBMs were integrated into the BQDH genetic map, and four of them were located in QTL intervals. Interestingly, two GBMs developed for the *Arabidopsis* B transporter genes were associated with putative QTLs detected under LB condition. ATBOR1-BrS3a, developed from *AtBOR1* [8], was mapped to the confidence interval of *SWLB-A1b* on A1, and NIP5;1-Bn4b developed from *NIP5;1* [13], was mapped to the confidence interval of *SWLB-C6* on C6. Two GBMs (NIP5;1-Br3a and NIP5;1-Br3b) were associated with four epistatic interactions, three of which were for SYLB and the BEC (Table S3). These results could be beneficial for identifying and cloning genes and could provide potential markers for MAS to produce B-efficient rapeseed cultivars.

Materials and Methods

Plant materials

Two double haploid populations TNDH and BQDH were employed in this study for mapping QTLs for yield and yield-related traits under different B conditions. The TNDH population was developed from a cross between Tapidor and Ningyou7 (NY7) [34]. Using the TNDH population, a genetic linkage map was constructed with 621 markers, including restriction fragment length polymorphisms (RFLPs), simple sequence repeats (SSRs) and single nucleotide polymorphisms (SNPs). The map covered 2060 cM on 19 chromosomes with an average interval between two adjacent markers of 3.3 cM [33]. The BQDH population, comprising 200 DH lines, was developed from one F_1 progeny derived from a cross between QY10 (B-efficient) and Bakow (B-inefficient) [21] via the microspore culture technique.

Field trials and traits investigation

For the TNDH population, a two-year field trial was conducted at our laboratory's boron fertilizer experiment base, which has sandy paddy soil, in Qichun county, Hubei Province (N 115°45′ E 30°19′), during the 2003–2004 and 2004–2005 crop seasons. No specific permits were required for the field trial. The average hot-water-soluble B in the plough layer soil before fertilization was 0.069 mg kg^{-1}. In both trials, two B treatments were employed: the LB treatment, 1.5 kg ha^{-1} borax, and the NB treatment, 15 kg ha^{-1} borax. The NB treatment was the control, and the application of N-, P-, K-containing fertilizers for each treatment was according to the following nutrient rates: 180 kg N ha^{-1}, 90 kg P_2O_5 ha^{-1}, and 150 kg K_2O ha^{-1}. K as potassium chloride and P as ordinary superphosphate were applied as base fertilizers, and N as urea was divided into 120 kg before transplanting and 60 kg at the bolting stage. The seeds of the 202 DH lines, together with the two parents, were firstly sown in seedbed in the middle of September, and then the uniform seedlings were transplanted to B-treated field plots 30 days later. The plants were harvested at the beginning of the following May. The planting was conducted in a complete randomized block design with three replicates. Every block for a line contained two rows and the interval between adjacent rows was 25.6 cm, and 10 plants were planted in a row with an interval of 18 cm between adjacent plants. The seeds were sown by hand, and the field management followed standard agricultural practice. In each replicate, ten representative individuals of each block were harvested at physiological maturity. Seed yield was investigated for all the lines, and then the BEC for each DH line was calculated as the ratio of mean values of seed yield in three replicates under low B level to that under normal B level. Seed yield was recorded as the average seed dry weight of the harvested individuals.

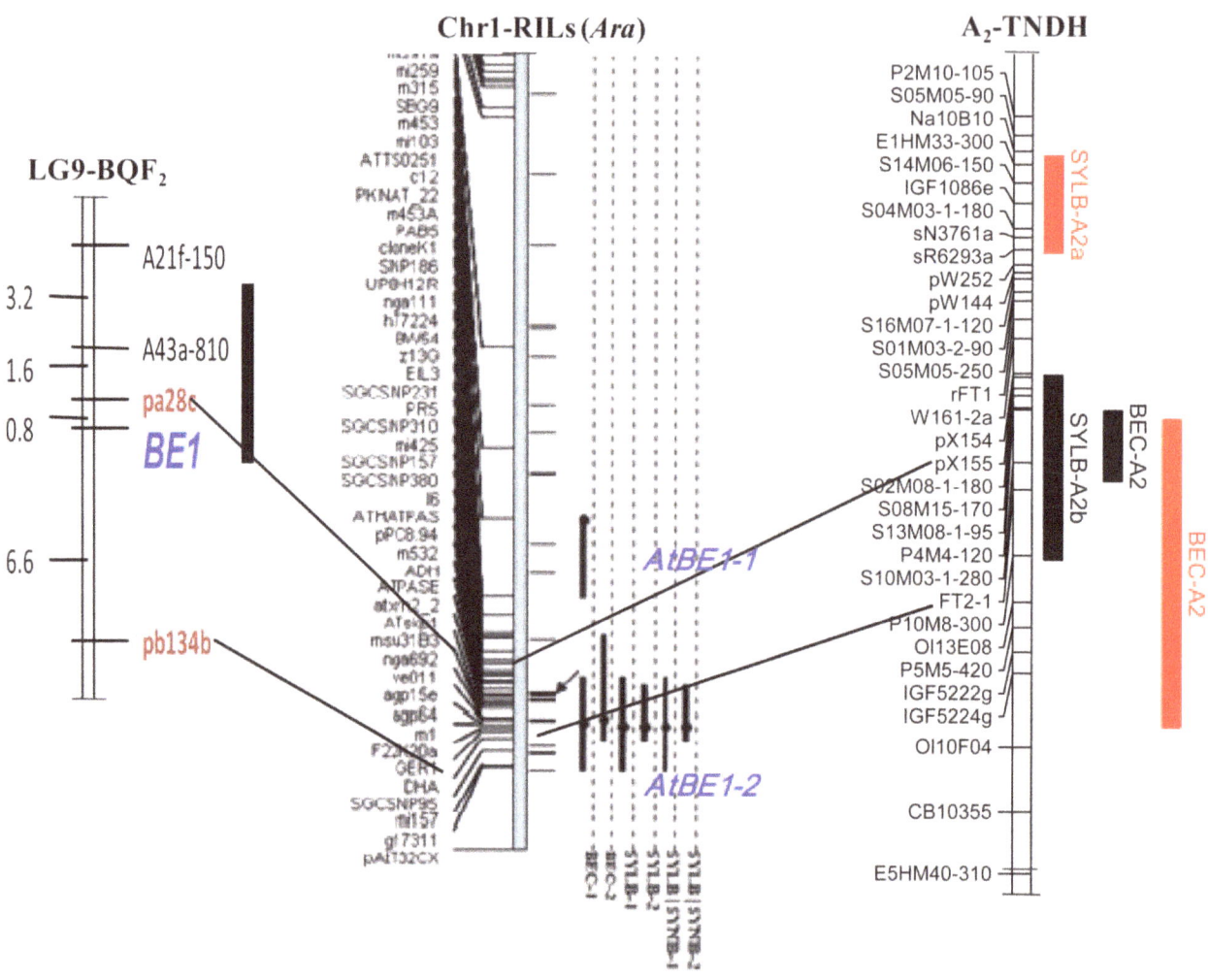

Figure 4. QTLs for seed yield at low B condition (SYLB) and B efficiency coefficient (BEC) in A2 linkage group of the TNDH population compared to B efficiency loci in LG9 linkage group of the BQ F$_2$ population and *Arabidopsis* Chromosome 1. Black and red blocks means QTLs detected in the TNDH population in 2004 and 2005, respectively.

For the BQDH population, a three-year field trial was carried out at our laboratory's boron fertilizer experiment base, which has sandy paddy soil, in Qichun county, Hubei Province (N 115°45′ E 30°19′), during the 2008–2009, 2009–2010 and 2010–2011 crop seasons. No specific permits were required for the field trial. The average hot-water-soluble B in the plough layer soil before

Table 7. Candidate genes in *B.rapa* in the homologous region of the intervals of *SYLB-A2* and *BEC-A2* for B efficiency in *B. napus*.

Gene in *B.rapa*	Start	Stop	Functional annotation	Homologous gene in *Arabidopsis*	Gene in *AtBE1-2*
Bra020592	24377992	24383065	auxiliary transport protein activity;	AT5G27360	AT1G08930
Bra020593	24358466	24363733	transport accessory protein activity;	AT5G27350	
Bra020595	24340611	24343166	small molecule transport;	AT5G27350	
			solute:solute exchange;		
			small-molecule carrier or transporter;		
			cellular component.		
Bra020609	24233008	24236011	cellular component;	AT5G28470	AT1G08930
			oligopeptide transport;		
			small-molecule carrier or transporter.		
Bra033181	17297210	17300039	small-molecule carrier or transporter	AtNIP5;1	AtNIP5;1

Table 8. Sequence information for the polymorphic GB-SSR primers developed from the four candidate genes in *B. rapa*.

Locus	Repeat motif	Repeat start position (bp)	Primer sequence (5'-3')	Expected size (bp)
Bra020592-1	$(TA)_{12}$	24379844	GTGGCTGAACATCGGAAGAT	199
			CCTGCTTGGTACATCCATCA	
Bra020592-2	$(AT)_{10}$	24382294	AGCGTCTGGGATGTGTTTGT	205
			TTGCATTTTCTGCTCGTACC	
Bra020592-3	$(TA)_{10}$	24382840	GTTTGAATGGAGCACTCAAGG	202
			CCGGAACGAGAAACCAAATA	
Bra020593-1	$(AT)_{9}$	24359107	CAAGTGTGGTCTTGTGGGAAT	196
			GTACGTTCGCATTCGGATACT	
Bra020593-2	$(TA)_{11}$	24363010	GGCTCGCCTTCACATTACA	257
			TGCAAGATACAGCTGCCAAG	
Bra020595-1	$(CT)_{12}$	24333492	CCTATTTGAGTTCCTTAAGCGATG	221
			GCTCCAAAAGCCCTTCTTCT	
Bra020595-2	$(TC)_{12}$	24333517	CCCACTATCACAAAAACATAGCTC	205
			ACAGCCTGAAGCGTCACTTT	
Bra020595-3	$(AT)_{9}$	24355967	GGTTGCATTGATGGCAAATAG	204
			GGCAAAGCATATGACAAAGC	
Bra020609-1	$(A)_{14}$	24235380	TGGTACAAAAACCCCATGCT	174
			GGGCTAAAACCTGTTTTGGATT	

fertilization was 0.09~0.11 mg kg^{-1}, and two B treatments were employed in the three trials. For the first trial, the LB treatment was 1.0 kg ha^{-1} of boric acid, which was sprayed as fertilizer before the bolting stage. For the second trial, no B fertilizer was applied throughout the season. For the third trial, 0.75 kg ha^{-1} of borax was applied as the base fertilizer. The NB treatment was the control, and the application of N-, P- and K-containing fertilizers, the trial block design and the agronomical management were the same as TNDH population field trial.

Six traits, including seed yield (SY), plant height (PH), branch number (BN), pod number (PN), seed number (SN) and seed weight (SW), were measured. One derived trait, the BEC, was calculated as the ratio of the seed yield under low B levels to that under normal B levels. The six traits were measured according to the methods described in [40].

Analysis of polymorphic loci

Genomic DNA was extracted from plant leaf tissues by the cetyltrimethylammonium bromide (CTAB) method [54]. The genotypes of the BQDH lines were analyzed using simple sequence repeat (SSR), sequence-related amplified polymorphism (SRAP), and gene-based markers (GBMs) for the construction of a genetic linkage map.

Primer sequences of the SSR markers were obtained from various sources: UK prefixed by OL and Na (http://www. brasscia.bbsrc.ac.uk/BrassicaDB), Australia prefixed by sA (http://www.hornbill.cspp.latrobe.edu.au), Canada prefixed by sR and sN (http://www.brassica.agr.gc.ca/index_e.shtml), Japan prefixed by BRMS [55], and France prefixed by BRAS, CB and MR [56]. A total of 171 *B.rapa* BAC sequences and/or BAC-end sequences (BES) primers [57], 698 SSR primers designated as BnGMS [58] and 1398 SSR primers designated as BoGMS [59] were used. Additionally, due to the release of the genome sequence of *B.rapa* [60], we developed 50 SSR primers designated as

BeA2ssr according to the A2 sequences to improve the A2linkage group in the BQDH genetic linkage map.

The analysis of the SRAP markers has been previously described [61], and the polymorphic primer pairs were named by combining the names of the forward and reverse primers (e.g., em5me24).

The gene-based markers (GBMs) were functional genes related to B transporters and channel proteins in *Arabidopsis* (http://www. arabidopsis.org/). The primers were designed based on the conserved sequences between the homologous genes from *Arabidopsis* and *Brassica*. The GBMs were named by using the suffix Bn or Br and the gene name in *Arabidopsis*, for example AtNIP5;1-Bn. The analysis of GBM followed the protocol described by [57].

All PCR products were separated by PAGE and stained with AgNO$_3$. If a primer pair showed more than one polymorphic loci, the different loci were distinguished by small letters after the name of the marker. For example, the primer pair CB10079 generated two polymorphic loci, which were named CB10079a and CB10079b and were distinguished by the product size in increasing order.

Linkage analysis and map construction

Linkage analysis and map construction were performed using the JoinMap software Version 4.0 [62]. The threshold for goodness-of-fit was set to ≤5.0 with a recombination frequency of <0.4 and a minimum logarithm of odds score of 1.0. Markers with a χ^2 value of >3.0 were excluded in all genetic groups. Recombination frequencies were converted to centimorgans (cM) using Kosambi's method for map distance calculation [63].

QTLs, interaction detection and *in silico* mapping

QTL detection was carried out by the composite interval method (CIM) [64], using WinQTLcart 2.5 software [65]. CIM was performed using Model 6 after scanning the genetic map and estimating the likelihood of a QTL and its corresponding effect every 1 cM. The number of control markers and the window size were set to 5 and 10 cM, respectively. For each trait, the threshold for detection of a significant QTL (P<0.05) was estimated by 1,000 permutations [66]. The estimated additive effect and the percentage of phenotypic variation explained by each putative QTL were obtained using the software with the CIM model. QTL support intervals were determined by 2-LOD intervals surrounding the QTL peak. When QTLs for the same trait during the same B treatment across two or three years had overlapping support intervals, they were assumed to be consistent.

Epistatic interactions and environmental interactions were detected using the software program QTLmapper 2.0 (http://www.cab.zju.edu.cn/ics/faculty/zhujun.htm) (for the TNDH population) and the software program QTLNetwork2.0 (for the BQDH population) [67], which are all based on mixed linear model approaches [68]. The 1D search for QTL×environment interval effects was carried out with a 10 cM testing window, a 1 cM walking speed, and a 5 cM window size. Both 1D and 2D genome scans were conducted with $p<0.05$ significance threshold based on 1,000 permutations.

The in silico mapping of the BQDH population was carried out according to the method described in [49]. Genes were identified by comparative mapping between the B. napus linkage groups and the A. thaliana genome in each syntenic block of A. thaliana and then associated with each putative QTL. If the position of an aligned gene(s) was located in the support interval of a QTL, the orthologous gene(s) was considered to be associated with the target QTL.

Statistical analysis

Statistical analysis for all traits was conducted using SAS8.1 (SAS Institute, Cary, NC, USA). Histograms and normality tests (Pearson chi-square test) were used to describe the variation of the phenotypic traits. The Pearson's phenotypic correlation coefficients among seven traits across all environments were calculated to examine their phenotypic association using the SAS PROC CORR. ANOVA was conducted using the SAS general linear model (GLM) procedure.

Supporting Information

Figure S1 QTL projection from other populations onto the BQDH genetic linkage map, via a map projection using BioMercator 2.1 software (Arcade et al. 2004) based on common markers.

Figure S2 Chromosomal locations of putative QTLs for yield and yield-related traits in Brassica napus BQDH population.

Table S1 The BQDH genetic linkage map and its syntenic segmental alignment with the Arabidopsis genome.

Table S2 Pearson's correlation analysis between traits in TNDH (A) and BQDH (B) populations.

Table S3 Epistatic interactions for six yield related traits under both B conditions and BEC in BQDH population of Brasscia napus.

Table S4 Orthologous genes associated with yield and yield-associated QTLs detected in BQDH population by in-silico mapping between A.thaliana and B.napus.

Acknowledgments

We appreciate the helps from Dr. Kede Liu with providing two groups of SSR primers designated as BnGMS and BoGMS for analysis of polymorphic loci in the BQDH population, and Dr. Ruiyuan Li with compiling program for comparative mapping between Brassica napus and Arabidopsis, Brassica napus and Brassica rapa.

Author Contributions

Conceived and designed the experiments: ZZ FX. Performed the experiments: ZZ FN TS. Analyzed the data: ZZ GD FX DZ LS JM. Contributed reagents/materials/analysis tools: ZZ LW TS JM. Wrote the paper: ZZ FX GD LS.

References

1. Warington K (1923) The effect of boric acid and borax on the broad bean and certain other plants. Ann Bot 37:629–672.
2. O'Neill MA, Warrenfeltz D, Kates K, Pellerin P, Doco T, et al. (1996) Rhamnogalacturonan-II, a pectic polysaccharide in the walls of growing plant cell, forms a dimer that Is covalently cross-linked by a borate ester. Journal of Biological Chemistry 271: 22923–22930.
3. Ishii T, Matsunaga T, Hayashi N (2001) Formation of rhamnogalacturonan II-borate dimer in pectin determines cell wall thickness of pumpkin tissue. Plant Physiol 126: 1698–1705.
4. Marschner P (1995) Marschner's mineral nutrition of higher plants, Second Edition. 2nd ed. Academic Press. 889 p.
5. González-Fontes A, Rexach J, Navarro-Gochicoa MT, Herrera-Rodríguez MB, Beato VM, et al. (2008) Is boron involved solely in structural roles in vascular plants? Plant Signal Behav 3: 24–26.
6. Raven JA (1980) Short- and long-distance transport of boric acid in plants. New Phytologist 84: 231–249.
7. Dannel F, Pfeffer H, Römheld V (2000) Characterization of root boron pools, boron uptake and boron translocation in sunflower using the stable isotopes ^{10}B and ^{11}B. Functional Plant Biol 27: 397–405.
8. Takano J, Noguchi K, Yasumori M, Kobayashi M, Gajdos Z, et al. (2002) Arabidopsis boron transporter for xylem loading. Nature 420: 337–340.
9. Takano J, Miwa K, Yuan L, von Wirén N, Fujiwara T (2005) Endocytosis and degradation of BOR1, a boron transporter of Arabidopsis thaliana, regulated by boron availability. Proc Natl Acad Sci USA 102: 12276–12281.
10. Nakagawa Y, Hanaoka H, Kobayashi M, Miyoshi K, Miwa K, et al. (2007) Cell-type specificity of the expression of OsBOR1, a rice efflux boron transporter gene, is regulated in response to boron availability for efficient boron uptake and xylem loading. Plant Cell 19: 2624–2635.
11. Sun J, Shi L, Zhang C, Xu F (2011) Cloning and characterization of boron transporters in Brassica napus. Mol Biol Rep 39: 1963–1973.
12. Kaya A, Karakaya HC, Fomenko DE, Gladyshev VN, Koc A (2009) Identification of a novel system for boron transport: Atr1 is a main boron exporter in yeast. Mol Cell Biol 29: 3665–3674.
13. Takano J, Wada M, Ludewig U, Schaaf G, von Wirén N, et al. (2006) The Arabidopsis major intrinsic protein NIP5;1 is essential for efficient boron uptake and plant development under boron limitation. Plant Cell 18: 1498–1509.
14. Tanaka M, Wallace IS, Takano J, Roberts DM, Fujiwara T (2008) NIP6;1 is a boric acid channel for preferential transport of boron to growing shoot tissues in Arabidopsis. Plant Cell 20: 2860–2875.
15. Kasajima I, Ide Y, Yokota Hirai M, Fujiwara T (2010) WRKY6 is involved in the response to boron deficiency in Arabidopsis thaliana. Physiol Plant 139: 80–92.
16. Miwa K, Takano J, Fujiwara T (2006) Improvement of seed yields under boron-limiting conditions through overexpression of BOR1, a boron transporter for xylem loading, in Arabidopsis thaliana. Plant J 46: 1084–1091.
17. Kato Y, Miwa K, Takano J, Wada M, Fujiwara T (2009) Highly boron deficiency-tolerant plants generated by enhanced expression of NIP5;1, a boric acid channel. Plant Cell Physiol 50: 58–66.
18. Meyer M (2009) Rapeseed oil fuel - the crisis-proof home-made eco-fuel. Agrarforschung 16: 262–267.
19. Shorrocks V (1997) The occurrence and correction of boron deficiency. Plant and soil 193: 121–148.
20. Yan X, Liao H, Beebe SE, Blair MW, Lynch JP (2004) QTL mapping of root hair and acid exudation traits and their relationship to phosphorus uptake in common bean. Plant and Soil 265: 17–29.
21. Xu F, Wang Y, Meng J (2001) Mapping boron efficiency gene(s) in Brassica napus using RFLP and AFLP markers. Plant Breeding 120: 319–324.

22. Paran I, Zamir D (2003) Quantitative traits in plants: beyond the QTL. Trends Genet 19: 303–306.
23. Cooper M, van Eeuwijk FA, Hammer GL, Podlich DW, Messina C (2009) Modeling QTL for complex traits: detection and context for plant breeding. Curr Opin Plant Biol 12: 231–240.
24. Xue W, Xing Y, Weng X, Zhao Y, Tang W, et al. (2008) Natural variation in Ghd7 is an important regulator of heading date and yield potential in rice. Nat Genet 40: 761–767.
25. El-Din El-Assal S, Alonso-Blanco C, Peeters AJM, Raz V, Koornneef M (2001) A QTL for flowering time in *Arabidopsis* reveals a novel allele of CRY2. Nat Genet 29: 435–440.
26. Quijada PA, Udall JA, Lambert B, Osborn TC (2006) Quantitative trait analysis of seed yield and other complex traits in hybrid spring rapeseed (*Brassica napus* L.): 1. Identification of genomic regions from winter germplasm. Theor Appl Genet 113: 549–561.
27. Chen W, Zhang Y, Liu X, Chen B, Tu J, et al. (2007) Detection of QTL for six yield-related traits in oilseed rape (*Brassica napus*) using DH and immortalized F(2) populations. Theor Appl Genet 115: 849–858.
28. Radoev M, Becker HC, Ecke W (2008) Genetic analysis of heterosis for yield and yield components in rapeseed (*Brassica napus* L.) by quantitative trait locus mapping. Genetics 179: 1547–1558.
29. Fan C, Cai G, Qin J, Li Q, Yang M, et al. (2010) Mapping of quantitative trait loci and development of allele-specific markers for seed weight in *Brassica napus*. Theor Appl Genet 121: 1289–1301.
30. Zhao J, Becker HC, Zhang D, Zhang Y, Ecke W (2006) Conditional QTL mapping of oil content in rapeseed with respect to protein content and traits related to plant development and grain yield. Theor Appl Genet 113: 33–38.
31. Shi L, Wang Y, Nian F, Lu J, Meng J, et al. (2009) Inheritance of boron efficiency in oilseed rape. Pedosphere 19 (3) 403–408.
32. Zhao H, Shi L, Duan X, Xu F, Wang Y, et al. (2008) Mapping and validation of chromosome regions conferring a new boron-efficient locus in Brassica napus. Mol Breeding 22: 495–506.
33. Long Y, Shi J, Qiu D, Li R, Zhang C, et al. (2007) Flowering time quantitative trait loci analysis of oilseed *Brassica* in multiple environments and genomewide alignment with *Arabidopsis*. Genetics 177: 2433–2444.
34. Qiu D, Morgan C, Shi J, Long Y, Liu J, et al. (2006) A comparative linkage map of oilseed rape and its use for QTL analysis of seed oil and erucic acid content. Theor Appl Genet 114: 67–80.
35. Shi L, Nian F, Zhao H, Xu F, Meng J, et al. (2004) Responses to boron deficiency in 7 varieties of rape (*Brassica napus* L.). Chin J Oil Crop Sci 26:50–53 (in Chinese with English abstract).
36. Masclaux-Daubresse C, Daniel-Vedele F, Dechorgnat J, Chardon F, Gaufichon L, et al. (2010) Nitrogen uptake, assimilation and remobilization in plants: challenges for sustainable and productive agriculture. Ann Bot 105: 1141–1157.
37. Xu G, Fan X, Miller AJ (2011) Plant nitrogen assimilation and use efficiency. Annu Rev Plant Biol 2012.63:5.1–5.30.
38. Zeng C, Han Y, Shi L, Peng L, Wang Y, et al. (2008) Genetic analysis of the physiological responses to low boron stress in *Arabidopsis thaliana*. Plant Cell Environ 31: 112–122.
39. Xu Y (1997) Quantitative Trait Loci: Separating, Pyramiding, and Cloning: 85–139.
40. Shi J, Li R, Qiu D, Jiang C, Long Y, et al. (2009) Unraveling the complex trait of crop yield with quantitative trait loci mapping in *Brassica napus*. Genetics 182: 851–861.
41. Wang L, Zhao J, Xu F, Liu R, Meng J (2002) Integration of DNA clones related to important economic traits of *Brassica napus* onto *Arabidopsis* genetic map. Acta Genetica Sinica 29(8): 741–746.
42. Arcade A, Labourdette A, Falque M, Mangin B, Chardon F, et al. (2004) BioMercator: integrating genetic maps and QTL towards discovery of candidate genes. Bioinformatics 20: 2324–2326.
43. Ding G, Zhao Z, Liao Y, Hu Y, Shi L, et al. (2012) Quantitative trait loci for seed yield and yield-related traits, and their responses to reduced phosphorus supply in *Brassica napus*. Ann Bot doi:10.1093/aob/mcr323.
44. Yu S, Li J, Xu C, Tan Y, Gao Y, et al. (1997) Importance of epistasis as the genetic basis of heterosis in an elite rice hybrid. Proc Natl Acad Sci U S A 94: 9226–9231.
45. Liu J, Yang J, Li R, Shi L, Zhang C, et al. (2009) Analysis of genetic factors that control shoot mineral concentrations in rapeseed (*Brassica napus*) in different boron environments. Plant and Soil 320: 255–266.
46. Wang Z, Wang Z, Shi L, Wang L, Xu F (2010) Proteomic alterations of *Brassica napus* root in response to boron deficiency. Plant Mol Biol 74: 265–278.
47. Quirino BF, Reiter WD, Amasino RD (2001) One of two tandem Arabidopsis genes homologous to monosaccharide transporters is senescence-associated. Plant Mol Biol 46: 447–457.
48. Qin Y, Leydon AR, Manziello A, Pandey R, Mount D, et al. (2009) Penetration of the stigma and style elicits a novel transcriptome in pollen tubes, pointing to genes critical for growth in a pistil. PLoS Genet 5: e1000621.
49. Parkin IAP, Gulden SM, Sharpe AG, Lukens L, Trick M, et al. (2005) Segmental structure of the *Brassica napus* genome based on comparative analysis with *Arabidopsis thaliana*. Genetics 171: 765–781.
50. Schranz ME, Lysak MA, Mitchell-Olds T (2006) The ABC's of comparative genomics in the Brassicaceae: building blocks of crucifer genomes. Trends Plant Sci 11: 535–542.
51. Yang M, Ding G, Shi L, Feng J, Xu F, et al. (2010) Quantitative trait loci for root morphology in response to low phosphorus stress in *Brassica napus*. Theor Appl Genet 121: 181–193.
52. Kobayashi M, Mutoh T, Matoh T (2004) Boron nutrition of cultured tobacco BY-2 cells. IV. Genes induced under low boron supply. J Exp Bot 55: 1441–1443.
53. Wang Z, Wang Z, Chen S, Shi L, Xu F (2011) Proteomics reveals the adaptability mechanism of *Brassica napus* to short-term boron deprivation. Plant Soil 347: 195–210.
54. Doyle JI (1990) Isolation of plant DNA from fresh tissue. Focus 12: 13–15.
55. Suwabe K, Iketani H, Nunome T, Kage T, Hirai M (2002) Isolation and characterization of microsatellites in *Brassica rapa* L. Theor Appl Genet 104: 1092–1098.
56. Piquemal J, Cinquin E, Couton F, Rondeau C, Seignoret E, et al. (2005) Construction of an oilseed rape (*Brassica napus* L.) genetic map with SSR markers. Theor Appl Genet 111: 1514–1523.
57. Ding G, Liao Y, Yang M, Zhao Z, Shi L, et al. (2011) Development of gene-based markers from functional Arabidopsis thaliana genes involved in phosphorus homeostasis and mapping in *Brassica napus*. Euphytica 181: 305–322.
58. Cheng X, Xu J, Xia S, Gu J, Yang Y, et al. (2009) Development and genetic mapping of microsatellite markers from genome survey sequences in *Brassica napus*. Theor Appl Genet 118: 1121–1131.
59. Li H, Chen X, Yang Y, Xu J, Gu J, et al. (2010) Development and genetic mapping of microsatellite markers from whole genome shotgun sequences in *Brassica oleracea*. Mol Breeding 28: 585–596.
60. Wang X, Wang H, Wang J, Sun R, Wu J, et al. (2011) The genome of the mesopolyploid crop species *Brassica rapa*. Nat Genet 43: 1035–1039.
61. Li, Quiros CF (2001) Sequence-related amplified polymorphism (SRAP), a new marker system based on a simple PCR reaction: its application to mapping and gene tagging in *Brassica*. Theor Appl Genet 103: 455–461.
62. Van Ooijen JW (2006) JoinMap®4.0: software for the calculation of genetic linkage maps in experimental populations. Kyazma BV, Wageningen, Netherlands
63. Kosambi DD (1943) The estimation of map distances from recombination values. Annals of Human Genetics 12: 172–175.
64. Zeng Z (1994) Precision mapping of quantitative trait loci. Genetics 136: 1457–1468.
65. Wang S, Bastern J, Zeng Z (2006) Windows QTL Cartographer 2.5. Department of Statistics, North Carolina State University, Raleigh, NC, USA.
66. Churchill GA, Doerge RW (1994) Empirical threshold values for quantitative trait mapping. Genetics 138: 963–971.
67. Yang J, Hu C, Hu H, Yu R, Xia Z, et al. (2008) QTLNetwork: mapping and visualizing genetic architecture of complex traits in experimental populations. Bioinformatics 24: 721–723.
68. Yang J, Zhu J, Williams RW (2007) Mapping the genetic architecture of complex traits in experimental populations. Bioinformatics 23: 1527–1536.

14

Effects of Tillage and Nitrogen Fertilizers on CH_4 and CO_2 Emissions and Soil Organic Carbon in Paddy Fields of Central China

Li Cheng-Fang[1]⁹, Zhou Dan-Na[2]⁹, Kou Zhi-Kui[1], Zhang Zhi-Sheng[1], Wang Jin-Ping[1], Cai Ming-Li[1], Cao Cou-Gui[1]*

1 College of Plant Science and Technology, Huazhong Agricultural University, Wuhan, Hubei, China, 2 Institute of Animal Husbandry and Veterinary Science, Hubei Academy of Aguicultural Sciences, Wuhan, Hubei, China

Abstract

Quantifying carbon (C) sequestration in paddy soils is necessary to help better understand the effect of agricultural practices on the C cycle. The objective of the present study was to assess the effects of tillage practices [conventional tillage (CT) and no-tillage (NT)] and the application of nitrogen (N) fertilizer (0 and 210 kg N ha^{-1}) on fluxes of CH_4 and CO_2, and soil organic C (SOC) sequestration during the 2009 and 2010 rice growing seasons in central China. Application of N fertilizer significantly increased CH_4 emissions by 13%–66% and SOC by 21%–94% irrespective of soil sampling depths, but had no effect on CO_2 emissions in either year. Tillage significantly affected CH_4 and CO_2 emissions, where NT significantly decreased CH_4 emissions by 10%–36% but increased CO_2 emissions by 22%–40% in both years. The effects of tillage on the SOC varied with the depth of soil sampling. NT significantly increased the SOC by 7%–48% in the 0–5 cm layer compared with CT. However, there was no significant difference in the SOC between NT and CT across the entire 0–20 cm layer. Hence, our results suggest that the potential of SOC sequestration in NT paddy fields may be overestimated in central China if only surface soil samples are considered.

Editor: Kurt O. Reinhart, USDA-ARS, United States of America

Funding: The study was supported by the National Technology Project for High Food Yield of China (2011BAD16B02), National Natural Science Foundation of China (31100319), Fundamental Research Funds for the Central Universities (2010QC032) and the Foundation of Hubei Key Laboratory of Animal Embryo Engineering and Molecular Breeding (2011ZD152). The funders had no role in study design, data collection and analysis, decision to publish, or preparation of the manuscript.

Competing Interests: The authors have declared that no competing interests exist.

* E-mail: ccgui@mail.hzau.edu.cn

⁹ These authors contributed equally to this work.

Introduction

Global surface temperatures have increased by 0.88°C since the late nineteenth century [1]. The observed climate changes are caused by the emission of greenhouse gases (GHGs) mainly through anthropogenic activities. Methane and CO_2 are the most important GHGs, respectively contributing 15% and 60% to the anthropogenic GHG effect [2]. Rice paddies are an important source of atmospheric CH_4. The amount of CH_4 emitted from wetland paddy fields accounts for 10% to 20% of the total CH_4 emissions (i.e. 50 Tg yr^{-1} to 100 Tg yr^{-1}) [2]. The rice production of China exceeds that of any other country, accounting for 30% of the world total [3]. Agricultural activity affects CH_4 and CO_2 emissions, contributing 39% of the excess CH_4 and 1% of the excess CO_2 to global emissions [4]. Hence, CH_4 emissions from paddy fields under different agricultural management practices in China are relevant to the discussion of the global C cycle and climate changes.

The entire process of CH_4 emission from rice fields, including production, oxidation, and transport into the atmosphere is influenced by agricultural management practices, such as tillage and N fertilizer use [5–7]. Tillage affects a range of biological,

chemical, and physical properties, thereby affecting the release of CH_4 [8]. No-tillage (NT) has been reported to reduce CH_4 emissions from paddy soils because rice straw is placed on the soil surface under NT and the soil conditions are more oxidative than those of conventional tillage (CT) [7,9]. CH_4 emissions from paddy fields are reportedly affected by the form and amount of N fertilizer applied [10]. Overall, the effects of N fertilizer application on CH_4 fluxes from paddy fields are mostly unclear. Therefore, more research on the effects of N addition on CH_4 emissions is needed.

Tillage practices can affect soil biochemical and physical properties, consequently influencing the release of CO_2 [8]. However, there is no consensus on the differences in the soil CO_2 emissions between NT- and CT-treated paddy fields. Some authors have reported similar soil CO_2 fluxes from NT- and CT-treated paddy fields [7]. However, Liang et al. [9] reported higher soil CO_2 emissions from CT-treated paddy fields than from the NT paddy fields. Nitrogen supplied by commercial fertilizers can be expected to affect soil CO_2 flux by increasing the C input from enhanced plant productivity and crop residues returned to the soil [11]. However, studies on the effects of N fertilizer on soil CO_2 emissions reveal diverse results [12]. Within the past few years, Iqbal et al. [13] and Xiao et al. [14] observed increased CO_2 emissions from paddy soils because of a positive effect of N fertilization on plant biomass. However, Burton et al. [15] and DeForest et al. [16] found that the use of N reduced extracellular enzymatic activities and fungal populations, resulting in decreased soil CO_2 flux. The effect of N fertilization on variation in CO_2 emission under anaerobic conditions in paddy soils remains unknown.

Land management practices are increasingly thought to affect soil carbon levels and may partially ameliorate CO_2 emissions and climate change [17,18]. Studies have indicated that NT can increase C sequestration in paddy soils compared with CT [19–21]. In 2007, Tang et al. [20] indicated that the NT could sequester 112.3 kg C ha^{-1} yr^{-1} in the top 20 cm of purple paddy soil in the Beipei district of Chongqing City, China. In a 12-year study, Gao et al. [21] reported that NT could sequester 26.68 kg C ha^{-1} yr^{-1} in gray fluvoaguic paddy soils to a depth of 30 cm in Zhangjiagang City, Jiangsu Province, China. However, Six et al. [22] and Su [23] indicated that the effects of NT on SOC sequestration depend on the soil type. In a 5-year study, He et al. [24] indicated that NT did not increase the SOC sequestration of paddy fields in the 20 cm layer of sandy silty loam in Ningxiang country, Hunan Province. However, Angers and Eriksen-Hamel [25] reviewed the related literature and concluded that soil variables do not affect the tillage effects on soil C sequestration. Hence, further research is needed to clarify the effects of soil type on C sequestration in NT-treated soils.

No-tillage may influence SOC accumulation when soil surface layers are considered, but the effect may not be detected more deeply [22]. The influence of NT on SOC sequestration is still unclear. Hence, Baker et al. [26] analyzed sampling strategies on the potential of SOC sequestration under conservation tillage and indicated that SOC sequestration under this tillage varied with soil depth. Thus, shallow sampling may not be sufficient to assess the differences in SOC sequestration between NT- and CT-treated soils, and further research on the effects of deeper soil sampling on SOC sequestration in NT-treated soils should be performed.

Application of N fertilizer may play a significant role in the soil C sequestration [17]. Application of N fertilizer affects the soil C stock in two ways. These compounds can increase the crop biomass and influence the microbial decomposition of crop residues by affecting the N availability [27]. However, a meta-analysis of 111 studies covering 12 soil types of divergent ecosystems indicated that the effects of N fertilizer application on soil C content vary with the soil type although N fertilizer application consistently increases the crop biomass [28]. For example, Tong et al. [29] found in a 17-year study published in 2009 that the use of chemical N fertilizers did not increase the SOC content in a hydromorphic paddy soil in Hunan Province compared with no fertilizer use. By contrast, Shang et al. [30] found in the same province that increased N fertilization increased the SOC sequestration in paddy soils derived from quaternary red clay.

Central China is one of the major rice-producing regions in the country, comprising 28% of the total area cultivated with rice in China [31]. Recently, NT practices have become increasingly popular in this region. However, to our knowledge, relatively few studies have been performed on the effects of tillage and N fertilizer on CH_4 and CO_2 emissions as well as on SOC sequestration in the paddy fields in this region. We hypothesized that tillage practices and N fertilizer use affect CH_4 and CO_2 emissions as well as soil C sequestration in hydromorphic paddy fields in this region. We specifically tested the effects of tillage practices and N fertilizer use on SOC in soils from 0 cm to 5 cm,

as well as from 0 cm to 20 cm, during the 2009 and 2010 rice growing seasons. This paper also aimed to evaluate the effects of tillage and N fertilizer on CH_4 and CO_2 emissions during the rice growing seasons.

Results

Temperature

The air temperature in the experimental site is shown in Table 1. The mean monthly air temperature ranged from 21.4°C to 28.9°C and from 19.7°C to 29.8°C during the 2009 and 2010 rice growing season, respectively. The mean monthly air temperature during rice growing seasons in 2009 was slightly lower than that in 2010. The mean air temperature from June to September, except for August, was significantly higher ($P<0.05$) in 2010 than in 2009.

CH_4 and CO_2 Emissions

The pattern of seasonal CH_4 emission fluxes was similar across NT and CT treatments during the 2009 and 2010 rice growing seasons (Fig. 1). In both years, the CH_4 emission fluxes in the four treatment groups were all initially low, increased gradually, and then peaked in mid-July (about 4–5 weeks after sowing). Thereafter, the CH_4 emission fluxes declined gradually and remained relatively low until harvesting when the CH_4 emission fluxes were lowest.

Application of N fertilizer significantly increased CH_4 emissions by 13%–66% in 2009 and 2010 ($P<0.05$) (Table 2). Tillage significantly affected CH_4 emissions, where NT significantly decreased CH_4 emissions by 10%–36% compared with CT ($P<0.05$). No significant effect of tillage×fertilizer on the cumulative CH_4 emissions was observed in 2009 or 2010. The cumulative CH_4 emissions in 2010 were 1.39–2.45 times those recorded in 2009.

Tillage treatments exhibited clear seasonal variations in soil CO_2 fluxes in the 2009 and 2010 rice growing seasons (Fig. 2). The soil CO_2 fluxes remained relatively low for the first two weeks after tillage, increased rapidly, stayed relatively high until about the middle 10 days of July, and then decreased to relatively low levels. Just one day after tillage (June 9, 2009 and June 13, 2010), the soil CO_2 fluxes from CT were 1.40–4.60 times higher than those from NT ($P<0.05$).

The cumulative CO_2 emissions from NT were 1.30–1.33 times those of CT ($P<0.05$) (Table 2). The application of N fertilizer had no significant effect on cumulative CO_2 emissions. We observed a significant effect of tillage×fertilizer on CO_2 emissions in 2009 ($P<0.05$) but not in 2010. In addition, cumulative CO_2 emissions in 2010 were 2.44–2.93 times those in 2009.

Table 1. Mean monthly air temperature during rice growing season in the experimental site/°C.

Time	2009	2010
June	26.1 b	27.1 a
July	28.9 a	29.1 a
August	28.0 b	28.8 a
September	24.7 b	25.6 a
October	21.4 a	19.7 a
Mean air temperature during the rice growing season	26.7 a	27.4 a

Different letters in a line mean significant differences at the 5% level.

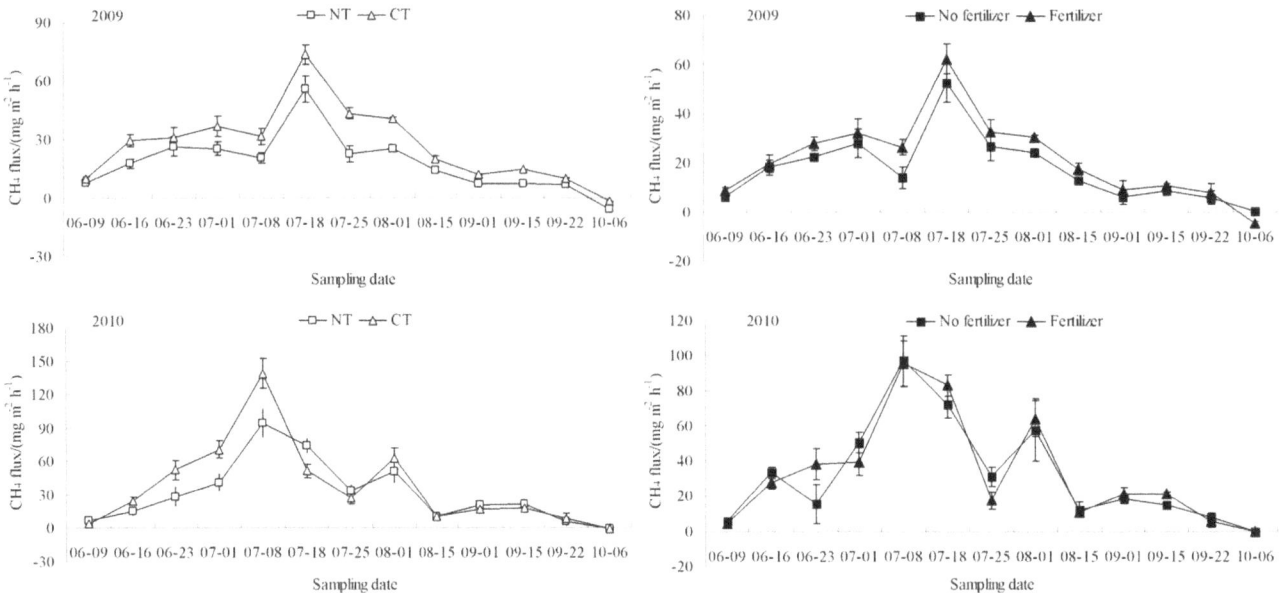

Figure 1. Changes in CH₄ emission fluxes from paddy fields under different management practices during the 2009 and 2010 rice growing seasons. The vertical bars are standard deviations of the mean, n = 3.

Soil Organic C and Bulk Density

As shown in Table 3, neither tillage nor N fertilizer application had any significant effect on bulk density before tilling fields or at harvesting in either year irrespective of the soil sampling depth. The SOC contents were significantly higher at 0–5 cm depth than at 0–20 cm depth under NT. In both years, application of N fertilizer significantly increased the SOC content by 4%–9% at harvesting and the SOC at the end of the growing seasons in 2009 and 2010 (21–94%) irrespective of soil sampling depths. Though NT had slightly higher SOC at the end of the growing seasons at

0–20 cm depth than CT in 2009 and 2010, we observed no significant effect of tillage or tillage×fertilizer in either year. However, across both years, tillage affected the SOC at 0–5 cm depth at harvesting, where NT significantly increased SOC contents by 12%–15% and SOC sequestration by 102%–270% than CT.

Based on the SOC content and bulk density at harvesting in the plow layer (0–20 cm; Table 3), we estimated SOC at harvesting in the plow layer to be 27.0–29.5 t C ha^{-1} in 2009 and 2010. Correspondingly, annual SOC accumulation rate in the plow layer

Table 2. Cumulative CH₄ and CO₂ emissions (g m^{-2}) from different tillage treatments in the 2009 and 2010 rice growing seasons, n = 3.

Tillage	N fertilizer	Cumulative CH₄ emissions		Cumulative CO₂ emissions	
		2009	2010	2009	2010
NT	No fertilizer	2.74 (0.57)	6.72 (0.91)	125.7 (10.6)	326.1 (15.6)
	Fertilizer	4.54 (0.44)	7.56 (1.02)	140.1 (6.6)	386.8 (10.5)
CT	No fertilizer	4.28 (0.27)	7.49 (0.33)	103.4 (7.2)	252.8 (12.2)
	Fertilizer	6.76 (0.40)	9.40 (0.60)	100.3 (4.3)	293.8 (14.1)
Analysis of variance					
T		*	*	**	*
F		*	*	NS	NS
T×F		NS	NS	*	NS

T, tillage;
F, application of N fertilizer;
*, significant at the 0.05 probability level;
**, significant at the 0.01 probability level;
NS, not significant;
The values in brackets are standard deviations of the mean.

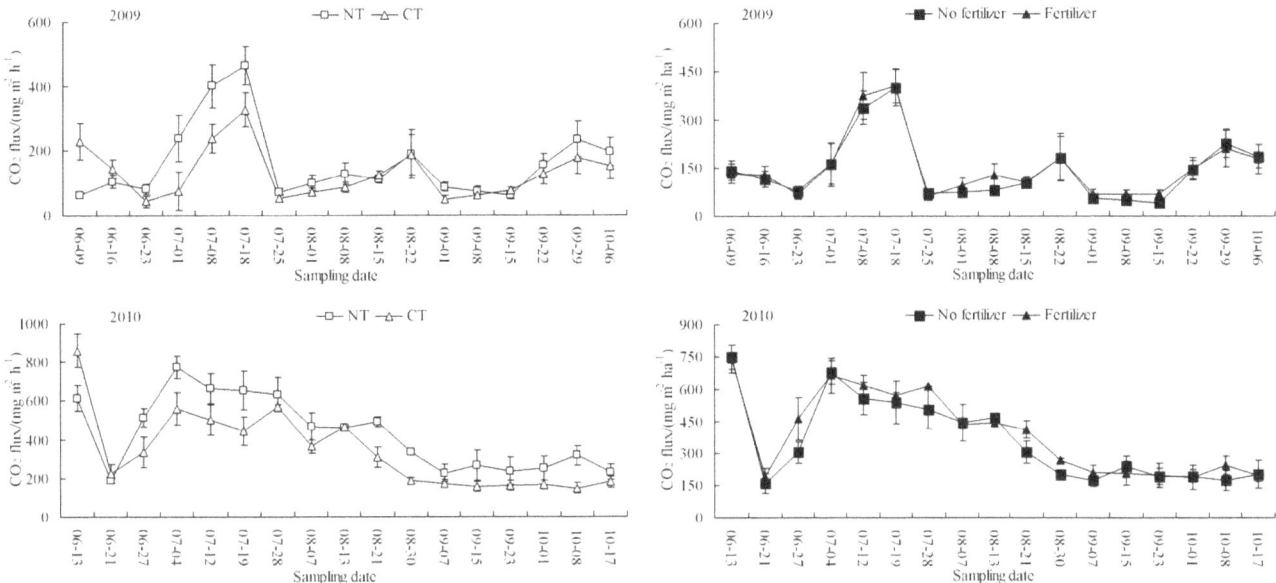

Figure 2. Changes in CO_2 emission fluxes from paddy fields under different management practices during the 2009 and 2010 rice growing seasons. The vertical bars are standard deviations of the mean, n = 3.

was estimated to be 0.06–0.14 t C ha^{-1} yr^{-1} for no fertilizer treatments and 0.25–0.47 t C ha^{-1} yr^{-1} for fertilizer treatments, with an average of 0.23 t C ha^{-1} yr^{-1} over the period 2009–2010.

Discussion

CH_4 Emission

Application of N fertilizer in the present study increased CH_4 emissions from paddy fields because of the promotion of rice growth, providing additional C sources and emission pathways [32]. Lindau and Bollich [33], in a study on a Louisiana rice field, which also had a humid subtropical climate, reported similar results from silt loam soil. However, Wassmann et al. [34] and Lu et al. [35] indicated no significant effect of N fertilizer application on CH_4 emissions from paddy fields in Zhejiang Province, China. Schütz et al. [36] found that the application of urea significantly decreased CH_4 emissions from paddy fields in Italy. Results varied among studies because of the differences in soil texture or climate. These findings show that further study is needed to understand the functioning of these complex and dynamic systems.

No-tillage significantly decreased CH_4 emissions relative to CT in the present study. This is in accordance with the findings reported by Harada et al. [7] and Liang et al. [9]. The decrease in CH_4 emissions under NT may be attributed to the differences regarding the size and activity of the methanotrophic community between tillage treatments [37]. Tillage also affects gaseous diffusivity and the rate of supply of atmospheric CH_4 [38]. By contrast, NT improves macroporosity and maintains its continuity [39]. The improvement probably allows greater air diffusion, increasing CH_4 uptake and decreasing CH_4 emissions.

CO_2 Emissions

Application of N fertilizer increases plant biomass production, stimulating soil biological activity, and consequently, CO_2 emission [40]. Wilson and Al-Kaisi [41], as well as Iqbal et al. [13], observed increased CO_2 emissions caused by N fertilizer

application. By contrast, Burton et al. [15] and DeForest et al. [16] indicated that reduced extracellular enzyme activities and fungal populations resulting from N fertilizer application resulted in decreased soil CO_2 emissions. We observed no significant effect of N fertilizer application on cumulative CO_2 emissions (Table 2), consistent with the results reported by Almaraz et al. [42]. This finding may be due to the fact that CO_2 is reduced to CH_4 under anaerobic conditions, thus leading to significant differences in CH_4 emissions rather than in CO_2 emissions between fertilized and unfertilized treatment areas (see Table 2).

We observed greater CO_2 emissions from NT than from CT during the 2009 and 2010 rice growing seasons (Table 2). Similar results were obtained by Liu et al. [43] and Oorts et al. [8]. The differences between the soil CO_2 emissions under the tillage treatments may have been caused by variation in soil C mineralization. Our own previously published work and those of other researchers indicated greater soil C mineralization under NT [7,8,44]. Increased SOC (Table 3) and higher microbial activity on the soil surface under NT [39] also resulted in greater soil CO_2 emissions for NT than CT. However, CT is generally reported to increase CO_2 emissions by exposing organic matter to more oxidizing conditions of the topsoil and accelerating the decomposition of aggregate-associated soil organic matter [38,45]. The increased levels of surface crop residues in NT probably serve as a barrier for CO_2 emissions from soil, decreasing the decomposition of crop residues because of reduced soil temperature and minimum soil-residue contact [46]. The inconsistent tillage effects on soil CO_2 fluxes suggest that tillage is not the only factor affecting CO_2 flux and that other factors are also involved. As suggested by Mosier et al. [47], CO_2 emissions caused by NT may be similar or slightly lower than those caused by CT if entire growing and fallow seasons are considered.

We observed a significant effect of tillage×N fertilizer on cumulative CO_2 emissions in 2009, in accordance with the results reported by Roberson et al. [48]. The cumulative CO_2 emissions during the rice growing seasons in the present study were 1003–1401 kg C ha^{-1} in 2009 and 2528–3868 kg C ha^{-1} in 2010. These values were greater than 363–371 and 506–926 kg C ha^{-1}

Table 3. SOC contents (g kg^{-1}) and bulk density (g cm^{-3}) before tillage and at harvesting, and SOC sequestration (kg C ha^{-1}) based on soil sampling depths from different tillage treatments in the 2009 and 2010 rice growing seasons, n = 3.

2009

Tillage	N fertilizer	0–20 cm					0–5 cm				
		Bulk density before tillage	Bulk density at harvesting	SOC contents before tillage	SOC contents at harvesting	SOC sequestration	Bulk density before tillage	Bulk density at harvesting	SOC contents before tillage	SOC contents at harvesting	SOC sequestration
NT	No fertilizer	1.19 (0.02)	1.19 (0.01)	18.40 (0.78)	20.16 b (1.12)	2318 (129)	1.20 (0.02)	1.22 (0.03)	18.92 (1.06)	22.55 a (1.22)	1389 (115)
	Fertilizer	1.19 (0.05)	1.17 (0.03)	18.89 (0.91)	21.77 b(1.00)	3439 (271)	1.21 (0.04)	1.23 (0.02)	19.47 (1.23)	23.90 a (4.15)	1685 (162)
CT	No fertilizer	1.17 (0.03)	1.18 (0.04)	18.10 (0.85)	19.70 a (0.89)	2187 (148)	1.18 (0.03)	1.20 (0.06)	18.32 (1.73)	19.68 a (2.11)	559 (90)
	Fertilizer	1.18 (0.04)	1.20 (0.05)	18.56 (1.02)	20.59 a (1.14)	3146 (347)	1.19 (0.05)	1.19 (0.06)	18.85 (1.15)	20.81 a (2.94)	835 (117)
Analysis of variance											
T		–	NS	–	NS	NS	–	NS	–	**	**
F		–	NS	–	*	*	–	NS	–	*	**
T×F		–	NS	–	NS	NS	–	NS	–	NS	NS

2010

Tillage	N fertilizer	0–20 cm					0–5 cm				
		Bulk density before tillage	Bulk density at harvesting	SOC contents before tillage	SOC contents at harvesting	SOC sequestration	Bulk density before tillage	Bulk density at harvesting	SOC contents before tillage	SOC contents at harvesting	SOC sequestration
NT	No fertilizer	1.19 (0.04)	1.25 (0.07)	18.76 (0.88)	19.18 b (1.21)	2102 (123)	1.19 (0.03)	1.25 (0.06)	19.96 (1.25)	21.92 a (1.42)	1032 (66)
	Fertilizer	1.18 (0.04)	1.25 (0.04)	18.85 (1.06)	19.81 b (1.44)	3630 (310)	1.21 (0.04)	1.23 (0.08)	19.93 (1.25)	23.87 a (1.64)	1492 (122)
CT	No fertilizer	1.18 (0.06)	1.24 (0.06)	18.63 (0.76)	18.81 a (1.35)	1949 (251)	1.19 (0.04)	1.21 (0.07)	19.05 (0.97)	19.53 a (1.54)	279 (57)
	Fertilizer	1.17 (0.03)	1.27 (0.04)	18.80 (1.11)	19.33 b (1.51)	2877 (346)	1.20 (0.05)	1.21 (0.07)	19.70 (1.33)	21.15 a (1.71)	542 (115)
Analysis of variance											
T		–	NS	–	NS	NS	–	NS	–	*	**
F		–	NS	–	*	**	–	NS	–	**	**
T×F		–	NS	–	NS	NS	–	NS	–	NS	NS

T, tillage; F, application of N fertilizer;
*, significant at the 0.05 probability level;
**, significant at the 0.01 probability level; NS, not significant; SOC, soil organic C.
Different letters in a year at different depths mean significant differences at the 5% level.
The values in brackets are standard deviations of the mean.

of cumulative CO_2 emissions from different rice tillage systems at an Ogata farm (Japan) and the Hailun Experimental Station of Ecology (Heilongjiang Province, China), respectively [7,9]. The differences in the emissions are possibly related to the dissimilar climates. The experimental field (humid mid-subtropical monsoon climate) in the present study is located at a lower latitude than those of the aforementioned studies.

We observed only one peak of CH_4 or CO_2 emission at the complete tillering stage, in contrast to the two or three emission peaks observed by other researchers [7,31,32]. The discrepancies are likely related to the different rice cropping systems (e.g. single, early, or late rice cropping), field pre-cropping management (e.g. rape and wheat), soil properties, weather conditions, and the use of N fertilizer [31]. The peak of CH_4 emission in the present study may be attributed to (1) the higher availability of substrates through root exudation or decayed plant residues for methanogenic bacteria in the rice rhizosphere [49,50] and (2) vigorous respiration by rice plants during this stage [51]. These processes promote CH_4 emission because most of the CH_4 is emitted through plants [52]. The peak of soil CO_2 emission might be attributed to the increased availability of substrates from root exudation or microbial decomposition of left-over plant residues at the active vegetative growth stage.

Higher cumulative CH_4 and CO_2 emissions (Table 2) were observed during the rice growing season in 2010 than in 2009. Similar interannual differences between CH_4 and CO_2 emissions have been found by other researchers [31,53]. Although these interannual differences in emissions are difficult to explain, discrepancies in climatic conditions and pre-crop residue management are probably involved. Residues of rapeseed were burnt before the experiment was started in 2009, which may be an important reason for the significantly lower emissions observed in this year. Higher mean air temperatures from June to August in 2010 than 2009 may be another important factor that led to higher cumulative CH_4 and CO_2 emissions.

Soil Organic C

In the present study, N fertilizer application had a positive effect on SOC (Table 3). This is attributed to more rice biomass and in turn more residue input to soil under the N fertilized treatments [19]. Others [19,30,54,55] also reported similar results. However, there were other reports indicating that application of chemical N fertilizers caused no significant or even negative effects on SOC [56–59]. The inconsistent results might depend on differences in the climatic and soil conditions, crop residue management, tillage regime, and experimental duration [60].

Here topsoil SOC (27.0–29.5 t C ha^{-1}) was comparable to the results of Pan et al. (27.9–30.9 t C ha^{-1}) [61], but lower than previous estimates of SOC of double-rice paddy soils reported by Shang et al. (36.4–48.2 t C ha^{-1}) [30] and Wang et al. (32.7–41.9 t C ha^{-1}) [62]. The SOC accumulation rate averaged 0.23 t C ha^{-1} yr^{-1} over the period 2009–2010 in the present study, generally lower than previous estimates in some double-rice paddy soils under short- or long-term chemical N fertilizer application [30,61,62]. However, it falls within the SOC sequestration rate range of 0.13–2.20 t C ha^{-1} yr^{-1} estimated by Pan et al. [63]. Lower levels of SOC in the present study could be attributed to differences in crop rotation systems. The decomposition rate of SOC in the single rice paddy-upland rotation system was higher than double rice-cropping paddy soils primarily dominated by surface waterlogging [30].

NT significantly increased SOC contents relative to CT at 0–5 cm depth but not at 0–20 cm depths (Table 3). A possible reason could be the return of moderately higher residues and root

biomass to the soil surface, instead of migrating deeper into the soil under NT. CT incorporates residues into a greater soil volume [64,65], resulting in relatively high SOC contents at deeper depths than NT [44]. Consequently, the lower SOC content at deeper depths under NT may weaken the tillage effects on SOC contents in the 20 cm layer. Similar observations were reported by other researchers [66,67].

The present results indicate that tillage has different effects on SOC sequestration based on the soil sampling depth (Table 3). NT significantly increased SOC compared with CT only at 0–5 cm but not at 0–20 cm. This result is likely caused by the residue accumulation on the soil surface. Similar results were observed by Wright et al. [65] and Wright and Hons [68]. Our results were in contrast to the results reported by other researchers [69–71]. Nyamadzawo et al. [69] found that NT had more SOC than CT at 0–20 cm depth. Deen and Kataki [70] reported that, compared to CT, NT increased SOC storage only for the surface layer (0–5 cm) but had significantly lower SOC for the entire soil profile (0–40 cm). However, Christopher et al. [71] found that NT had similar amounts of SOC to CT across the entire soil profile (0–60 cm). We can speculate that the potential of SOC sequestration under NT paddy fields in the present study may be overestimated at deeper soil depths (>20 cm). Further research is needed to understand the sequestration of SOC under CT and NT systems based on different soil sampling depths.

Materials and Methods

Site Description

The experimental site is situated at an experimental farm in Zhonggui Country, Dafashi Town, Wuxue City, Hubei Province, China (29°55′ N, 115°30′ E). This region has a humid mid-subtropical monsoon climate, an average annual temperature of 16.8°C, and a mean annual precipitation of 1360.6 mm. Rainfall mostly occurred between April and August in the past 5 years. The paddy field soil is a hydromorphic paddy soil, which is silty clay loam (3% sand, 50% silt, and 47% clay) derived from quaternary yellow sediment. The main soil properties (0–20 cm depth) of the site are as follows: pH (extracted by H_2O; soil: water = 1:2.5), 6.58; organic C, 18.29 g kg^{-1}; total N, 1.05 g kg^{-1}; NO_3^-–N, 4.37 mg kg^{-1}; NH_4^+–N, 2.43 mg kg^{-1}; total P, 0.70 g kg^{-1}; Bray-P, 3.65 mg kg^{-1}; and available K (extracted by CH_3COONH_4), 111 mg kg^{-1}.

The rice variety planted was *Liangyoupeijiu* (*Oryza sativa* L.), a mid-season rice variety. The experimental site was cultivated with a rape (*Brassica napus*)–rice (*Oryza sativa* L.) rotation. Rice was directly seeded from May to October each year and rape was planted from October to May the following year for the past 30 years.

Experimental Design

Implementation of NT was initiated in 2006. Treatments were established following a split-plot design of a randomized complete block with standard tillage practices in the main plot and N fertilizers in the sub-plots. Each treatment had three replications. Each plot was isolated with a plastic film driven to a depth of 40 cm along the inner edge of the field ridge (30 cm at the base and 30 cm in height) in order to prevent lateral water movement caused by either leakage or permeable lateral flow. Each plot had an area of 45 m^2 and an inlet for irrigation as well as an outlet for drainage. Two water meters were installed at the inlet and outlet to record water flow.

The weeds were controlled by spraying 36% glyphosate at 3 L ha^{-1} on June 4, 2009 and June 10, 2010. The field was then flooded on June 5, 2009 and June 12, 2010, respectively. Thereafter, the CT treatments were cultivated to 8–10 cm depth by hoeing, and were subsequently mouldboard ploughed twice to 20 cm depth before sowing. There was no tillage in the NT-treated subplots. Before sowing, rice seeds were soaked in water for 12 h and mixed with Dry-Raised Nurse (provided by Yangzhou Lvyuan Biochemial Co., LTD), a biological seed coat agent that can promote rice seed germination at a ratio of 1:3 ratio. Rice seeds were sown manually at a rate of 22.5 kg ha^{-1} on June 8, 2009 and June 12, 2010. The crops were then harvested on October 8, 2009 and October 17, 2010, respectively. Commercial inorganic N–phosphorus (P)–potassium (K) fertilizer (15% N, 15% P_2O_5, 15% K_2O), urea (46% N), single superphosphate (16% P_2O_5) and potassium chloride (60% K_2O) were used to furnish 210 kg N ha^{-1}, 135 kg P_2O_5 ha^{-1} and 240 kg K_2O ha^{-1} during the rice growing season. Nitrogen fertilizers were broadcast at a rate of 84 kg N ha^{-1} as basal fertilizers immediately after sowing. The P and K fertilizers were only used as basal fertilizers immediately after seeding. The remaining N fertilizers were split into three doses of 42 kg N ha^{-1} on June 24, July 19 and August 12, 2009, as well as June 25, July 21, and August 14, 2010. The irrigation and application of pesticide were the same in all experimental treatments. According to local conventional irrigation-drainage practices, the plots were irrigated immediately upon the germination of rice seeds. Thereafter, the plots were reirrigated to a depth of 10 cm whenever that the water depth decreased to 1 cm to 2 cm above the soil surface during the growing season. The fields were not flooded for the entire 2 weeks before the rice was harvested.

Methane Emission

Closed steel cylinders with diameters of 58 cm and height of 110 cm were used to quantify the CH_4 fluxes from all plots during the rice growing seasons [72]. CH_4 gas samples were collected from June 9 to October 8, 2009 and from June 12 to October 17, 2010. Two permanent rings were placed below water level to create a seal in each treatment plot and chambers were temporarily placed on these rings to measure the gas fluxes. Fans installed on the tops of the chambers were run for 1 min to mix the air within the chamber before each gas sample was taken. Then the gases in the chamber were drawn off with a syringe and immediately transferred into a 20 ml vacuum glass container. Three gas samples from the chamber headspace were collected at 8 min intervals using 25 ml plastic syringes during a half-hour period. Measurements of CH_4 fluxes were conducted twice a day in the morning (9:00 to 11:00) and afternoon (15:00 to17:00). The morning and afternoon measurements from each plot were then averaged and considered as representative of that plot. The gas samples were collected 1 day after each N fertilizer application, and weekly.

We measured CH_4 concentrations with gas chromatograph meter (Shimadzu GC-14B), fitted with a 6′ to 1/8′ stainless steel column (Porapack N, length×inner diameter: 3 m×2 mm) and a flame ionization detector as previously presented [73]. For determination of CH_4, N_2 (flow rate: 330 ml min^{-1}), H_2 (flow rate: 30 ml min^{-1}) and zero air (flow rate: 400 ml min^{-1}) were used as the carrier, fuel, and supporting gas, respectively. The temperatures of the column, injector, and detector were set at 55, 100, and 200°C, respectively. The changes in CH_4 concentrations remained linear throughout the sampling period. The gas emission flux was calculated from the difference in the gas concentration according to the equation given by Zheng et al. [74]:

$$F = \rho \times h \times dC/dt \times 273 \div (273 + T)$$

where F is the gas emission flux (mg m^{-2} h^{-1}), ρ is the gas density at the standard state, h is the height of the chamber above the soil (m), C is the gas mixing ratio concentration (mg m^{-3}), and T is the mean air temperature inside the chamber during sampling.

Carbon Dioxide Emission

The soil CO_2 flux was measured using the soil respiration method described by Parkinson [75]. In this method, a cylinder static chamber of 20 cm in diameter and 30 cm in height was placed on the soil and the rate of increase in CO_2 concentration within the chamber was monitored using a LI–6400 portable photosynthesis analyzer (Li–Cor Inc., Lincoln, NE). We measured soil fluxes from 2 h measurements between 9:00 and 11:00 (a representative time of daily averages in this region described by Lou et al. [76]). The soil CO_2 flux in the present study was measured in this way. The soil CO_2 fluxes were measured 17 times at weekly intervals from June 9 to October 6, 2009, and 17 times at a 7–10 day interval from June 13 to October 17, 2010.

Each soil CO_2 flux was determined every 1 min for 20 min. Three measurements were performed for each plot on each sampling day, and soil CO_2 flux was the average of three individual measurements. Meteorological data were collected from the weather station in Wuxue City, 1 km from the experimental site.

Changes in the concentration with the sampling time were used to calculate the soil CO_2 flux rate. The flux rate was calculated by simple linear regression when the concentration of gas inside the chamber varied linearly over time. Otherwise, the rate flux was calculated by nonlinear regression [77]. For the nonlinear regression, a model based on Fick's law was fitted to the chamber data:

$$C(t) = C_{\max} - (C_{\max} - C_0) \times \exp^{(-k \times t)}$$

where the regression parameter C_0 is the air concentration at time $t = 0$; $C_{\max}$ is the maximum concentration that can be reached in the chamber, and k is a rate constant. The values of $C_{\max}$, C_0 and k were estimated iteratively using the observed concentration versus time data. Methane and CO_2 fluxes were both expressed as mg m^{-2} h^{-1}.

The cumulative CH_4 and CO_2 emissions were calculated for each plot by linearly interpolating the gas emissions between sampling dates under the assumption that the measured fluxes represented the average daily fluxes. The cumulative emissions were calculated according to the following equation:

$$CE = \sum \left\{ (F_i + F_{i+1}) \div 2 \times 10^{-3} \times t \times 24 \right\}$$

where CE is the cumulative emissions (g m^{-2}), F_i and F_{i+1} are the measured fluxes of two consecutive sampling days (mg m^{-2} h^{-1}), and t is the number of days between two consecutive sampling days (d).

Sampling and Analytical Methods

Paddy soil samples (0–5 or 0–20 cm depth) were collected using a soil sampler with a diameter of 5 cm at eight random positions in each plot 1 day before the field was tilled and immediately after rice was harvested. The SOC were determined by dichromate oxidation and titration with ferrous ammonium sulfate [78]. The soil bulk density was determined by the method as described by Bao [78]. Soil bulk density samples for 0–5 or 0–20 cm soil layers were collected from each plot using metallic cores of 5.3 cm in diameter and 5.0 cm tall or 5.3 cm in diameter and 20 cm tall. Three soil cores were collected from each plot at 0–5 cm depth. The soil bulk density was computed as the weight to volume ratio of oven-dried (105°C) soil. Each measurement was replicated thrice. The SOC density (kg C ha^{-1}) at the soil depth was evaluated by the methods described by Lu et al. [19]. The SOC density was calculated as follows:

$$SOCD = SOCC \times BD \times H$$

where *SOCD* and *SOCC* are the SOC density (kg C ha^{-1}) and SOC concentration (g kg^{-1}), respectively; *BD* is the soil bulk density, and *H* is the soil sampling depth in the paddy field.

Statistical Analysis of Data

The SPSS 16.0 analytical software package was used for all statistical analyses. All data (mean±SE, $n = 3$) were checked for normal distribution. Statistical analysis was performed by two–way ANOVA to analyze the effects of N fertilizer and tillage on the CH$_4$ and CO$_2$ flux, as well as other C indices, using the SPSS general linear model procedure. The least significant difference (LSD) was calculated only when the ANOVA F-test was found to be significant at the $P<0.05$ probability level.

Author Contributions

Conceived and designed the experiments: LCF CCG. Performed the experiments: ZDN KZK ZZS. Analyzed the data: LCF ZDN KZK CCG. Contributed reagents/materials/analysis tools: ZDN WJP CML. Wrote the paper: LCF ZDN CCG.

References

1. IPCC (2007) Climate change 2007. impacts Climatechange, ed. adaptation and vulnerability. Working Group II. Geneva, Switzerland: IPCC.
2. Reiner W, Milkha SA (2000) The role of rice plants in regulating mechanisms of methane missions. Biol Fertil Soils 31: 20–29.
3. IRRI, International Rice Research Institute (2004) Rice Stat. Database, Los Banõs, Philippines. Available: http://www.irri.org/science/ricestat/index.asp. Accessed 2011 Sep 13.
4. OECD (2000) Environmental indicators for agriculture methods and results. Executive Summary, Paris.
5. Chu H, Hosen Y, Yagi K (2007) NO, N$_2$O, CH$_4$ and CO$_2$ fluxes in winter barley field of Japanese Andisol as affected by N fertilizer management. Soil Biol Biochem 39: 330–339.
6. Guo J, Zhou C (2007) Greenhouse gas emissions and mitigation measures in Chinese agroecosystems. Agric Forest Meteorol 142: 270–277.
7. Harada H, Kobayashi H, Shindo H (2007) Reduction in greenhouse gas emissions by no-tilling rice cultivation in Hachirogata polder, northern Japan: life-cycle inventory analysis. Soil Sci Plant Nutr 53: 668–677.
8. Oorts K, Merckx R, Gréhan E, Labreuche J, Nicolardot B (2007) Determinants of annual fluxes of CO$_2$ and N$_2$O in long-term no-tillage and conventional tillage systems in northern France. Soil Till Res 95: 133–148.
9. Liang W, Shi Y, Zhang H, Yue J, Huang GH (2007) Greenhouse gas emissions from northeast China rice fields in fallow season. Pedosphere 17(5): 630–638.
10. Minami K (1995) The effect of nitrogen fertilizer use and other practices on methane emission from flooded rice. Fertil Res 40: 71–84.
11. Paustian K, Collins HP, Paul EA (1997) Management controls on soil carbon. In: Paul EA, Paustian K, Elliot ET, Cole CV (Eds.) *Soil Organic Matter in Temperate Agroecosystems - Long-term Experiments in North America*, CRC Press, Boca Raton, FL, 15–49.
12. Lee DK, Doolittle JJ, Owens VN (2007) Soil carbon dioxide fluxes in established switch grass land managed for biomass production. Soil Biol Biochem 39: 178–186.
13. Iqbal J, Hu RG, Lin S, Hatano R, Feng ML, et al. (2009) CO$_2$ emission in a subtropical red paddy soil (Ultisol) as affected by straw and N fertilizer applications: a case study in Southern China. Agric Ecosyst Environ 131: 292–302.
14. Xiao Y, Xie G, Lu G, Ding X, Lu Y (2005) The value of gas exchange as a service by rice paddies in suburban Shanghai, PR China. Agric Ecosyst Environ 109: 273–283.
15. Burton AJ, Pregitzer KS, Crawford JN, Zogg GP, Zak DR (2004) Simulated chronic NO$_3$-deposition reduces soil respiration in Northern hardwood forests. Global Change Biol 10: 1080–1091.
16. DeForest JL, Zak DR, Pregitzer KS, Burton AJ (2004) Atmospheric nitrate deposition, microbial community composition, and enzyme activity in Northern hardwood forests. Soil Sci Soc Am J 68: 132–138.
17. Lal R (2004) Soil carbon sequestration impacts on global climate change and food security. Science 304: 1623–1627.
18. DeLuca TH, Zabinski CA (2011) Prairie ecosystems and the carbon problem. Front Ecol Environ 9: 407–413.
19. Lu F, Wang XK, Han B, Ouyang ZY, Duan XN, et al. (2009) Soil carbon sequestrations by nitrogen fertilizer application, straw return and no-tillage in China's cropland. Global Change Biol 15: 281–305.
20. Tang XH, Shao JA, Gao M, Wei CF, Xie DT, et al. (2007) Effects of conservational tillage on aggregate composition and organic carbon storage in purple paddy soil. Chin J Appl Ecol 18: 1027–1032. (in Chinese).
21. Gao YJ, Zhu PL, Huang DM, Wang ZM (2000) Long-term impact of different soil management on organic matter and total nitrogen in rice-based cropping system. Soil Environ Sci 9: 27–30. (in Chinese).
22. Six J, Feller C, Denef K, Ogle SM, Moraes Sa JC, et al. (2002) Soil organic matter, biota and aggregation in temperate and tropical soils – effects of no-tillage. Agronomie 22: 755–775.
23. Su YZ (2007) Soil carbon and nitrogen sequestration following the conversion of cropland to alfalfa forage land in northwest China. Soil Till Res 92: 181–189.
24. He YY, Zhang HL, Sun GF, Tang WG, Li Y, et al. (2010) Effect of different tillage on soil organic carbon and the organic carbon storage in two-crop paddy field. J Agro-Environ Sci 29(1): 200–204. (in Chinese).
25. Angers DA, Eriksen-Hamel NS (2008) Full-inversion tillage and organic carbon distribution in soil profiles: a meta-analysis. Soil Sci Soc Am J 72: 1370–1374.
26. Baker JM, Ochsner TE, Venterea RT, Griffis TJ (2007) Tillage and soil carbon sequestration – what do we really know? Agric Ecosyst Environ 118: 1–5.
27. Green CJ, Blackmer AM, Horton R (1995) Nitrogen effects on conservation of carbon during corn residue decomposition in soil. Soil Sci Soc Am J 59: 453–459.
28. Alvarez R (2005) A review of nitrogen fertilizer and conservation tillage effects on soil organic carbon storage. Soil Use Manage 21: 38–52.
29. Tong CL, Xiao HA, Tang GY, Wang HQ, Huang TP, et al. (2009) Long-term fertilizer effects on organic carbon and total nitrogen and coupling relationships of C and N in paddy soils in subtropical China. Soil Till Res 106: 8–14.
30. Shang QY, Yang XX, Gao CM, Gao CM, Wu PP, et al. (2011) Net annual global warming potential and greenhouse gas intensity in Chinese double rice-cropping systems: a 3-year field measurement in long-term fertilizer experiments. Global Change Biol 17: 2196–2210.
31. Wang MX, Li J (2002) CH$_4$ emission and oxidation in Chinese rice paddies. Nutr Cy Agroecosyst 64: 43–55.
32. Neue HU, Roger PA (2000) Rice agriculture: factors controlling emissions. In: Khalil MAK (ed.), Atmospheric Methane. Its Role in the Global Environment, 134–169.
33. Lindau CW, Bollich PK (1993) Methane emissions from Louisiana first and Ratoon crop rice. Soil Sci 156: 42–48.
34. Wassmann R, Schüetz H, Papen H, Rennenberg H, Seiler W, et al. (1993) Quantification of methane emissions from Chinese rice fields (Zhejiang Province) influenced by fertilizer treatment. Biogeochemistry 20: 83–101.
35. Lu WF, Chen W, Duan WM, Lu Y, Lantin RS, et al. (2000) Methane emission and mitigation options in irrigated rice fields in southeast China. Nutr Cy Agroecosyst 58: 65–73.
36. Schütz H, Holzapfel-Pschorn A, Conrad R, Rennenberg H, Seiler W (1989) A 3-year continuous record on the influence of daytime, season and fertilizer treatment on methane emission rate from an Italian rice paddy. J Geophys Res 94: 16406–16416.
37. Ussiri DAN, Lal R, Jarecki MK (2009) Nitrous oxide and methane emissions from long term tillage under a continuous corn cropping system in Ohio. Soil Till Res 104: 247–255.
38. Hütsch BW (1998) Tillage and land use effects on methane oxidation rates and their vertical profiles in soil. Biol Fertil Soils 27: 284–292.
39. Ball BC, Scott A, Parker JP (1999) Field N$_2$O, CO$_2$ and CH$_4$ fluxes in relation to tillage, compaction and soil quality in Scotland. Soil Till Res 53: 29–39.
40. Dick RP (1992) A review: long term effects of agricultural systems on soil biochemical and microbial parameters. Agric Ecosyst Environ 40: 25–36.
41. Wilson HM, Al-Kaisi MM (2008) Crop rotation and nitrogen fertilization effect on soil CO$_2$ emissions in central Iowa. Appl Soil Ecol 39: 264–270.
42. Almaraz JJ, Zhou XM, Mabood F, Madramootoo C, Rochette P, et al. (2009) Greenhouse gas fluxes associated with soybean production under two tillage systems in southwestern Quebec. Soil Till Res 104: 134–139.
43. Liu XJ, Mosier AR, Halvorson AD, Zhang FS (2006) The impact of nitrogen placement and tillage on NO, N$_2$O, CH$_4$ and CO$_2$ fluxes from a clay loam soil. Plant Soil 280: 177–188.

44. Li CF, Kou ZK, Yang JH, Cai ML, Wang JP, et al. (2010) Soil CO_2 fluxes from direct seeding rice fields under two tillage practices in central China. Atmos Environ 44: 2696–2704.

45. Reicosky DC, Dugas WA, Torbert HA (1997) Tillage–induced carbon dioxide loss from different cropping systems. Soil Till Res 41: 105–118.

46. Omonode RA, Vyn TJ, Smith DR, Hegymegi P, Gál A (2007) Soil carbon dioxide and methane fluxes from long–term tillage systems in continuous corn and corn–soybean rotations. Soil Till Res 95: 182–195.

47. Mosier AR, Halvorson AD, Peterson GA, Robertson GP, Sherrod L (2005) Measurement of net global warming potential in three agroecosystems. Nutr Cy Agroecosyst 72: 67–76.

48. Roberson T, Reddy KC, Reddy SS, Nyakatawa EZ, Raper RL, et al. (2008) Carbon dioxide efflux from soil with poultry litter applications in conventional and conservation tillage systems in northern Alabama. J Environ Qual 37: 535–541.

49. Xu H, Cai ZC, Jia ZJ, Tsuruta H (2000) Effect of land management in winter crop season on CH_4 emission during the following flooded and rice growing period. Nutr Cy Agroecosyst 58: 12–18.

50. Mitra S, Aulakh MS, Wassmann R, Olk DC (2005) Triggering of methane production in rice soils by root exudates: effects of soil properties and crop management. Soil Sci Soc Am J 69: 563–570.

51. Zhan M, Cao CG, Wang JP, Jiang Y, Cai ML, et al. (2011) Dynamics of methane emission, active soil organic carbon and their relationships in wetland integrated rice-duck systems in Southern China. Nutr Cy Agroecosyst 89: 1–13.

52. Butterbach-Bahl K, Papen H, Rennenberg H (1997) Impact of gas transport through rice cultivars on methane emission from rice paddy fields. Plant Cell Environ 20: 175–1183.

53. Drury CF, Reynolds WD, Tan CS, Welacky TW, Calder W, et al. (2006) Emissions of nitrous oxide and carbon dioxide: influence of tillage type and nitrogen placement depth. Soil Sci Soc Am J 70: 570–581.

54. Nayak P, Patel D, Ramakrishnan B, Mishra AK, Samantaray RN (2009) Long-term application effects of chemical fertilizer and compost on soil carbon under intensive rice–rice cultivation. Nutr Cy Agroecosyst 83: 259–269.

55. Wang CJ, Pan GX, Tian YG, Li LQ, Zhang XH, et al. (2010) Changes in cropland topsoil organic carbon with different fertilizations under long-term agro-ecosystem experiments across mainland China. Sci China Life Sci, 53: 858–867.

56. López-Bellido RJ, Fontán JM, López-Bellido FJ, López-Bellido L (2010) Carbon sequestration by tillage, rotation, and nitrogen fertilization in a Mediterranean Vertisol. Agron J 102: 310–318.

57. Li JT, Zhang B (2007) Paddy soil stability and mechanical properties as affected by long-term application of chemical fertilizer and animal manure in subtropical China. Pedosphere 17(5): 568–579.

58. Halvorson AD, Wienhold BJ, Black AL (2002) Tillage, nitrogen, and cropping system effects on soil carbon sequestration. Soil Sci Soc Am J 66: 906–912.

59. Khan SA, Mulvaney RL, Ellsworth TR, Boast CW (2007) The myth of nitrogen fertilization for soil carbon sequestration. J Environ Qual 36: 1821–1832.

60. Lou YL, Xu MG, Wang W, Sun XL, Zhao K (2011) Return rate of straw residue affects soil organic C sequestration by chemical fertilization. Soil Till Res 113: 70–73.

61. Pan GX, Zhou P, Li ZP, Smith P, Li LQ, et al. (2009) Combined inorganic/organic fertilization enhances N efficiency and increases rice productivity through organic carbon accumulation in a rice paddy from the Tai Lake region, China. Agr Ecosyst Environ 131: 274–280.

62. Wang SX, Liang XQ, Luo QX, Fan F, Chen YX, et al. Fertilization increases paddy soil organic carbon density. J Zhejiang Uni Sci B Doi:10.1631/jzus.B1100145, in press.

63. Pan G, Li L, Wu L, Zhang X (2003) Storage and sequestration potential of topsoil organic carbon in China's paddy soils. Global Change Biol 10: 79–92.

64. Six J, Elliot ET, Paustian K (1999) Aggregate and soil organic matter dynamics under conventional and no-tillage systems. Soil Sci Soc Am J 63: 1350–1358.

65. Wright AL, Hons FM, Lemon RG, McFarland ML, Nichols RL (2008) Microbial activity and soil C sequestration for reduced and conventional tillage cotton. Appl Soil Ecol 38: 168–173.

66. Causarano HJ, Franzluebbers AJ, Reeves DW, Shaw JN (2006) Soil organic carbon sequestration in cotton production systems of the southeastern United States: a review. J Environ Qual 35: 1374–1383.

67. Dolan MS, Clapp CE, Allmaras RR, Baker JM, Molina JAE (2006) Soil organic carbon and nitrogen in a Minnesota soil as related to tillage, residue and nitrogen management. Soil Till Res 89: 221–231.

68. Wright AL, Hons FM (2005) Carbon and nitrogen sequestration and soil aggregation under sorghum cropping sequences. Biol Fertil Soils 41: 95–100.

69. Nyamadzawo G, Chikowo R, Nyamugafata P, Nyamangara J, Giller KE (2008) Soil organic carbon dynamics of improved fallow-maize rotation systems under conventional and no-tillage in Central Zimbabwe. Nutr Cy Agroecosyst 81: 85–93.

70. Deen K, Kataki PK (2003) Carbon sequestration in a long-term conventional versus conservation tillage experiment. Soil Till Res 74: 143–150.

71. Christopher S, Lal R, Mishra U (2009) Long-term no-till effects on carbon sequestration in the Midwestern U.S. Soil Sci Soc Am J 73: 207–216.

72. Crill PM, Bartlett KB, Harriss RC, Gorham E, Verry ES, et al. (1988) Methane flux from Minnesota peatlands. Global Biogeochem Cy 2: 371–384.

73. Li CF, Cao CG, Wang JP, Zhan M, Yuan WL, et al. (2009) Nitrous oxide emissions from wetland rice–duck cultivation systems in southern China. Arch Environ Contam Toxicol 56: 21–29.

74. Zheng XH, Wang MX, Wang YS, Shen RX, Li J (1998) Comparison of manual and automatic methods for measurement of methane emission from rice paddy fields. Adv Atmos Sci 15(4): 569–579.

75. Parkinson KJ (1981) An improved method for measuring soil respiration in the field. J Appl Ecol 18: 221–228.

76. Lou YS, Li ZP, Zhang TL (2003) Carbon dioxide flux in a subtropical agricultural soil of China. Water Air Soil Pollut 149: 281–293.

77. Kroon PS, Hensen A, van den Bulk WCM, Jongejan PAC, Vermeulen AT (2008) The importance of reducing the systematic error due to non-linearity in N_2O flux measurements by static chambers. Nutr Cy Agroecosyst 82: 175–186.

78. Bao SD (2000) Analytical Method for Soil and Agricultural Chemistry. Beijing: Chinese Agriculture Press, 42–56. (in Chinese).

15

Testing the Paradox of Enrichment along a Land Use Gradient in a Multitrophic Aboveground and Belowground Community

Katrin M. Meyer[1,2]*, Matthijs Vos[3,4], Wolf M. Mooij[4,5], W. H. Gera Hol[1], Aad J. Termorshuizen[6], Wim H. van der Putten[1,7]

1 Department of Terrestrial Ecology, Netherlands Institute of Ecology NIOO-KNAW, Wageningen, The Netherlands, 2 Ecosystem Modelling, Faculty of Forest Sciences and Forest Ecology, University of Goettingen, Göttingen, Germany, 3 Department of Ecology & Ecosystem Modeling, Institute of Biochemistry and Biology, University of Potsdam, Potsdam, Germany, 4 Department of Aquatic Ecology, Netherlands Institute of Ecology NIOO-KNAW, Wageningen, The Netherlands, 5 Aquatic Ecology and Water Quality Management, Department of Environmental Sciences Group, Wageningen University, Wageningen, The Netherlands, 6 BLGG Research, Wageningen, The Netherlands, 7 Laboratory of Nematology, Wageningen University, Wageningen, The Netherlands

Abstract

In the light of ongoing land use changes, it is important to understand how multitrophic communities perform at different land use intensities. The paradox of enrichment predicts that fertilization leads to destabilization and extinction of predator-prey systems. We tested this prediction for a land use intensity gradient from natural to highly fertilized agricultural ecosystems. We included multiple aboveground and belowground trophic levels and land use-dependent searching efficiencies of insects. To overcome logistic constraints of field experiments, we used a successfully validated simulation model to investigate plant responses to removal of herbivores and their enemies. Consistent with our predictions, instability measured by herbivore-induced plant mortality increased with increasing land use intensity. Simultaneously, the balance between herbivores and natural enemies turned increasingly towards herbivore dominance and natural enemy failure. Under natural conditions, there were more frequently significant effects of belowground herbivores and their natural enemies on plant performance, whereas there were more aboveground effects in agroecosystems. This result was partly due to the "boom-bust" behavior of the shoot herbivore population. Plant responses to herbivore or natural enemy removal were much more abrupt than the imposed smooth land use intensity gradient. This may be due to the presence of multiple trophic levels aboveground and belowground. Our model suggests that destabilization and extinction are more likely to occur in agroecosystems than in natural communities, but the shape of the relationship is nonlinear under the influence of multiple trophic interactions.

Editor: Martin Heil, Centro de Investigación y de Estudios Avanzados, Mexico

Funding: This work was supported by the ALW-biodiversity programme of the NWO (Project ALWPB/05-02) and by the State of Lower Saxony (Ministry of Science and Culture; Cluster of Excellence "Functional Biodiversity Research"). The funders had no role in study design, data collection and analysis, decision to publish, or preparation of the manuscript.

Competing Interests: The authors have declared that no competing interests exist.

* E-mail: kmeyer5@uni-goettingen.de

Introduction

Land use change currently is the most important cause of terrestrial biodiversity loss [1]. One of the most outspoken changes is that nutrient-poor natural ecosystems are transformed into highly fertilized agroecosystems. Thus, predicting plant community responses to land use change requires understanding how plant species [2] and their associated aboveground and below-ground trophic networks [3] respond to changes in resource availability and vegetation structure. The paradox of enrichment predicts that fertilization leads to a destabilization of population dynamics. The ensuing boom-bust cycles may lead to the extinction of predators and/or their prey [4]. Testing the paradox of enrichment in a complex community with multiple trophic levels requires species exclusion experiments that are difficult to implement in the field or greenhouse, but are highly amenable to modeling approaches [5]. In line with claims for more integration between models and field studies in aboveground-belowground

interactions research [3], [6], we use a successfully validated simulation model [7] to test the paradox of enrichment along a land use intensity gradient including multiple trophic levels aboveground and belowground.

Predator-prey theory predicts that stable coexistence of predators and preys is possible if their population dynamics are not too closely coupled [8]. The stability of a predator-prey system can depend on factors such as searching efficiency of the predator or resource availability to the prey. According to the paradox of enrichment, high resource availability leads to an explosion of prey populations causing a sudden increase in predator populations which may first drive the prey and subsequently the predators to extinction. It is also possible that the predator goes locally extinct following the major population decline of its prey, after which the prey population explodes back up, now unchecked by predator-induced mortality. Stochastic processes determine whether only the predator or both predator and prey go extinct. This implies that enrichment increases uncertainty in how the relative performance of herbivores and natural enemies will turn out to affect the fate of plant populations. During land use change, fertilization often goes along with increased insect searching efficiencies, because searching for food plants and host species is facilitated in structurally poor agricultural monocultures (e.g. [9–11]). The Rosenzweig-MacArthur model predicts that high predator searching efficiencies will destabilize predator-prey systems [12], because high predator efficiency increases the coupling between predator and prey population dynamics. In the simple bi-trophic Rosenzweig-MacArthur model, this is the case when the predator zero-growth isocline is located at very low prey densities. Also in more complex food chain models [13], predator efficiency is an important factor affecting stability. We may therefore expect enhanced destabilization in agricultural systems where both fertilization and searching efficiencies are high. We also expect that destabilization will increase smoothly, in a non-abrupt way, if land use change is smooth and linear.

Predator-prey theory has been tested for two trophic levels, and occasionally for tritrophic systems [13–14]. However, plants are involved in both aboveground and belowground multitrophic level interactions [15] and much is unknown about the applicability of predator-prey theory to multiple trophic levels aboveground and belowground. Results obtained in a simple predator-prey system or food chain will not necessarily hold for more complex communities and food webs. In a trophic network with plants at the first trophic level, herbivores at the second, parasitoids at the third and hyperparasitoids at the fourth level, the natural enemies at the third trophic level can enhance agricultural production as biological control agents. The third trophic level is also central in the debate of top-down versus bottom-up control in natural food webs [16]. The efficiency of biological control depends on the relative susceptibility of herbivores and natural enemies to land use changes.

Disentangling the effects of aboveground and belowground control [17–18] could help to design more efficient and better targeted biological control schemes. Most studies comparing aboveground with belowground interactions showed dominant aboveground effects and weak belowground effects on plants [19–24]. On the other hand, a few studies indicated strongest trophic effects belowground [7], [25–26]. Predator-prey theory predicts that the impact of aboveground and belowground natural enemies of herbivores is strongly related to the tendency of these populations to cycle. Population fluctuations may be larger aboveground, because air is more permeable to movement than soil, so that insect searching efficiencies can be much higher. This may result in larger increases and declines in population density. This effect would be enhanced by fertilization, so that we expect an increased variability in aboveground versus belowground interactions with increasing land use intensity. This increased variability may increase the maximum effect that aboveground interactions can have on plant performance.

The overall aim of the present study was to investigate whether the predictions of the paradox of enrichment hold when multiple aboveground and belowground trophic levels and their searching efficiencies are included, and how the results relate to plant performance. We used herbivore-induced plant mortality as a proxy for system instability and assessed it along a land use intensity gradient from low to high fertilization levels and insect searching efficiencies. We derived three predictions from predator-prey theory and tested them using an individual-based simulation model of aboveground-belowground multitrophic interactions [7]: (1) Fertilization and increasing searching efficiencies lead to destabilization of the system, i.e. to increased plant mortalities, (2)

destabilization is stronger for aboveground than for belowground interactions, and (3) destabilization is changing smoothly, i.e. non-abruptly, with land use intensity. We found support for the first two predictions that destabilization increases with land use intensity and that it is stronger aboveground than belowground. However, contradicting the third prediction there were abrupt changes in destabilization with land use intensity.

Materials and Methods

The Model

The individual- and rule-based ABove-BElowground interactions model ABBE was developed to investigate plant performance in a food web of plant shoots, herbivores, parasitoids and hyperparasitoids aboveground and plant roots, herbivores and their antagonists belowground [7]. The model also included earthworms that have positive effects on plants without being part of the plant-based food web. Aboveground interactions with the plant are modeled on an individual-by-individual basis, while belowground interactions are considered at the level of populations [7]. One single plant individual is modelled at the centre of aboveground and belowground interactions in each simulation to allow comparisons with the experimental conditions in Soler et al. [27], which provided most of the parameter values for ABBE.

The plant is simulated according to the wild cruciferous plant *Brassica nigra* L. (Brassicaceae), the aboveground herbivores are represented by the specialist chewing larvae of *Pieris brassicae* L. (Lepidoptera: Pieridae) and the belowground herbivores are represented by the specialized root-feeding larvae of the cabbage root fly *Delia radicum* L. (Diptera: Anthomyiidae). Aboveground, the third and fourth trophic levels are parameterized according to the gregarious koinobiont endoparasitoid *Cotesia glomerata* L. (Hymenoptera: Braconidae) and its secondary hyperparasitoid, the solitary idiobiont *Lysibia nana* Gravenhorst (Hymenoptera: Ichneumonidae). As belowground third trophic level, we added the coleopteran egg predator *Aleochara billineata* Gyll. (Coleoptera: Staphylinidae) to the system of Soler et al. [27]. The earthworm component of the community was parameterized with data from the species *Aporrectodea caliginosa* Savigny (Lumbricidae) and *Octolasion tyrtaeum* Savigny (Lumbricidae) (see [7]).

The timeframe of the simulations is one growing season, the general time step is six weeks, space is not considered explicitly. The simulated processes in any one time step are nutrient pool replenishment, plant growth, earthworm action, shoot herbivore mortality, parasitoid mortality and reproduction, shoot biomass reduction by herbivores, parasitism of herbivores by parasitoids, hyperparasitoid mortality and reproduction, parasitism of parasitoids by hyperparasitoids, plant mortality due to shoot herbivory, belowground antagonist mortality and reproduction, root herbivore mortality due to antagonist attack, root herbivore natural mortality, root biomass reduction by root herbivores, plant mortality due to root herbivory, shoot and root herbivore reproduction [7]. We assume that the only cause of plant mortality is excessive herbivory and that shoot herbivore efficiency depends on parasitism, whereas parasitoid efficiency is independent of parasitism by hyperparasitoids [7]. Shoot herbivore efficiency is the proportional shoot biomass decrease per individual and parasitoid efficiency is parasitizing success probability (Table 1). Shoot herbivore presence increases shoot quality [27] and induces shoot volatile emission, increasing parasitism efficiency. Shoot quality is standardized to range from 0 to 1 and shoot volatile emission can only be turned on or off in the model. Individual herbivore body size depends on shoot quality while body sizes of higher trophic levels depend on the body size of their hosts. Egg

Table 1. Composition of the land use intensity gradient from low to high fertilization and searching efficiencies.[a]

Trophic level	Property	Model parameter [unit]	Parameter value	
			Low	High
Abiotic	Fertilization	Nutrient supply [proportion of 100% nutrient supply]	0	1
AG parasitoid	Efficiency	Parasitizing success probability [probability]	0.092	0.788
BG herbivore	Efficiency	Proportional root biomass decrease per individual [proportion]	0.0002	0.0667
BG antagonist	Efficiency	Maximum number of BG herbivore eggs killed per antagonist couple lifetime [numbers]	260	1438
AG herbivore	Load	Initial number of AG herbivores [numbers]	0	2580
AG herbivore	Mortality	AG herbivore natural mortality [probability]	0.9975	0.785
Plant	Quality increase due to herbivory	Shoot quality increase due to defense induction by AG herbivores [proportion of 100% quality]	0.93	0.0275

[a]The parameters plant quality and aboveground herbivore mortality are linearly decreasing with increasing land use intensity, the others are linearly increasing. All parameters were changed at the same time. AG – aboveground, BG – belowground.

load of aboveground trophic levels depends on body size. ABBE has successfully been validated [7] against independent experimental data (e.g. [27]). Further details on structure and parameterization of ABBE can be obtained from [5],[7].

Land Use Intensity Gradient

Plant mortality and biomass was investigated along a simulated increase in land use intensity consisting of fertilization and searching efficiency parameters (Table 1). These parameters represented a gradient from structurally complex, unfertilized natural systems to fertilized agricultural monocultures (see e.g. [9–11]). Greater searching efficiencies of insects were simulated by increasing insect numbers and efficiencies at all trophic levels, decreasing herbivore natural mortalities, and reducing the positive impact of herbivory on shoot quality (Table 1; the positive impact of herbivory on shoot quality was originally reported by [27] for nutrient-poor conditions). We included parameters of all trophic levels with the exception of hyperparasitoids and earthworms for which knowledge was too limited. We specified a meaningful range for each parameter (Table 1) and divided these ranges linearly into 40 steps to represent small changes from low to high land use intensity. The ranges were based on expert knowledge provided by the authors and their colleagues at the Department of Terrestrial Ecology at the NIOO-KNAW. During the simulations, all gradient parameters were changed at the same time to acknowledge the simultaneous change of fertilization levels and searching efficiencies along real land use gradients.

Simulated Removal Experiments

To investigate the direct and indirect influences of different trophic levels on plant performance along the land use intensity gradient, we designed removal scenarios. In each removal scenario, one trophic level was removed from the simulations (including earthworm removal). In total, this yielded seven removal scenarios including the standard scenario in which all levels were present. Removal can also be seen as a proxy for the application of a very effective pesticide. We performed simulated experiments with 1000 replicate runs per removal scenario and determined plant performance at the end of a growing season for each of the 40 steps along the land use intensity gradient. We measured plant performance as median shoot and root biomass of surviving plants at the end of a growing season and used plant mortality as an inverse proxy of plant performance and as

a measure of instability. Plant mortality was calculated as the proportion of replicate runs in which the plant did not survive until the end of the growing season.

Sensitivity Analysis of the Gradient Parameters

We conducted a sensitivity analysis to identify the gradient parameters with the greatest contribution to plant performance and to assess their relative contribution to the land use intensity gradient. Plant performance was measured as shoot biomass, root biomass and – inversely – as plant mortality. We ran 1000 replicate simulations each for every possible combination of plant performance measure and gradient parameter, varying the parameters one after the other within the ranges specified for the parameters. For each combination, we separately applied a linear regression model of the form performance ~ gradient parameter value × removal scenario. We accounted for the removal scenario in the model formula to obtain sensitivity estimates that are independent of removal scenario. Sensitivity of the respective plant performance measure was calculated as standardized regression coefficient of the respective parameter value. Regression coefficients were standardized through division by the respective standard error.

Statistical Analysis

To quantify the effects of trophic level removal on plant performance along the land use intensity gradient, we compared plant performance before and after removal with Pearson's Chi-squared test statistic in the case of plant mortality and Mann-Whitney U-tests in the cases of root and shoot biomass. We used non-parametric tests and calculated the medians of root and shoot biomass, because non-normal errors and heterogeneous variance precluded calculating parametric statistics based on averages. We calculated the test statistics separately for each removal scenario (always comparing with the standard scenario) and for each of the 40 steps along the land use gradient. All statistical analyses were carried out with the software package R 2.5.1 (R Development Core Team 2007).

Results

The sensitivity analysis showed that plant root and shoot biomass were by far most sensitive to the parameter nutrient supply (Fig. 1). Plant mortality was also sensitive to nutrient supply,

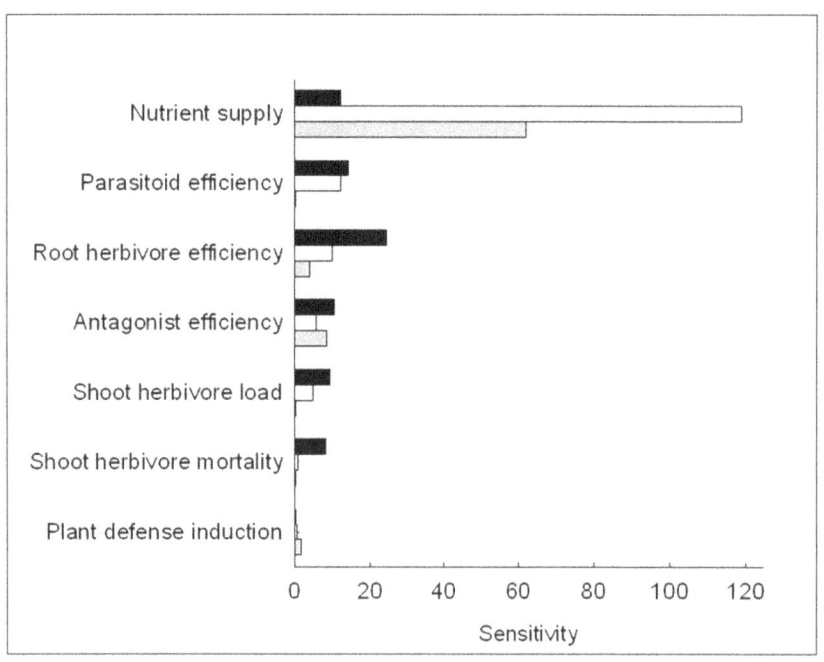

Figure 1. Sensitivity of model outputs to parameters of the land use intensity gradient. The model outputs were plant mortality (solid bars), average shoot biomass (empty bars), and average root biomass (grey bars). Sensitivity was derived from linear models of the form output ~ gradient parameter * removal scenario for all possible combinations of outputs and gradient parameters. Sensitivity was calculated as the regression slope of the parameter divided by the corresponding standard error for standardization.

but much more so to root herbivore efficiency and parasitoid efficiency. The sensitivity of all plant performance measures to all other tested parameters was relatively small.

Plant mortality as a proxy of instability spanned the whole range from 0 to 1 (Fig. 2a). Plant mortality increased steeply and non-linearly with increasing land use intensity, approaching a plateau of 100% mortality in the quarter representing the most fertilized agroecosystems (Fig. 2a). Apart from this quarter, shoot biomass increased steeply with land use intensity and root biomass first remained stable at a low level and then decreased towards medium levels of land use intensity (Figs. 2b, c). Beyond the plant mortality level of 100%, root and shoot biomass reached zero, because the plant had died by the end of the simulations (see Fig. 2a). The transition to 100% plant mortality was sharp for shoot biomass but fuzzier for root biomass where some extreme values occurred (Fig. 2c). The interquartile ranges showed that the variability of shoot and root biomass was greatest at medium land use intensity (Figs. 2b, c).

As expected, the removal of the second trophic level significantly reduced plant mortality at most land use intensities, i.e. herbivores diminished plant performance aboveground and belowground (see points below the zero-line in Figs. 3a, d). The removal of the third trophic level, the parasitoids and antagonists, significantly increased plant mortality aboveground and belowground (Figs. 3b, e), i.e. presence of the third trophic level increased plant performance. The removal of the fourth trophic level, the hyperparasitoids, and the earthworms, did not have a strong effect on plant mortality (Figs. 3c, f). Shoot herbivores were the only trophic level whose removal significantly reduced plant mortality below hundred per cent in highly fertilized agroecosystems (Figs. 3a). Aboveground herbivores had the strongest effect on plant mortality in these agricultural conditions, whereas belowground herbivores affected plant mortality more in natural to intermediate environments (compare position of

horizontal bars along the gradient in Figs. 3a, d). Neither aboveground nor belowground third trophic level removal affected plant mortality at the 100% plant mortality plateau (Figs. 3b, e). However, the effect of aboveground parasitoids was more prominent at intermediate conditions while a significant effect of belowground antagonists also occurred under more natural conditions (Figs. 3b, e).

The effects of the removal of trophic levels on shoot biomass generally were in line with the effects on plant mortality, only their direction was opposite (Fig. 4). For instance, as expected, removal of herbivores had a significant positive effect on shoot biomass (Figs. 4a, d), whereas removal of the third trophic level had a significant negative effect (Figs. 4b, e). Earthworms did not have a strong effect on plant mortality, and their removal had a small negative effect on shoot biomass (Fig. 4f). The effect of shoot herbivores on shoot biomass was much greater than that of belowground herbivores (Figs. 4a, d). This revealed the potential biomass level that the plant could reach without aboveground herbivory in agroecosystems (Fig. 4a). The significantly positive effect of aboveground natural enemies of herbivores on shoot biomass was limited to intermediate land use intensities (Fig. 4b). For plant mortality, this effect was significant also at lower land use intensities (Fig. 3b).

In contrast to shoot biomass, root biomass was almost exclusively affected by belowground species (Fig. 5). Removal of belowground herbivores had a strong significantly positive effect on root biomass, and this positive effect increased from natural to intermediate conditions; of course this effect disappeared at 100% plant mortality (Fig. 5d). The effect of removing the belowground third trophic level was smaller with respect to root biomass than to shoot biomass (Figs. 4e and 5e). Removal of earthworms had a small negative effect on root biomass that was constant over the land use intensity gradient and disappeared at 100% plant mortality (Fig. 5f). Aboveground parasitoids did not affect root

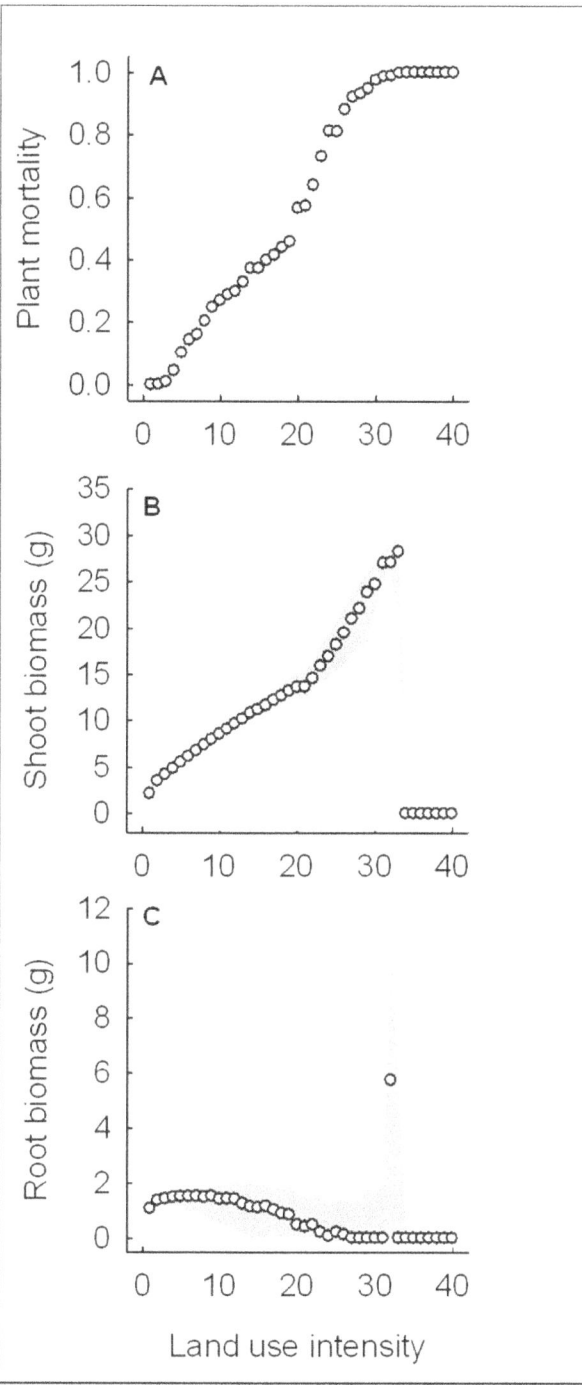

Figure 2. Plant performance with increasing land use intensity in the standard scenario. (a) Mortality and median (b) shoot and (c) root biomass of the model plant were based on 1000 simulation runs of the standard scenario. The land use intensity gradient was constructed by increasing nutrient supply, aboveground herbivore load, aboveground parasitoid efficiency, belowground herbivore efficiency, and belowground antagonist efficiency and decreasing plant defense induction and aboveground herbivore mortality in 40 steps within a meaningful range (Table 1). Mortality is the proportion of simulation runs in which the plant did not survive until the end of the growing season. Median shoot and root biomass are based on surviving plants only. Shaded areas indicate the interquartile range, i.e. contain 50% of the simulated data points. Note that the scales of the y-axes in (b) and (c) differ.

biomass (except for two single significant steps along the gradient, Fig. 5b).

Discussion

Using an individual-based simulation model, we tested predictions derived from predator-prey theory [12] and specifically the paradox of enrichment [4] in a range of systems from natural communities to agroecosystems including multiple trophic interactions aboveground and belowground. Plant mortality increased steeply with increasing land use intensity, supporting our first prediction that systems should destabilize when fertilization and predator searching efficiencies are increased under changes in land use. We also found support for our second prediction, because belowground effects were more often significant under natural conditions and aboveground effects more often at high land use intensity. We found abrupt, non-linear responses of plant performance along the land use intensity gradient, rejecting the third prediction that changes in instability should be smooth if the underlying land use intensity gradient is smooth.

Consistent with our first prediction, plant mortality as a measure of instability increased towards highly fertilized agroecosystems in the presence of all aboveground and belowground trophic levels. Hence, the paradox of enrichment is still applicable in a system with multiple trophic levels in aboveground and belowground subsystems. Our finding implies that parasitoids and antagonists were not able to offset the negative effect of herbivores on plant performance. This was not a trivial outcome, because the gradient parameters included increased efficiencies of both herbivores and their natural enemies and the removal experiments showed negative effects of herbivores on plant performance and positive effects of natural enemies. It seems that shoot herbivores alone were responsible for the 100% plant mortality plateau (Fig. 3a compared to panels b–f), confirming their great potential for damaging highly fertilized crops. Our results on increasing herbivory with increasing fertilization and insect searching efficiencies are in line with previous findings. With respect to searching efficiencies, it has been shown that decreasing proportions of non-crop area in agricultural landscapes leads to increasing plant damage by rape pollen beetles [28] which may be due to greater insect searching efficiencies in crop-dominated areas. With respect to fertilization, phytophagous arthropod abundances have been found to increase with increasing land use intensity in tropical agroecosystems [29]. Hence, control of herbivores by their natural enemies seems to fail in highly fertilized or structurally poor environments.

Aboveground parasitoid removal gave more detailed insight into the efficiency of control of herbivores by their natural enemies along the land use intensity gradient: Aboveground biological control failed at high land use intensity and under natural conditions, but was successful in reducing plant mortality at intermediate land use intensities. With respect to shoot biomass loss, aboveground biological control was effective within an even smaller range located even further away from natural conditions (compare horizontal bars in Figs. 3b and 4b). This agrees well with the finding that parasitism of rape pollen beetles is at its maximum at 50% of non-crop area, corresponding to intermediate land use intensity [28] and supports the notion that strong aboveground top-down control is probably not a natural phenomenon (*sensu* [30]). In contrast, belowground natural enemies were effective at both low and intermediate land use intensities, indicating that belowground top-down control is probably largely independent of environmental context as long as land use intensity is not extremely high. This corresponds to greater control of herbivores

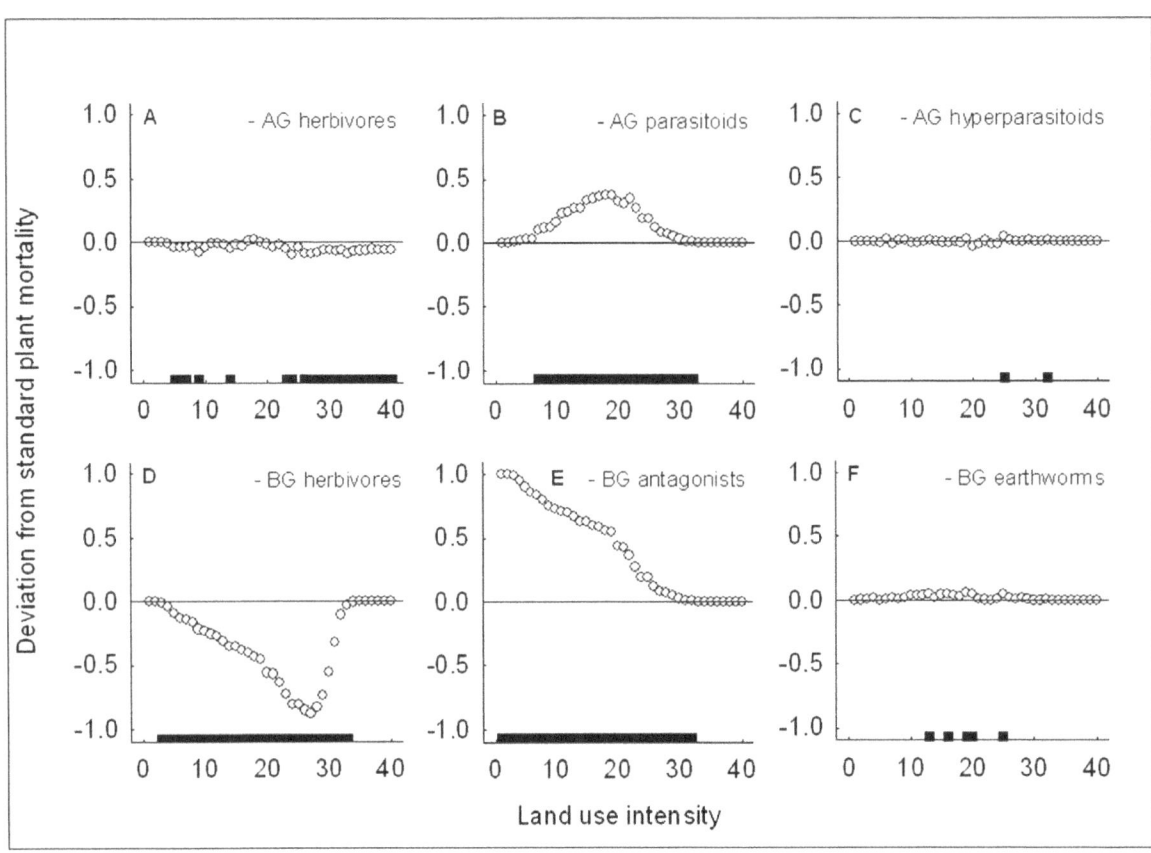

Figure 3. Plant mortality and multitrophic interactions with increasing land use intensity. Deviation from plant mortality in the standard scenario from low to high land use intensity after removal of aboveground (top row) and belowground (bottom row) trophic levels and earthworms, i.e. removal of aboveground (a) herbivores, (b) parasitoids, or (c) hyperparasitoids, or belowground (d) herbivores, (e) antagonists, or (f) earthworms. AG – aboveground, BG – belowground. The horizontal line at 0.0 represents the values of the standard scenario. Points indicate the difference in plant mortality between the respective removal scenario and standard scenario, and horizontal bars at the bottom of each graph indicate for each of the 40 points whether this difference is significant based on Pearson's Chi-squared test statistic ($P<0.05$). Points above the 0.0-line show a positive effect of removal of the respective trophic level on plant mortality, points below the line a negative one. The horizontal bars show that belowground effects are more important under natural conditions while aboveground effects are stronger towards agroecosystems.

exerted by soil organisms found in less managed systems [31]. This outcome may be accelerated if the attraction of natural enemies by volatiles emitted from damaged roots is considered [32–35] which may be integrated into future model versions. Our findings highlight that fertilization can to a certain extent be beneficial for successful biological control, especially aboveground, as long as it does not reach a level at which the herbivores and natural enemies are out of balance, allowing herbivores to dominate the system. Much of modern agriculture is characterized by such extremely high levels of fertilization, often necessitating the application of pesticides to correct for biological control failure if high yields are to be maintained.

Our results supported the second prediction that the strength of aboveground effects on system instability should increase towards agroecosystems. Effects of belowground trophic levels on plant mortality and biomass were more important at low land use intensities whereas aboveground trophic levels had a greater impact on plant performance at intermediate to high land use intensities. This was a general trend for all components of the community (compare the location of the horizontal bars in panels a versus d, b versus e and c versus f in Figs. 3, 4, and 5). Particularly the effect of shoot versus root herbivores is interesting and can be explained in terms of differences in population fluctuations. Shoot herbivores showed strong "boom-bust" pop-

ulation cycles with scramble competition [7], which can lead to rapid reduction of plant biomass in phases of booming population growth, but also rapid herbivore population crashes when resources are limited or parasitism gets too strong. In this case, the plant may survive while the herbivore population goes extinct. This is a typical case of too close coupling of predator and prey populations leading to instability and species extinction (*sensu* [8]). In contrast, root herbivores had much lower-amplitude population fluctuations. They contributed less to plant performance and mortality at high land use intensities. As an exception to this pattern, sedentary endoparasitic nematodes can produce great instability in agroecosystems by causing patches of decline in sugar beet and potato [36]. Nonetheless, the greater importance of belowground herbivores under more natural conditions is in line with findings of Blossey and Hunt-Joshi [37] and with evidence on the crucial role of root herbivores during early succession [38] which corresponds to low land use intensities.

Removal of the fourth trophic level, the hyperparasitoids, and of the earthworms did not have marked effects on plant performance. Only earthworm removal decreased shoot biomass with increasing land use intensity, which can be explained by model architecture: Presence of earthworms increases root and shoot biomass by a certain proportion of the current biomass and current biomass increased with increasing resource supply.

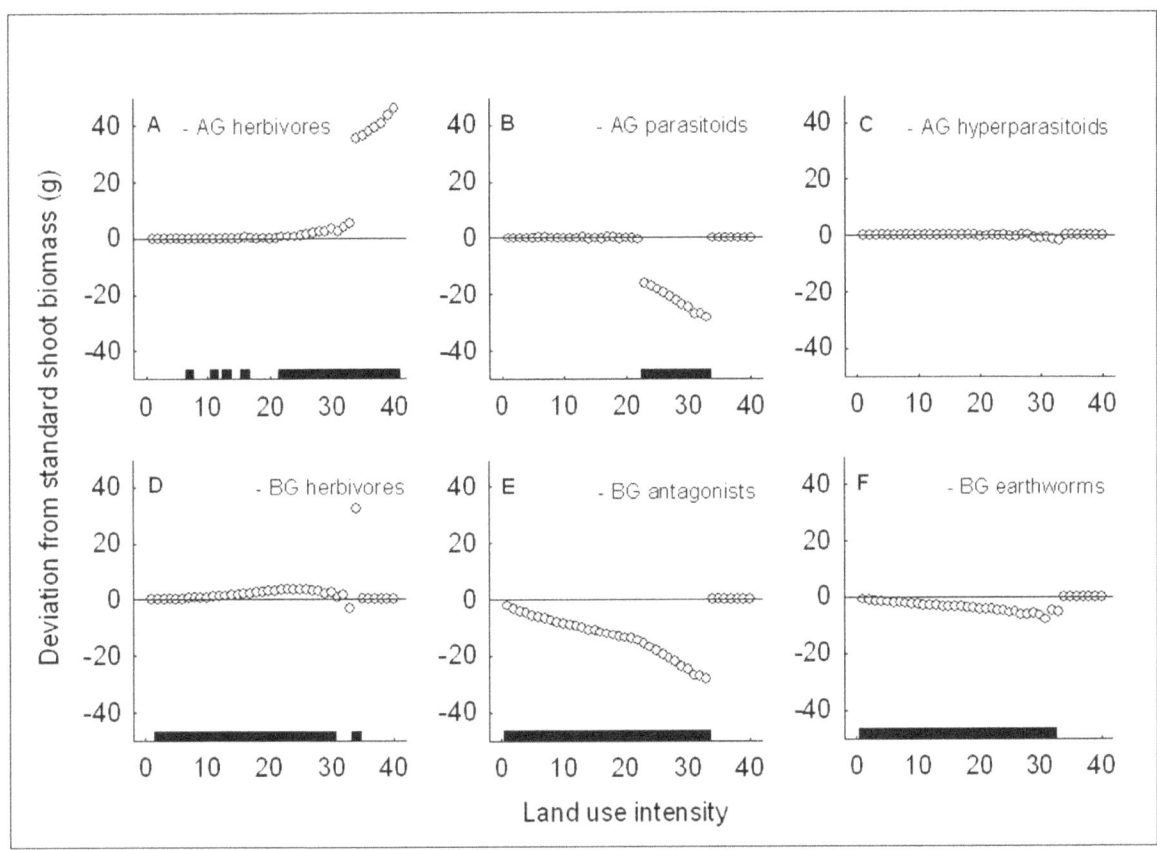

Figure 4. Shoot biomass and multitrophic interactions with increasing land use intensity. Deviation from median shoot biomass in the standard scenario from low to high land use intensity after removal of aboveground (top row) and belowground (bottom row) trophic levels and earthworms, i.e. removal of aboveground (a) herbivores, (b) parasitoids, or (c) hyperparasitoids, or belowground (d) herbivores, (e) antagonists, or (f) earthworms. AG – aboveground, BG – belowground. The horizontal line at 0.0 represents the values of the standard scenario. Points indicate the difference in shoot biomass between the respective removal scenario and standard scenario, and horizontal bars at the bottom of each graph indicate for each of the 40 points whether this difference is significant based on the Mann-Whitney U-test ($P<0.05$). Points above the 0.0-line show a positive effect of the removal of the respective trophic level on shoot biomass, points below the line a negative one. The horizontal bars show that belowground effects are more important under natural conditions while aboveground effects are stronger towards agroecosystems.

Our model analysis did not support our third prediction that the imposed smooth land use intensity gradient should result in smooth plant responses to trophic interactions. We observed abrupt non-linear effects of trophic level removal across all land use intensities and for all trophic levels. The most pronounced non-linearity was the rather suddenly reached 100% mortality plateau. Such enrichment-induced mortality and extinction have earlier been shown for plants in North America [39] and Great Britain [40] and for forest trees [41]. The few extreme values and the associated large variation in biomass that occurred at the transition to the 100% mortality plateau can be explained by model-inherent stochasticity that was inflated by the boom-bust dynamics of the herbivores. Just before reaching the 100% plant mortality plateau, these dynamics led to a broad range of possible biomass outcomes, also reflected in the few outliers at the same transition point in Figs. 4d and 5a–f. These observations are examples of alternative transient states [42] that are possible in dynamic systems under the same environmental conditions. Beyond the 100% plant mortality plateau, the great increase of shoot biomass in the absence of shoot herbivores revealed the immense potential growth that the plant could reach without aboveground herbivory at high land use intensity. This underlines how powerful aboveground herbivores with a boom-bust-population dynamical behavior can be as soon as fertilization exceeds

a certain limit. It confirms the notion that the predictability and reliability of plant performance will decrease under enrichment. However, it also shows the great potential of aboveground biological control for crop yields, if natural enemies are supported to act as powerful control agents.

In conclusion, our model suggests that destabilization and extinction are more likely in fertilized agroecosystems than in natural communities, confirming the predictions of the paradox of enrichment also for communities with multiple aboveground and belowground trophic levels. Abrupt destabilization of the multi-trophic system at high land use intensities implies that land use gradients as well as multiple trophic levels need to be considered to make reliable predictions on plant responses to land use change. These predictions will gain generality if more complex food webs with greater species diversities are considered in future modelling approaches. Based on our current predictions, it should be possible to tune land use in multitrophic agroecosystems such that the positive effect of natural enemies on plant performance is maximized while the risk of sudden transitions to high plant mortality is minimized.

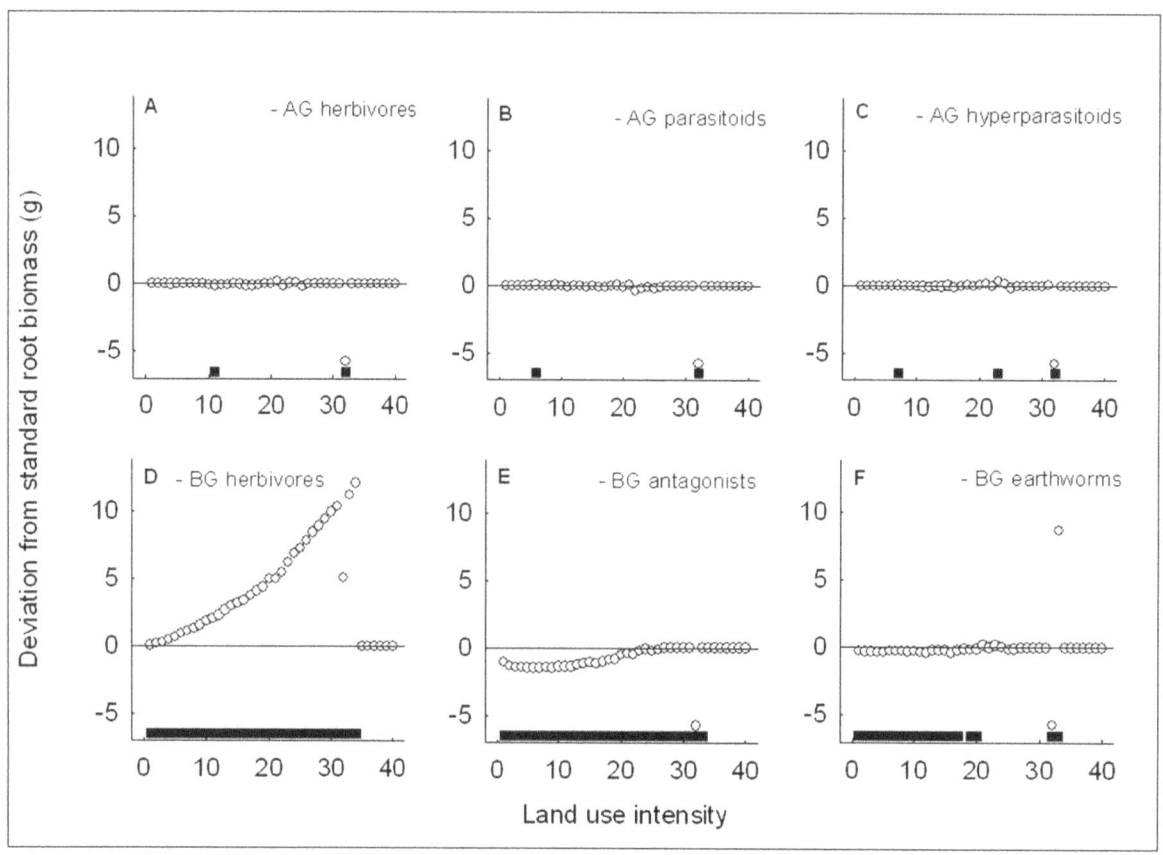

Figure 5. Root biomass and multitrophic interactions with increasing land use intensity. Deviation from median root biomass in the standard scenario from low to high land use intensity after removal of aboveground (top row) and belowground (bottom row) trophic levels and earthworms, i.e. removal of aboveground (a) herbivores, (b) parasitoids, or (c) hyperparasitoids, or belowground (d) herbivores, (e) antagonists, or (f) earthworms. AG – aboveground, BG – belowground. The horizontal line at 0.0 represents the values of the standard scenario. Points indicate the difference in root biomass between the respective removal scenario and standard scenario, and horizontal bars at the bottom of each graph indicate for each of the 40 points whether this difference is significant based on the Mann-Whitney U-test ($P<0.05$). Points above the 0.0-line show a positive effect of the removal of the respective trophic level on root biomass, points below the line a negative one. The horizontal bars show that belowground effects are dominating the effect on root biomass while aboveground effects are virtually absent.

Acknowledgments

We thank the members of the Multitrophic Interactions (now Terrestrial Ecology) Department for fruitful discussions. This is NIOO publication No. 5354.

Author Contributions

Conceived and designed the experiments: KMM MV WMM WHGH AJT WHvdP. Performed the experiments: KMM MV. Analyzed the data: KMM MV WMM WHGH AJT WHvdP. Wrote the paper: KMM MV WMM WHGH AJT WHvdP.

References

1. Sala OE, Chapin FS, Armesto JJ, Berlow E, Bloomfield J, et al. (2000) Biodiversity - Global biodiversity scenarios for the year 2100. Science 287: 1770–1774. Doi: 10.1126/science.287.5459.1770.

2. Temperton VM, Hobbs RJ, Nuttle T, Halle S (2004) Assembly Rules and Restoration Ecology - Bridging the Gap Between Theory and Practice. Washington: Island Press.

3. Van der Putten WH, Bardgett RD, de Ruiter PC, Hol WHG, Meyer KM, et al. (2009) Empirical and theoretical challenges in aboveground-belowground ecology. Oecologia 161: 1–14. Doi: 10.1007/s00442-009-1351-8.

4. Rosenzweig ML (1971) The Paradox of Enrichment. Science 171: 385–387.

5. Meyer KM, Mooij WM, Vos M, Hol WHG, van der Putten WH (2009) The power of simulating experiments. Ecol Model 220: 2594–2597. Doi: 10.1016/j.ecolmodel.2009.06.001.

6. Van Dam NM, Heil M (2011) Multitrophic interactions below and above ground: *en route* to the next level. J Ecol 99: 77–88. Doi: 10.1111/j.1365-2745.2010.01761.x.

7. Meyer KM, Vos M, Mooij WM, Hol WHG, Termorshuizen AJ, et al. (2009) Quantifying the impact of above- and belowground higher trophic levels on plant and herbivore performance by modeling. Oikos 118: 981–990. Doi: 10.1111/j.1600-0706.2009.17220.x.

8. Vandermeer J (1993) Lose coupling of predator-prey cycles: entrainment, chaos, and intermittency in the classic MacArthur consumer-resource equations. Am Nat 141: 687–716.

9. Verschoor BC, Pronk TE, De Goede RGM, Brussaard L (2002) Could Plant-Feeding Nematodes Affect the Competition between Grass Species during Succession in Grasslands under Restoration Management? J Ecol 90: 753–761. Doi: 10.1046/j.1365-2745.2002.00710.x.

10. Bukovinszky T, van Veen FJF, Jongema Y, Dicke M (2008) Direct and indirect effects of resource quality on food web structure. Science 319: 804–807. Doi: 10.1126/science.1148310.

11. Hoehn P, Steffan-Dewenter I, Buchori D, Tscharntke T (2009) Spatiotemporal density patterns of the pest predator *Phynchium haemorrhoidale* (F.) along a land-use gradient in cacao agroforestry systems. Agroforestry Systems 76: 163–171. Doi: 10.1007/s10457-008-9154-1.

12. Rosenzweig ML, MacArthur RH (1963) Graphical representation and stability conditions of predator-prey interactions. Am Nat 97: 209–223.

13. Vos M, Kooi BW, DeAngelis DL, Mooij WM (2004) Inducible defences and the paradox of enrichment. Oikos 105: 471–480. Doi: 10.1111/j.0030-1299.2004.12930.x.

14. Verschoor AM, Vos M, van der Stap I (2004) Inducible defenses prevent strong population fluctuations in bi- and tritrophic food chains. Ecol Lett 7: 1143–1148. Doi: 10.1111/j.1461-0248.2004.00675.x.

15. Van der Putten WH, Vet LEM, Harvey JA, Wackers FL (2001) Linking above- and belowground multitrophic interactions of plants, herbivores, pathogens, and their antagonists. Trends Ecol Evol 16: 547–554. Doi:/10.1016/S0169-5347(01)02265-0.

16. Walker M, Jones TH (2001) Relative roles of top-down and bottom-up forces in terrestrial tritrophic plant-insect herbivore-natural enemy systems. Oikos 93: 177–187. Doi: 10.1034/j.1600-0706.2001.930201.x.

17. Maron JL (1998) Insect herbivory above- and belowground: Individual and joint effects on plant fitness. Ecology 79: 1281–1293.

18. Morrien E, Engelkes T, van der Putten WH (2011) Additive effects of aboveground polyphagous herbivores and soil feedback in native and range-expanding exotic plants. Ecology 92: 1344–1352. Doi: 10.1890/10–1937.1.

19. Moran NA, Whitham TG (1990) Interspecific competition between root-feeding and leaf-galling aphids mediated by host-plant resistance. Ecology 71: 1050–1058. Doi: 10.2307/1937373.

20. Masters GJ (1995) The effect of herbivore density on host-plant mediated interactions between two insects. Ecol Res 10: 125–133. Doi: 10.1007/BF02347934.

21. Mueller-Schaerer H, Brown VK (1995) Direct and indirect effects of aboveground and belowground insect herbivory on plant-density and performance of Tripleurospermum perforatum during early plant succession. Oikos 72: 36–41. Doi: 10.1078/1439-1791-00133.

22. Hufbauer RA, Root RB (2002) Interactive effects of different types of herbivore damage: Trirhabda beetle larvae and Philaenus spittlebugs on goldenrod (Solidago altissima). American Midland Naturalist 147: 204–213. Doi: 10.1674/0003-0031(2002)147[0204:IEODTO]2.0.CO;2.

23. Rudgers JA, Hoeksema JD (2003) Inter-annual variation in above- and belowground herbivory on a native, annual legume. Plant Ecol 169: 105–120.

24. Hunt-Joshi TR, Blossey B, Root RB (2004) Root and leaf herbivory on Lythrum salicaria: implications for plant performance and communities. Ecol Appl 14: 1574–1589. Doi: 10.1890/03-5181.

25. Gerber E, Hinz HL, Blossey B (2007) Interaction of specialist root and shoot herbivores of Alliaria petiolata and their impact on plant performance and reproduction. Ecol Entomol 32: 357–365. Doi: 10.1111/j.1365-2311.2007.00875.x.

26. Wurst S, van der Putten WH (2007) Root herbivore identity matters in plant-mediated interactions between root and shoot herbivores. Basic Appl Ecol 8: 491–499. Doi: 10.1016/j.baae.2006.09.015.

27. Soler R, Bezemer TM, van der Putten WH, Vet LM, Harvey J (2005) Root herbivore effects on above-ground herbivore, parasitoid and hyperparasitoid performance via changes in plant quality. J Anim Ecol 74: 1121–1130. Doi: 10.1111/j.1365-2656.2005.01006.x.

28. Thies C, Tscharntke T (1999) Landscape structure and biological control in agroecosystems. Science 285: 893–895. Doi: 10.1126/science.285.5429.893.

29. Klein AM, Steffan-Dewenter I, Tscharntke T (2002) Predator-prey ratios on cocoa along a land-use gradient in Indonesia. Biodiv Conserv 11: 683–693. Doi: 10.1023/A:1015548426672.

30. Hawkins BA, Mills NJ, Jervis MA, Price PW (1999) Is the biological control of insects a natural phenomenon? Oikos 86: 493–506.

31. Sylvain ZA, Wall DH (2011) Linking soil biodiversity and vegetation: implications for a changing planet. American Journal of Botany 98: 517–527. Doi: 10.3732/ajb.1000305.

32. Erb M, Flors V, Karlen D, De Lange E, Planchamp C, et al. (2009) Signal signature of aboveground induced resistance upon belowground herbivory in maize. Plant Journal 59: 292–302.

33. Erb M, Robert CAM, Hibbart BE, Turlings TCJ (2011) Sequence of arrival determines plant-mediated interactions between herbivores. J Ecology 99: 7–15.

34. Rasmann S, Kollner TG, Degenhardt J, Hiltpold I, Toepfer S, et al. (2005) Recruitment of entomopathogenic nematodes by insect-damaged maize roots. Nature 434: 732–737.

35. Royer L, Boivin G (1999) Infochemicals mediating the foraging behaviour of Aleochara bilineata Gyllenhal adults: sources of attractants. Entomologia Experimentalis et Applicata 90: 199–205.

36. Steele AE (1986) Nematode parasites of sugar beet. In: Whitney ED, Duff JE, editors. Compendium of beet diseases and insects. St. Paul: American Phytopathological Society, APS Press. 33–36.

37. Blossey B, Hunt-Joshi TR (2003) Belowground herbivory by insects: influence on plants and aboveground herbivores. Annual Review of Entomology 48: 521–547. Doi: 10.1146/annurev.ento.48.091801.112700.

38. Schädler M, Jung G, Brandl R, Auge H (2004) Secondary succession is influenced by belowground insect herbivory on a productive site. Oecologia 138: 242–252. Doi: 10.1007/s00442-003-1425-y.

39. Sunding KN, Collins SL, Gough L, Clark C, Cleland EE, et al. (2005) Functional- and abundance-based mechanisms explain diversity loss due to N fertilization. PNAS 102: 4387–4392.

40. McClean CJ, van den Berg LJL, Ashmore MR, Preston CD (2011) Atmospheric nitrogen deposition explains patterns of plant species loss. Global Change Biology 17: 2882–2892.

41. Fenn ME, Poth MA, Aber JD, Baron JS, Bormann BT, et al. (1998) Nitrogen excess in American ecosystems: Predisposing factors, ecosystem responses and management strategies. Ecol Appl 8: 706–733.

42. Fukami T, Nakajima M (2011) Community assembly: alternative stable states or alternative transient states? Ecol Lett 14: 973–984. Doi: 10.1111/j.1461-0248.2011.01663.x.

16

Real-Time Cytotoxicity Assay for Rapid and Sensitive Detection of Ricin from Complex Matrices

Diana Pauly, Sylvia Worbs, Sebastian Kirchner, Olena Shatohina, Martin B. Dorner, Brigitte G. Dorner*

Center for Biological Security - Microbial Toxins, Robert Koch-Institut, Berlin, Germany

Abstract

Background: In the context of a potential bioterrorist attack sensitive and fast detection of functionally active toxins such as ricin from complex matrices is necessary to be able to start timely countermeasures. One of the functional detection methods currently available for ricin is the endpoint cytotoxicity assay, which suffers from a number of technical deficits.

Methodology/Findings: This work describes a novel online cytotoxicity assay for the detection of active ricin and *Ricinus communis* agglutinin, that is based on a real-time cell electronic sensing system and impedance measurement. Characteristic growth parameters of Vero cells were monitored online and used as standardized viability control. Upon incubation with toxin the cell status and the cytotoxic effect were visualized using a characteristic cell index–time profile. For ricin, tested in concentrations of 0.06 ng/mL or above, a concentration-dependent decrease of cell index correlating with cytotoxicity was recorded between 3.5 h and 60 h. For ricin, sensitive detection was determined after 24 h, with an IC50 of 0.4 ng/mL (for agglutinin, an IC50 of 30 ng/mL was observed). Using functionally blocking antibodies, the specificity for ricin and agglutinin was shown. For detection from complex matrices, ricin was spiked into several food matrices, and an IC50 ranging from 5.6 to 200 ng/mL was observed. Additionally, the assay proved to be useful in detecting active ricin in environmental sample materials, as shown for organic fertilizer containing *R. communis* material.

Conclusions/Significance: The cell-electrode impedance measurement provides a sensitive online detection method for biologically active cytotoxins such as ricin. As the cell status is monitored online, the assay can be standardized more efficiently than previous approaches based on endpoint measurement. More importantly, the real-time cytotoxicity assay provides a fast and easy tool to detect active ricin in complex sample matrices.

Editor: Nicholas J. Mantis, Wadsworth Center, New York State Dept. Health, United States of America

Funding: This work was supported by grants from the Federal Ministry of Education and Research to BGD (BiGRUDI project, 13N9601; http://www.bmbf.de/en/index.php). The funders had no role in study design, data collection and analysis, decision to publish or preparation of the manuscript.

Competing Interests: The authors have declared that no competing interests exist.

* E-mail: DornerB@rki.de

Introduction

Ricin, one of the most poisonous toxins known, is a glycoprotein derived from the seeds of the castor plant *Ricinus communis*. This cytotoxin is highly toxic both to humans and animals [1,2]. The plant is cultivated as an ornamental and industrial flower all over the world. As by-product during castor oil production, ricin is mass-produced above 1 million tons per year [3]. On the basis of its availability, toxicity, ease of preparation and the current lack of medical countermeasures, ricin has gained attention as potential biological warfare agent and is listed as category B agent of potential bioterrorism risk by the Centers for Disease Control and Prevention (CDC, Atlanta, GA, USA; [4,5]).

Functionally active ricin consists of two ~32-kDa subunits, the A-chain and the B-chain, which are linked by a disulfide bond. Both chains are needed for toxic action *in vivo*. The B-chain acts as a lectin, which binds to terminal galactose residues on the eukaryotic cell surface and mediates ricin cell entry by endocytosis [6]. The A-chain is an RNA-specific *N*-glycosidase that hydrolytically removes a specific adenine from the 28S ribosomal subunit, thereby inhibiting the protein biosynthesis and ultimately leading to cell death [7]. As well as ricin, castor seeds also contain a second

lectin, *Ricinus communis* agglutinin (abbreviated in the text as agglutinin), which is highly homologous to ricin, but less toxic [8]. Agglutinin consists of a dimer of two associated ricin-like molecules, each of which contains A- and B-chains. The homology at the amino acid level between agglutinin and ricin is around 93% for the A-chains and around 84% for the B-chains [9].

The mortality in ricin poisoning is dependent on the route of administration. In mice, the median lethal doses (LD50) for injection, inhalation or ingestion are reported as 2–10 µg ricin/kg body weight, 3–5 µg/kg or 20 000–30 000 µg/kg, respectively. In humans, the oral LD50 of 1 000–20 000 µg ricin/kg body weight is estimated from accidental ingestion of castor beans [10,11]. For agglutinin, it is approximated from animal studies that the toxicity is about two orders of magnitude less than that for ricin [12,13].

Similar to other toxins, ricin acts in the absence of the producing plant and its genetic information. Therefore, it is necessary to detect the protein itself, not only the plant's nucleic acid. Currently, the analysis of ricin is mainly based on immunological methods [14,15], mass spectrometry analysis [16,17], or functional *in vitro* and *in vivo* assays (for an overview of the latter see Table S1). In the case of an intentional release of ricin into the environment, the discrimination of functionally active and denatured ricin is important, especially with regard to emergency operating schedules, forensic analysis and therapy. This information can only be obtained from functional assays, which can be principally differentiated into assays detecting the A-chain activity, the B-chain activity, or both. *In vitro* assays, like adenine-release assays or cell-free translation assays based on rabbit reticulocyte lysate, analyze the enzymatic activity of the A-subunit [18,19,20,21,22,23,24]. Glycan binding of the ricin B-subunit is detected by enzyme-linked lectin assays [25]. However, the detection of the activity of the isolated subchains provides no information on the activity of the intact 64 kDa ricin molecule. Therefore, the detection of active ricin requires *in vitro* or *in vivo* assays for both subchains. *In vivo* assays have the advantage that whole-organism responses can be monitored but, on the basis of different species and strain susceptibilities that have been reported for ricin, animal bioassays seem to be difficult to standardize and raise ethical concerns [26,27,28]. An alternative are *in vitro* immunocapture assays combined with adenine release measurement [17,29], or cell-based functional assays. Current cell assays use different endpoint read-outs of cell death via biochemical, fluorescent or radioactive detection [30,31,32,33]. The detection limits for ricin analysis in cell-based bioassays have been described as being between 0.01 ng/mL and 0.8 ng/mL from complex matrices. Cytotoxicity is detected at the end of the assay after different cell cultivation times (ranging from 4 to 28 h), in order to obtain a "snapshot" of ricin action. Taking into account the duration of the assay required to reach high sensitivity, time-dependent toxicity values might be more informative than endpoint measurements. Another drawback of the current cell-based assays is represented by a lack of online and internal viability control. It is only known at the end of the assay if the cells adhere, grow and die, according to a standardized procedure. Therefore cell-based assays usually show a higher degree of variability than other methods. In terms of reproducibility of cell-based assays, it is important to standardize all growth parameters and to include specificity controls (i.e. functionally blocking antibodies) in order to avoid experimental artefacts.

In the present study, we describe an online functional ricin cytotoxicity assay based on a real-time cell electronic sensing (RT-CES) system. The cell proliferation and toxin-induced cell death of African green monkey (Vero) cells is monitored online in the RT-CES system. This system uses an impedance sensor technology to noninvasively and label-free quantify cell viability, based on cell number, morphology and adhesion in real-time [34]. The cells are seeded in E-plates into which microelectrodes are integrated. Low-voltage application leads to the generation of an electric field, which is differentially modulated by the cells. The higher the number of cells attached to the plate surface, the higher the impedance monitored by the RT-CES system as a read-out for cell viability.

The aim of this report was to detect and quantify biologically active ricin in a real-time cytotoxicity assay. The assay duration, sensitivity and specificity was tested for ricin and agglutinin, as compared to other types of lectins. This method allowed for the detection of low concentrations of active ricin in different food matrices and in organic fertilizer, without significant interfering matrix effects. The real-time impedance measurement of adhering, proliferating and, ultimately, dying cells turned out to be highly reproducible, thus opening the door to standardized cell-based cytotoxicity assays.

Results

Real-time monitoring of Vero cell growth pattern

In order to set up a standardized cell-based cytotoxicity assay, the culture conditions and assay procedures were thoroughly optimized. First, the Vero cell growth pattern was dynamically monitored using the RT-CES system, which detects cellular impedance as measure of cell number, morphology and adherence [34]. To this end, we seeded different Vero cell concentrations, from 390 to 50 000 cells/well, into an E-plate and monitored the cell proliferation online for up to 60 h. Cell growth was recorded as cell index (CI), which corresponds to the electrical impedance of a well measured by the RT-CES system [34]. Depending on the number of cells seeded within an E-well, the CI of proliferating Vero cells ranged from 0.5 to 10 (Figure 1). Cells seeded in concentrations below 3 125 cells/well did not proliferate in the observation period shown. Vero cells seeded with 12 500 cells/ well or above showed vigorous growth in the first 12 h post-seeding (Figure 1) and reached a plateau after 42 h hours, up to a CI of 10 (data not shown). Characteristic cell proliferation parameters were observed for the optimal cell concentration of 12 500 cells/well, which was chosen for further experiments: The attachment phase of the cells to the plate was completed after 1 h (CI increased from 0 to 0.5), the lag-phase lasted from 1 to 2 h (CI remained at 0.5), the start of proliferation was recorded after 2 h (CI increased over 5), and after 14 h the proliferation of the cells decelerated into the confluent phase (CI remained at 7; Figure 1).

Measurement of ricin toxicity in the real-time cell system

We used the optimal cell concentration of 12 500 cells/well to analyze the toxic effect of serial dilutions of ricin (230 000 to 0.023 ng/mL) or agglutinin (4 600 to 0.046 ng/mL) in cell culture medium, respectively. Immediately after the seeding of cells into E-plates, ricin or agglutinin was added to the cells (without prior attachment of the cells to the plate). Electrical impedance of cells within an E-well was monitored using the RT-CES system over 24 h, where attachment and proliferation were visualized as a rise in CI (correlating with rise in impedance) and detachment and cell death as a drop in CI (correlating with drop in impedance). As shown in Figures 2A and 2B, all cells showed a characteristic

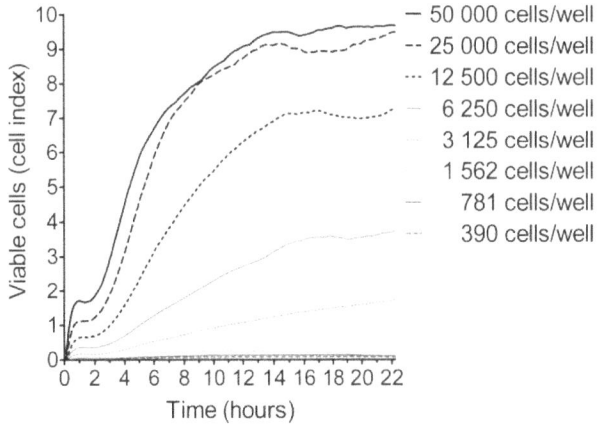

Figure 1. Dynamic monitoring of Vero cell proliferation. Serial dilutions of Vero cells were seeded at indicated densities of 50 000 to 390 cells/well in a 96-well E-plate. The attachment phase, lag-phase and proliferation phase were dynamically monitored every 15 min for 22 h, as indicated in the text. Data shown are representative of three independent experiments showing similar results.

attachment and lag phase of cell growth, independent of toxin concentration. Notably, the duration of the proliferation phase was dependent on the toxin concentration. The growth curves of Vero cells treated with a toxin concentration of 0.23 ng/mL ricin or 46 ng/mL agglutinin (or higher), respectively, did not reach confluence compared to untreated cells (Figures 2A, 2B). Time-to-inhibition of cell proliferation was dependent on the toxin dose, with the earliest inhibition visible after 3 to 4 hours, at the highest concentrations tested. For ricin tested in concentrations of 2.3 ng/mL or above (or 460 ng/mL agglutinin), a concentration-dependent decrease of CI was recorded after 3.5 to 10 h, reaching a CI below 2 after 24 h. Toxin concentrations of 0.23 ng/mL ricin or 46 ng/mL agglutinin also showed an inhibition of cell growth, albeit after more than 13 h of incubation time (Figures 2A, 2B). Based on these data, Figures 2C and D show dose–response curves at selected time points of the real-time cytotoxicity assay. As expected, the toxic effect of ricin and agglutinin on Vero cells increased with toxin concentration and incubation time. *In vitro* cytotoxicity, where 50% of the cells were alive (IC50), was detected at the earliest after 8 h for 880 ng/mL ricin (Table 1). For agglutinin, the IC50 value at 8 h was higher (2 400 ng/mL). For ricin, sensitive detection was determined after 24 h, with an IC50 of 0.4 ng/mL. At the same time point, agglutinin showed a toxicity that was 75 times lower than ricin (IC50 of 30 ng/mL; Table 1). Upon prolonged incubation for 42 h, the most sensitive detection of ricin was reached with an IC50 of 0.1 ng/mL (and 6.7 ng/mL for agglutinin). Longer incubation times did not significantly result in higher sensitivities (Table 1).

Since the real-time cytotoxicity assay allows for monitoring the cell proliferation in real-time, the assay can be better standardized than conventional cell-based cytotoxicity tests: even slight changes in cell culture conditions can be visualized and used to optimize the test, as shown in Figure S1: parameters like the cell density prior to seeding cells into E-plates (Fig. S1 A) or the method used to detach cells from the culture flask (Fig. S1 B) quite strongly influence the growth characteristics of the cells, similar to seeding of different cell numbers (Fig. 1).

The precision of the optimized real-time cytotoxicity assay was evaluated by the measurement of cytotoxicity of serial dilutions of ricin after 24 h or 42 h, respectively: for within-run precision serial dilutions of ricin (1 ng/mL to 0.03 ng/mL) were performed in four replicates and measured on one day yielding within-run CVs between 13% and 2% after 24 h incubation (Table S2). For between-run precision serial dilutions of ricin were performed on four different days and the results were used for calculation of CVs. At the 24 h time point, the between-run CVs were determined between 30% and 6% (Table S2).

Comparison of RT-CES and MTT cytotoxicity assay

In order to compare the performance of the online cytotoxicity assay using the RT-CES system with the classical endpoint cell viability test, we performed parallel assays using ricin or agglutinin. Serial dilutions of ricin or agglutinin were incubated with Vero cells, and cell growth was determined either in real-time format over 24 h (RT-CES system) or as endpoint measurement after 20 h proliferation using an MTT assay. Both techniques resulted in sigmoidal dose-dependent response curves (Figure 3). The IC50 for ricin in the RT-CES system and the MTT assay were similar (0.4 ng/mL and 1.5 ng/mL, respectively). For agglutinin, the IC50 values were in the same concentration range using both assay systems (30 ng/ml for RT-CES system, 50 ng/mL for MTT assay). The duration of the full assay, starting from cell seeding to sensitive read-out, was 24 h for the RT-CES system and 45 h for the MTT assay. On a quantitative level, the RT-CES system

reached a similar sensitivity as the MTT assay, but was significantly faster. At the same time the overall precision of the RT-CES system was similar to the precision of the MTT assay (Table S3).

Specificity of the real-time cytotoxicity assay

In order to demonstrate the specificity of the assay, ricin (in concentrations ranging from 230 000 to 0.023 ng/mL) was preincubated with chicken anti-ricin polyclonal antibodies (IgY). The action of the toxin on cells was then monitored online for 23 h. Undisturbed cell proliferation, equivalent to a complete block of ricin's functional activity, was observed up to a ricin concentration of 23 000 ng/mL (Figure 4A). For the highest ricin concentration tested (230 000 ng/mL), the IgY concentration was not sufficient to block ricin activity, resulting in cell death.

We further analyzed the specificity of the real-time cytotoxicity assay by comparing the cytotoxic effects of ricin and agglutinin with other plant lectins (*Abrus precatorius* abrin [abrin], *Dolichos biflorus* agglutinin [DBA] and *Triticum vulgaris* agglutinin [TVA]) in the absence or presence of anti-ricin IgY (incubation period of 21 h). Ricin, agglutinin and abrin caused cell death, whereas the plant lectins DBA and TVA showed no significant toxic effects (Figure 4B). As expected, anti-ricin IgY were able to specifically block the functional activity of ricin and agglutinin, but not the activity of abrin. Only functionally active ricin induced cell death, since heat-inactivated ricin had no effect on cell proliferation (Figure 4C).

Detection of ricin activity in complex matrices

In order to detect ricin in the presence of complex food matrices, we performed a series of experiments to find out how much food matrix would be tolerated by the cells (data not shown). In the final protocol the food matrices were diluted 1:14 in medium and a clarified homogenate was added to the cells. As shown in Figure 5A, cells treated with 1:14-diluted carrot juice and milk showed a similar proliferation pattern as untreated cells. Diluted baby food extract, however, interfered more strongly with cell proliferation leading to a decelerated growth. To determine the ability of the real-time cytotoxicity assay to detect ricin from complex food matrices, serial dilutions of ricin were spiked into homogenized food extracts (from milk, carrot juice and baby food) and incubated with the cells for 24 h (Figure 5B) and 42 h (Figure 5C). Functionally active toxin was detectable from all complex matrices spiked with ricin after 24 h and 42 h of incubation. The sigmoidal dose-dependent response curve for ricin spiked into diluted carrot juice was very similar to the medium control, while the response curve for ricin spiked into diluted milk significantly shifted to higher concentrations (Figure 5B and 5C). Based on these data, the IC50 for ricin spiked into diluted carrot juice was 0.4 ng/mL (corresponding to 5.6 ng/mL in the undiluted matrix) and for diluted milk 14.3 ng/mL (corresponding to 200 ng/mL in the undiluted matrix, respectively; Table S4). For ricin spiked into the difficult matrix baby food first results could be visualized as drop in CI within 24 h yielding an IC50 of about 0.4 ng/mL. However, the spreading of the dose-dependent response curve was marginal. In this case incubation for 42 h resulted in a reasonable dose-dependent response curve and delivered an IC50 of 0.1 ng/mL for the diluted matrix (corresponding to 1.4 ng/mL in the undiluted matrix; Table S4). As shown in Table S4, for all ricin-spiked food matrices the IC50 obtained after 24 h is about 3–4 times higher than the IC50 obtained after 42 h.

In order to show that the real-time cytotoxicity assay can be used to screen for active ricin in environmental samples, we tested

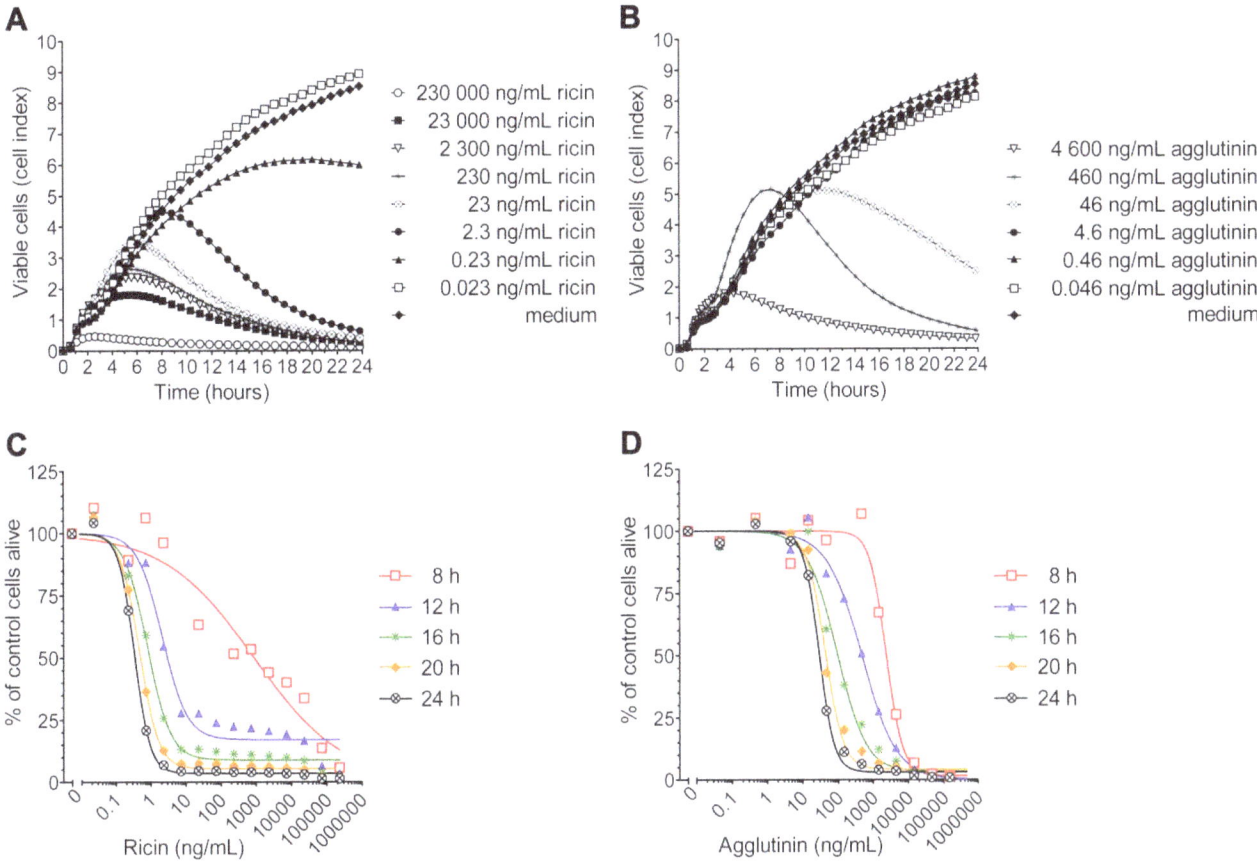

Figure 2. Real-time measurement of cytotoxicity and time-dependent standard curves for ricin and agglutinin. Vero cells were seeded in a 96-well E-plate (12 500 cells/well). Immediately after seeding, cells were exposed to the indicated concentrations of ricin (A, C), agglutinin (B, D), or medium (control). Cell proliferation was dynamically monitored every 15 min for 24 h. Figures A and B show the time-dependent alteration of the CI for different ricin or agglutinin concentrations. Figure C and D display the percentage of viable cells plotted against toxin concentrations at selected time points (conversion of the data from figure [A, B] to [C, D] is described in material and methods). Data shown are representative of five (A, C) or three (B, D) independent experiments with similar results.

different lots of organic fertilizer containing *Ricinus communis* material (castor meal). During the industrial production of castor oil, ricin is obtained as a by-product in the castor bean meal and

Table 1. IC50 values for ricin and agglutinin at different time points of the real-time cytotoxicity assay.

time [h]	IC50 ricin [ng/mL]	IC50 agglutinin [ng/mL]
8	880	2 400
12	3.0	470
16	1.0	100
20	0.5	44
24	0.4	30
30	0.36	26
36	0.17	11
42	0.10	6.7
48	0.08	4.8
54	0.06	4.2
60	0.06	4.1

has to be thoroughly heat-inactivated. However, in the past there have been reports of dog poisoning, presumably caused by the accidental ingestion of fertilizer that was insufficiently heat-treated [2,35]. Having been involved in a similar case of dog poisoning in Germany, we tested extracts of the suspected *Ricinus communis*-containing fertilizer using the real-time cytotoxicity assay. An example of one fertilizer, Figure 5D, shows that different dilutions of the fertilizer extract induced cytotoxicity in Vero cells. Using a 1:14 dilution of the buffered fertilizer extract, about 37% of the Vero cells were alive after 21 h. Preincubation of this dilution with anti-ricin IgY (+Ab) prevented cell death. Similarly, heat treatment of the extract for 30 min at 95°C also prevented cytotoxicity, both results indicate that the toxic effect on the cells is caused by functionally active ricin contained in the fertilizer extract (for guidance, different concentrations of purified ricin were tested in parallel).

Discussion

We report a sensitive and rapid real-time cytotoxicity assay for the detection and quantification of functionally active ricin and agglutinin based on impedance sensor technology. This specific method was applicable for the detection of *Ricinus communis* material from complex matrices, e.g. food and organic fertilizer.

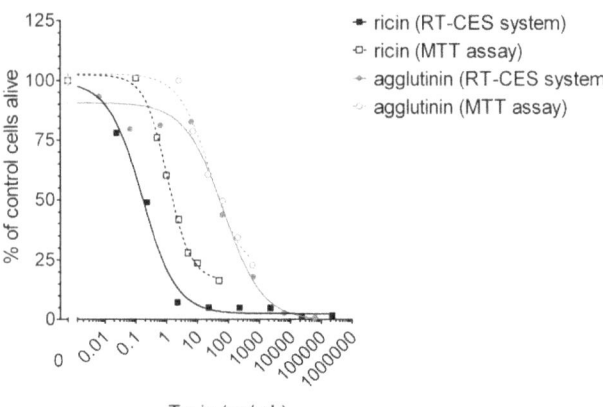

Figure 3. Comparison of ricin and agglutinin cytotoxicity in RT-CES system and MTT assay. Vero cells (12 500 cell/well RT-CES system, 10 000 cells/well MTT assay) were seeded in a 96-well E-plate (RT-CES system) or 96-well cell culture plate (MTT assay), respectively. In the RT-CES system (filled symbols), serial dilutions of ricin (grey) and agglutinin (black) were incubated immediately after cell seeding, and cell proliferation was monitored online for 24 h. For the MTT assay (open symbols), cells were cultivated for 18 h and incubated afterwards with ricin or agglutinin. After 2 h cells were washed and cultured for a further 20 h, before the MTT reagent was used to determine cell viability. Data shown are representative of two independent experiments showing similar results.

Classical functional assays for ricin include animal bioassays, endpoint cytotoxicity assays, adenine release assays, cell-free translation assays or enzyme-linked lectin assays (for an overview, see Table S1). For AB toxins like ricin, it has been shown previously that neither the isolated A-chain nor the isolated B-chain are toxic alone [36,37]. Furthermore, there are hints that under certain denaturing conditions the sugar-binding property of the B-chain is destroyed, whereas the enzymatic activity of the A-chain is retained, even in the presence of an intact disulfide bond between the A- and B-chain [38]. This indicates the necessity for functional assays that show the presence AND the functional activity of both subunits. According to this criterion, several of the assays known are suitable to detect active ricin, e.g. animal bioassays and cell-based cytotoxicity assays. However, these assays have certain limitations, because animals are required [26,28,39], because of long cell cultivation and assay times [33], the need for transfection to generate fluorescently labeled reporter cells [32], or the use of radioactive materials [31]. The ricin detection based on the RT-CES system works label-free, and does not require radioactivity, enzymatic assays or transfection steps. Depending on the toxin concentration and the matrix used, active ricin can be detected between 3.5 to 60 h, which is comparable to previously described assays and faster than most endpoint cytotoxicity assays. High sensitivity, however, is reached after 24 h, with an IC50 of 0.4 ng/mL which is in the range of endpoint assays (Table S1). Compared to endpoint assays, the most important advantage of the impedance measurement is the online monitoring of cellular status, which offers the possibility to standardize the assay. Impedance read-outs in the RT-CES system depend on cell number, cell size and morphology, as well as adhesion characteristics of the cell line chosen. Ricin action was recorded online after cell attachment and lag-phase, and the kinetic control of the cellular status before and after ricin action revealed continuous information about growth, morphological changes and cell death. Therefore, abnormal first growth phases were an indicator of

problems in assay performance. Even slight changes in cell culture conditions could be visualized and were used to optimize the assay. The CI values for the attachment and lag-phase were used as quality parameters to reduce assay variability. Based on this, the intra-assay variability was routinely below 13% for ricin tested at 1 ng/mL or below and the inter-assay variability was between 6 and 30% (24 h incubation, Table S2).

In recent years, highly sophisticated mass-spectrometry-based methods for detection and quantification of ricin have been introduced [29,40,41]. These methods combine an immunoaffinity enrichment of ricin, e.g. via its B-chain, followed by detection and/or quantification of adenine release by the A-chain. Strictly speaking, these assays detect the presence of the B-chain (not its activity), plus the activity of the A-chain. The advantage of these assays is their high precision – they are able to discriminate and quantify the highly homologous ricin and agglutinin down to a few fmol/mL, even in food matrices [40]. This data cannot be obtained by the cellular assays described or by conventional immunological assays. Since these technologies require high-end mass spectrometric equipment and specialised technical expertise, they might not be useful for broad application in routine laboratories. The impedance technology described, however, is easy to use and reduces cell culture work-load to a minimum, combined with low operative cost and effort. Therefore, this technology might complement standard routine ricin detection approaches.

The impedance technology might also be useful for screening of functionally blocking anti-ricin monoclonal antibodies or small molecule inhibitors, since multiple samples can be analyzed in parallel in a 96-well or 384-well format, thereby further improving endpoint cytotoxicity screening approaches [42,43]. Combined with automated liquid handling platforms, high-throughput screening and objective quantitative data analysis is possible within a minimal amount of time.

On a qualitative basis, the impedance technology allows the comparison of the toxicity of substances, as shown here for ricin and agglutinin. At the time points of highest sensitivity (24–48 h), ricin was 60- to 75-fold more toxic than the closely related agglutinin, and the data are in accordance with previous data in the literature [12,13]. For related plant AB toxins, abrin and abrin agglutinin, it was shown that the reduced toxicity of abrin agglutinin is associated with amino acid substitutions in a conserved region of ribosome inactivating proteins, resulting in a modified three-dimensional structure, which prevents an affine substrate binding [44]. With respect to ricin and agglutinin, the difference in toxicity is not understood on a molecular basis.

Since the ricin-producing plant *R. communis* is used on an industrial scale for the production of castor oil, there is concern that as a by-product, ricin could be used to deliberately contaminate the food supply chain. Furthermore, the castor meal itself is used as a cheap additive in organic fertilizers, since it is a rich source of nitrate. Therefore, it was important to show the applicability of the impedance technology on food matrices and fertilizer samples. Ricin was detected in concentrations between 5.6 ng/mL in carrot juice or baby food and 200 ng/mL in milk. It is known that the functional activity of ricin is inhibited by various sugars, e.g. lactose and galactose [45]. The predominant carbohydrate in milk is lactose, which interacts with the ricin B-subunit and therefore interferes with ricin detection in functional assays. Generally, the real-time cytotoxicity assay described here showed only minor interference with matrix compounds from the food tested. The assay was also applicable for the detection of ricin from fertilizer samples and actually turned out to be useful for forensic analysis. Having been involved in a case of dog poisoning

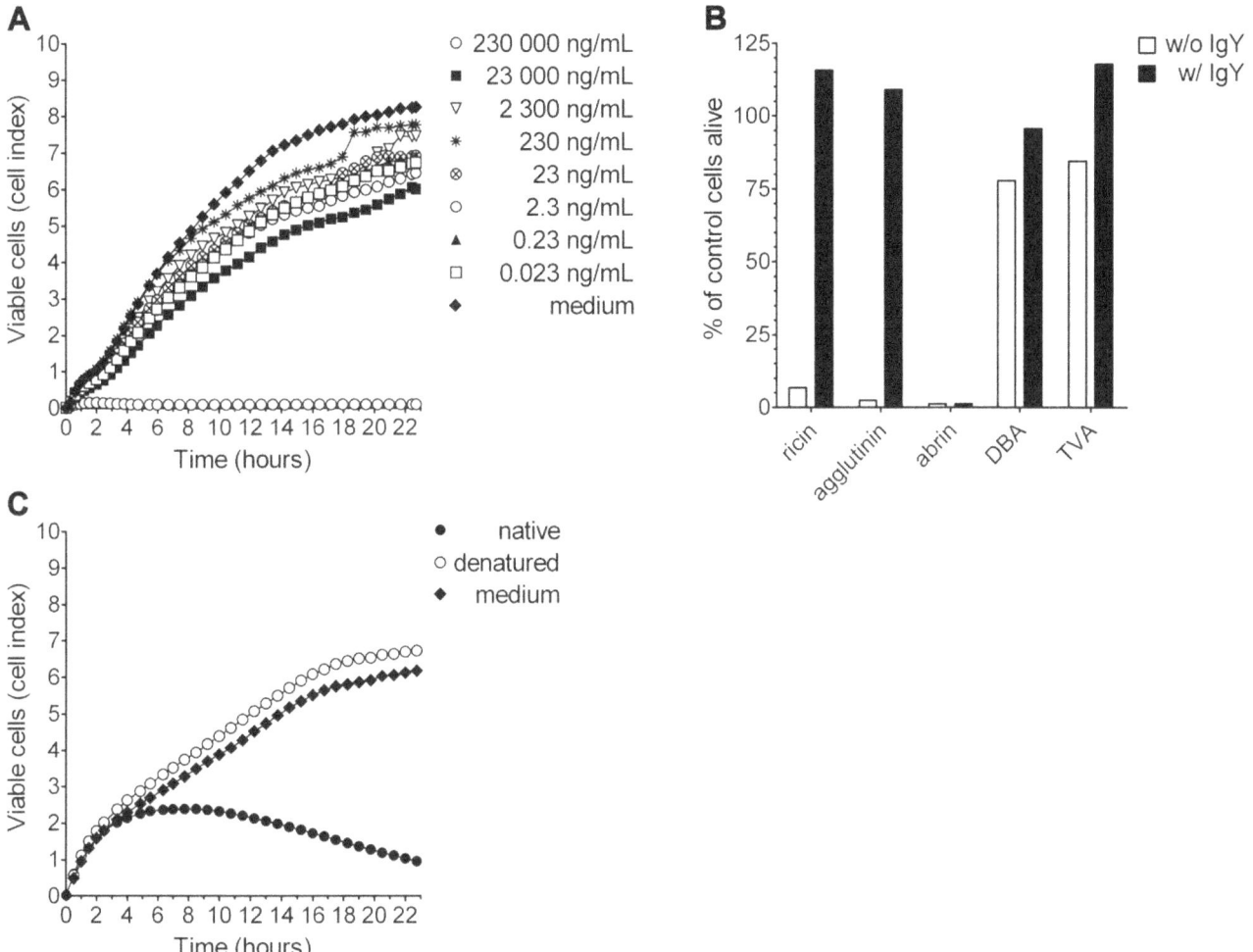

Figure 4. Specificity of the cytotoxicity test. Vero cells (12 500 cells/well) were seeded in a 96-well E-plate. (A) Serial dilutions of ricin were preincubated with polyclonal anti-ricin IgY for 1.5 h at 37°C, and then added to the cells. Cell proliferation was dynamically monitored every 15 min for 23 h. Data shown are representative of five independent experiments showing similar results. (B) Vero cells were exposed to 10 μg/mL ricin, agglutinin, abrin, DBA or TVA, respectively (white columns). In order to show the specificity of the assay, the different lectins were preincubated with anti-ricin IgY as above (black columns). The viability of the cells after 21 h is depicted as percentage of the viability of untreated control cells (100%). Data shown are exemplary data out of two independent experiments showing similar results. (C) Ricin (20 000 ng/mL) was heated in PBS for 30 min at 95°C (denatured, white circles), or was left untreated (native, black circles) and then added to the cells. Cell proliferation was dynamically monitored every 15 min for 23 h. In parallel, cell growth was monitored in medium only (negative control, black diamond). Data shown are representative of two independent experiments showing similar results.

in Germany in 2010, we were able to show that the deceased animal had taken up fertilizer containing significant amounts of active ricin which was obviously not correctly heat-inactivated during the production process [46]. Similar cases have been documented worldwide [2,35,47].

In conclusion, the impedance sensor technology presented offers a fast and label-free real-time monitoring of the functional activity of the ricin holo-toxin, using cell death as read-out as a consequence of both the sugar-binding and the enzymatic activity of the molecule. Combined with ELISA and precise mass-spectrometry assays, it will be useful to complement the information obtained from analyzing real sample materials. The method is easy to use in routine laboratories, requires minimum hands-on time and can be automated for high-throughput screening for anti-ricin inhibitory substances.

Materials and Methods

Material

Ricin (purity 98%) was purified along with agglutinin (purity 96%) either from seeds of *Ricinus communis carmencita* or from an unknown cultivar according to standard procedures [48]. Abrin was purchased from Toxin Technology (Sarasota, FL, USA). *Dolichos biflorus* agglutinin (DBA) and *Triticum vulgaris* agglutinin (TVA) were obtained from Sigma-Aldrich (Munich, Germany). Milk (UHT-milk, 3.5% fat, pH 7, brand: Tip, Metro, Düsseldorf, Germany), carrot juice (with lemon juice and ascorbic acid, pH 5, brand: Drink, Kaiser's Tengelmann, Viersen, Germany) and baby food (blueberry and apple dessert, pH 3, brand: Hipp, Pfaffenhofen, Germany) were purchased from a local retail store. Anti-ricin-specific polyclonal chicken IgY has been described elsewhere [49].

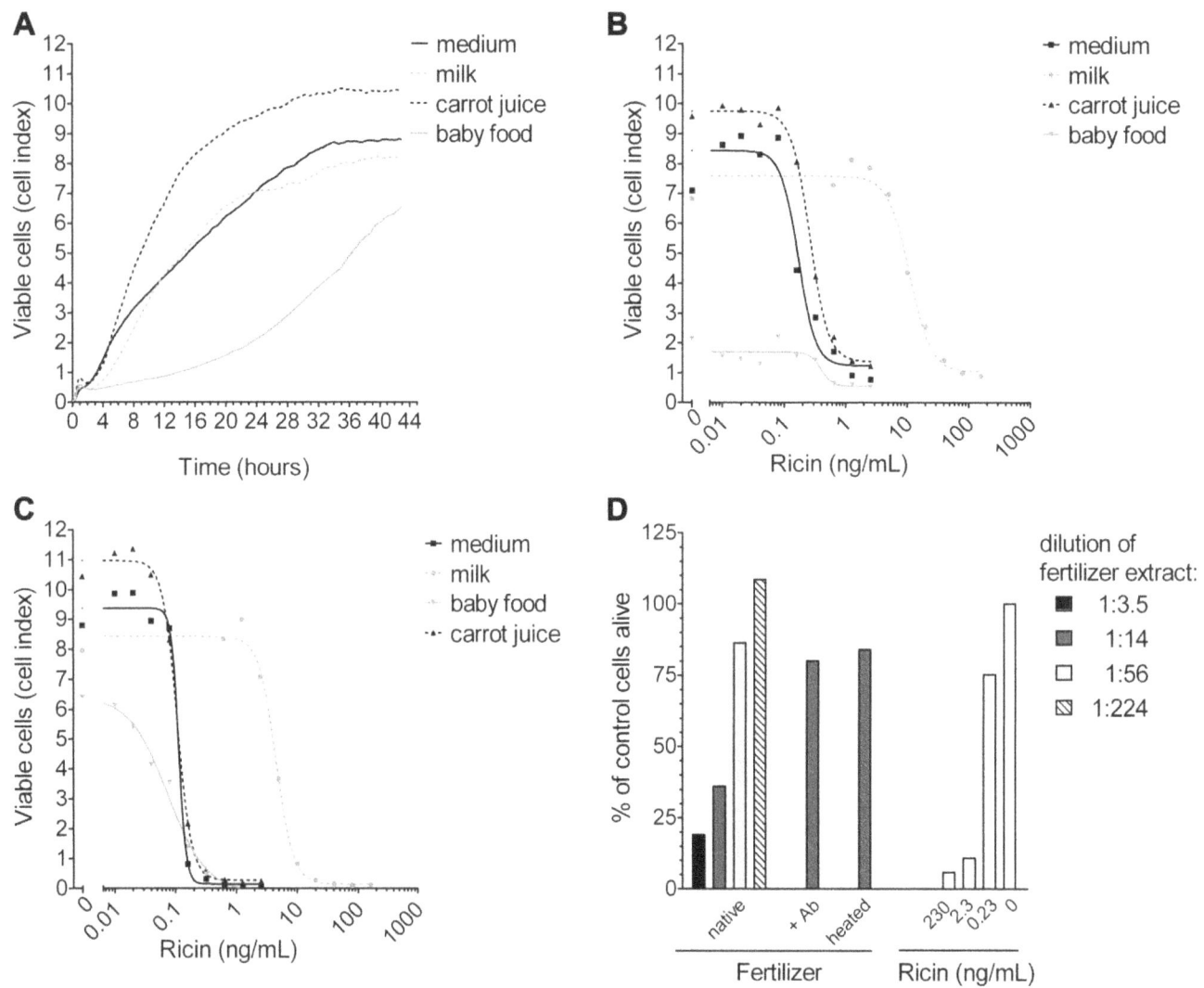

Figure 5. Detection of functionally active ricin in complex matrices. (A) Vero cells (12 500 cells/well) were treated immediately after seeding into E-plates with 1:14-diluted food extracts from milk (dotted grey line), carrot juice (dotted black line), baby food (grey line) or medium (black line). Characteristic growth phases of the cells were dynamically monitored every 15 min for 43 h. (B) Vero cells were exposed to ricin spiked into milk, carrot juice, baby food or medium, respectively. The indicated toxin concentrations are post-dilution concentrations. The viability of the cells is depicted as percentage of viable cells plotted against toxin concentrations in the different food matrices, measured after 24 h. (C) Vero cells were treated as described in (B). The viability of the cells is depicted as percentage of viable cells plotted against toxin concentrations in the different food matrices, measured after 42 h. (D) Vero cells were incubated with different dilutions of *Ricinus communis*-containing fertilizer extract (1:3.5, black; 1:14 dark grey; 1:56 light grey; 1:224, hatched) either without treatment (native), preincubated with 6 µg polyclonal anti-ricin IgY for 1.5 h at 37°C (+Ab) or heated for 30 min at 95°C (heated). For guidance, Vero cells were treated in parallel with different concentrations of purified ricin (white bars, 230 ng/mL, 2.3 ng/mL, 0.23 ng/mL). The viability of the cells after 21 h is depicted as percentage of the viability of the untreated control cells (100%). Data shown are exemplary data out of two independent experiments showing similar results.

Cell culture conditions

African green monkey (Vero) cells were acquired from the American Type Culture Collection (ATCC, Manassas, VA, USA). Cells grown to 50–80% confluent monolayers (~25 000 cells/cm²), at passages 153 to 163, were used for the experiments. Before each assay, Vero cells were trypsinized with Trypsin/EDTA solution (0.2%/0.02% (w/v)) in PBS, centrifuged and resuspended in the appropriate medium volume and counted using a hemacytometer. Cells were cultured in Dulbecco's Modified Eagle Medium (DMEM, Biochrom, Berlin, Germany) containing 10% fetal bovine serum (Invitrogen, Karlsruhe, Germany), L-glutamine (2 mM, Biochrom, Berlin, Germany), 100 IU/mL penicillin and 100 µg/mL streptomycin (PAA

Laboratories, Pasching, Austria). Cells were incubated at 37°C in a humidified 5% CO_2 atmosphere.

Cell proliferation test

For real-time monitoring of Vero cell proliferation, the baseline impedance of each sensor well in a 96-well E-plate (Roche Diagnostics, Mannheim, Germany) was measured by addition of 105 µl cell culture medium. Vero cells were adjusted to the appropriate concentrations (serial dilutions from 666 666 to 5 208 cells/mL) in cell culture medium, and the baseline medium in the E-plate well was substituted by 105 µl of cell suspension. Empty wells were filled with 105 µl PBS to avoid desiccation. Edge wells were not used in order to reduce variability resulting from edge effects [50]. The CI was automatically determined every 15 min by the RT-CES system (Roche Diagnostics, Mannheim, Germany), over a period up to 90 h. CI as a quantitative measure of the status of the cells in an electrode-containing well is calculated by the software according to [34]:

$$CI = \max_{i=1,...,N} \left[\frac{R_{cell}(f_i)}{R_b(f_i)} - 1 \right]$$

Where R_{cell} stands for resistance of the electrode with attached cells, R_b stands for resistance of the electrode without attached cells and N is the number of the frequency points at which the impedance is measured.

RT-CES cytotoxicity assay

After baseline measurement of the E-plate, a Vero cell suspension containing 12 500 cells/well in a volume of 75 μl was seeded into the E-plates. Immediately after the seeding of cells into the wells (i.e. without prior attachment of the cells onto the plate), ricin (either native or denatured), agglutinin, complex matrices, extracts of *R. communis*-containing fertilizer or other plant lectins at given concentrations were added onto the cells in a volume of 30 μl. Each sample was measured at least in duplicate. The CI was automatically determined every 15 min by the RT-CES system (Roche Diagnostics, Mannheim, Germany), over a period of up to 24–42 h. During the incubation, only live cells attached onto the plate and showed a vigorous proliferation (equivalent to an increase in impedance and CI). Depending on the toxin concentration present on the cells, the proliferation was terminated after different time points, followed by detachment and cell death (equivalent to a drop of impedance and CI). To specifically block ricin and agglutinin cytotoxicity, the test compounds were preincubated with 880 μg anti-ricin IgY for 1.5 h at 37°C on a shaker.

MTT cytotoxicity assay

To determine ricin and agglutinin cytotoxicity in a colorimetric endpoint assay, an MTT (3-[4,5-Dimethylthiazol-2-yl]-2,5-diphenyltetrazolium bromide) assay was performed as described previously [14,51]. Briefly, Vero cells (10 000 cells/well) were cultured in a 96-well plate for 18 h. Cells were treated with toxin dilutions for 2 h, washed and further incubated for 20 h in medium. The endpoint viability of the Vero cells was quantified using the CellTiter96 Non-Radioactive Cell Proliferation Assay (Promega, Madison, WI, USA).

Heat inactivation of ricin and analysis of *R. communis*-containing fertilizer

Ricin was diluted in PBS to a concentration of 20 000 ng/ml. In order to guarantee optimal heat transfer, 100 μl solution were incubated in a thin-wall PCR tube for 30 min at 95°C in a PCR cycler, cooled on ice and then used for the experiments, as described for our recent experiments on the stability of botulinum neurotoxins [52].

Ricinus communis-containing fertilizer was ground with a coffee grinder. 2 g of ground fertilizer were mixed with 20 ml of PBS and rotated for 2 h at room temperature followed by centrifugation. The supernatant was filtered through a 70 μm sieve and stored at 4°C. For heat inactivation of the *R. communis*-containing fertilizer, 100 μl of the extract was heated for 30 min at 95°C, as indicated above.

Analysis of complex food samples

Serial dilutions of ricin were spiked into milk, carrot juice or baby food. Spiked and unspiked food samples were diluted 1:4 (v/v) in cell culture medium, centrifuged and filtered through a 0.45 μm filter.

Data analysis

All calculations and figures were obtained using GraphPad Prism software 5.01 (GraphPad, San Diego, CA, USA). The curve fitting of the standard curves was a nonlinear regression: log(inhibitor) vs. response-variable slope (four parameters). Cell viability was either depicted as CI value of the RT-CES system over time, or converted into percent (%) of the control cells alive over toxin concentration. To this end, the CI value of nontreated cells at several time points was set to 100%, and for a given time point the ratio of CI values of toxin-treated cells to nontreated cells was calculated. *In vitro* cytotoxicity at 50% (IC50) was defined as the toxin concentrations required to reduce cell viability by 50% compared to untreated control cells at various time points. Within- and between-run precision was evaluated by the measurement of cytotoxicity of serial dilutions of ricin after 24 h or 42 h, respectively: for within-run precision serial dilutions of ricin (1 ng/mL to 0.03 ng/mL) were performed in four replicates and measured on one day; for between-run precision serial dilutions of ricin were performed on four different days and the results were used for calculation. The coefficient of variation (CV) equals the standard deviation of the concentration-dependent CI values divided by the mean of the CI values.

Supporting Information

Figure S1 Dynamic monitoring of Vero cell proliferation depending on different cell culture conditions. (A) To illustrate the different growth characteristics of Vero cells depending on culture conditions prior to the cytotoxicity assay, the cells were grown in different densities in culture flasks at 10 000 cells/cm^2 (dashed line), 25 000 cells/cm^2 (black line) and 100 000 cells/cm^2 (dotted line). Then Vero cells were trypsinized and seeded in a 96-well E-plate at 12 500 cells/well. Cell proliferation was dynamically monitored every 15 min for 22 h. (B) Vero cells were grown at a density of 25 000 cells/cm^2 in culture flasks before the cytotoxicity assay and removed by trypsinization with either Trypsin (0.2%) containing EDTA (0.02%, black line) or EDTA (0.07%, dotted line). Then the cells were seeded in a 96-well E-plate at 12 500 cells/well and proliferation was dynamically monitored every 15 min for 22 h.

Table S1 Comparison of real-time ricin cytotoxicity assay with other functional ricin detection methods. The table depicts information on different functional assays for ricin detection highlightening their detection principle, measurement parameters, assay time, specificity and the application to detect ricin from complex matrices.

Table S2 Within-run and between-run precision for ricin detection using the novel real-time cytotoxicity assay. The precision of the optimized real-time cytotoxicity assay was evaluated by the determination of the coefficient of variation (CV) analyzing the cytotoxicity data obtained by measuring serial dilutions of ricin on Vero cells after 24 h or 42 h, respectively: For within-run precision serial dilutions of ricin (1 ng/mL to 0.03 ng/mL) were performed in four replicates and measured on one day; for between-run precision serial dilutions of ricin were performed on four different days. The CV near the IC50 value is highlighted in grey.

Table S3   Within-run and between-run precision for ricin detection using the conventional endpoint cytotoxicity assay (MTT assay).
The precision of the MTT assay was evaluated by the determination of the coefficient of variation (CV) analyzing the cytotoxicity data obtained by measuring serial dilutions of ricin on Vero cells after 45 h: For within-run precision serial dilutions of ricin (100 ng/mL to 0.4 ng/mL) were performed in four replicates and measured on one day; for between-run precision serial dilutions of ricin were performed on four different days. The CV near the IC50 value is highlighted in grey.

Table S4   IC50 values for ricin spiked into different food matrices.
Vero cells were exposed to ricin spiked into milk, carrot juice, baby food or medium, respectively. The IC50 values of serial dilutions of ricin in the complex matrices (see Figure 5) after 24 h and 42 h are shown.

Acknowledgments

We thank R. Schade and B. Diemar for helpful support with the IgY preparation. We are grateful to B. Brembs for fruitful discussions.

Author Contributions

Conceived and designed the experiments: DP OS SW BGD. Performed the experiments: DP OS SW. Analyzed the data: DP SW MBD BGD. Contributed reagents/materials/analysis tools: SW SK MBD BGD. Wrote the paper: DP BGD.

References

1. Bradberry SM, Dickers KJ, Rice P, Griffiths GD, Vale JA (2003) Ricin poisoning. Toxicol Rev 22: 65–70.
2. Hong IH, Kwon TE, Lee SK, Park JK, Ki MR, et al. (2011) Fetal death of dogs after the ingestion of a soil conditioner. Exp Toxicol Pathol 63: 113–117.
3. Balint GA (1974) Ricin: the toxic protein of castor oil seeds. Toxicology 2: 77–102.
4. Drociuk D, Fabian GT (2009) It can happen here: ricin in a South Carolina postal facility-2003. J S C Med Assoc 105: 94–96.
5. Khan AS, Levitt AM, Sage MJ (2000) Biological and Chemical Terrorism: Strategic Plan for Preparedness and Response. MMWR Recomm Rep 49: 1–14.
6. Lord MJ, Jolliffe NA, Marsden CJ, Pateman CS, Smith DC, et al. (2003) Ricin. Mechanisms of cytotoxicity. Toxicol Rev 22: 53–64.
7. Endo Y, Tsurugi K (1988) The RNA N-glycosidase activity of ricin A-chain. Nucleic Acids Symp Ser. pp 139–142.
8. Sweeney EC, Tonevitsky AG, Temiakov DE, Agapov II, Saward S, et al. (1997) Preliminary crystallographic characterization of ricin agglutinin. Proteins 28: 586–589.
9. Lord JM, Roberts LM, Robertus JD (1994) Ricin: structure, mode of action, and some current applications. FASEB J 8: 201–208.
10. Audi J, Belson M, Patel M, Schier J, Osterloh J (2005) Ricin poisoning: a comprehensive review. JAMA 294: 2342–2351.
11. Griffiths GD, Lindsay CD, Upshall DG (1994) Examination of the toxicity of several protein toxins of plant origin using bovine pulmonary endothelial cells. Toxicology 90: 11–27.
12. Lin JY, Liu SY (1986) Studies on the antitumor lectins isolated from the seeds of Ricinus communis (castor bean). Toxicon 24: 757–765.
13. Zhan J, Zhou P (2003) A simplified method to evaluate the acute toxicity of ricin and ricinus agglutinin. Toxicology 186: 119–123.
14. Pauly D, Kirchner S, Stoermann B, Schreiber T, Kaulfuss S, et al. (2009) Simultaneous quantification of five bacterial and plant toxins from complex matrices using a multiplexed fluorescent magnetic suspension assay. Analyst 134: 2028–2039.
15. Thullier P, Griffiths G (2009) Broad recognition of ricin toxins prepared from a range of Ricinus cultivars using immunochromatographic tests. Clin Toxicol (Phila) 47: 643–650.
16. Kull S, Pauly D, Störmann B, Kirchner S, Stämmler M, et al. (2010) Multiplex detection of microbial and plant toxins by immunoaffinity enrichment and matrix-assisted laser desorption/ionization mass spectrometry. Anal Chem 82: 2916–2924.
17. Kalb SR, Barr JR (2009) Mass spectrometric detection of ricin and its activity in food and clinical samples. Anal Chem 81: 2037–2042.
18. Bevilacqua VL, Nilles JM, Rice JS, Connell TR, Schenning AM, et al. (2010) Ricin activity assay by direct analysis in real time mass spectrometry detection of adenine release. Anal Chem 82: 798–800.
19. Hines HB, Brueggemann EE, Hale ML (2004) High-performance liquid chromatography-mass selective detection assay for adenine released from a synthetic RNA substrate by ricin A chain. Anal Biochem 330: 119–122.
20. Keener WK, Rivera VR, Young CC, Poli MA (2006) An activity-dependent assay for ricin and related RNA N-glycosidases based on electrochemiluminescence. Anal Biochem 357: 200–207.
21. Roday S, Sturm MB, Blakaj D, Schramm VL (2008) Detection of an abasic site in RNA with stem-loop DNA beacons: application to an activity assay for Ricin Toxin A-Chain. J Biochem Biophys Methods 70: 945–953.
22. He X, Lu S, Cheng LW, Rasooly R, Carter JM (2008) Effect of food matrices on the biological activity of ricin. J Food Prot 71: 2053–2058.
23. Langer M, Rothe M, Eck J, Mockel B, Zinke H (1996) A nonradioactive assay for ribosome-inactivating proteins. Anal Biochem 243: 150–153.
24. Hale ML (2001) Microtiter-based assay for evaluating the biological activity of ribosome-inactivating proteins. Pharmacol Toxicol 88: 255–260.
25. Vang O, Larsen KP, Bog-Hansen TC (1986) A new quantitative and highly specific assay for lectin binding activity. In: Bog-Hansen TC, van Driessche E, eds. Lectins: Biology, Biochemistry, Clinical Biochemistry. Berlin: Walter de Gruyter & co. pp 637–644.
26. Beyer NH, Kogutowska E, Hansen JJ, Engelhart Illigen KE, Heegaard NH (2009) A mouse model for ricin poisoning and for evaluating protective effects of antiricin antibodies. Clin Toxicol (Phila) 47: 219–225.
27. Griffiths GD, Phillips GJ, Holley J (2007) Inhalation toxicology of ricin preparations: animal models, prophylactic and therapeutic approaches to protection. Inhal Toxicol 19: 873–887.
28. Fodstad Ø, Olsnes S, Pihl A (1976) Toxicity, distribution and elimination of the cancerostatic lectins abrin and ricin after parenteral injection into mice. Br J Cancer 34: 418–425.
29. Becher F, Duriez E, Volland H, Tabet JC, Ezan E (2007) Detection of functional ricin by immunoaffinity and liquid chromatography-tandem mass spectrometry. Anal Chem 79: 659–665.
30. Brzezinski JL, Craft DL (2007) Evaluation of an in vitro bioassay for the detection of purified ricin and castor bean in beverages and liquid food matrices. J Food Prot 70: 2377–2382.
31. Colombatti M, Johnson VG, Skopicki HA, Fendley B, Lewis MS, et al. (1987) Identification and characterization of a monoclonal antibody recognizing a galactose-binding domain of the toxin ricin. J Immunol 138: 3339–3344.
32. Halter M, Almeida JL, Tona A, Cole KD, Plant AL, et al. (2009) A mechanistically relevant cytotoxicity assay based on the detection of cellular GFP. Assay Drug Dev Technol 7: 356–365.
33. Oda T, Komatsu N, Muramatsu T (1997) Cell lysis induced by ricin D and ricin E in various cell lines. Biosci Biotechnol Biochem 61: 291–297.
34. Atienza JM, Zhu J, Wang X, Xu X, Abassi Y (2005) Dynamic monitoring of cell adhesion and spreading on microelectronic sensor arrays. J Biomol Screen 10: 795–805.
35. Roels S, Coopman V, Vanhaelen P, Cordonnier J (2010) Lethal ricin intoxication in two adult dogs: toxicologic and histopathologic findings. J Vet Diagn Invest 22: 466–468.
36. Vervecken W, Kleff S, Pfüller U, Bussing A (2000) Induction of apoptosis by mistletoe lectin I and its subunits. No evidence for cytotoxic effects caused by isolated A- and B-chains. Int J Biochem Cell Biol 32: 317–326.
37. Olsnes S, Refsnes K, Pihl A (1974) Mechanism of action of the toxic lectins abrin and ricin. Nature 249: 627–631.
38. Pfüller U, Franz H, Pfüller K, Junghahn I, Bielka H (1988) Selective inactivation of mistletoe lectin I and ricin using ethylammonium nitrate - a molten salt liquid at room temperature. Lectins - Biology, Biochemistry, Clinical Biochemistry. St. Louis, Missouri USA: Sigma Chemical Company. pp 299–304.
39. Garber EA (2008) Toxicity and detection of ricin and abrin in beverages. J Food Prot 71: 1875–1883.
40. McGrath SC, Schieltz DM, McWilliams LG, Pirkle JL, Barr JR (2011) Detection and quantification of ricin in beverages using isotope dilution tandem mass spectrometry. Anal Chem 83: 2897–2905.
41. Schieltz DM, McGrath SC, McWilliams LG, Rees J, Bowen MD, et al. (2011) Analysis of active ricin and castor bean proteins in a ricin preparation, castor bean extract, and surface swabs from a public health investigation. Forensic Sci Int 209: 70–79.
42. Wahome PG, Bai Y, Neal LM, Robertus JD, Mantis NJ (2010) Identification of small-molecule inhibitors of ricin and shiga toxin using a cell-based high-throughput screen. Toxicon 56: 313–323.

43. Pelat T, Hust M, Hale M, Lefranc MP, Dubel S, et al. (2009) Isolation of a human-like antibody fragment (scFv) that neutralizes ricin biological activity. BMC Biotechnol 9: 60.

44. Bagaria A, Surendranath K, Ramagopal UA, Ramakumar S, Karande AA (2006) Structure-function analysis and insights into the reduced toxicity of Abrus precatorius agglutinin I in relation to abrin. J Biol Chem 281: 34465–34474.

45. Jang HY, Kim JH (1993) Isolation and Biochemical Properties of Ricin from Ricinus communis. Korean Biochem J 26: 98–104.

46. Worbs S, Köhler K, Pauly D, Avondet MA, Schaer M, et al. (2011) Ricinus communis Intoxications in Human and Veterinary Medicine-A Summary of Real Cases. Toxins 3: 1332–72.

47. Albretsen JC, Gwaltney-Brant SM, Khan SA (2000) Evaluation of castor bean toxicosis in dogs: 98 cases. J Am Anim Hosp Assoc 36: 229–233.

48. Lin TT, Li SL (1980) Purification and physicochemical properties of ricins and agglutinins from Ricinus communis. Eur J Biochem 105: 453–459.

49. Pauly D, Dorner M, Zhang X, Hlinak A, Dorner B, et al. (2009) Monitoring of laying capacity, immunoglobulin Y concentration, and antibody titer development in chickens immunized with ricin and botulinum toxins over a two-year period. Poult Sci 88: 281–290.

50. Lundholt BK, Scudder KM, Pagliaro L (2003) A simple technique for reducing edge effect in cell-based assays. J Biomol Screen 8: 566–570.

51. McGuinness CR, Mantis NJ (2006) Characterization of a novel high-affinity monoclonal immunoglobulin G antibody against the ricin B subunit. Infect Immun 74: 3463–3470.

52. Weingart OG, Schreiber T, Mascher C, Pauly D, Dorner MB, et al. (2010) The case of botulinum toxin in milk: experimental data. Appl Environ Microbiol 76: 3293–3300.

17

Understanding Dry Matter and Nitrogen Accumulation with Time-Course for High-Yielding Wheat Production in China

Qingfeng Meng[1], Shanchao Yue[1], Xinping Chen[1], Zhenling Cui[1]*, Youliang Ye[2], Wenqi Ma[3], Yanan Tong[4], Fusuo Zhang[1]

1 Center for Resources, Environment and Food Security, China Agricultural University, Beijing, China, 2 College of Resources and Environmental Sciences, Henan Agricultural University, Zhengzhou, China, 3 College of Resources and Environmental Sciences, Hebei Agricultural University, Baoding, China, 4 College of Resources and Environmental Sciences, Northwest Sci-Tech University of Agriculture and Forestry, Yangling, China

Abstract

Understanding the time-course of dry matter (DM) and nitrogen (N) accumulation in terms of yield–trait relationships is essential to simultaneously increase grain yield and synchronize N demand and N supply. We collected 413 data points from 11 field experiments to address patterns of DM and N accumulation with time in relation to grain yield and management of winter wheat in China. Detailed growth analysis was conducted at the Zadok growth stages (GS) 25 (regreening), GS30 (stem elongation), GS60 (anthesis), and GS100 (maturity) in all experiments, including DM and N accumulation. Grain yield averaged 7.3 Mg ha^{-1}, ranging from 2.1 to 11.2 Mg ha^{-1}. The percent N accumulation was consistent prior to DM accumulation, while both DM and N accumulation increased continuously with growing time. Both the highest and fastest DM and N accumulations were observed from stem elongation to the anthesis stage. Significant correlations between grain yield and DM and N accumulation were found at each of the four growth stages, although no positive relationship was observed between grain yield and harvest index or N harvest index. The yield increase from 7–9 Mg ha^{-1} to >9 Mg ha^{-1} was mainly attributed to increased DM and N accumulation from stem elongation to anthesis. Although applying more N fertilizer increased N accumulation during this stage, DM accumulation was not improved, indicating that N fertilizer management and related agronomic management should be intensified synchronously across the wheat growing season to simultaneously achieve high yields and match N demand and N supply.

Editor: Mingliang Xu, China Agricultural University, China

Funding: The authors thank the National Basic Research Program of China (973, Program: 2009CB118606), and Special Fund for Agriculture Profession (200803030). The funders had no role in study design, data collection and analysis, decision to publish, or preparation of the manuscript.

Competing Interests: The authors have declared that no competing interests exist.

* E-mail: cuizl@cau.edu.cn

Introduction

As the largest wheat producer and consumer in the world, China produced around 115 million t wheat with a planting area of about 24.3 million ha in 2010 [1]. To meet the projected demands of population growth and increasing consumption, cereal production must increase by 70–100% until 2050 [2,3]. However, as one of major cereal, yield growth of wheat in China has dropped from 9% in 1970s to <3% annually in 2000s [1]. Because of the great difficulty of improving yield potential over the short-term through genetic improvement [4], great efforts should be made to improve agronomical management to enhance average farm yields. Knowledge of DM and nutrient accumulation associated with yield–trait relationships in crops would supply an efficient tool to improve management efficiency and thus increase yield.

Grain yield has usually been positively correlated with total DM production and nutrient accumulation in crops [5]. Meanwhile, DM accumulation and nutrient accumulation vary with growth stage of crops [6]. Many previous studies indicated that DM and N accumulation for wheat occurred mainly pre-anthesis, and that grain yields greatly depended on the translocation of pre-anthesis assimilates and N accumulation [7–10]. However, in practice, a significant correlation has also been found between grain yield and DM production post-anthesis in high-yield systems in China [11,12]. Because few studies have focused on the relationship between grain yield and DM and N accumulation with time-course (e.g., [13]), the pattern of DM and N accumulation pre- and post-anthesis remains unclear.

Due to lack of proper knowledge, most Chinese farmers tried to increase wheat DM and N accumulation as high as possible at earlier stage to pursue high yield in practice. For example, farmers applied a large amount of irrigation and N fertilizer at early stage (e.g., regreening stage, GS25) [14,15], which leaded to higher DM through more spring tillering. However, higher tiller population would lead to more non-reproductive tillers [16], and more risk of lodging, pests and diseases when it is more than optimum effective plant population. Meanwhile, the non-reproductive tillers would have a negative effect as competitors for assimilates and plant nutrients [17,18]. Thus, further studies on DM and N accumulation with time-course are needed, which could help to identify the critical stage of optimal spring tillering through management to improve grain yield.

N management also influences the dynamics of DM and N accumulation by crops [19]. For example, N overuse leads to higher plant N concentrations due to so-called "luxury accumulation" by crops, while N shortage decreases plant N concentrations and sacrifices some grain yield [20,21]. Recently, N recommendations in crop production systems emphasized the need for greater synchrony between crop N demand and the N supply from all sources throughout the growing season [22,23]. However, most of previous studies have been made to quantify total N accumulation at harvest through yield goals and crop models [20,24–26]. Due to the limited understanding dynamics of N accumulation, N supply in current N management practices often does not match crop demand. For instance, farmers typically apply a large amount of N fertilizer for wheat before sowing (often >50% total N fertilizer), which results in large N losses to the environment because the plant capacity for N accumulation is small during this period [27].

Thus, much effort should be made for dynamic DM and N accumulations associated with yield–trait relationship to improve management efficiency and simultaneously achieve high yields and synchronize N demand and N supply. In this study, we collected 413 field-year data at nine sites in major wheat area in China, to investigate the relationship between DM and N accumulation and grain yield, the pattern of DM and N accumulation for different yield levels with time, and the dynamics of DM and N accumulation with different N management practices.

Materials and Methods

Eleven field experiments at nine sites in four key winter wheat domains of China were conducted in Dongbeiwang (DBW) in Beijing; Xiangyun (XY), Zhaobao (ZB), and Lankao (LK) in Henan Province; Dingzhou (DZ) and Quzhou (QZ) in Hebei Province; and Yangling (YL) in Shannxi Province. A typical winter wheat–summer maize rotation system was adopted at all sites. No specific permissions were required for these locations. The field studies did not involve endangered or protected species.

The amount and distribution pattern of precipitation varied widely from year to year, as affected by the continental monsoon climate. Wheat was irrigated two to four times based on soil water content at about 100 mm at each time point. More details such as experimental year, soil texture, organic matter content, total N content, Olsen-P, and NH_4OAc-K in the different experimental locations were shown in Table S1.

Experimental Design

For the eleven field experiments, the details of treatments, including varieties and N application rates were listed in Table S2. At least three N treatments were included for all eight N level experiments: no N as a control (N-0), an optimal N rate (N-opt), and the farmer's N practice (FNP). At XY (2008–2009, 2009–2010), LK, DZ and QZ sites, N treatments at sub- or supra-optimal N rates were also applied in experiments with five or six N treatments, including 40%, 50%, 70% and 75% of N-opt, 125%, 130%, and 150% of N-opt according to the experimental design at each site.

The optimal N rate was based on either in-season root zone N management (DBW and QZ sites) or yield goal-based N recommendation (XY, ZB, LK and DZ sites). For the in-season root zone N management (IRNM), wheat growing season was divided into two periods: from sowing to the stem elongation stage, and from the stem elongation stage to the maturity. The amount of

N fertilizer applied at the beginning of each growing period was determined by deducting the amount of soil N_{min} (NH_4^+–N+NO_3^-–N) in the root-zone from the target N value, which was estimated based on the yield target and crop N accumulation. Detailed descriptions of the target N value and the soil N_{min} measurement have been previously reported [27,28]. The yield goal-based N recommendation was derived from the mass balance approach [29]. In this method, optimal N rate was estimated as total N accumulation of the grain less all of the other sources of grain N and adjusted for inefficiencies in the ability of the crop to recover fertilizer N from the soil [30]:

$$\text{Optimal N rate} = \left(\begin{array}{l} \text{N accumulation in the harvested grain} \\ - \text{ Soil N in the harvest grain} \end{array} \right) \quad (1)$$
$$/ \text{ Fertilizer N use efficiency}$$

where optimal N rate was estimated for a selected yield goal, fertilizer N use efficiency was the proportion of fertilizer N applied to the soil that was recovered in the grain. For the FNP treatment, the N rate varied from 300 to 360 kg ha^{-1}, mostly with a split application (50%:50%) as a broadcast application at pre-sowing, and as a side-dress application near the stem elongation stage.

For other three cropping systems experiments (DZ, QZ and YL sites), details of the system design were in the Text S1. In this study, the current farmers' system (FP) and higher yield system from agronomists (HY) were considered N-over systems, while the optimized system (HYHR) and integrated soil–crop management system (ISSM) were considered N-opt systems.

All of the experiments consisted of a randomized complete block design with three to four replications. The plot area varied from 20 m^2 at XY (2007–2008) to 1800 m^2 at QZ (2007–2010). Urea as the main N fertilizer has been used for all field experiments. Phosphorus (P), potassium (K) and other nutrients were applied as needed according to soil tests. No organic manure was applied. At all sites, winter wheat was seeded in early to mid-October and harvested the following June. During the study year, no obvious water, weed, pest, or disease stresses were observed during the wheat growing season at any site. No obvious winter killing happened for all field experiments.

Sampling and Laboratory Procedures

Before sowing, soil samples from the 0–30 cm soil layer were air-dried and sieved to remove un-decomposed plant material. Then the sieved samples were used to measure soil organic matter, total N, Olsen-P, and ammonium acetate-extractable K. To determine DM and N content, plant samples were collected four times: at the regreening (GS25), stem elongation (GS30), anthesis (GS60), and maturity (GS100) stages. Plant samples were dried at 60°C to determine the DM, and sub-samples were taken to measure the N content using the Kjeldahl method. At harvest, areas ranging from 6 m^2 at XY (2007–2008) to 18 m^2 at QZ (2007–2010) were harvested manually and dried in an oven at 60°C to calculate grain yield. The grain yield value was adjusted to 14% moisture. The number of grains per square meter was counted manually for at least two rows within 1.0 m, and the number per ear was counted manually to estimate a mean value from more than 20 ears. For the thousand kernel weight (TGW), 1000 grains were randomly counted and weighed.

At XY (2007–2008, 2008–2009, 2009–2010), ZB (2007–2008), LK (2008–2009), QZ (2008–2009), DZ (2008–2010), and YL (2008–2009), the total tillers in the N-opt treatment were investigated before tillering (GS20), before winter (GS23), and at stem elongation, anthesis, and maturity.

Data Analysis

According to grain yield, all data from the nine sites were divided into three groups: <7, 7–9, and >9 Mg ha^{-1}. According this yield range, wheat DM and N accumulation were analyzed with time -course. According N application, all data were also divided into three groups: N-0, N-opt and N-over (125%, 130%, and 150% of N-opt, and FNP) for DM and N accumulation analysis. The data were then further categorized into nine groups according to the yield ranges (1) <7 Mg ha^{-1}, N-0; (2) <7 Mg ha^{-1}, N-opt; (3) <7 Mg ha^{-1}, N-over; (4) 7–9 Mg ha^{-1}, N-0; (5) 7–9 Mg ha^{-1}, N-opt; (6) 7–9 Mg ha^{-1}, N-over; (7) >9 Mg ha^{-1}, N-0; (8) >9 Mg ha^{-1}, N-opt, (9) >9 Mg ha^{-1}, N-over. The investigated stems dynamics in the N-opt treatment at six sites (XY (2007–2008, 2008–2009, 2009–2010), ZB (2007–2008), LK (2008–2009), QZ (2008–2009), DZ (2008–2010), and YL (2008–2009)) were also grouped according to grain yield (<7, 7–9, and >9 Mg ha^{-1}).

For all collected data, the correlation coefficients (r) between grain yield and average yield components, DM production, and N accumulation at different growth stages, were analyzed. At harvest, the amount of DM production translocation was calculated as the difference between dry DM in the stover at anthesis and at harvest. N accumulation translocation was calculated as the difference between N accumulation in the stover at anthesis and at harvest. The rate of fertile tillers from stem-elongation to anthesis was calculated according the result that the number of total ears at anthesis was divided by the tiller number at stem-elongation stage. N accumulation was calculated as:

$$N\ accumulation = Dry\ matter \times N\ concentration \quad (2)$$

At regreening, stem elongation and anthesis stage, N accumulation was calculated as straw DM multiplying straw N concentration. At maturity, N accumulation was the sum of straw and grain N accumulation.

Results

Grain Yield, DM and N Accumulation, and N Concentration

Overall, wheat grain yield averaged 7.3 Mg ha^{-1}, ranging from 2.1 Mg ha^{-1} to 11.2 Mg ha^{-1} (Table 1). According to published statistical data, the average wheat yield was 4.7 Mg ha^{-1} for China and 3.0 Mg ha^{-1} for the whole world in 2009 [1]; thus, the mean yield observed in the present study is 155% and 243% higher than the national and world yields, respectively. However, this is still lower than that in high-yield wheat production areas of Europe, which produce more than 8.0 Mg ha^{-1} [1].

Sixty-four percent of total DM accumulated pre-anthesis, while the other 36% accumulated post-anthesis (Figure S1). Before anthesis, 10% and 20% of DM accumulated in the regreening and stem-elongation stages, respectively. From stem elongation to anthesis, DM accumulated the most and the fastest across all four stages. More than 6.3 Mg ha^{-1} DM accumulated from stem elongation to anthesis, with an average accumulation rate of 253 kg ha^{-1} day^{-1} (Figure 1A). A total of 1.5 Mg ha^{-1} DM accumulated from sowing to regreening, 1.4 Mg ha^{-1} DM from regreening to stem elongation, and 5.1 Mg ha^{-1} DM accumulated post-anthesis, while the associated daily accumulation rate averaged 21, 38, and 130 kg ha^{-1} day^{-1}, respectively.

The percent N accumulation was consistent prior to DM accumulation (Figure S1). Unlike DM accumulation, 80% of N accumulated pre-anthesis; 27%, 20%, and 33% of N accumulated from sowing to regreening, from regreening to stem elongation, and from stem elongation to anthesis, respectively. Similar to DM accumulation, the highest and fastest N accumulation was observed from stem elongation to anthesis. During this period, 2.44 kg N ha^{-1} accumulated per day and 61 kg N ha^{-1} accumulated in total (Figure 1B). Moreover, 50 kg ha^{-1} N, with 0.7 kg N ha^{-1} per day, accumulated from sowing to regreening; 38 kg ha^{-1} N, with 1.1 kg N ha^{-1} per day, accumulated from regreening to stem elongation; and 37 kg ha^{-1} N, with 1.0 kg N ha^{-1} per day, accumulated post-anthesis.

The relationships between grain yield and DM production, N accumulation, and N concentration for different growth stages and yield components were shown in Table 2. For yield components, grains per ear was significantly correlated with grain yield ($r = 0.54$). Significant correlations between DM and grain yield were observed in each stage (0.58–0.95). Grain yield was significantly correlated with DM accumulation from stem elongation to anthesis and post-anthesis. No positive correlation was found between grain yield and harvest index.

Similar to DM production, significant correlations between grain yield and N accumulation were found at each of stage ($r = 0.54$–0.81; Table 2). The amount of N translocation from stover to grain was important for yield increase ($r = 0.47$), although DM translocation was not significant. Between grain yield and N concentration, no significant relationship was found with an exception of grain N concentration at maturity (Table 2). Moreover, no significant correlation was found between grain yield and N harvest index.

Dynamics of DM and N Accumulation with Different Yield Levels

To further understand the relationship between yield and temporal dynamics of DM and N accumulation, the collected data were further categorized into three groups according to the yield range (Table 1). In total, 179, 112, and 122 data points were grouped according to the yield ranges <7 Mg ha^{-1}, 7–9 Mg ha^{-1}, and >9 Mg ha^{-1}, respectively. The grain yield averaged 5.3 Mg ha^{-1} (2.1–7.0 Mg ha^{-1}), 8.1 Mg ha^{-1} (7.0–9.0 Mg ha^{-1}), and 9.7 Mg ha^{-1} (9.0–11.2 Mg ha^{-1}), respectively. According to the yield components, the yield difference was mainly related to the grain number per square meter (Table S3), which was determined by ear number and grains per ear. The grain number per square meter averaged 24,000 m^{-2} for 7–9 Mg ha^{-1} and 28,000 m^{-2} for >9 Mg ha^{-1}, an increase of 33% and 56% compared to the yield range of <7 Mg ha^{-1} (18,000 m^{-2}). Similar grain weight per 1000 grains (40–42 g) was observed among all three yield levels.

Across the whole wheat season, wheat DM and N accumulation of the yield for <7 Mg ha^{-1} were consistently lower than those for 7–9 Mg ha^{-1} and >9 Mg ha^{-1} (Figure 2). This difference started from the beginning of the growth stage. From the sowing to the regreening stage, 2.2 Mg ha^{-1} DM accumulated in 7–9 Mg ha^{-1}, 314% higher than that in <7 Mg ha^{-1} (0.7 Mg ha^{-1}) (Figure 2A). Similarly, the 71 kg ha^{-1} of N that accumulated in 7–9 Mg ha^{-1} is substantially higher than the 27 kg ha^{-1} of N that accumulated in <7 Mg ha^{-1} during this stage (Figure 2B). At regreening stage, the percentage of DM (7%) and N accumulation (20%) to total in <7 Mg ha^{-1} was 47% and 43% lower compared with that in 7–9 Mg ha^{-1} yield range (14% and 34%), respectively. The lower DM and N accumulation starting at sowing to regreening and continuing through the subsequent growth stages in yield range

Table 1. Descriptive statistics of yield for total samples, three yield ranges, and nitrogen levels (Mg ha^{-1}).

	n[a]	Mean	SD[b]	Minimum	25%Q[c]	Median	75%Q	Maximum
Total	413	7.3	2.1	2.1	5.8	7.5	9.2	11.2
Yield category								
<7 Mg ha^{-1}	179	5.3	1.3	2.1	4.2	5.6	6.4	7
7–9 Mg ha^{-1}	112	8.1	0.6	7	7.5	8.1	8.6	9
>9 Mg ha^{-1}	122	9.7	0.5	9	9.4	9.6	9.8	11.2
N-level								
N-0	79	6.6	2.5	2.1	4.1	7.3	8.7	10.4
N-opt	126	7.6	2	3.1	6.1	7.5	9.4	11.2
N-over	154	7.5	1.8	3.1	6.2	7.7	9.3	10.6

[a]n = number of observation.
[b]SD = standard deviation.
[c]Q = quartile.

<7 Mg ha^{-1}, suggested poor agronomic management (e.g., sowing quality, late sowing) may be the main cause, as the yield potential and weather conditions were similar at all sites.

Compared to the yield of 7–9 Mg ha^{-1}, DM and N accumulation at a yield of >9 M ha^{-1} was similar before the stem-elongation stage (Figure 2). The greatest difference between these two yield levels for DM and N accumulation took place from stem elongation to anthesis. During this stage, DM accumulation increased from 6.3 Mg ha^{-1} in 7–9 Mg ha^{-1} to 8.7 Mg ha^{-1} in >9 Mg ha^{-1} and was associated with an increase in N accumulation from 55 kg ha^{-1} to 83 kg ha^{-1}. Meanwhile, the 46% of DM and 35% of N accumulation to total from stem elongation stage to anthesis in >9 Mg ha^{-1} was 15% and 30% higher compared with that in 7–9 Mg ha^{-1} (40% and 27%), respectively. These results showed that DM and N accumulation from stem elongation to anthesis was very important to achieve the

>9 Mg ha^{-1} grain yield. Post-anthesis, 5.6 Mg ha^{-1} DM and 45 kg ha^{-1} N accumulated in 7–9 Mg ha^{-1} while DM and N accumulation averaged 6.8 Mg ha^{-1} and 50 kg ha^{-1} in >9 Mg ha^{-1}, respectively.

The yield range <7 Mg ha^{-1} group always had the highest N concentration for all growth stages. At regreening stage, the N concentration in <7 Mg ha^{-1} averaged 3.72%, 13–16% higher than that in both 7–9 Mg ha^{-1} and >9 Mg ha^{-1} yield ranges (3.21–3.29%). At stem elongation stage, the N concentration in <7 Mg ha^{-1} group was 3.32%, 6–16% higher than that in other two yield ranges (2.85–3.13%). At anthesis, 1.75% of N concentration in <7 Mg ha^{-1} was 7–13% higher than in other two yield ranges (1.55–1.63%). At maturity, N concentration in <7 Mg ha^{-1} averaged 1.36%, 4–8% higher than that in other two yield ranges (1.26–1.31%). This may be because of the lower DM accumulation in <7 Mg ha^{-1}. For yield ranges 7–9 Mg ha^{-1} and

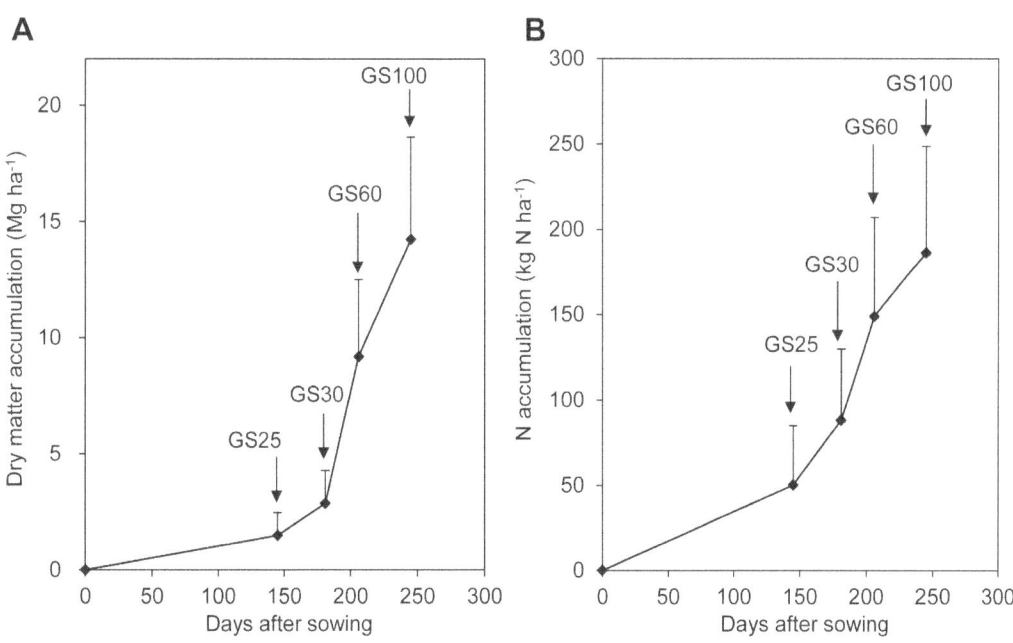

Figure 1. Dynamics of dry matter (A) and nitrogen (B) accumulation from sowing to maturity (n = 413). (GS25, GS30, GS60, and GS100 are the regreening, stem elongation, anthesis, and maturity stages, respectively.).

Table 2. Average yield components, dry matter and nitrogen accumulation in different growth stages, and correlative coefficient (r) between grain yield and these parameters for all samples (n = 413).

		Mean	r
Yield components	Ears number (10^4 ha^{-1})	600	0.36
	Grains per ear	39	0.54**
	Grain weight (g 1000^{-1})	41	0.10
Dry matter accumulation	Regreen (Mg ha^{-1})	1.5	0.60**
	Regreen to stem elongation (Mg ha^{-1})	1.4	0.29
	Stem elongation (Mg ha^{-1})	2.9	0.58**
	Stem elongation to anthesis (Mg ha^{-1})	6.3	0.65**
	Anthesis (Mg ha^{-1})	9.2	0.82**
	Post-anthesis (Mg ha^{-1})	5.1	0.70**
	Maturity (Mg ha^{-1})	14.2	0.95**
	Translocation (Mg ha^{-1})	1.3	0.29
N accumulation	Regreen (kg ha^{-1})	50	0.54**
	Regreen to stem elongation (kg ha^{-1})	38	0.24
	Stem elongation (kg ha^{-1})	88	0.59**
	Stem elongation to anthesis (kg ha^{-1})	61	0.37
	Anthesis (kg ha^{-1})	149	0.71**
	Post-anthesis (kg ha^{-1})	37	0.23
	Maturity (kg ha^{-1})	186	0.81**
	Translocation (kg ha^{-1})	97	0.47**
N concentration	Regreening-straw (%)	3.45	−0.29
	Stem elongation-straw (%)	3.14	0.00
	Anthesis-straw (%)	1.66	−0.14
	Maturity-straw (%)	0.63	0.24
	Maturity-grain (%)	2.15	−0.43**
HI and NHI	Harvest index	0.45	0.02
	N harvest index	0.74	−0.38

**p < 0.01.

>9 Mg ha^{-1}, N concentration was similar for most stages but not at stem elongation stage. At stem elongation stage, N concentration in >9 Mg ha^{-1} was 3.13%, 10% higher than that in 7–9 Mg ha^{-1} (2.85%).

To further understand DM and N accumulation with time in the three yield levels, wheat stem dynamics was investigated at six sites with optimized N treatment (Figure 3). Although the basic seedling was similar among the three yield ranges, total stems differed before winter. Total stems in <7 Mg ha^{-1} was consistently lower compared to that in 7–9 Mg ha^{-1} and >9 Mg ha^{-1} from wintering to maturity. This further verified that sowing quality or other related agronomic management practices were poor and did not supply the optimal growth conditions for wheat. The final ear number was 0.77 million ha^{-1} higher in 9 Mg ha^{-1} than in 7–9 Mg ha^{-1}, although total tillers was 2.74 million ha^{-1} higher in 7–9 Mg ha^{-1} in the stem-elongation stage. As a result, earbering tiller percentage averaged 44% (from 16.69 to 7.36 million ha^{-1}) in >9 Mg ha^{-1} from the stem-elongation stage to maturity while it averaged only 34% (from 19.43 to 6.59 million ha^{-1}) in 7–9 Mg ha^{-1}. This suggested that the final ear number was determined by tiller quality, not total number, in earlier growth stages. In the stem-elongation stage, the N concentration in the stover for >9 Mg ha^{-1} averaged 3.27 g kg^{-1}, 7% higher than the 3.07 g kg^{-1} for 7–9 Mg ha^{-1} (Table 3). The higher N concentration in the plant in the stem-elongation stage may lead to higher stem quality (e.g., higher tiller weight), and thus more stems would survive from stem elongation to anthesis.

Dynamics of DM and N Accumulation with Different N Management Regimes

To further compare the partitioning of DM and N accumulation among different N application and yield levels, pooled data from all nine sites were grouped into three N application categories: N-0, N-opt, and N-over (Table 1). Accordingly, 79 samples with a mean yield of 6.6 Mg ha^{-1} (2.1–10.4 Mg ha^{-1}), 126 samples with a mean yield of 7.6 Mg ha^{-1} (3.1–11.2 Mg ha^{-1}), and 154 samples with a mean yield of 7.5 Mg ha^{-1} (3.1–10.6 Mg ha^{-1}) were grouped according to these three N application levels. Compared to the N-0 treatment, DM and N accumulation increased for the N-opt and N-over treatments in each growth stage for each yield range (Figure 4). Between N-opt and N-over treatments, DM accumulation was similar at each growth stage (Figure 4A). N accumulation was also similar between N-opt and N-over treatments in most growth stages but not at anthesis (Figure 4B). N accumulation averaged around 55 kg ha^{-1} at regreening stage and 100 kg ha^{-1} at stem elongation stage for both N-opt and N-over treatments. At anthesis, N accumulation in N-opt averaged 159 kg ha^{-1}, 6% lower compared with that in N-over (169 kg ha^{-1}). At maturity, N accumulation averaged about 200 kg ha^{-1} for both N-opt and N-over treatments. However, the N rate in all N-opt management averaged 179 kg ha^{-1} (n = 126), just 59% of the 304 kg ha^{-1} N rate in N-over (n = 154). This showed that more N fertilizer input over the rational level did not lead to more DM and N accumulation except N accumulation from stem elongation to anthesis, and not resulted in a higher grain yield.

For further comparison, the data for each N level were categorized according to the three yield ranges (<7, 7–9, and >9 Mg ha^{-1}). DM and N accumulation in the <7 Mg ha^{-1} yield was consistently lower than those in the 7–9 and >9 Mg ha^{-1} (Figure 5). When N was deficient, the difference in N accumulation occurred throughout the whole growing season in the 7–9 and >9 Mg ha^{-1} yield ranges, and the difference in DM accumulation was mainly post-anthesis (Figure 5A, D). Post-anthesis, DM accumulation in >9 Mg ha^{-1} averaged 6.5 Mg ha^{-1} (10.9–17.4 Mg ha^{-1}), 20% higher than that in 7–9 Mg ha^{-1} (9.9–15.3 Mg ha^{-1}) in the N-0 treatment (Figure 5A). When N was applied rationally, the differences in DM and N accumulation between the yield ranges of 7–9 and >9 Mg ha^{-1} occurred from stem elongation to maturity, particularly during the stem elongation to anthesis stage (Figure 5B, E). During this stage, 9.0 Mg ha^{-1} DM accumulation in >9 Mg ha^{-1} was 67% higher than that in 7–9 Mg ha^{-1} (5.4 Mg ha^{-1}), while 40 kg ha^{-1} more N accumulated in >9 Mg ha^{-1}. Post-anthesis, 0.7 Mg ha^{-1} more DM accumulated in >9 Mg ha^{-1} than in 7–9 Mg ha^{-1}, while the same 45 kg ha^{-1} N accumulated in both yield ranges. Unlike N-opt, the N accumulation was similar between yield ranges of 7–9 and >9 Mg ha^{-1} from sowing to anthesis when N was overused (Figure 5F). At anthesis, N accumulation averaged around 190 kg ha^{-1} for both 7–9 Mg ha^{-1} and >9 Mg ha^{-1}, which was similar to the N accumulation in >9 Mg ha^{-1} in the same stage in the N-

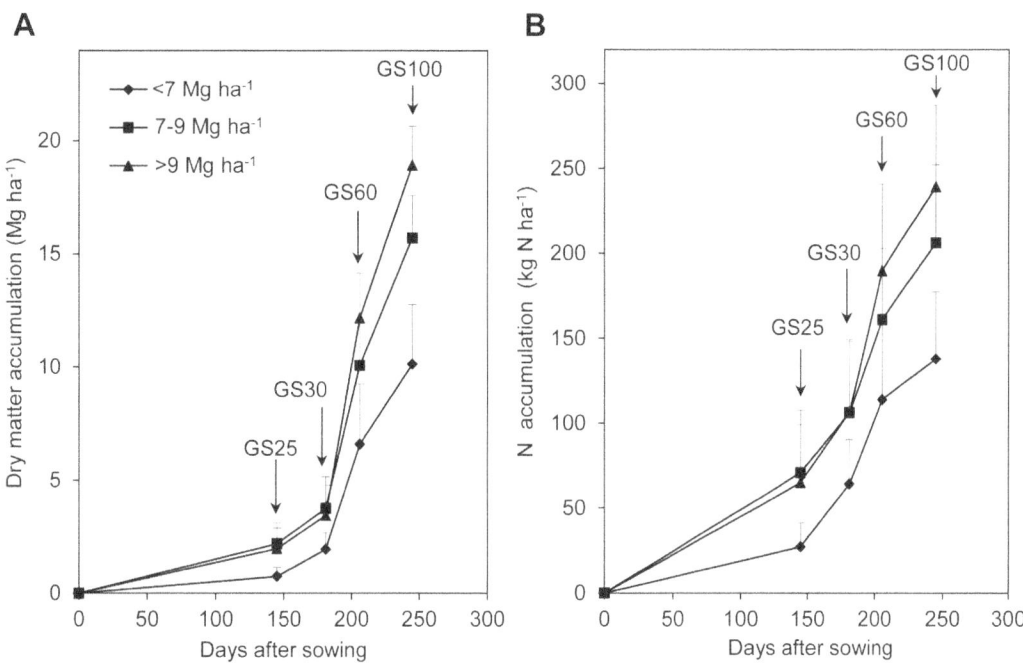

Figure 2. Total dry matter (A) and nitrogen (B) accumulation dynamics at yield ranges of <7 Mg ha^{-1} (n = 179), 7–9 Mg ha^{-1} (n = 112), and >9 Mg ha^{-1} (n = 122). (GS25, GS30, GS60, and GS100 are the regreening, stem elongation, anthesis, and maturity stages, respectively.).

opt treatment. Moreover, DM accumulation was similar at this stage in >9 Mg ha^{-1} for both N-opt and N-over (Figure 5B, C). However, DM in 7–9 Mg ha^{-1} in this stage was 21% lower than that in >9 Mg ha^{-1}, while DM for both of these was around 4.0 Mg ha^{-1} in stem-elongation stage in the N-over treatment (Figure 5C). These results indicated that higher N accumulation did not lead to more DM accumulation synchronously when N fertilizer was overused and thus N luxury accumulation occurred during growth, especially from the stem-elongation stage to anthesis.

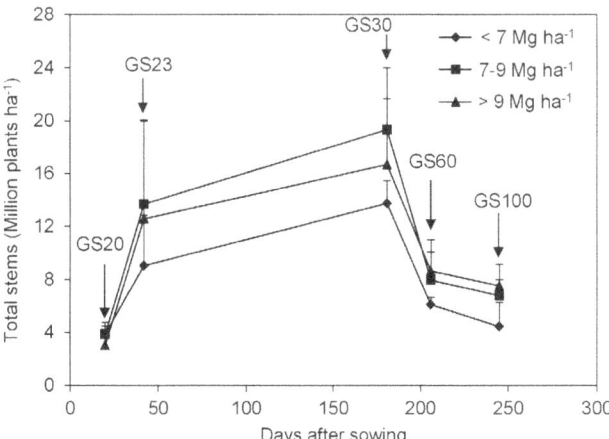

Figure 3. Wheat stem dynamics at grain yields of <7 Mg ha^{-1} (n = 19), 7–9 Mg ha^{-1} (n = 21), and >9 Mg ha^{-1} (n = 33) with optimized N treatment. (GS20, GS23, GS30, GS60, and GS100 are the pre-tillering, before winter, stem elongation, anthesis, and maturity stages, respectively.).

Discussion

We found that wheat yield improvement greatly depended on an increase in DM and N accumulation across the whole growing season. This finding contradicts some previous results, which showed that grain yield was mainly associated with the pre-anthesis assimilate contribution to grain filling and greater DM translocation efficiency [31,32]. This is because improving the harvest index (by introducing dwarfing genes into new varieties) has traditionally been considered the most important process for increasing the wheat yield potential of modern wheat varieties [33–35]. However, the utility of this strategy is limited because of the need to maintain sufficient leaf area and stem dry matter for interception of solar radiation, physical support, and storage of assimilates and N used in grain filling [34]. In the present study, the harvest index averaged 0.45, which was not significantly correlated with grain yield (Table 2). Given the relatively small possibility of further increasing yield by improving the harvest index, the greater yield potential must come from increases in net primary productivity (DM) [36]. The view that increasing DM and N accumulation over the entire growing season is important to achieve high grain yield, is also consistent with recent genetic gains in wheat breeding found in some studies [37,38].

We found a positive correlation between grain yield and DM accumulation post-anthesis (Table 2), in agreement with other research on modern wheat production in China [11,12]. Furthermore, yield increase was mainly attributable to DM and N accumulation from stem elongation to anthesis and post-anthesis (Figure 2). The stage from stem elongation to anthesis accumulated the most DM and N, and the accumulation rate was also highest among the four growth seasons (Figure 1). In plant physiology, the period a few weeks before anthesis (from terminal spikelet initiation to anthesis) is of paramount importance in determining the number of fertile florets at anthesis and final grain yield [39,40].

Table 3. Stover nitrogen concentration (%) from stem elongation to maturity and grain N concentration at maturity for the N-opt treatment for yield ranges of <7 Mg ha^{-1} (n = 19), 7–9 Mg ha^{-1} (n = 21), and >9 Mg ha^{-1} (n = 33), in which total stem dynamics was investigated.

	Stover			Grain
	Stem elongation (GS30)	Anthesis (GS60)	Maturity (GS100)	Maturity (GS100)
<7 Mg ha^{-1}	3.40±0.48[a]	1.74±0.18	0.50±0.07	2.30±0.14
7–9 Mg ha^{-1}	3.07±0.45	1.67±0.32	0.58±0.21	2.20±0.20
>9 Mg ha^{-1}	3.27±0.54	1.56±0.20	0.62±0.18	1.99±0.31

[a]average±SD.

The final ear number was similar or even higher in 9 Mg ha^{-1} than in 7–9 Mg ha^{-1}, although total stems was less in >9 Mg ha^{-1} than in 7–9 Mg ha^{-1} in the stem-elongation stage (Figure 3). This indicates that the higher ear number at maturity is likely attributable to population quality, not tiller number. In the stem-elongation stage, the N concentration in >9 Mg ha^{-1} was significantly higher than that in 7–9 Mg ha^{-1} (Table 3). The higher N concentration in plants the stem-elongation stage might lead to higher stem quality (e.g., higher tiller weight) and thus more stems would survive from stem elongation to anthesis. In practice, this suggests that tiller quality is of more importance than total tiller number.

DM accumulation was fastest (253 kg ha^{-1} day^{-1}) from stem elongation to anthesis, and then decreased significantly with time (Figure 1A). This differs from wheat production in most European countries. For example, a study in Germany showed that the DM accumulation rate reached 200 kg ha^{-1} day^{-1} during this stage, and then consistently increased to more than 250 kg ha^{-1} per day in the early grain-filling stage [41]. Meanwhile, the grain-filling stage for wheat is longer in Germany than in China because of

cool temperatures during this stage. The dry and hot winds in May and June in China's wheat production area accelerate wheat maturation [42]. The higher DM accumulation rate during the earlier grain-filling stage and the longer grain-filling stage in Germany would lead to higher wheat yields.

In China, pursuing high grain yield has been the top priority in policy and in practice. The typical N rate applied by winter wheat farmers in the North China plain is around 369 kg N ha^{-1} [15], whereas results from region-wide experiments have demonstrated the optimal N rate to be 128 kg N ha^{-1} [27]. Most Chinese farmers apply large amounts of chemical fertilizer at a uniform rate as an "insurance" against low yields. Affected by the Green Revolution, most farmers still believe that more fertilizer and higher grain yield are synonymous. Our results showed applying more N fertilizer than the rational didn't result in higher wheat yield (Table 1).

On the other hand, being restricted by old knowledge and habits, farmers often apply large amounts of N fertilizer before sowing or during early growth stages (~50% of total N application). Our present study clearly demonstrated that this

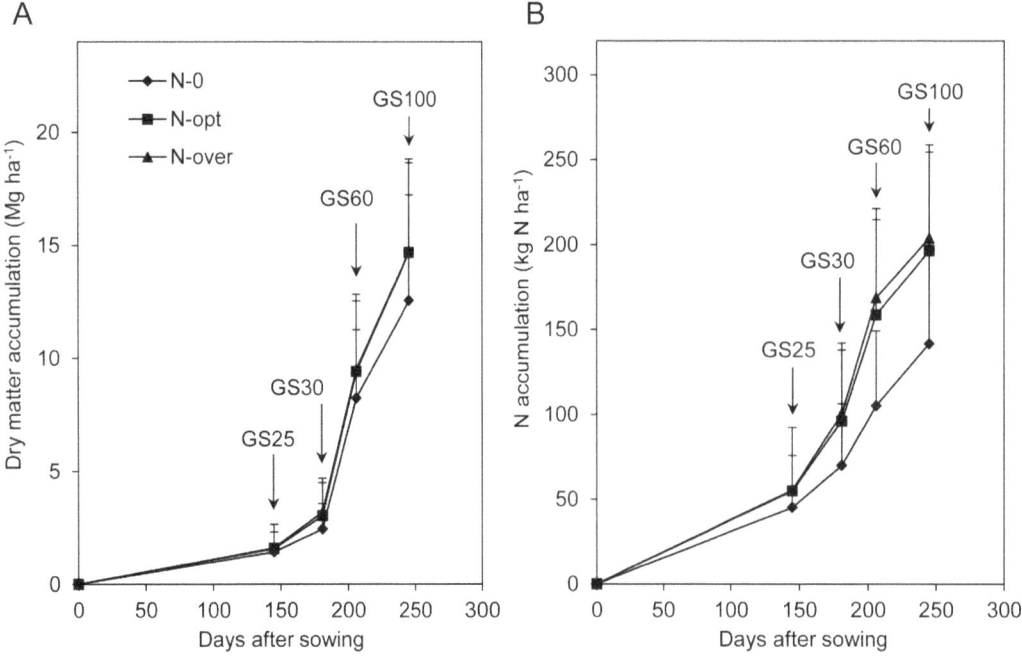

Figure 4. Total dry matter (A) and nitrogen (B) accumulation dynamics with N-0 (n = 79), N-opt (n = 126) and N-over (n = 154) treatments. (GS25, GS30, GS60, and GS100 are the regreening, stem elongation, anthesis, and maturity stages, respectively.).

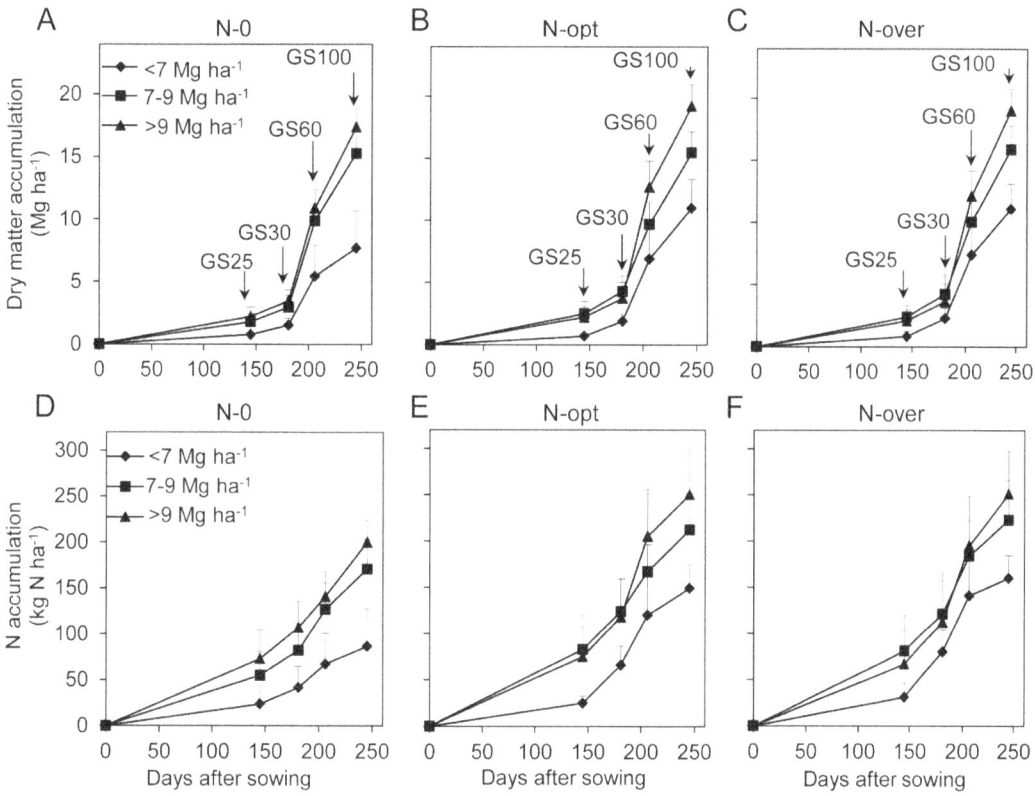

Figure 5. Total dry matter and nitrogen accumulation dynamics in the yield ranges <7, 7–9, and >9 Mg ha^{-1} with N-0 (n = 32, 34, and 13) (A and D), N-opt (n = 56, 29, and 41) (B and E), and N-over treatments (n = 67, 41, and 46) (C and F). (GS25, GS30, GS60, and GS100 are the regreening, stem elongation, anthesis, and maturity stages, respectively.).

large amount of N fertilization before wheat planting was completely unnecessary, and was likely to move beyond the root zone, particularly under irrigated conditions. In addition, surplus N at early stage would lead to more non-reproductive tillers [16], and more risk of lodging, pests and diseases when it is more than optimum effective plant population. To achieve high yield and N use efficiency at the same time, 60–70% of total N fertilizer application should be applied during the rapid crop-growth period, such as stem elongation stages for wheat.

Conclusions

A lack of information on DM and N accumulation dynamics with different yield ranges and N management regimes has resulted in N overuse and misuse and various environment problems in China and worldwide. We found significant correlations between grain yield and DM and N accumulation in each growth stage. Both the highest and fastest DM and N accumulations were observed from stem elongation to anthesis. During this stage, increased DM and N accumulation in >9 Mg ha^{-1} yield plots was key for a final higher yield compared with 7–9 Mg ha^{-1}. Applying more N could increase the N accumulation during this stage; however, DM accumulation was not further improved and thus final yield was not increased. This suggests that N management alone is insufficient to increase grain yield and that associated agronomic management practices should be intensified. Moreover, we found that higher final ear number at harvest was determined by higher tiller quality (e.g., higher tiller weight) in earlier stages, not by tiller number. Thus, more attention should be paid to increasing wheat tiller quality.

Supporting Information

Figure S1 Changes in dry matter and nitrogen accumulation expressed as a percentage of the levels at maturity (n = 413). (GS25, GS30, GS60, and GS100 are the regreening, stem elongation, anthesis, and maturity stages, respectively).

Table S1 Location, year, soil texture, and selected chemical properties in the top 30-cm soil layer at nine sites in intensive wheat production areas in China.

Table S2 Location, year, variety and treatment number, treatment and N rate for the eleven field experiments.

Table S3 Ear number, grains per ear, grain number per square meter and grain weight per thousand grains for grain yield ranges of <7 Mg ha^{-1} (n = 179), 7–9 Mg ha^{-1} (n = 112), and >9 Mg ha^{-1} (n = 122).

Text S1 Design of the system experiment.

Author Contributions

Conceived and designed the experiments: ZC XC FZ. Performed the experiments: QM SY YY WM YT. Analyzed the data: QM ZC. Contributed reagents/materials/analysis tools: ZC XC FZ YY WM YT. Wrote the paper: QM ZC.

References

1. FAO FAOSTAT–Agriculture Database. Available: http://faostat.fao.org/site/339/default.aspx. Accessed 2013 Jan 6.
2. Tilman D, Cassman KG, Matson PA, Naylor R, Polasky S (2002) Agricultural sustainability and intensive production practices. Nature 418: 671–677.
3. Tester M, Langridge P (2010) Breeding technologies to increase crop production in a changing world. Science 327: 818–822.
4. Tollenaar M, Lee EA (2002) Yield potential, yield stability and stress tolerance in maize. Field Crops Res 75: 161–169.
5. Rhoads FM, Stanley RL (1981) Fertilizer scheduling, yield, and nutrient uptake of irrigated corn. Agron J 73: 971–974.
6. Fageria NK, Baligar VC (2005) Enhancing nitrogen use efficiency in crop plants. Adv Agron 88: 97–185.
7. Cox MC, Qualset CO, Rains DW (1985) Genetic-variation for nitrogen assimilation and translocation in wheat.1. dry-matter and nitrogen accumulation. Crop Sci 25: 430–435.
8. Cox MC, Qualset CO, Rains DW (1985) Genetic-variation for nitrogen assimilation and translocation in wheat.2. nitrogen assimilation in relation to grain-yield and protein. Crop Sci 25: 435–440.
9. Heitholt JJ, Croy LI, Maness NO, Nguyen HT (1990) Nitrogen partitioning in genotypes of winter-wheat differing in grain N-concentration. Field Crops Res 23: 133–144.
10. Papakosta DK, Gagianas AA (1991) Nitrogen and dry-matter accumulation, remobilization, and losses for Mediterranean wheat during grain filling. Agron J 83: 864–870.
11. Pan QM, Yu ZW, Wang YF, Tian QZ (1999) Studies on uptake and distribution of nitrogen in wheat at the level of 9000 kg per hectare. (In Chinese). Acta Agron Sin 25: 541–547.
12. Ye Y, Wang G, Huang Y, Zhu Y, Meng Q, et al. (2011) Understanding physiological processes associated with yield–trait relationships in modern wheat varieties. Field Crops Res 124: 316–322.
13. Malhi SS, Johnston AM, Schoenau JJ, Wang ZH, Vera CL (2007) Seasonal biomass accumulation and nutrient uptake of pea and lentil on a black chernozem soil in Saskatchewan. J Plant Nutr 30: 721–737.
14. Wang ZM, Wang P, Lan LW, Zhou DX (2003) A water-saving and high-yielding cultivation system for bread wheat in Huang-Huai-Hai area of China. (In Chinese). Chinese Agric Sci Bullet 19: 22–43.
15. Cui ZL (2005) Optimization of the nitrogen fertilizer management for a winter wheat-summer maize rotation system in the North China Plain-from field to regional scale. China Agricultural University. Ph.D. Dissertation.
16. Weisz R, Crozier CR, Heiniger RW (2001) Optimizing nitrogen application timing in no-till soft red winter wheat. Agron J 93: 435–442.
17. Langer RHM, Dougherty CT (1976) Physiology of grain yield in wheat. Persp Exp Biol 2: 59–70.
18. Sharma RC (1995) Tiller mortality and its relationship to grain yield in spring wheat. Field Crops Res 41: 55–60.
19. Latiri-Souki K, Nortcliff S, Lawlor DW (1998) Nitrogen fertilizer can increase dry matter, grain production and radiation and water use efficiencies for durum wheat under semi-arid conditions. Eur J Agron 9: 21–34.
20. Greenwood DJ, Verstraeten LMJ, Draycott A (1987) Response of winter- wheat to N-fertilizer-quantitative relations for components of growth. Fert Res 12: 119–137.
21. Pask AJD, Sylvester-Bradley R, Jamieson PD, Foulkes MJ (2012) Quantifying how winter wheat crops accumulate and use nitrogen reserves during growth. Field Crops Res 126: 104–118.
22. Cassman KG, Dobermann A, Walters DT (2002) Agroecosystems, nitrogen-use efficiency, and nitrogen management. Ambio 31: 132–140.
23. Shanahan JF, Kitchen NR, Raun WR, Schepers JS (2008) Responsive in-season nitrogen management for cereals. Comput Electron Agric 61: 51–62.
24. Stanford G, Legg JO (1984) Nitrogen and yield potential. In Hauck RD, ed. Nitrogen in crop production. ASA, CSSA, and SSSA: Madison. 263 p.
25. Meisinger JJ, Randall GW (1991) Estimating nitrogen budgets for soil-crop systems. In Follett RF, Keeney DR, Cruse RM, eds. Managing nitrogen for groundwater quality and farm profitability. SSSA: Madison. 85 p.
26. Gastal F, Lemaire G (2002) N uptake and distribution in crops: an agronomical and ecophysiological perspective. J Exp Bot 53: 789–799.
27. Cui ZL, Zhang FS, Xinping CP, Miao YX, Li JL, et al. (2008) On-farm evaluation of an in-season nitrogen management strategy based on soil Nmin test. Field Crops Res 105: 48–55.
28. Zhao RF, Chen XP, Zhang FS, Zhang HL, Schroder J, et al. (2006) Fertilization and nitrogen balance in a wheat-maize rotation system in North China. Agron J 98: 938–945.
29. Stanford G (1973). Rational for optimum fertilization in corn produc tion. J Environ Qual 2: 159–166.
30. Lory JA, Scharf PC (2003) Yield goal versus delta yield for predicting fertilizer nitrogen need in corn. Agron J 95: 994–999.
31. Shearman VJ, Sylvester-Bradley R, Scott RK, Foulkes MJ (2005) Physiological processes associated with wheat yield progress in the UK. Crop Sci 45: 175–185.
32. Álvaro F, Isidro J, Villegas D, García del Moral LF, Royo C (2008) Breeding effects on grain filling, biomass partitioning and remobilization in Mediterranean durum wheat. Agron J 100: 361–370.
33. Austin RB, Ford MA, Morgan CL (1989) Genetic-improvement in the yield of winter-wheat-a further evaluation. J Agric Sci 112: 295–301.
34. Donmez E, Sears RG, Shroyer JP, Paulsen GM (2001) Genetic gain in yield attributes of winter wheat in the great plains. Crop Sci 41: 1412–1419.
35. Acreche MM, Briceño-Félix G, Sánchez JAM, Slafer GA (2008) Physiological bases of genetic gains in Mediterranean bread wheat yield in Spain. Eur J Agron 28: 162–170.
36. Cassman KG (1999) Ecological intensification of cereal production systems: Yield potential, soil quality, and precision agriculture. Proc Natl Acad Sci USA 96: 5952–5959.
37. Singh RP, Huerta-Espino J, Rajaram S, Crossa J (1998) Agronomic effects from chromosome translocations 7DL.7Ag and 1BL.1RS in spring wheat. Crop Sci 38: 27–33.
38. Reynolds MP, Rajaram S, Sayre KD (1999) Physiological and genetic changes of irrigated wheat in the post-green revolution period and approaches for meeting projected global demand. Crop Sci 39: 1611–1621.
39. Fischer RA (1985) Number of kernels in wheat crops and the influence of solar-radiation and temperature. J Agric Sci 105: 447–461.
40. Slafer GA, Satorre EH, Andrade FH (1994) Increases in grain yield in bread wheat from breeding and associated physiological changes. In Slafer GA, ed. Genetic improvement of field crops. Marcel Dekker Inc: New York. 1 p.
41. Diekmann F, Fischbeck G (2005) Differences in wheat cultivar response to N supply. I: Differences in grain yield formation. J Agron Crop Sci 191: 351–361.
42. Sun HY, Liu CM, Zhang XY, Shen YJ, Zhang YQ (2006) Effects of irrigation on water balance, yield and WUE of winter wheat in the North China Plain. Agric Water Manage 85: 211–218.

18

Growth but Not Photosynthesis Response of a Host Plant to Infection by a Holoparasitic Plant Depends on Nitrogen Supply

Hao Shen[1]*, Shu-Jun Xu[1,2], Lan Hong[3], Zhang-Ming Wang[1], Wan-Hui Ye[1]

1 Key Laboratory of Vegetation Restoration and Management of Degraded Ecosystems, South China Botanical Garden, Chinese Academy of Sciences, Guangzhou, Guangdong, PR China, **2** College of Life Sciences, University of Chinese Academy of Sciences, Beijing, PR China, **3** College of Horticulture and Landscape Architecture, Zhongkai University of Agriculture and Engineering, Guangzhou, Guangdong, PR China

Abstract

Parasitic plants can adversely influence the growth of their hosts by removing resources and by affecting photosynthesis. Such negative effects depend on resource availability. However, at varied resource levels, to what extent the negative effects on growth are attributed to the effects on photosynthesis has not been well elucidated. Here, we examined the influence of nitrogen supply on the growth and photosynthesis responses of the host plant *Mikania micrantha* to infection by the holoparasite *Cuscuta campestris* by focusing on the interaction of nitrogen and infection. *Mikania micrantha* plants fertilized at 0.2, 1 and 5 mM nitrate were grown with and without *C. campestris* infection. We observed that the infection significantly reduced *M. micrantha* growth at each nitrate fertilization and more severely at low than at high nitrate. Such alleviation at high nitrate was largely attributed to a stronger influence of infection on root biomass at low than at high nitrate fertilization. However, although *C. campestris* altered allometry and inhibited host photosynthesis, the magnitude of the effects was independent of nitrate fertilizations. The infection reduced light saturation point, net photosynthesis at saturating irradiances, apparent quantum yield, CO_2 saturated rate of photosynthesis, carboxylation efficiency, the maximum carboxylation rate of Rubisco, and maximum light-saturated rate of electron transport, and increased light compensation point in host leaves similarly across nitrate levels, corresponding to a similar magnitude of negative effects of the parasite on host leaf soluble protein and Rubisco concentrations, photosynthetic nitrogen use efficiency and stomatal conductance across nitrate concentrations. Thus, the more severe inhibition in host growth at low than at high nitrate supplies cannot be attributed to a greater parasite-induced reduction in host photosynthesis, but the result of a higher proportion of host resources transferred to the parasite at low than at high nitrate levels.

Editor: Gabriele Sorci, CNRS, Université de Bourgogne, France

Funding: This work was funded by the National Key Technologies R&D Program of China (2012BAC07B04). The funders had no role in study design, data collection and analysis, decision to publish, or preparation of the manuscript.

Competing Interests: The authors have declared that no competing interests exist.

* E-mail: shenhao@scbg.ac.cn

Introduction

Parasitic plants are a taxonomically diverse group of organisms that obtain some or all of their nutrients and other resources, such as water, carbon and phytohormones, from their host plants via haustoria [1]. Interactions between them and their hosts are one of the key research topics in parasitic plant biology [2,3]. Press *et al.* [1] indicated that the extent to which parasites compete with their hosts for nutrients depends on the relative sink strength and the degree of autotrophy of the parasites. In hemiparasitic plants, nutrient transfer and resource acquisition from the hosts are facilitated by the parasite maintaining high transpiration rates, high leaf conductance and low water potentials, and in holoparasitic plants, by high osmotic potentials [3]. Furthermore, parasitic plants can affect the photosynthesis of their hosts at the leaf and/or whole plant level [4]. These processes can adversely affect the hosts, and such negative effects depend on resource availability: they might be negligible when resources are abundant but when resources are limiting they can be severe, ranging from reduction of growth and development to death of the hosts [3].

The influence of nitrogen on host-parasite associations has been investigated in the economically important root hemiparasite *Striga hermonthica* [5,6] and the stem holoparasite *Cuscuta reflexa* [7,8]. *Striga hermonthica*-infected C_4 sorghum had lower rates of photosynthesis than uninfected plants, but the difference in both growth and photosynthesis between uninfected and infected sorghum plants was lower or even negligible when high nitrogen concentrations were supplied [5]. In contrast, high nitrogen supply did not result in an alleviation of the effects of the parasite on the host C_3 rice to the same degree that *S. hermonthica* did on the sorghum host, as reflected by similar growth and photosynthesis in uninfected and infected plants at high nitrogen supply [6].

Among the species in *Cuscuta* (Convolvulaceae), nitrogen relations in the parasitic associations of *C. reflexa* and its leguminous or non-leguminous hosts have been studied [7–9]. Modelling the solute transfer between *C. reflexa* and its leguminous host *Lupinus albus* [9] indicated that the massive demand of the parasite led to resource losses of the host, particularly nitrogen from leaves and roots. As a result of such highly competitive sink activity of the parasite, net photosynthesis of *L. albus* appeared to be stimulated. However, *C. reflexa* infection increased tissue nitrogen levels in the non-N_2-fixing hosts *Ricinus communis* [8] and *Coleus blumei* [7]. Growth and development of *C. reflexa* were restricted similarly with those of the hosts when fed with different concentrations of nitrate, suggesting a fine tuning of the parasite sink strength with the source capacity of both hosts [7,8]. In these associations, *C. reflexa* led to a substantial sink-dependent stimulation of the host's photosynthesis and, under N-limiting conditions, to an increase in the host's tissue nitrogen concentrations. The reason for the different effects of *C. reflexa* on the symbiotically N-fed *L. albus* and on nitrate-fed *R. communis* and *C. blumei* was attributed to the overriding competition between *C. reflexa* and *L. albus* in the tripartite association *L. albus*-Rhizobium-*C. reflexa*, whilst this additional factor was absent in the associations with *R. communis* and *C. blumei* [7,8]. Although the holoparasite *C. reflexa* substantially decreased growth of both *R. communis* [8] and *C. blumei* [7] regardless of nitrate supply, the inhibition in growth of infected *R. communis* was exacerbated at low N supply, but in contrast, the inhibition in growth of infected *C. blumei* was similar at low and high N supply.

In our previous studies, we investigated the influence of another *Cuscuta* species, *C. campestris*, on growth, biomass allocation and photosynthesis of an invasive weed, *Mikania micrantha* H.B.K. (Asteraceae). We found different growth and photosynthesis influence patterns from those of *C. reflexa*. *Cuscuta campestris* significantly reduced the total biomass, changed the biomass allocation patterns and completely inhibited the flowering of *M. micrantha* plants [10]. In addition to direct resource capture by *C. campestris*, the parasite also reduced the stomatal conductance, and carboxylation and light use efficiencies of the host, resulting in reduced growth of infected plants [11]. We also observed that the total biomass of the parasite plus its host was significantly less than that of uninfected hosts [10], and the parasite suppressed photosynthesis of the hosts [11]. However, Jeschke and Hilpert [8] and Jeschke *et al.* [7] observed that the total biomass of *C. reflexa* plus its hosts was similar to that of the uninfected and *C. reflexa* led to a sink-dependent stimulation of host photosynthesis. Thus, it is of interest to study if the nitrogen relations are also different between *C. campestris*-host and *C. reflexa*-host associations.

In the present project we investigated the nitrogen relations in the *M. micrantha*-*C. campestris* host-parasite association by focusing on the interaction of nitrogen and infection. We hypothesized that both growth and photosynthesis responses in *M. micrantha* to *C. campestris* infection would be more affected by parasitism at low than high nitrogen supply.

Materials and Methods

Study Species

Mikania micrantha H.B.K. is a fast-growing climbing perennial vine of the family Asteraceae, native to Central and South America [12]. In its palaeotropic exotic range, it is a notorious invasive weed, severely damaging forestry and plantation crops [13]. In South China, it grows in poor to fertile soils with total nitrogen 0.14–1.62 g kg^{-1} [13]. In the field, the generalist stem parasite *Cuscuta campestris* Yuncker infects *M. micrantha* and it has been one of the most effective means of biologically controlling *M. micrantha* in South China [10,13]. *Cuscuta campestris* is the most widespread species in the genus and the only parasitic weed of North America that has spread to the Old World [14]. It is a holoparasite and draws all nutrients from its host. It is a very powerful sink for host photosynthates, severely suppressing host growth, preventing flowering and fruiting, and even resulting in host death [10,14]. It can infect many herbaceous plants and

results in damage to horticultural and agricultural crops, and it is the worst pest of alfalfa and other legumes [14].

Plant Culture and Growth Conditions

The experiment was carried out during the July 2011–January 2012 growing season in an unheated greenhouse with natural light at the same field station of South China Botanical Garden as in our previous study [10]. On 26 July 2011, whole *M. micrantha* plants were collected from a *M. micrantha* population near the station. Two-node segments, similar in size, were obtained from the middle of the stems. The segments were planted in containers filled with washed moist sand, with the low nodes buried below and the upper about 5 cm above the sand surface. The upper nodes began to sprout 5 days later. On 20 August, 90 healthy sprouts about 20 cm long were transplanted into 8.36 L pots filled with washed moist sand, one per pot, and the pots were placed in the glasshouse with a temperatures range 12–28°C, mean 17.8°C, and relative humidity range 50–90%, mean 70% during August 2011–January 2012. The plants were watered twice daily at 06:00 h and 18:00 with distilled water during the first week after transplanting. From then on to the end of the experiment, they were watered at 06:00 h with distilled water and at 18:00 h with modified Hoagland solutions containing 0.2, 1 or 5 mM nitrate with 200 ml per pot and 30 pots per nitrate concentration. The pots were thoroughly rinsed with water once a week.

On 7 October when the *M. micrantha* plants had been treated with nitrate for 41 days, half of them within each nitrate treatment were randomly chosen and inoculated with *C. campestris* filaments about 5 cm in length, one per plant, and the rest were left as control. To ensure simultaneous attachment, excised and previously twined shoot cuttings of *C. campestris* were allowed to attach to the lowest two *M. micrantha* stem internodes. By 14 October, all the inoculated *M. micrantha* plants had become infected with *C. campestris* stems as indicated by renewed vigorous growth of the filaments. Thus, this day was considered day 0 after parasitization (DAP). To prevent *M. micrantha* from climbing from one pot to another, a bamboo cane was placed vertically in each pot for *M. micrantha* to climb on. The experiment ended on 14 January 2012, 90 DAP or 172 days after planting, when the uninfected *M. micrantha* plants fertilized at 5 mM nitrate were in full bloom.

Growth Measurements and Observations

During the experiment, both destructive and nondestructive measurements of growth were made. Height from the base of the stem to the apex of the shoot and number of visible leaves per *M. micrantha* plant were recorded on 0, 15, 40, 60, 90 DAP. Flowering times of *M. micrantha* and *C. campestris* plants were also recorded.

Mikania micrantha plants on 0 DAP, and the uninfected and infected and parasite plants on 90 DAP were randomly sampled and harvested, five per treatment. We measured the leaf area using a LI-3000C portable laser area meter (LI-COR Inc., Lincoln, NE, USA), removed the dead material and counted the number of dead leaves of the sampled *M. micrantha* plants, but the number of dead leaves was not used in the growth analyses. We separated the living parts of the sampled plants into stems, leaves, reproductive organs (if present) and roots. Roots were soaked in tap water, washed and separated carefully in running water over a 2-mm mesh sieve. Stems, tendrils and reproductive organs of *C. campestris* were carefully dissected from stems and leaves of *M. micrantha* plants.

All plant material was oven dried at 70°C until constant weights were achieved, and they were used to obtain tissue C and N concentrations and dry weights. For the *M. micrantha* plants harvested on 90 DAP, specific leaf area (SLA, the ratio of leaf area to dry mass) and shoot-to-root dry weight ratio (S/R), relative growth rate (RGR, the dry weight increase per plant per day), leaf area ratio (LAR, the ratio of leaf area to dry weight per plant) and unit leaf rate (ULR, dry weight production per m^2 leaf area per day) were calculated according to Hunt and Parsons [15].

Measurements of Photosynthesis

In situ gas exchange measurements were made on *M. micrantha* leaves using a LI-6400 portable photosynthesis system with a standard 6 cm^2 leaf chamber (LI-COR Inc., Lincoln, NE, USA) on 30 and 80 DAP, at around 10:00 h, and photosynthetic parameters were calculated based on von Caemmerer and Farquhar [16]. To ensure that leaves measured were similar in age and developmental stage, only the youngest fully expanded mature sun leaves were sampled, one leaf per plant, from five randomly selected *M. micrantha* plants per treatment. Conditions inside the leaf chamber during the measurements were controlled as follows. Irradiance was provided by an integrated red-blue light-emitting diode source (model 6400-02B, LI-COR, Inc.) at photosynthetic photon flux density (PPFD) of 1000 μmol photons $m^{-2} s^{-1}$ except for the light response study, CO_2 concentration (C_a) was controlled at 360 μmol mol^{-1} with a CO_2 mixer except for the leaf internal CO_2 concentration (C_i) response study, flow rate was set at 500 μmol s^{-1}, and leaf temperature (T_l) was controlled at 20°C on 80 DAP and at 30°C on 30 DAP. Net photosynthetic rate (P_n), stomatal conductance (g_s, mol H_2O $m^{-2} s^{-1}$), rate of transpiration (E, mmol H_2O $m^{-2} s^{-1}$), intercellular CO_2 concentration (C_i), C_a, air temperature (T_a), T_l, air relative humidity (RH), and PPFD were recorded after equilibration to a steady state with a coefficient of variation ≤1% at each measurement had been reached. Water use efficiency (WUE, μmol CO_2 mmol H_2O^{-1}) was calculated as P_n/E for each measurement. Methods and conditions used to obtain photosynthesis light and C_i response curves were the same as described in the above paragraph unless specified otherwise.

Determination of Chlorophyll and Carotenoid Concentrations

Leaf chlorophyll concentrations were measured on the leaves used for the measurements of photosynthesis. Leaf pigments were extracted from about 70 mg of leaf sample put in 10 mL 80% acetone for 72 hours in the dark, and carotenoid and chlorophyll *a* and *b* concentrations were determined spectrophotometrically at 663, 645 and 470 nm according to Arnon [17].

Light Response Curves

To construct light response curves, on two clear days, 80–81 DAP, photosynthesis measurements were made between 08:00 h and 11:00 h. Leaf temperature in the leaf chamber was maintained at 20°C. When a leaf in the chamber had acclimated to a PPFD of 500 μmol photons $m^{-2} s^{-1}$ for 20 min, photosynthesis measurements were taken at PPFD in the following order: 500, 800, 1000, 1500, 1800, 2000, 200, 100, 50, 20, 0 μmol photons $m^{-2} s^{-1}$. For each measurement, apparent quantum yield (Φ, mol CO_2 mol^{-1} photons), dark respiration rate (R_d, μmol CO_2 $m^{-2} s^{-1}$) and light compensation point (LCP) were obtained by linear regression using data obtained at PPFD of 0, 20 and 50 μmol photons $m^{-2} s^{-1}$ [18]. The entire photosynthetic light response curves were fitted using Photosynthesis Work Bench (LI-COR Inc., Lincoln, NE, USA). Maximum leaf light-saturated photosynthetic rate (P_{max}) and light saturating point (LSP) were estimated.

C_i Response Curves

To study the relationship between P_n and leaf internal CO_2 concentration C_i, photosynthesis was measured on two clear days, 75–76 DAP. Leaf temperature inside the leaf chamber was maintained at 20°C, and PPFD, at 1000 μmol photons $m^{-2} s^{-1}$. P_n was measured at C_a in the following order: 400, 250, 150, 100, 50, 0, 400, 400, 600, 800, 1000 and 1200 μmol mol^{-1} provided by a CO_2 mixer. Sigma Plot for Windows 10.0 was used to fit the P_n/C_i response curves using an exponential function [19]:

$$P_n = a(1 - e^{-bx}) + c,$$

where P_n is leaf net photosynthetic rate and x is C_i. Using this equation, the CO_2 saturated rate of photosynthesis (P_{sat}) was calculated as $a+c$, and the carboxylation efficiency (CE), as the slope at $P_n = 0$ or b(a+c).

Maximum carboxylation rate of Rubisco (V_{cmax}) and maximum light-saturated rate of electron transport (J_{max}) were determined using Photosynthesis Assistant software (Version 1.1, Dundee Scientific, Dundee, UK) according to Farquhar *et al.* [20], modified by Harley and Sharkey [21] and Harley *et al.* [22].

Soluble Protein and Rubisco Contents

The leaves used for light and C_i response curves were collected to determine soluble protein and Rubisco content. Approximately 0.5 g of fresh leaf material per sample with the mid-vein removed was ground in liquid nitrogen to a fine powder with 10 mg of PVPP. Extraction buffer [50 mM sodium phosphate buffer pH 7.8, 10% (v/v) glycerol, 1% (v/v) β-mercaptoethanol] was added at 3 ml g^{-1} fresh weight. The homogenate was centrifuged at 16,000 g for 15 min at 4°C. Protein concentration of the supernatant was estimated by the protein dye-binding method of Bradford [23] using bovine serum albumin (BSA) as the standard.

Rubisco content was determined following the protocol of Makino *et al.* [24] modified by Irving and Robinson [25]. Briefly, equal amounts of protein and extraction buffer were mixed and boiled for 2 min. Proteins in the extracts together with bovine serum albumin standards were separated using SDS-PAGE following the method of Laemmli [26] using 12% acrylamide resolving and 5% acrylamide stacking gels and the Mini-PROTEAN 3 System (Bio-Rad Laboratories, Richmond, CA, USA). Gels were stained using 1% (w/v) Coomassie Brilliant Blue R250 for 3 hours, the Rubisco containing band was excised, and the protein concentration was determined spectrophotometrically at 595 nm after elution of the stain in formamide at 50°C for 12 hours.

Carbon and Nitrogen Analysis

Tissue C and N concentrations in *M. micrantha* and in stems and flowers of *C. campestris* plants harvested on 90 DAP were assayed by GC using a Vario EL CHNS elemental analyzer (Elementar Analysensysteme GmbH, Hanau, Germany). They were also determined for the leaves used to measure photosynthesis. Photosynthetic nitrogen use efficiency (PNUE) was calculated as P_{max}/N_{area}.

Data Analysis

All statistical tests were carried out at α = 0.05 level using SPSS (version 11.5, SPSS Inc., Chicago, IL, USA). Two-way analysis of variance (ANOVA) was performed to evaluate the effects of nitrate supply, *C. campestris* infection, and their interaction on the growth and physiological traits. Repeated measures ANOVA was conducted to test the main effects, their interactions and measuring times (0, 15, 40, 60 and 90 DAP) on the number of leaves. One-way ANOVA was performed to test the effects of nitrate treatments on parasite biomass. Treatment means of the significant ANOVA effects were compared at $\alpha = 0.05$ level using the least significant difference (LSD) analysis. Correlation analysis was conducted to test the relationships between P_{sat} or CE and leaf nitrogen concentrations for *M. micrantha* plants. To satisfy the assumptions of ANOVA, some data were square-root transformed; however, untransformed data are presented in tables and figures.

Results

There were no differences in the flowering initiation of the uninfected or infected *M. micrantha* among the three nitrate treatments. In both uninfected and infected plants, compared with 5 mM nitrate fertilization, the other two delayed the further development of inflorescence after the inflorescence had formed and such delay was more at 0.2 than at 1 mM nitrate (data not shown), and they also reduced the number of florets. At all nitrate levels, the uninfected started to develop terminal inflorescences on 15 DAP, but the infected, on 40 DAP. From 15 to 60 DAP, *C. campestris* grew vigorously with a lot of branching. It started flowering on 20 DAP at 0.2 mM, and on 25 DAP at 1 or 5 mM nitrate treatments.

Number of Leaves

Repeated measures ANOVA indicated there were significant ($P<0.001$) infection, nitrate and their interaction effects on the number of leaves of *M. micrantha* over the measurement times. From 0 to 15 DAP, the number of leaves increased regardless of infection and fertilization treatments (Figure 1). From 15 to 90 DAP, the number of leaves of the infected *M. micrantha* was smaller than that of the control, and the differences between them became greater as nitrate fertilization levels increased from 0.2 to 1 to 5 mM. The number of leaves of the infected decreased continuously from 15 to 90 DAP, and that of the control increased from 0 to 60 DAP and then decreased slightly from 60 to 90 DAP. At harvest on 90 DAP, infected plants had 61%, 58% and 34% of the number of leaves of uninfected plants at 0.2, 1 and 5 mM nitrate supplies, respectively.

Plant Biomass Components

By 90 DAP, the dry mass of the infected system (host plus parasite) was significantly less than that of uninfected *M. micrantha* across all nitrate treatments (Table 1). *Mikania micrantha* total biomass and its components were significantly reduced by *C. campestris* infection at all nitrate treatments, and the magnitude of the reduction was dependent on nitrate fertilization levels as indicated by significant nitrate × *Cuscuta* interaction (Table 1). The infection reduced *M. micrantha* root biomass by about 71%, 73% and 61%, flower biomass by about 91%, 79% and 71% and total biomass by about 70%, 64% and 59% at 0.2, 1 and 5 mM nitrate fertilizations, respectively. These proportional decreases in biomass with the increases in nitrate supply occurred although the infected plants supported significantly higher parasite biomass at high than at low nitrate fertilizations (Figure 2A; Table 1). However, the parasite was always a similar proportion of the infected system (host plus parasite) at all nitrate levels (Figure 2B).

RGR and Leaf Area

RGR was affected significantly by nitrate and infection, but not by their interaction (Table 2). Significant decreases in RGR occurred in the infected *M. micrantha* plants at each nitrate fertilization level, and generally as nitrate supplies increased, RGR increased within each infection treatment. Infection significantly reduced leaf area of *M. micrantha*, and this negative effect was greater at 5 than at 0.2 or 1 mM nitrate (Table 2).

Biomass Allocation

Biomass allocation parameters, except the percentage of total biomass allocated to flowers, of *M. micrantha* were all significantly affected by infection, but not by the interaction of nitrate and infection (Tables 2, 3).

Generally, *C. campestris* infection significantly increased LAR, SLA and shoot:root ratios (S:R), but it reduced ULR of *M. micrantha* plants, and its effects on these traits were independent of nitrate treatments (Table 2). Within each nitrate treatment, the infection effects were more negative on root than on shoot growth (Figure 2A), resulting in higher S:R in infected plants than in control plants. The ratio of above to below-ground biomass in the host-parasite system was 2.3–3.8 times that of the uninfected plants among the nitrate treatments (Table 2).

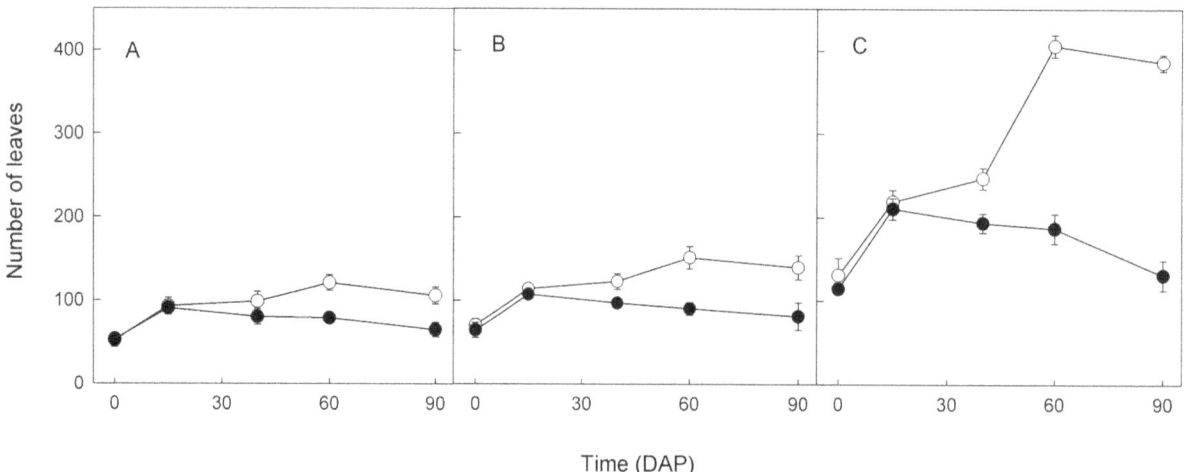

Figure 1. Means (±SE, *n* = 5) of number of leaves of uninfected (○) and infected (●) *M. micrantha* plants by *C. campestris* on different days after parasitization (DAP) at 0.2 (A), 1 (B) and 5 (C) mM nitrate fertilizations.

Table 1. ANOVA results and means (±SE, n = 5) of plant dry biomass (g) components based on data collected on 90 DAP for uninfected and infected *M. micrantha* plants by *C. campestris* at 0.2, 1 and 5 mM nitrate fertilization concentrations.

Treatments	M. micrantha					Infected system (host+parasite)	C. campestris
	Roots	Stem	Leaves	Flowers	Total		
0.2 mM nitrate							
Uninfected	22.64±2.17c	8.50±0.76de	3.86±0.49c	1.17±0.50bd	36.16±3.25c	36.16±3.25d	
Infected	6.46±1.91d	2.66±0.41c	1.63±0.14d	0.11±0.11c	10.86±2.28d	16.44±2.12e	5.58±0.22b
1 mM nitrate							
Uninfected	32.95±3.37b	12.70±0.75bd	6.44±0.64b	2.88±0.53b	54.97±4.20b	54.97±4.20c	
Infected	8.77±1.78d	7.12±1.69ce	3.35±0.65cd	0.60±0.30cd	19.83±4.27d	29.04±3.23d	9.21±1.33b
5 mM nitrate							
Uninfected	53.10±3.51a	35.35±5.22a	14.94±0.86a	13.87±3.25a	117.26±5.61a	117.26±5.61a	
Infected	20.49±2.96c	17.06±2.69b	6.31±1.01b	3.96±1.32b	47.81±4.63bc	75.39±4.55b	27.57±2.82a
Source of variation	F values from ANOVA						
Nitrate (N)	35.89***	36.77***	70.12***	18.72***	109.36***	170.00***	42.76***[†]
Infection (I)	121.16***	22.86***	68.59***	13.55**	161.39***	80.36***	
N × I	4.61*	4.11*	12.77***	5.33*	15.42***	4.11*	

[†]F value from one-way ANOVA, and the rest F values are from two-way ANOVA.
ns, $P>0.05$; *$P<0.05$; **$P<0.01$, ***$P<0.001$. Means in the same column not sharing a common letter are significantly different according to LSD analysis at $p = 0.05$ level. The same apply to Tables 2, 3, 4, 5, 6 and 7.

Cuscuta campestris infection increased biomass allocation to stems and leaves and reduced them to roots and flowers of *M. micrantha* plants although the effects within all nitrate levels were not always significant (Table 3). The interaction of nitrate and infection was not significant in the allocations to these biomass components of *M. micrantha* plants.

P_n of *M. micrantha* Leaves

The interaction of nitrate and infection had no significant effects on P_n and related parameters of *M. micrantha* leaves on 30 and 80

DAP (Table 4). The infected plants had lower leaf P_n, g_s, E and WUE, but higher C_i than the uninfected plants at each nitrate fertilization level. Mostly, nitrate treatment did not result in significant changes in P_n, g_s, E, C_i and WUE measured on 80 DAP but led to no consistent changes in these traits on 30 DAP within infection treatments (Table 4).

Photosynthesis in Response to Light

Cuscuta campestris infection had significant effects on LSP, P_{max}, LCP and Φ, but not on R_d of *M. micrantha* leaves in response to

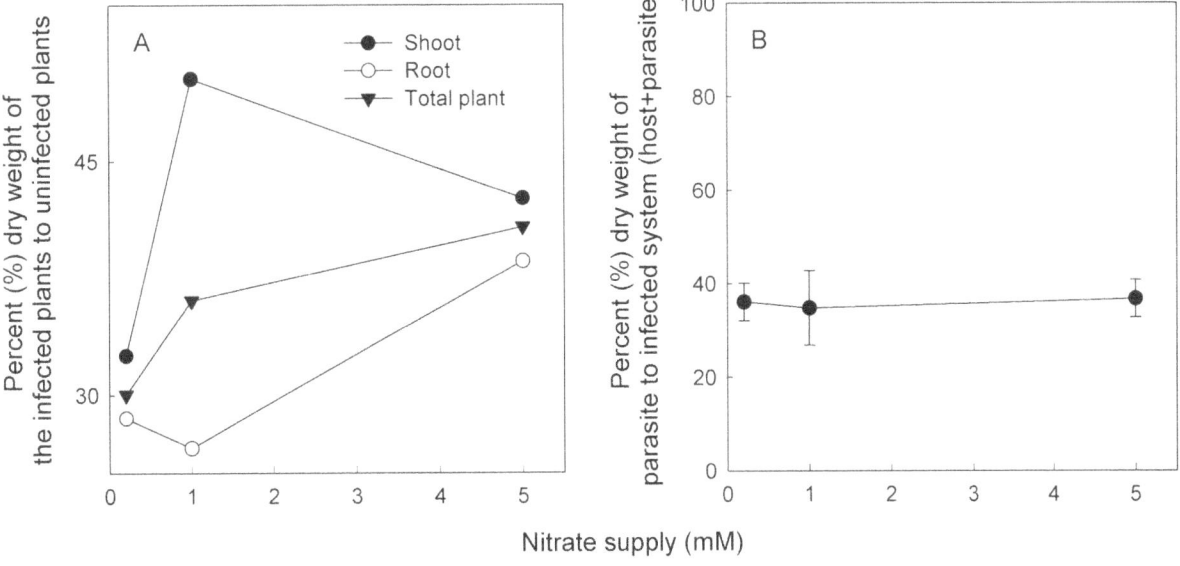

Figure 2. Percent (%) dry weight of the infected to the uninfected *M. micrantha* plants (A) and of the parasite to infected system (host plus parasite) (B) in the association *M. micrantha-C. campestris* fertilized at 0.2, 1 and 5 mM nitrate fertilizations.

Table 2. ANOVA results and mean ($\pm$SE, $n = 5$) of relative growth rate (RGR, g g^{-1}d^{-1}), leaf area (m^2), leaf area ratio (LAR, m^2 kg^{-1} of plant, specific leaf area (SLA, m^2 kg^{-1} of leaves), unit leaf rate (ULR, g m^{-2} d^{-1}), shoot:root dry weight ratio (S:R, g g^{-1}), and (host shoot+*Cuscuta*):host root ((H+C)/HR, g g^{-1}) for *M. micrantha* plants infected and uninfected by *C. campestris* and fertilized at 0.2, 1 and 5 mM nitrate.

Treatments	RGR	Leaf area	LAR	SLA	ULR	S:R	(H+C)/HR
0.2 mM nitrate							
Uninfected	0.045±0.001b	0.12±0.02c	4.12±0.28c	32.00±0.93ab	11.22±0.78a	0.61±0.05c	0.61±0.05b
Infected	0.031±0.002d	0.06±0.01d	6.05±0.88ab	33.63±2.57ab	5.79±1.10b	0.87±0.20bc	2.03±0.47a
1 mM nitrate							
Uninfected	0.052±0.001a	0.18±0.02b	4.16±0.13c	29.04±1.78b	12.55±0.46a	0.69±0.05c	0.69±0.05b
Infected	0.040±0.002c	0.12±0.02c	6.65±0.67a	34.54±2.63ab	6.24±0.82b	1.24±0.11ab	2.63±0.44a
5 mM nitrate							
Uninfected	0.056±0.001a	0.45±0.03a	4.89±0.15bc	30.27±1.92ab	11.42±0.36a	1.22±0.05ab	1.22±.05b
Infected	0.046±0.001b	0.22±0.04b	5.80±0.48ab	35.06±1.88a	8.07±0.66b	1.41±0.22a	2.84±0.32a
Source of variation	*F* values from ANOVA						
Nitrate (N)	32.79***	50.31***	0.22 ns	0.15 ns	1.51 ns	9.28**	2.90 ns
Infection (I)	94.98***	34.23***	18.02***	5.75*	69.48***	9.27**	47.48***
N × I	0.80 ns	6.63**	1.24 ns	0.51 ns	2.11 ns	1.01 ns	0.39 ns

light (Figure S1; Table 5). Leaves of uninfected plants had higher LSP, P_{max} and Φ but lower LCP than infected plants at each nitrate treatment. However, nitrate and its interaction with infection had no significant effects on these parameters (Table 5).

Photosynthesis in Response to C_i

Leaves of uninfected plants had significantly higher CE, P_{sat}, V_{cmax} and J_{max} than infected plants at each nitrate treatment (Figure S2; Table 6). CE and P_{sat} were higher at 5 than at 0.2 and 1.0 mM nitrate in both the infected and uninfected.

Chlorophyll and Carotenoid

The concentrations of total chlorophyll, chlorophyll *a* and *b*, and carotenoid of *M. micrantha* leaves were significantly affected by nitrate, infection and their interaction (Table 7). There was a greater reduction in chlorophyll concentration of infected plants at 5 mM nitrate than at 0.2 or 1 mM nitrate (Table 7). The chlorophyll *a*:*b* ratio was not significantly affected by infection at each nitrate treatment.

Proteins and Rubisco Contents

Nitrate treatment had a significant influence on total soluble protein content, but not on Rubisco content (Table 7). Higher

Table 3. ANOVA results and means ($\pm$SE, $n = 5$) of the percentages (%) of total biomass allocated to roots, stems, leaves and flowers of the uninfected and infected *M. micrantha* plants by *C. campestris* on 90 DAP under 0.2, 1 and 5 mM nitrate fertilization treatments.

Treatments	Roots	Stems	Leaves	Flowers
0.2 mM nitrate				
Uninfected	62.55±2.34a	23.69±1.36b	10.66±0.74b	3.10±1.08b
Infected	55.93±5.58a	25.44±1.57b	17.11±3.18a	1.53±1.53b
1 mM nitrate				
Uninfected	59.47±1.82a	23.35±1.34b	11.73±0.84b	5.44±1.14bc
Infected	45.00±2.18b	35.27±0.82a	17.25±1.11a	2.48±1.19b
5 mM nitrate				
Uninfected	45.16±1.16b	29.86±3.24ab	12.74±0.45b	12.23±2.92a
Infected	42.81±3.85b	35.87±4.04a	12.93±1.08ab	8.38±1.93ac
Source of variation	*F* values from ANOVA			
Nitrate (N)	11.58***	6.21**	0.60 ns	9.14**
Infection (I)	9.10**	11.53**	10.57**	3.00 ns
N × I	1.88 ns	2.33 ns	2.45 ns	0.17 ns

Table 4. ANOVA results and means ($\pm$SE, $n = 5$) of the net photosynthetic rate (P_n), stomatal conductance (g_s), transpiration rate (E), water use efficiency (WUE), intercellular CO_2 concentration (C_i) of the youngest fully expanded mature leaves of the uninfected and infected *M. micrantha* plants by *C. campestris* on different days after parasitization (DAP) under different nitrate fertilization treatments.

Treatments	P_n (μmol CO_2 m^{-2} s^{-1})		g_s (mol H_2O m^{-2} s^{-1})		E (mmol H_2O m^{-2} s^{-1})		C_i (μmol mol^{-1})		WUE (μmol CO_2 mmol^{-1} H_2O)	
	30 DAP	80 DAP	30 DAP	80 DAP	30 DAP	80 DAP	30 DAP	80 DAP	30 DAP	80 DAP
0.2 mM nitrate										
Uninfected	5.15±0.42bd	4.47±0.42a	0.11±0.01bc	0.09±0.02a	1.64±0.15b	1.10±0.21ab	278.52±4.67ab	259.68±6.96bc	3.16±0.14b	4.38±0.45b
Infected	2.23±0.28c	1.75±0.34b	0.09±0.02c	0.07±0.02ab	1.53±0.37b	0.71±0.13bc	303.14±8.16a	316.51±8.26a	1.65±0.30ac	2.48±0.27c
1 mM nitrate										
Uninfected	6.22±0.85b	4.07±0.54a	0.16±0.03a	0.06±0.01ab	1.48±0.19b	0.71±0.13ac	287.72±8.21a	240.37±8.87c	4.30±0.45a	5.99±0.43ab
Infected	4.08±0.35cd	1.08±0.27b	0.12±0.01ac	0.04±0.01b	1.46±0.17b	0.46±0.11c	290.68±12.95a	289.95±21.27ab	3.04±0.55bc	2.50±0.72c
5 mM nitrate										
Uninfected	8.44±0.97a	4.48±0.66a	0.15±0.03ab	0.07±0.01ab	2.71±0.42a	0.73±0.14ac	255.01±12.06b	231.18±13.66c	3.28±0.38ab	6.41±0.84a
Infected	3.51±0.85cd	1.21±0.58b	0.09±0.02c	0.04±0.01b	1.66±0.30b	0.40±0.11c	289.19±6.47a	295.72±11.37ab	2.09±0.24c	2.59±0.53c
Source of variation	*F* values from ANOVA									
Nitrate (N)	5.80**	0.60 ns	3.57*	2.71 ns	3.59*	3.55*	2.53 ns	2.36 ns	6.51**	1.92 ns
Infection (I)	36.11***	56.44***	7.85*	4.51*	2.83 ns	7.79*	7.46*	30.33***	19.31***	43.23***
N × I	2.26 ns	0.16 ns	0.43 ns	0.11 ns	1.96 ns	0.14 ns	1.50 ns	0.17 ns	0.10 ns	1.62 ns

Table 5. ANOVA results and means ($\pm$SE, $n = 5$) of photosynthesis parameter estimates from the light response curves for the youngest fully expanded mature leaves of the uninfected and infected *M. micrantha* by *C. campestris* under different nitrate fertilization concentrations.

| Treatments | Parameter estimates of photosynthesis | | | | |
	LSP (μmol photons m^{-2} s^{-1})	P_{max} (μmol CO$_2$ m^{-2} s^{-1})	LCP (μmol photons m^{-2} s^{-1})	Φ (mol CO$_2$ mol^{-1} photons)	R_d (μmol m^{-2} s^{-1})
0.2 mM nitrate					
Uninfected	376.00$\pm$53.11bc	4.96$\pm$0.41a	9.73$\pm$0.73c	0.042$\pm$0.002a	0.40$\pm$0.03
Infected	293.60$\pm$23.38c	2.19$\pm$0.43b	21.09$\pm$3.84ab	0.021$\pm$0.004b	0.40$\pm$0.07
1 mM nitrate					
Uninfected	516.00$\pm$65.62ab	5.17$\pm$0.70a	16.16$\pm$3.43abc	0.042$\pm$0.005a	0.62$\pm$0.07
Infected	361.40$\pm$38.89bc	2.79$\pm$0.48b	25.39$\pm$2.80ab	0.015$\pm$0.001b	0.39$\pm$0.06
5 mM nitrate					
Uninfected	586.40$\pm$75.82a	5.33$\pm$0.77a	14.07$\pm$2.66bc	0.034$\pm$0.001a	0.46$\pm$0.08
Infected	273.40$\pm$33.00c	2.24$\pm$0.52b	27.96$\pm$5.45a	0.019$\pm$0.004b	0.42$\pm$0.04
Source of variation	F values from ANOVA				
Nitrate (N)	2.49 ns	0.26 ns	1.68 ns	0.83 ns	1.43 ns
Infection (I)	18.88***	35.10***	16.61***	45.17***	2.99 ns
N $\times$ I	2.61 ns	0.20 ns	0.23 ns	1.20 ns	1.83 ns

LSP, light saturation point; P_{max}, net photosynthesis at LSP; LCP, light compensation point; Φ, apparent quantum yield; R_d, dark respiration rate.

nitrate supply resulted in higher soluble protein content. Infection significantly reduced both total soluble protein and Rubisco contents, and the response to infection was similar across nitrate levels (Table 7).

Leaf Nitrogen, Nitrogen Partitioning and PNUE

Cuscuta campestris infection significantly reduced *M. micrantha* plant leaf nitrogen content and its effect depended on nitrate supply (Table 7). The infected plants had significantly reduced leaf nitrogen contents at 1 and 5 mM nitrate fertilizations, but not at 0.2 mM. The nitrogen concentrations in *C. campestris* were not significantly different among the three nitrate fertilizations; 14.4$\pm$0.66, 15.2$\pm$0.76 and 16.9$\pm$0.60 mg g^{-1} at 0.2, 1 and 5 mM nitrate, respectively. There was a significant positive linear correlation between P_{sat} or CE and leaf nitrogen content for uninfected *M. micrantha* plants (Figure 3). *Cuscuta campestris* infection

Table 6. ANOVA results and means ($\pm$SE, $n = 5$) of the photosynthesis parameter estimates from the P_n-C_i response curves for the youngest fully expanded mature leaves of the uninfected and infected *M. micrantha* by *C. campestris* under different concentrations of nitrate fertilization.

| Treatments | Parameter estimates of photosynthesis | | | |
	CE	P_{sat} (μmol CO$_2$ m^{-2} s^{-1})	V_{cmax} (μmol m^{-2} s^{-1})	J_{max} (μmol m^{-2} s^{-1})
0.2 mM nitrate				
Uninfected	0.038$\pm$0.005b	8.22$\pm$0.47a	18.8$\pm$1.08ab	78.64$\pm$4.97ab
Infected	0.010$\pm$0.001cd	3.24$\pm$0.38bc	8.63$\pm$0.87c	36.02$\pm$3.54cd
1 mM nitrate				
Uninfected	0.034$\pm$0.006b	8.66$\pm$1.02a	19.6$\pm$2.71ab	87.84$\pm$13.98ab
Infected	0.006$\pm$0.001c	2.15$\pm$0.13c	6.76$\pm$0.78c	28.78$\pm$2.70d
5 mM nitrate				
Uninfected	0.061$\pm$0.008a	10.52$\pm$1.18a	23.5$\pm$2.14a	104.26$\pm$10.64a
Infected	0.025$\pm$0.009bd	4.87$\pm$1.05b	13.39$\pm$4.21bc	58.94$\pm$18.79bcd
Source of variation	F values from ANOVA			
Nitrate (N)	8.06**	4.70*	3.13 ns	3.21 ns
Infection (I)	38.53***	74.95***	34.07***	30.60***
N $\times$ I	0.27 ns	0.45 ns	0.23 ns	0.33 ns

CE, carboxylation efficiency; P_{sat}, CO$_2$ saturated rate of photosynthesis; V_{cmax}, maximum rate of Rubisco carboxylase activity; J_{max}, maximum rate of photosynthetic electron transport.

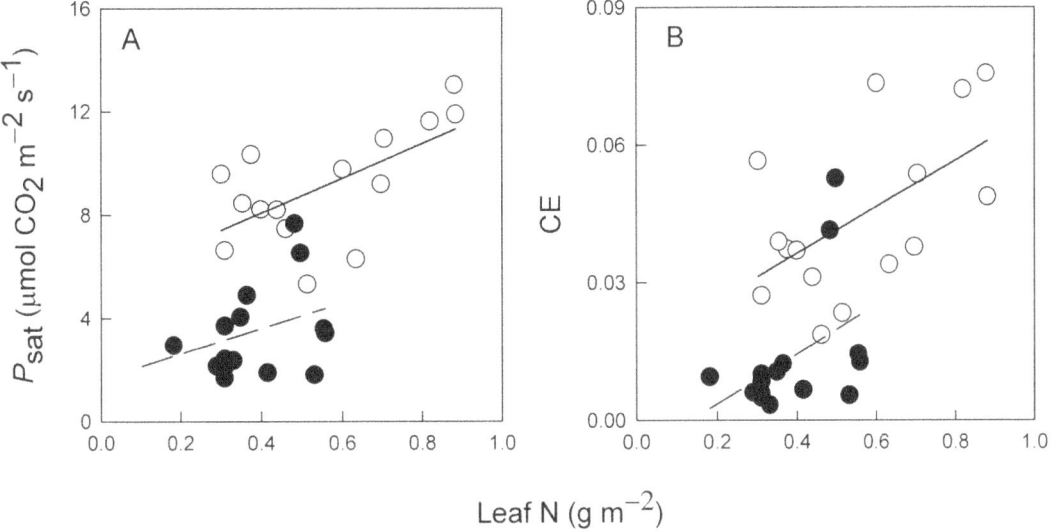

Figure 3. The relationships between leaf nitrogen concentrations (Leaf N) and P_{sat} (A) and CE (B) for *M. micrantha* **plants either uninfected (○; solid line) or infected (●; broken line) with** *C. campestris* **(Data from all nitrate treatments are included).** The correlation coefficients are 0.62 ($P<0.05$) and 0.40 ($P>0.05$) in (a), and 0.56 ($P<0.05$) and 0.44 ($P>0.05$) in (b) for the uninfected and infected plants, respectively.

significantly reduced photosynthetic nitrogen use efficiency (PNUE) of *M. micrantha* plants but the negative effects did not differ across the three nitrate treatments (Table 7).

Discussion

The present study shows that *C. campestris* infection had significant effects on most of the traits related to growth (biomass traits, number of leaves and leaf area), biomass allocation, photosynthesis and biochemical parameters of *M. micrantha* host plants. The extent of the negative effects of the parasite on host growth, and chlorophyll and leaf nitrogen content varied with the concentration of nitrate supplied to the host plants, as indicated by the significant nitrate × infection effects on these variables.

However, the effects of infection on biomass allocation and leaf photosynthesis related traits of *M. micrantha* were independent of nitrate supply.

Growth

In the present study, *C. campestris* infection reduced the number of leaves, leaf area, and biomass traits (total, root, stem, leaf and flower dry weights) of *M. micrantha* plants at each level of nitrate fertilization, and the negative impacts of the parasite on host growth were generally less severe at high than at low nitrate supplies. Such an alleviation of the impacts at high nitrogen was attributable to a more negative influence of the infection on root biomass at low than at high nitrogen fertilization. Alleviation of growth inhibition at high N supply has also been observed in *R.*

Table 7. ANOVA results and mean ($\pm$SE, $n=5$) concentrations (mg g^{-1}) of total chlorophyll (Chl), Chl *a*, Chl *b*, carotenoid (mg g^{-1}), Chl *a:b* ratio, soluble protein, Rubisco and nitrogen (g m^{-2}) of the youngest fully expanded mature leaves of uninfected and infected *M. micrantha* plants fertilized at 0.2, 1 and 5 mM nitrate concentrations.

Treatments	Total Chl	Chl *a*	Chl *b*	Chl *a/b* ratio	Carotenoid	Soluble protein	Rubisco	Nitrogen	PNUE
0.2 mM nitrate									
Uninfected	0.73±0.04b	0.52±0.03b	0.21±0.01b	2.51±0.05b	0.057±0.003b	3.39±0.22b	1.08±0.14a	0.36±0.03c	12.73±1.56a
Infected	0.29±0.05c	0.20±0.04c	0.09±0.01c	2.25±0.20b	0.038±0.005c	1.27±0.17c	0.59±0.14b	0.36±0.06c	5.21±1.01bcd
1 mM nitrate									
Uninfected	0.70±0.12b	0.50±0.08b	0.20±0.03b	2.44±0.12b	0.055±0.009b	4.50±0.19a	1.42±0.13a	0.55±0.07b	7.58±1.01b
Infected	0.22±0.04c	0.15±0.03c	0.07±0.01c	2.25±0.17b	0.031±0.006c	1.41±0.41c	0.52±0.13b	0.31±0.01c	3.54±0.89cd
5 mM nitrate									
Uninfected	1.47±0.15a	1.10±0.11a	0.37±0.040a	2.97±0.04a	0.105±0.008a	5.12±0.42a	1.53±0.14a	0.76±0.06a	6.00±1.09bc
Infected	0.35±0.07c	0.26±0.05c	0.09±0.02c	3.10±0.23a	0.046±0.007bc	2.08±0.44c	0.50±0.24b	0.49±0.03b	2.52±0.82d
Source of variation	*F* values from ANOVA								
Nitrate (N)	15.61***	18.02***	9.33**	12.92***	13.67***	7.38**	0.71 ns	16.55***	9.94**
Infection (I)	88.26***	87.35***	87.06***	0.75 ns	37.70***	104.26***	38.34***	19.58***	31.65***
N × I	9.38**	9.98**	7.48**	0.93 ns	5.28*	1.35 ns	1.61 ns	5.07*	2.01 ns

communis infected by *C. reflexa* [8] and in sorghum [5] or rice [6] infected by *S. hermonthica*. However, as N supply increased, inhibition in the growth of *C. blumei* infected by *C. reflexa* increased [7], which might be due to the strangling effect exerted by the haustorial coil of *C. reflexa*. Such strangling was not found in *C. reflexa*-infected *R. communis* [8] or in *C. campestris*-infected *M. micrantha* in this study.

Inhibitions in the reproductive growth of infected plants also became more severe at lower nitrate supply with fewer resources for reproduction as growth became severely inhibited. It has been reported that flowering was delayed and the number of florets was reduced at high N supply in *C. reflexa*-infected *C. blumei* [7], *R. communis* [8], *Vicia faba* [27] and *L. albus* [9], and flowering of *M. micrantha* was completely inhibited by low N supply and by *C. campestris* infection [10]. We did not find complete flowering inhibition in the present study which contradicts results of our earlier study [10]. However, although the plants in both studies were of similar age at the time of infection treatments, the treatments were applied about 45 days later in the growing season in the present than in the previous study.

In this study, the total biomass of the infected system (host+parasite) was less than uninfected *M. micrantha*, with greater proportional reductions as nitrate concentration decreased. Similar results were observed in the same system [10] and in *S. hermonthica*-sorghum association [5]. In these cases, the reduced growth of infected plants resulted from resource capture by the parasite and the negative effects of the parasite on host photosynthesis. As infection reduced host photosynthesis in our present study at each level of nitrate supplies, and in our previous study [11], the response of the host to infection cannot be explained by a simple source-sink relation regardless of nitrogen treatment [1,7,8]. As nitrate supply increased, the biomass of infected hosts increased, as did the corresponding biomass of the parasite. However, the percentage of total biomass allocated to the parasite did not differ among the three nitrate treatments, indicating the growth of the parasite is dependent on or tuned to the size or carrying capacity of the host. Therefore, *C. campestris* growth on its host may be resource-dependent: resource uptake is not linear but eventually reaches a plateau. This is possible as *C. campestris* obtains all its resources from its host *M. micrantha*, and its host's physiological conditions would change directly with parasite densities; in turn, *C. campestris* somehow 'senses' these changes and then regulates its growth accordingly. A fine tuning of the sink power of the parasite in the association *R. communis*–*C. reflexa* [8] and the adaptation of *C. campestris* life cycle completion to the resource availability of its host *M. micrantha* [10] has been observed. Such sensing or tuning strategies can ensure the survival of the hosts for the normal growth and development of the parasites, and the biochemical and physiological mechanisms underlying them are unknown and merit future studies.

Biomass Allocation

Cuscuta campestris had more negative effects on host root than shoot growth. *Cuscuta campestris* infection resulted in greater biomass allocation to stems and leaves but lesser allocation to roots, thus resulting in increased shoot:root ratios of infected *M. micrantha* plants. Similar results were found in our previous studies [10,28]. *Cuscuta campestris* is a shoot parasite and competes for the resources that the host allocates to shoot and root growth. The host may allocate relatively more resources to shoots to compensate for the resources directly captured by the parasite, or transfer relatively fewer resources to roots, or a higher competitive demand and a stronger source demand from the shoot system resulting in greater transfer of resources to shoots. In

root parasites, reduced shoot:root ratios have been reported in *S. hermonthica*-infected rice and sorghum [5,6] and *Orobanche aegyptiaca*-infected tomato [29]. Therefore, the negative effects of shoot parasites may be more severe on the roots than on the shoots of their hosts, and the opposite may apply to root parasites. This requires further study.

Photosynthesis

Our previous studies [11,30] showed that *C. campestris* infection reduced leaf P_n of *M. micrantha* and speculated that this was due to the parasite's indirect adverse impacts on g_s and direct negative effects on the photosynthetic metabolism of *M. micrantha*. Our present study shows similar negative effects resulting from the lower light and CO_2 use efficiencies of leaves of infected plants than uninfected plants. Infection reduced LSP, P_{max}, Φ, P_{sat}, CE, V_{cmax} J_{max} and PNUE and increased LCP across all nitrate levels. Lower photosynthetic efficiency of infected plants was also caused by lower leaf nitrogen, chlorophyll *a* and *b*, soluble protein and Rubisco concentrations, and g_s than uninfected plants. Low nitrogen concentrations in infected leaves could accelerate leaf senescence, reducing leaf photosynthesis and the number of leaves. This chain of effects in infected plants explains the lower photosynthesis we observed at leaf level, resulting in lower total photosynthesis and growth at the plant level in comparison to uninfected plants.

The magnitude of the negative effects of *C. campestris* on *M. micrantha* photosynthesis was similar across all nitrate fertilization levels. Thus, the less severe inhibition in host growth at high than at low nitrate levels is not attributable to inhibition of host photosynthesis and hence leaf production. It has been shown that *Cuscuta* can form a strong sink to redirect the flow of host resources to itself [8,27], and *Cuscuta* species alter host physiology by acting as a stronger sink for photosynthates than any host organs [31]. Redirection of more resources by *C. campestris* at low than at high nitrate levels resulted in a greater reduction in infected *M. micrantha* total biomass and root biomass.

Uninfected *M. micrantha* plants made greater use of the extra nitrogen to produce 'greener' leaves than infected plants, as shown by the higher leaf chlorophyll content at 5 mM than at 1 and 0.2 mM nitrate in uninfected plants. However, greener leaves did not have increased P_{max} although there was a good relationship between N concentration and P_{sat}. The significant effect of nitrate and infection interaction on leaf chlorophyll content (per leaf area or leaf mass) resulted from the chlorophyll content of the leaves of infected plants being more reduced at 5 mM than at 0.2 or 1 mM nitrate. This was consistent with the variation pattern for leaf nitrogen content per unit leaf area. The reason for this might be that the total nitrogen absorption and supply capacities of the roots of infected plants were more reduced at high than at low nitrate level.

Infection reduced transpiration and g_s on 30 and 80 DAP at each nitrate treatment, which may have induced stomatal closure or reduced stomatal opening. Parallel reductions in leaf nitrogen and P_n were observed, and P_n decreases would reduce carbon production. It has been suggested that low leaf nitrogen often leads to high leaf abscisic acid (ABA) levels and increases in xylem translocation of ABA from root to shoot [32] and high leaf ABA induces stomatal closure [30]. *Cuscuta campestris* infection lowered leaf nitrogen in *M. micrantha*, which may increase leaf ABA and thus contribute to the stomatal closure of infected plants [30].

In this study, as in our previous studies, *C. campestris* reduced the number of leaves of *M. micrantha*, through a host response of reducing new leaf initiation and/or accelerating leaf senescence and abscission [10,11]. Leaf senescence is characterized by a decline in photosynthesis accompanied by the loss of Rubisco and chlorophyll/protein complexes and the decline in stomatal conductance [33–35]. Leaf chlorophyll and protein contents are often used as indicators of leaf senescence [36]. In the present study, the lower leaf P_n and g_s, leaf nitrogen, total soluble protein and chlorophyll concentrations in infected plants than in uninfected plants suggest that host leaf senescence is a response to *C. campestris* infection.

In summary, the results indicate that the negative effects of the holoparasite *C. campestris* on the growth of *M. micrantha* were dependent on nitrate supply to the host, and they were more severe at low than at high nitrate levels. The more severe inhibition in host growth at low than at high nitrate supplies is largely attributable to the transfer of more host resources to *C. campestris* at low than at high nitrate levels as the magnitude of inhibition in host photosynthesis was similar across nitrate levels. In addition, *C. campestris* seems able to sense the carrying capacity of the host and regulates its growth accordingly, indicating a synchronicity in growth and development between the parasite and its host.

Supporting Information

Figure S1 Mean net photosynthetic rates (P_n, ±SE, $n = 5$) at different photosynthetic photon flux densities (PPFD) for the youngest fully expanded mature leaves of the uninfected (○) and infected (•) *M. micrantha* plants by *C. campestris* at (a) 0.2, (b) 1 and (c) 5 mM nitrate fertilizations.

Figure S2 Response of net photosynthetic rates (P_n) to intercellular CO_2 concentrations (C_i) in the youngest fully expanded mature leaves of the uninfected (○) and infected (•) *M. micrantha* plants by *C. campestris* at (a) 0.2, (b) 1 and (c) 5 mM nitrate fertilizations. Data points are means ±SE ($n = 5$).

Acknowledgments

We thank Dr. Jane Prider for her constructive comments on the manuscript.

Author Contributions

Conceived and designed the experiments: HS. Performed the experiments: HS S-JX LH. Analyzed the data: HS S-JX Z-MW W-HY. Contributed reagents/materials/analysis tools: HS S-JX W-HY. Wrote the paper: HS S-JX LH Z-MW W-HY.

References

1. Press MC, Scholes JD, Watling JR (1999) Parasitic plants: physiological and ecological interactions with their hosts. In: Press MC, Scholes JD, Barker MG, editors. Physiological plant ecology: the 39th Symposium of the British Ecological Society. York: University of York. 175–197.

2. Press MC, Graves JD, Stewart GR (1990) Physiology of the interaction of angiosperm parasites and their higher plant hosts. Plant Cell and Environment 13: 91–104.

3. Shen H, Ye W, Hong L, Huang H, Wang Z, et al. (2006) Progress in parasitic plant biology: Host selection and nutrient transfer. Plant Biology 8: 175–185.

4. Watling JR, Press MC (2001) Impacts of infection by parasitic angiosperms on host photosynthesis. Plant Biology 3: 244–250.

5. Cechin I, Press MC (1993) Nitrogen relations of the sorghum-*Striga hermonthica* host-parasite association: growth and photosynthesis. Plant, Cell and Environment 16: 237–247.

6. Cechin I, Press MC (1994) Influence of nitrogen on growth and photosynthesis of a C₃ cereal, *Oryza sativa*, infected with the root hemiparasite *Striga hermonthica*. Journal of Experimental Botany 45: 925–930.

7. Jeschke WD, Baig A, Hilpert A (1997) Sink-stimulated photosynthesis, increased transpiration and increased demand-dependent stimulation of nitrate uptake: nitrogen and carbon relations in the parasitic association *Cuscuta reflexa-Coleus blumei*. Journal of Experimental Botany 48: 915–925.

8. Jeschke WD, Hilpert A (1997) Sink-stimulated photosynthesis and sink-dependent increase in nitrate uptake: nitrogen and carbon relations of the parasitic association *Cuscuta reflexa-Ricinus communis*. Plant, Cell and Environment 20: 47–56.

9. Jeschke WD, Bäumel P, Räth N, Czygan FC, Proksch P (1994) Modelling of the flows and partitioning of carbon and nitrogen in the holoparasite *Cuscuta reflexa* Roxb. and its host *Lupinus albus* L. II. Flows between host and parasite and within the parasitized host. Journal of Experimental Botany 45: 801–812.

10. Shen H, Ye WH, Hong L, Cao HL, Wang ZM (2005) Influence of the obligate parasite *Cuscuta campestris* on growth and biomass allocation of its host *Mikania micrantha*. Journal of Experimental Botany 56: 1277–1284.

11. Shen H, Hong L, Ye WH, Cao HL, Wang ZM (2007) The influence of the holoparasitic plant *Cuscuta campestris* on the growth and photosynthesis of its host *Mikania micrantha*. Journal of Experimental Botany 58: 2929–2937.

12. Holm LG, Plucknett DL, Pancho JV, Herberger JP (1977) The world's worst weeds: distribution and biology. Honolulu: University Press of Hawaii. 320–327 p.

13. Zhang LY, Ye WH, Cao HL, Feng HL (2004) *Mikania micrantha* H.B.K. in China - an overview. Weed Research 44: 42–49.

14. Dawson JH, Musselman LJ, Wolswinkel P, Dörr I (1994) Biology and control of *Cuscuta*. Review of Weed Science 6: 265–317.

15. Hunt R, Parsons IT (1974) A computer program for deriving growth-functions in plant growth-analysis. Journal of Applied Ecology 11: 297–307.

16. von Caemmerer S, Farquhar GD (1981) Some relationships between the biochemistry of photosynthesis and the gas exchange of leaves. Planta 153: 376–387.

17. Arnon DI (1949) Copper enzymes in isolated chloroplasts: polyphenoloxidase in *Beta vulgaris*. Plant Physiology 24: 1–15.

18. Hieke S, Menzel CM, Lüdders P (2002) Effects of light availability on leaf gas exchange and expansion in lychee (*Litchi chinensis*). Tree Physiology 22: 1249–1256.

19. Watling JR, Press MC (2000) Infection with the parasitic angiosperm *Striga hermonthica* influences the response of the C₃ cereal *Oryza sativa* to elevated CO₂. Global Change Biology 6: 919–930.

20. Farquhar GD, von Caemmerer S, Berry JA (1980) A biochemical model of photosynthetic CO₂ assimilation in leaves of C₃ species. Planta 149: 78–90.

21. Harley PC, Sharkey TD (1991) An improved model of C₃ photosynthesis at high CO₂: reversed O₂ sensitivity explained by lack of glycerate reentry into the chloroplast. Photosynthesis Research 27: 169–178.

22. Harley PC, Thomas RB, Reynolds JF, Strain BR (1992) Modeling photosynthesis of cotton grown in elevated CO₂. Plant Cell and Environment 15: 271–282.

23. Bradford MM (1976) A rapid and sensitive method for the quantitation of microgram quantities of proteins utilizing the principle of protein-dye-binding. Analytical Biochemistry 72: 248–254.

24. Makino A, Mae T, Ohira K (1986) Colorimetric measurement of protein stained with Coomassie Brilliant Blue R on sodium dodecyl sulfate-polyacrylamide gel electrophoresis by eluting with formamide. Agricultural and Biological Chemistry 50: 1911–1912.

25. Irving LJ, Robinson D (2006) A dynamic model of Rubisco turnover in cereal leaves. New Phytologist 169: 493–504.

26. Laemmli UK (1970) Cleavage of structural proteins during the assembly of the head of bacteriophage T4. Nature 227: 680–685.

27. Wolswinkel P (1974) Complete inhibition of setting and growth of fruits of *Vicia faba* L., resulting from the draining of the phloem system by *Cuscuta* species. Acta Botanica Neerlandica 23: 48–60.

28. Shen H, Hong L, Chen H, Ye WH, Cao HL, et al. (2011) The response of the invasive weed *Mikania micrantha* to infection density of the obligate parasite *Cuscuta campestris* and its implications for biological control of *M. micrantha*. Botanical Studies 52: 89–97.

29. Barker ER, Press MC, Scholes JD, Quick WP (1996) Interactions between the parasitic angiosperm *Orobanche aegyptiaca* and its tomato host: growth and biomass allocation. New Phytologist 133: 637–642.

30. Chen H, Shen H, Ye W, Cao H, Wang Z (2011) Involvement of ABA in reduced photosynthesis and stomatal conductance in *Cuscuta campestris* - *Mikania micrantha* association. Biologia Plantarum 55: 545–548.

31. Parker C, Riches CR (1993) Parasitic weeds of the world: biology and control. Wallingford: CAB International. 332 p.

32. Peuke AD, Jeschke WD, Hartung W (1994) The uptake and flow of C, N and ions between roots and shoots in *Ricinus communis* L. III. Long-distance transport of abscisic acid depending on nitrogen nutrition and salt stress. Journal of Experimental Botany 45: 741–747.

33. Navabpour S, Morris K, Allen R, Harrison E, A-H-Mackerness S, et al. (2003) Expression of senescence-enhanced genes in response to oxidative stress. Journal of Experimental Botany 54: 2285–2292.

34. Murchie EH, Chen YZ, Hubbart S, Peng SB, Horton P (1999) Interactions between senescence and leaf orientation determine in situ patterns of photosynthesis and photoinhibition in field-grown rice. Plant Physiology 119: 553–563.

35. Jiang CZ, Rodermel SR, Shibles RM (1993) Photosynthesis, Rubisco activity and amount, and their Regulation by transcription in senescing soybean leaves. Plant Physiology 101: 105–112.

36. Vanacker H, Sandalio LM, Jimenez A, Palma JM, Corpas FJ, et al. (2006) Roles for redox regulation in leaf senescence of pea plants grown on different sources of nitrogen nutrition. Journal of Experimental Botany 57: 1735–1745.

19

Effects of Fertilization and Clipping on Carbon, Nitrogen Storage, and Soil Microbial Activity in a Natural Grassland in Southern China

Zhimin Du[1,2], Yan Xie[1,2], Liqun Hu[1], Longxing Hu[1,2], Shendong Xu[3], Daoxin Li[3], Gongfang Wang[3], Jinmin Fu[1,2]*

1 Key Laboratory of Plant Germplasm Enhancement and Specialty Agriculture, Wuhan Botanical Garden, Chinese Academy of Sciences, Wuhan, Hubei, China, 2 Graduate University of Chinese Academy of Sciences, Beijing, Hebei, China, 3 National Dalaoling Forest Park, Yichang, Hubei, China

Abstract

Grassland managements can affect carbon (C) and nitrogen (N) storage in grassland ecosystems with consequent feedbacks to climate change. We investigated the impacts of compound fertilization and clipping on grass biomass, plant and soil (0–20 cm depth) C, N storage, plant and soil C: N ratios, soil microbial activity and diversity, and C, N sequestration rates in grassland *in situ* in the National Dalaoling Forest Park of China beginning July, 2011. In July, 2012, the fertilization increased total biomass by 30.1%, plant C by 34.5%, plant N by 79.8%, soil C by 18.8% and soil N by 23.8% compared with the control, respectively. Whereas the clipping decreased total biomass, plant C and N, soil C and N by 24.9%, 30.3%, 39.3%, 18.5%, and 19.4%, respectively, when compared to the control. The plant C: N ratio was lower for the fertilization than for the control and the clipping treatments. The soil microbial activity and diversity indices were higher for the fertilization than for the control. The clipping generally exhibited a lower level of soil microbial activity and diversity compared to the control. The principal component analysis indicated that the soil microbial communities of the control, fertilization and clipping treatments formed three distinct groups. The plant C and N sequestration rates of the fertilization were significantly higher than the clipping treatment. Our results suggest that fertilization is an efficient management practice in improving the C and N storage of the grassland ecosystem via increasing the grass biomass and soil microbial activity and diversity.

Editor: Ting Wang, Wuhan Botanical Garden, Chinese Academy of Sciences, Wuhan, China, China

Funding: Funding came from the "Strategic Priority Research Program - Climate Change: Carbon Budget and Relevant Issues" of the Chinese Academy of Sciences (No. XDA0505040704), and the National Natural Science Foundation of China (No. 31272194). The funders had no role in study design, data collection and analysis, decision to publish, or preparation of the manuscript.

Competing Interests: The authors have declared that no competing interests exist.

* E-mail: jfu@wbgcas.cn

Introduction

The grasslands in China cover an area of 3.92 million km^2 and provide 9% to 16% of the total C in the world grasslands [1,2,3]. Concerns about global warming has increased an attention to understand the role of potential C and nitrogen (N) sink in grasslands in mitigating the emission of greenhouse gases (i.e. CO_2 and N_2O) [4–6]. The C and N sequestration in terrestrial ecosystems constitutes a major mitigation strategy against the global warming [7]. China's grasslands make an important contribution to the world C and N storage and may have significant effects on C and N cycles worldwide [2]. Natural grasslands of southern China cover an area of 79.58 million km^2, and probably have a high yield owing to good hydrothermal conditions [8], which can be an important C and N pool.

The processes of C and N sequestration can be greatly affected by grassland managements [9], and good managements are critical for grasslands to enhance C and N sequestration [10–12]. Compound fertilizers or organic amendments affected grasslands C and N storage via increasing plant biomass [10,13,14]. Dersch and Böhm [15] reported that N, phosphorus (P), and potassium (K) fertilizers combined with farmyard manure application enhanced C storage to about 5.6 Mg ha^{-1} after 21 years in Australia. The N fertilization and cover cropping can increase soil organic C and total N by increasing the amount of plant residues returned to the soil [11,16]. Similarly, the application of manure can increase soil organic C and total N levels [17,18]. Clipping was found to affect the grassland C and N storage via reducing plant biomass [9] and changing grass species [19]. Particularly, the potentially dominant plants (i.e. usually larger than their neighbors) often lose a higher proportion of their biomass than their neighbors after clipping [9].

Soil microorganisms exert a dominant influence on the net C and N balance of terrestrial ecosystems by controlling soil organic matter (SOM) decomposition and plant nutrient availability [20,21]. The grassland SOM mainly derived from roots, senescent leaves and stems of the vegetations [22]. The processes and functions of breakdown of the plants residues in soil are greatly impacted by soil microorganisms [23]. Agricultural managements can affect soil microorganisms' condition and ultimately affect the C and N cycling in ecosystems [24,25]. Microbial populations were significantly increased in the soils amended with green manure throughout two-year experiment [26]. Soil microbial diversity and/or activity may be a sensitive indicator of ecosystem change, as it can be quickly affected by disturbances [27,28].

Zhong and Cai [29] demonstrated that soil microbial diversity and average well color development (AWCD) which reflects total microbial activity [30] in the NPK treatment were increased in response to fertilization. Soil microbial biomass, populations and diversity were increased by optimum and balanced fertilization [31,32]. On the other hand, the clipping significantly reduced soil microbial and respiration rate in both warmed and un-warmed plots of tallgrass prairie [33]. Above-ground biomass removal could significantly reduce C inputs from vegetation to soil and lead to significant N loss, resulting in substrate limitation to soil microorganisms [34,35].

Understanding the fate of stored C and N and their potential for anthropogenic manipulation is critically important to evaluate the future state of the atmosphere or terrestrial ecosystems and manage the foreseen global change [36,37]. However, the effects of management in relation to soil microorganisms on the redistribution and cycling of C and N within the plant-soil system were unclear. The objective of this study was to investigate the effects of compound fertilizer and clipping on C and N storage and distribution within a natural grassland ecosystem.

Materials and Methods

Site description and experimental design

The study was conducted in the National Dalaoling Forest Park near the dam of the Three Gorges Reservoir in China from July 2011 to September 2012. The experimental site located at approximately 110°56′E, 31°4′N and 1696 m asl. The climate in this region is of a northern subtropical type, with a warm, humid summer, and an obvious altitudinal change. Maximum, minimum, and mean annual temperature was 19.2°C in July, −2.7°C in January, and 8.5°C, respectively. The mean annual precipitation was 1446.8 mm. Although the majority of precipitation occurs in summer, there was still 179.6 mm in winter occurring as snow and sleet [38].

The experimental site had a vegetation coverage of more than 60%, composed of over 20 grass species, but was dominated by *Festuca arundinacea* Schreb (approximately 30% of total above-ground biomass), *Potentilla freyniana* (50% of total above-ground biomass) and *Lysimachia clethroides* Duby (10% of total above-ground biomass). In this humid ungrazed montane meadow, all grasses were shallow rooted in the depth of 0–20 cm, with the maximum density occurred in the 0–10 cm soil layer. The soil in the 0–20 cm depth zone had a pH of 5.8, 12.9 g organic matter per kg soil, 1.1 g total N per kg soil, and 0.4 g total P per kg soil.

The grassland was exposed to three treatments: (i) control; (ii) fertilization; and (iii) clipping. Five replicated plots were conducted for each of the three treatments and arranged in a randomized complete block design. Each plot was measured 10 m by 5 m and fenced on June 20, 2011 to prevent the rabbits or other animals from grazing. The grassland was untreated in the control. In the fertilization treatment, compound fertilizers (15-15-15, N-P_2O_5-K_2O) were applied on July 15, 2011 and May 15, 2012 (600 kg per ha for each time). In the clipping treatment, the grassland vegetations were clipped to 3 to 5 cm with sickles on July 15, 2011 and May 15, 2012, respectively. The clippings were left *in situ*.

Plant and soil sampling and analysis

Plant biomass was assessed five times from 2011 to 2012, in May (late spring 2012), July (middle summer 2011, 2012) and September (early autumn 2011, 2012). In each plot, five random 1 m×1 m quadrats were assigned. One quadrat was selected each time for plant sampling. Shoot including living and standing dead within the quadrat was collected. After the shoot was removed, the litter was picked up. Then, five soil sub-samples (7 cm in diam, and 0-5-10-20 cm depths) were collected from each quadrat using a soil auger and pooled together to be a composite sample. The roots (including roots and rhizomes) in the pooled soil cores were picked up and washed with deionized water three times to get rid of residual soil. Shoots, litters, and roots were killed at 105°C for 30 minutes and dried to constant weight at 80°C. The C and N concentrations of the plant samples of July 1, 2011 and July 18, 2012 were measured based on the methods described by Lu [39]. Each of the fresh composite soil samples of July 1, 2011 and July 18, 2012 was sieved (2 mm wire mesh) and divided into two sub-samples: one was kept in the refrigerator at 4°C until microbial analysis and the other was air-dried for the analysis of soil organic C and total N concentrations [39]. Plant and soil C, N storage was calculated as Post's and Tian's methods [40,41].

Soil microbial populations' analysis

Traditional culture techniques were used to determine the distribution of the main physiological groups [42]. Microbial populations of bacteria, fungi and actinomyces were determined by soil dilution plating on beef extract peptone medium, Martin's medium and Gause's No. 1 synthetic medium, respectively. All media were made up as the methods described by Dong et al. [43] and all microbes were cultivated in a 28°C incubator for 2, 3 and 7 d, respectively.

BIOLOG analysis and calculation of microbial activity and diversity indices

The soil microbial activity and functional diversity was evaluated using BIOLOG ECO microplate (BIOLOG Inc., Hayward, CA, USA). Soil sample kept in the refrigerator (equal to 10.0 g dried soil) was suspended in 90 mL of sterile 0.85% NaCl solution, shaken at 220 rpm for 30 min and held for 5 min. Then the suspensions were diluted to a final dilution of 10^{-3} with sterile 0.85% NaCl solution. Each well of BIOLOG ECO plates was inoculated with 125 μL of the diluted soil extracts and incubated at 25°C. Optical density of the wells was read with BIOLOG Micro Station reader (MicroLog release 4.20) at 590 nm every 12 h for 7 days. The optical density readings were corrected for the water controls in subsequent analysis. Negative readings after the correcting were adjusted to zero. Soil microbial activity measured as AWCD was calculated by the method described by Garland and Mills [30]. The substrate richness, Shannon's diversity index, Shannon's evenness index, McIntosh's diversity index, and McIntosh's evenness index were calculated using the data at 72 h, since the highest rate of microbial growth was observed at this incubation time [44,45]. Formulas used for the above indices calculations were described by Magurran [46] and Staddon et al. [47]. Principal component analysis (PCA) was performed on BIOLOG data divided by the AWCD [30].

C and N sequestration rates

Changes of C (or N) sequestration were estimated by calculating the difference of C (or N) storage between July, 2012 and July, 2011. C and N sequestration rates (CSR, g C m^{-2} yr^{-1} and NSR, g N m^{-2} yr^{-1}) were calculated using the following equations, respectively:

$$CSR = \frac{C_{rn} - C_{r0}}{n - 0} \tag{1}$$

$$NSR = \frac{N_{rn} - N_{r0}}{n - 0} \qquad (2)$$

where $C_{r0 \text{ and } n}$ is C storage (g C m^{-2}) under the certain management (the fertilization, clipping or the control) during the first and second years in which C storage was measured, respectively; $N_{r0 \text{ and } n}$ is N storage (g N m^{-2}) under the certain treatment (the fertilization, clipping or the control) during the first and second years in which N storage was measured, respectively; n is the number of years of duration of the experiment.

Data analysis

Management effects and interactions between the variables were determined by the analysis of variance using SPSS 13.0 (SPSS, Inc.). Significantly different means were separated using Fisher's protected least significant difference (LSD) test (p<0.05). PCA analysis was performed using the Canoco 4.5 software package [48]. The data in July, 2011 were not shown since no obvious difference was found in effect of clipping and fertilization on soil microbial activity and diversity.

Results

Biomass

The shoot, root, litter and total biomass of the three treatments varied considerably through different growing season in 2011 and 2012 (Fig. 1). There was no significant difference in biomass among the treatments in July, 2011 (before the implementation of fertilization and clipping). As time went by, the fertilization treatment increased the shoot, root, and total biomass than the control. The clipping treatment, on the contrary, had the less shoot, root, litter and total biomass compared to the control. The total biomass was much higher in middle summer (July) relative to late spring (May) and early autumn (September) for each year. The total biomass in the fertilization treatment in July, 2012 was 30.1% and 73.4% greater than that in the control and clipping treatments, respectively.

Plant and soil C, N storage and C: N ratios

Plant C storage ranged from 258.6 to 295.2 g C m^{-2}, and soil C storage ranged from 2086.9 to 2752.5 g C m^{-2} for the three treatments as measured in July, 2011. Plant N storage ranged from 5.8 to 7.0 g N m^{-2}, and soil N storage ranged from 199.8 to 223.0 g N m^{-2} for the three treatments as measured in July, 2011. No difference in C and N storage was found among the three treatments in July, 2011 (data were not presented).

The shoot and root C storage was the highest in the fertilization treatment, followed by the control and clipping in July, 2012 (Table 1). Shoot and root C storage in the fertilization treatment was 1.4 and 1.4 times more than that in the control, 2.2 and 1.9 times more than that in the clipping, respectively. Root C storage was similar between the control and clipping treatment. The shoot and root N storage of the fertilization treatment was 1.8 and 2.0 times more than the control, 3.0 and 2.8 times more than the clipping in July, 2012, respectively (Table 1). No difference in litter C and N storage was observed among the three treatments. The C: N ratios of all plant parts were decreased due to the fertilization treatment, but no difference was found between the clipping and the control (Table 1). Both plant C and N storage increased linearly as the plant biomass increased (both p<0.001; Fig. 2a, b).

Fertilized soil stored more C and N compared to the control and clipping treatments regardless soil depths in July, 2012 (Table 2). The soil had less C and N storage in the 0–5 and 10–20 cm zone

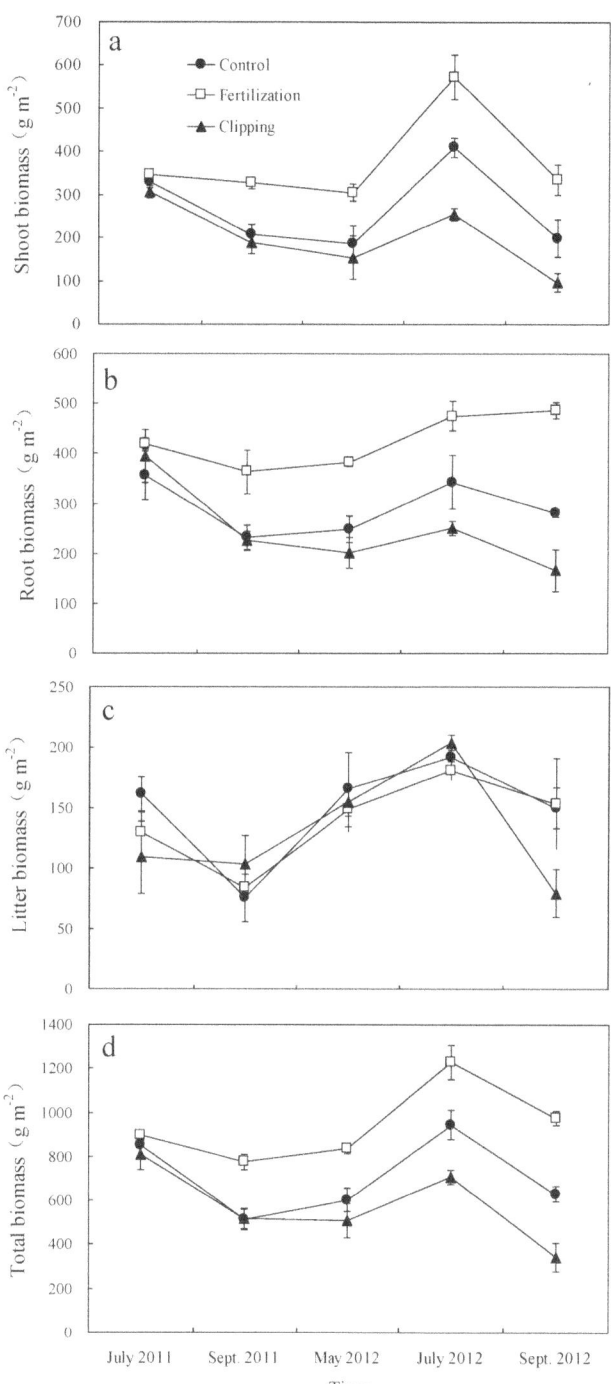

Figure 1. The shoot (a), root (b), litter (c) and total biomass (d) variation under the different treatments over time. Vertical bars represent standard error (SE). $n = 5$.

in the clipping treatment. The fertilization increased soil C by 18.8% and soil N by 23.8%, respectively, when compared to the control. Whereas the clipping decreased soil C and N by 18.5% and 19.4% compared to the control, respectively. No significant difference in soil C: N ratio among the three treatments was found (Table 2). Both soil C and N storage increased linearly as the plant biomass increased (both p<0.001; Fig. 2e, i). Furthermore, the plant C storage increased linearly as the soil C storage increased

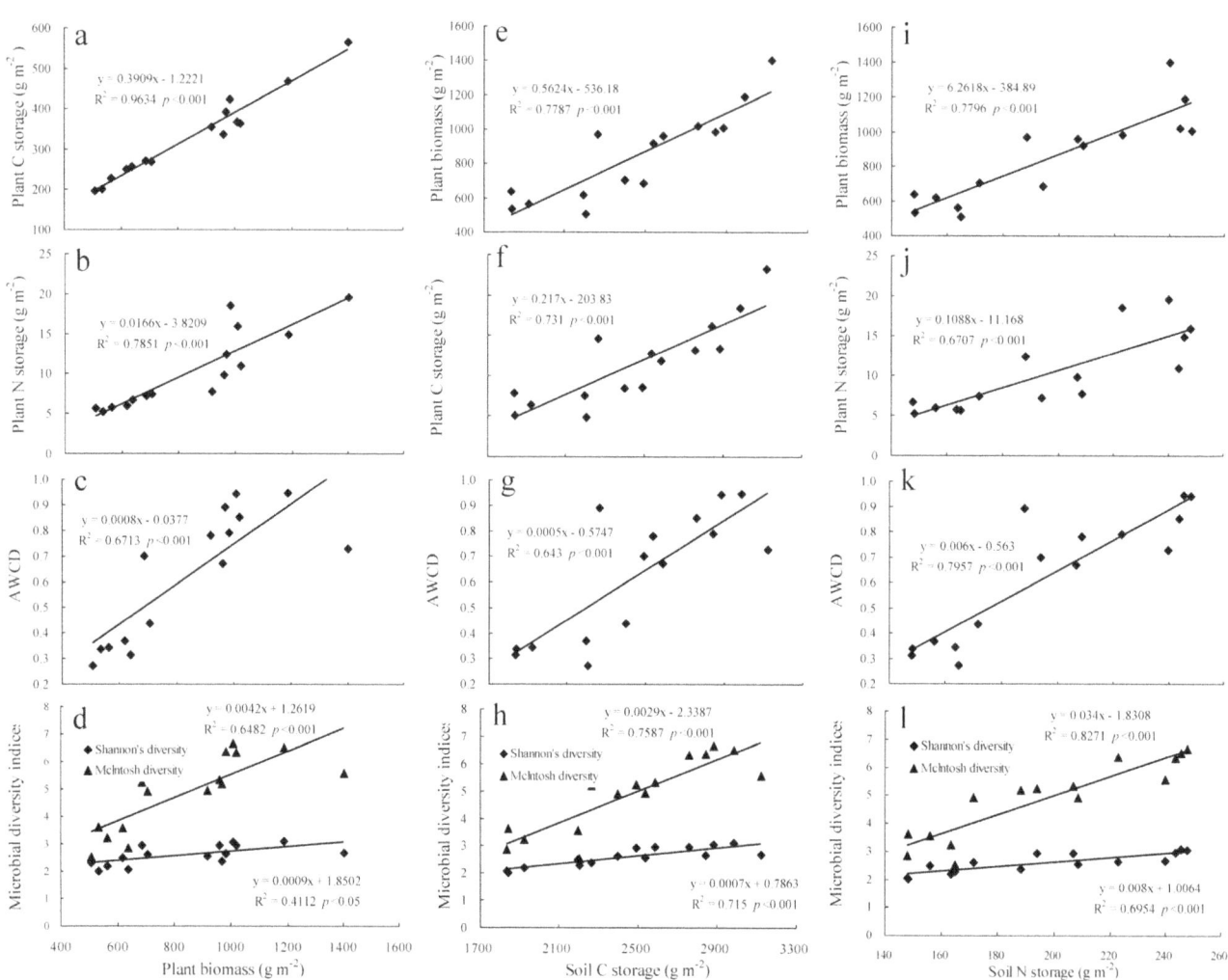

Figure 2. The relationships of plant biomass with plant C/N storage (a/b), AWCD (c) and soil microbial diversity indices (d), and the relationships of soil C/N storage with plant biomass (e/i), plant C/N storage (f/j), AWCD (g/k) and soil microbial diversity indices (h/l) across the three treatments in July, 2012.

Table 1. Carbon and nitrogen storage and C: N ratio among plant parts under different grassland treatments in July, 2012.

Treatments	Shoot	Root	Litter	Total
	-------C storage (g C m^{-2})--------			
Control	170.8±9.3b	139.3±21.8b	15.2±1.0a	325.2±24.1b
Fertilization	233.0±22.9a	195.8±19.3a	8.7±1.9a	437.5±37.0a
Clipping	108.0±7.1c	102.7±4.9b	16.0±2.6a	226.7±12.1c
	--------N storage (g N m^{-2}) ---------			
Control	5.5±0.6b	2.8±0.5b	0.6±0.0a	8.9±1.0b
Fertilization	10.0±1.1a	5.6±0.9a	0.4±0.1a	16.0±1.5a
Clipping	3.3±0.1b	2.0±0.2b	0.5±0.1a	5.8±0.2b
	---------C:N ratio----------			
Control	32.1±2.2a	50.1±2.8a	26.7±1.5a	37.3±2.5a
Fertilization	23.6±1.9b	37.3±4.2b	20.3±1.8b	27.9±2.2b
Clipping	32.8±2.0a	51.5±2.0a	30.3±1.8a	38.8±1.1a

The data represent means ± SE ($n = 5$). Letters a, b, c in a column indicate statistical significance base on Fisher's protected LSD test ($p<0.05$) among the three different treatments.

Table 2. Carbon and nitrogen storage and C: N ratio of soils at the 0–5, 5–10, and 10–20 cm depths under different grassland treatments in July, 2012.

Treatments	0–5 cm	5–10 cm	10–20 cm	0–20 cm
	––––––––C storage (g C m^{-2}) ––––––––			
Control	869.6±21.2b	573.1±26.7b	1014.1±47.5b	2456.9±56.5b
Fertilization	992.0±29.3a	732.9±39.1a	1193.8±29.5a	2918.8±62.5a
Clipping	690.3±29.1c	527.5±21.6b	784.4±52.4c	2002.2±82.7c
	––––––––N storage (g N m^{-2}) ––––––––			
Control	64.8±5.3b	44.3±3.2b	84.2±2.0b	193.9±6.7b
Fertilization	79.6±2.5a	62.2±3.4a	98.3±1.5a	240.1±4.5a
Clipping	50.0±2.3c	40.6±2.1b	66.1±3.5c	156.2±3.6c
	–––––––––C: N ratio––––––––			
Control	13.8±1.2a	13.1±0.9a	12.0±0.6a	12.7±0.4a
Fertilization	12.5±0.5a	11.9±0.9a	12.2±0.3a	12.2±0.3a
Clipping	13.9±0.9a	13.4±1.3a	11.9±0.4a	12.8±0.4a

The data represent means ± SE ($n = 5$). Letters a, b, c in a column indicate statistical significance base on Fisher's protected LSD test ($p < 0.05$) among the three different treatments.

($p < 0.001$; Fig. 2f), and the plant N storage increased linearly as the soil N storage increased ($p < 0.001$; Fig. 2j).

The high, medium and low values of total C and N storage (including plant and 0–20 cm soil zone) were observed in the fertilization, control and clipping grassland ecosystems, respectively (Tables 1, 2). The fertilization increased total C and N storage by 20.6% and 26.3%, respectively. The clipping reduced the total C and N storage by 19.9% and 20.1%, respectively. Furthermore, the plants had lower C and N storage and most C and N was stored in the soils. The 0–20 cm zone soils of the control, fertilization, and clipping treatments held 88%, 87% and 90% of the total C storage, and 96%, 94% and 96% of the total N storage, respectively.

Soil microbial populations, activity and diversity

There was no difference in soil microbial number, activity and diversity among the three treatments in July, 2011 (data were not presented). On July 18, 2012, the number of bacteria, fungi and actinomyces in the fertilization treatment was 1.5, 1.3 and 1.6 times more than the control, and 2.3, 2.0 and 2.4 times more than the clipping treatment, respectively (Table 3). There was no difference in bacteria and actinomyces numbers between the control and the clipping. However, the clipping had fewer fungi than the control. The soil microbial activity (measured as AWCD)

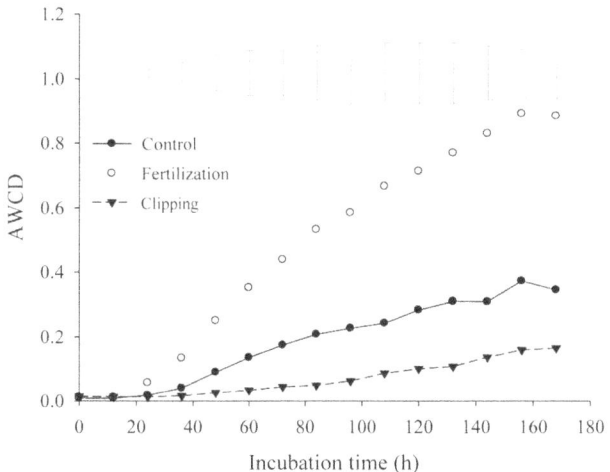

Figure 3. Change of average well color development (AWCD) of soil microbial community during the incubation time in July, 2012. Vertical bars represent Fisher's protected LSD ($p < 0.05$).

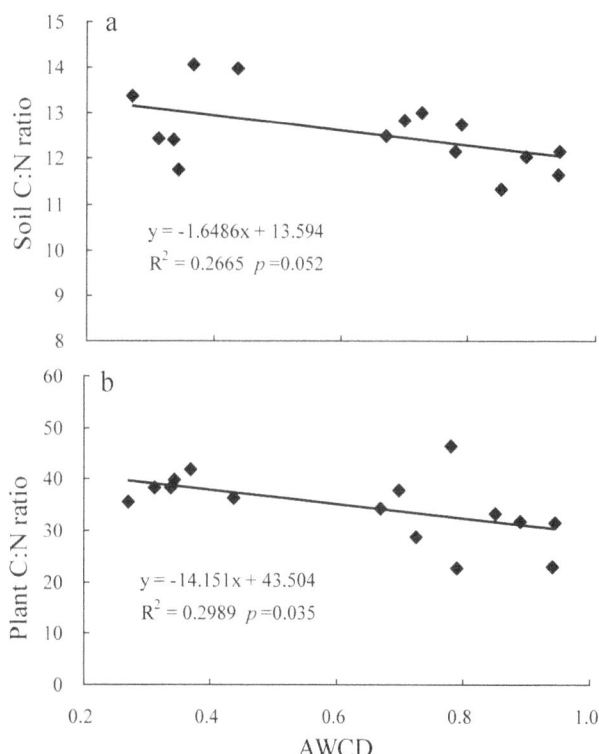

Figure 4. The relationships of AWCD of soil microbial community with soil C: N ratio (a) and plant C: N ratio (b) in

Table 3. The microbial community structure in grassland soils under different treatments in July, 2012. (CFU g^{-1} dry weight soil).

Treatments	bacteria number ($\times 10^6$)	fungi number ($\times 10^2$)	actinomyces number ($\times 10^4$)
Control	2.8±0.1b	2.7±0.2b	1.2±0.1b
Fertilization	4.3±0.2a	3.6±0.1a	1.9±0.1a
Clipping	1.9±0.1b	1.8±0.1c	0.8±0.0b

The data represent means ± SE ($n=5$). Letters a, b, c in a column indicate statistical significance base on Fisher's protected LSD test ($p<0.05$) among the three different treatments.

in July, 2012 was increased with incubation time for all the three treatments, and ranked in the order of fertilization > control > clipping (Fig. 3). The AWCD was positively associated with plant biomass, soil C and N storage ($p<0.001$, Fig. 2c, g, k). Both soil and plant C: N ratios decreased linearly with increasing AWCD ($p=0.052$, $p=0.035$ Fig. 4a, b). Soil microbial diversity, evenness indices and substrate richness in July, 2012 calculated from the BIOLOG data (72 h) were affected by the fertilization and clipping treatments (Table 4). The higher levels of substrate richness, Shannon's evenness index, and McIntosh diversity and evenness indices were detected in the fertilization treatment, but lower level in the clipping treatment, compared to that in the control. Both the Shannon's and McIntosh diversity indices were positively associated with plant biomass, soil C and N storage ($p<0.001$, Fig. 2d, h, l).

The PCA analysis indicated that the first two principal components accounted for 51.4% of the total variance (Fig. 5). The soil microbial community of the control, fertilization and the clipping treatments formed three distinct groups. The control was distinctly separated from the fertilization and clipping treatments by Factor 2. The fertilization and clipping treatments were distinctly separated by Factor 1. Furthermore, the factor loading plot also showed that the affinity of soil microbes for the substrates depended on the grassland treatments. The substrates including *L*-Arginine (A4), *L*-Phenylalanine (C4), N-Acetyl-D-Glucosamine (E2), Glucose-1-Phosphate (G2), Phenylethyl-amine (G4) and D-Malic Acid (H3) were favored by soil microbes of the fertilization treatment. The substrates including *L*-Threonine (E4), α-D-Lactose (H1), and D,L-α-Glycerol phosphate (H2) were favored by soil microbes of the clipping treatment, while the substrates including i-Erythritol (C2), 2-Hydroxy Benzoic Acid (C3), γ-Hydroxybutyric Acid (E3) and α-Ketobutyric Acid (G3) were favored by soil microbes of the control. Compared to the control, the fertilization treatment increased the utilization level of the substrates of amines (G4) and amino acids (A4 and C4), while the clipping treatment decreased the utilization level of the substrates of carboxylic acids.

Plant and soil C, N sequestration rates

The plant, soil (0–20 cm) and total C, N sequestration rates during July, 2011 to July, 2012 were ranked in descending order of fertilization > control > clipping (Table 5). There was no statistical difference in the grassland soil (0–20 cm) and the total C and N sequestration rates among the three treatments. The plant C and N sequestration rates of the fertilization treatment were significantly higher than the clipping treatment ($p<0.05$).

Discussion

This investigation was conducted on the natural grassland with vegetation coverage of more than 60%, composed of over 20 species of grass. In grassland ecosystems, the immobilization of C and N in the soil is the basic solution for C and N sequestration. Schleinger [49] reported that the below-ground C pool generally had much slower turnover rate than above-ground C. The data collected on July 1, 2011 exhibited that soil C and N (i.e. 15 days prior to imposing experiment) was similar among the three treatments. However, soil C and N on July 18, 2012 (i.e. at the end of the two-year experiment) increased by 18.8% and 18.5%, respectively, in the fertilization relative to the control. Previous studies investigated the effect of fertilizer on the grassland C and N storage and indicated that the accumulation of soil C and N were attributed to the increase of plant biomass [50–52].

As the main source of soil organic matter, the increase of grass biomass (including the shoots, roots, senescent litter) may be the first step to enhance the C and N sequestration in the soil [10,53]. The results of this study indicated that shoot and root biomass was greater in the fertilization treatment vs. the control, but less in the clipping treatment vs. the control. The increase in plant and soil C, N storage was significantly associated with a greater grass biomass (Fig. 2a, b, e, i). Data collected in July, 2012 indicated that the plant and soil C was significantly related to grass biomass ($r^2 = 0.9634$, 0.7787, both $p<0.001$). The correlation coefficient was 0.886 between plant N and biomass, and 0.883 between soil N and biomass (both $p<0.001$). Meanwhile, the plant C and N

Table 4. Effects of fertilization and clipping on microbial functional diversity as evaluated by substrate richness (*S*), Shannon's diversity index (*H'*), Shannon's evenness index (*E (S)*), McIntosh diversity index (*U*) and McIntosh evenness index (*E (M)*) in July, 2012 (72 h).

Treatments	S	H'	E (S)	U	E (M)
Control	24.6±0.5b	2.7±0.1a	0.8±0.0b	5.1±0.1b	0.8±0.0b
Fertilization	27.0±0.6a	2.9±0.1a	0.9±0.0a	6.3±0.2a	0.9±0.0a
Clipping	21.8±0.4c	2.2±0.1b	0.7±0.0c	3.2±0.2c	0.5±0.0c

The data represent means ± SE ($n=5$). Letters a, b, c in a column indicate statistical significance base on Fisher's protected LSD test ($p<0.05$) among the three different treatments.

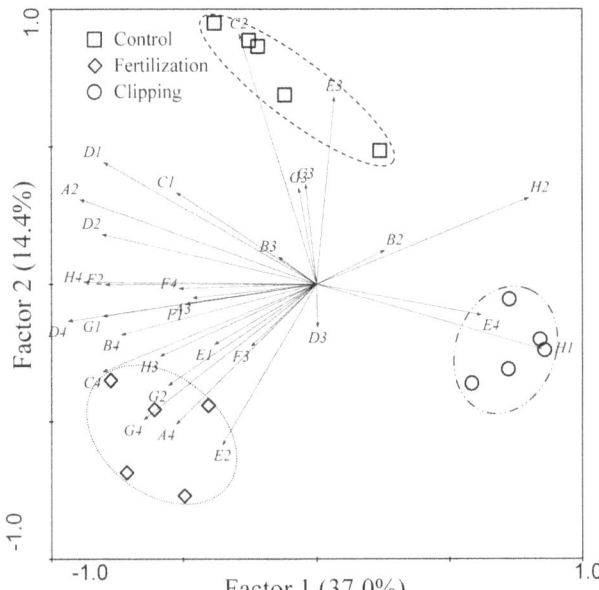

Figure 5. Principle components analysis of biological data in July, 2012 (72 h).

storage were positively associated with soil C and N storage (both $p < 0.001$). Because the C and N stored in the plant was ultimately transferred into the soil in the forms of plant residues [54]. The beneficial effect of the fertilization on plant biomass could be contributed to the input nutrient [14,55]. The compound fertilizer provided the essential elements of N, P, and K for plant growth, which improved the shoot, root, and total biomass. On the contrary, the clipping limited the plant growth by damaging photosynthesis organs, causing a slow-growing period and a decrease in grass biomass consequently [9]. In addition, the variation of grass biomass indicated that the biomass could be affected by seasonal variation, management types and management time. From the perspective of increasing biomass, spring fertilization could give better results compared with summer fertilization.

The nutrient for plant growth is mainly derived from decomposition of SOM and plant residues input to the soil [56]. Similar to previous studies [9,57], we found that the plant of the

fertilization treatment grew rapidly and sequestrated more C (carbohydrates) through photosynthesis and more N through passively and/or actively uptake, while the clipping treatment reduced the plant C and N uptake. Cheng et al [35] also found that the clipping decreased the plant N uptake in the tallgrass prairie. But Ruess et al [58] have reported that the clipping stimulated uptake rates of both ammonium (NH_4^+) and nitrate (NO_3^-), and ultimately accumulated more total plant N. This inconsistence was perhaps caused by the differences of soil nutrient conditions and plant species. The plant C: N ratio of the fertilization treatment was much lower than the control and the clipping treatments (Table 2). Chen [54] indicated that the SOM and plant residues with lower C: N ratios could be decomposed by microorganisms more easily due to improved soil microbial activity. We also found that the soil and plant C: N ratios decreased linearly with increasing AWCD ($p = 0.035$, $p = 0.052$). In addition, more humus was predicted for plant substrates with lower C: N ratios [59]. Thus the fertilization grassland could have a higher humification degree and sequestrated C and N in soil for a longer time than the control and the clipping treatments.

Soil microbial activity was involved in the mineralization of soil organic matter and plant residues [60,61]. Previous studies [56,62] indicated that the plant residues provided soil microorganisms with the major resource of nutrients and energy and controlled the soil microbial activity and composition. The results of this study exhibited that the fertilization and clipping treatments affected the grass biomass and ultimately changed the soil microbial activity, diversity and the C substrates utilized by soil microorganisms. The increase in soil microbial activity and diversity in the fertilization treatment was similar to Zhong and Cai's [29] and Marschner et al's study [14].

Our study suggested that the C and N cycles in the grassland ecosystem are determined not only by plant biomass, but also by soil microbial activity. The fertilization treatment, especially spring fertilization, improved the plant growth, increased the soil microbial activity, and ultimately increased the plant and soil C, N storage and sequestration rates. While the clipping treatment had the opposite effect and reduced the C, N storage and sequestration rates in grassland. Grassland management including fertilization was crucial to the grassland recovery from barren and overgrazing [63,64]. Other studies [65,66] indicated that the soil C may reach a new equilibrium in approximately decades after management changes, so continuous fertilization treatment was needed.

Table 5. Carbon and nitrogen sequestration rates of plant and soil (0–20 cm) under different grassland treatments from July, 2011 to July, 2012.

Treatments	Plant	Soil (0–20 cm)	Total
	-----C sequestration rate (g C m^{-2} yr^{-1}) ------		
Control	66.6±33.6ab	9.0±328.3	75.5±229.3
Fertilization	142.3±43.3a	166.3±380.3	308.6±379.8
Clipping	−31.9±39.2b	−84.7±194.6	−116.6±510.3
	-----N sequestration rate (g N m^{-2} yr^{-1}) ------		
Control	2.9±1.0ab	−5.7±43.6	−2.9±43.8
Fertilization	9.0±2.1a	28.2±15.7	37.2±16.2
Clipping	−0.4±0.7b	−11.1±9.4	−11.5±10.0

The data represent means ± SE ($n = 5$). Letters a, b, c in a column indicate statistical significance base on Fisher's protected LSD test ($p < 0.05$) among the three different treatments.

Conclusions

The fertilization and the clipping treatments exhibited remarkable effect on the grass biomass, C and N storage, and soil microbial activity and diversity of grassland ecosystems in National Dalaoling Forest Park. After the two-year experiment, the compound fertilizer increased grass biomass, improved soil microbial activity and diversity and increased C and N sequestration in grassland ecosystem. The clipped plots had a lower level of C, N storage, which were mainly attributed to less grass biomass. The soil C, N storage was increased linearly with increasing grass biomass, plant C and N storage and AWCD. The principal component analysis indicated that the soil microbial communities of the control, fertilization and the clipping treatments formed three distinct groups, respectively. Previous and our results suggested that the fertilization might improve soil C and N slowly. So, to improve soil C, continuous fertilization management of grassland is needed. Long-term *in situ* studies to extrapolate the effect of the plant biomass on C and N dynamics combined to soil microbial activity and diversity and soil microclimate might contribute to a better understanding of C and N cycling and mitigation of global warming.

Acknowledgments

The authors gratefully acknowledge the help of Dr Cheng Xiaoli in providing valuable revisions of the manuscript.

Author Contributions

Conceived and designed the experiments: ZD SX JF. Performed the experiments: ZD YX LiqunH. LongxingH. Analyzed the data: ZD. Contributed reagents/materials/analysis tools: DL GW JF. Wrote the paper: ZD.

References

1. Scurlock JMO, Hall DO (1998) The global carbon sink: a grassland perspective. Global Change Biol 4: 229–233.
2. Ni J (2002) Carbon storage in grasslands of China. J Arid Environ 50: 205–218.
3. Deng YL, Wang YK (2010) Assessment on water retention function of grassland ecosystems in the Upper Yangtze River Basin. Asian J Water Environ Pollut 7: 1–6.
4. Powlson DS, Whitmore AP, Goulding KWT (2011) Soil carbon sequestration to mitigate climate change: a critical re-examination to identify the true and the false. Eur J Soil Sci 62: 42–55.
5. Lal R (2003) Global potential of soil carbon sequestration to mitigate the greenhouse effect. Crit Rev Plant Sci 22: 151–184.
6. César Izaurralde R, Rosenberg NJ, Lal R (2001) Mitigation of climatic change by soil carbon sequestration: issues of science, monitoring, and degraded lands. Adv Agron 70: 1–75.
7. Dhillon RS, von Wuehlisch G (2013) Mitigation of global warming through renewable biomass. Biomass Bioenerg 48: 75–89.
8. Lv SH (2005) Status and development prospects of meadow resource in southern of China. J Sichuan Grassland 6: 37–41 (In Chinese with English abstract).
9. Klimeš L, Klimešová J (2001) The effects of mowing and fertilization on carbohydrate reserves and regrowth of grasses: do they promote plant coexistence in species-rich meadows? Evol Ecol 15: 363–382.
10. Smith P, Martino D, Cai Z, Gwary D, Janzen H, et al. (2008) Greenhouse gas mitigation in agriculture. Phil Trans R Soc B: Biol Sci 363: 789–813.
11. Sainju UM, Senwo ZN, Nyakatawa EZ, Tazisong IA, Reddy KC (2008) Soil carbon and nitrogen sequestration as affected by long-term tillage, cropping systems, and nitrogen fertilizer sources. Agr Ecosyst Environ 127: 234–240.
12. Qiu LP, Wei XR, Zhang XC, Cheng JM (2013) Ecosystem carbon and nitrogen accumulation after grazing exclusion in semiarid grassland. PLoS One 8: e55433.
13. Anindo DO, Potter HL (1994) Seasonal variation in productivity and nutritive value of Napier grass at Muguga, Kenya. E Afr Agric For J 59: 177–185.
14. Marschner P, Kandeler E, Marschner B (2003) Structure and function of the soil microbial community in a long-term fertilizer experiment. Soil Biol Biochem 35: 453–461.
15. Dersch G, Böhm K (2001) Effects of agronomic practices on the soil carbon storage potential in arable farming in Austria. Nutr Cycl Agroecosys 60: 49–55.
16. Sainju UM, Singh BP, Whitehead WF (2002) Long-term effects of tillage, cover crops, and nitrogen fertilization on organic carbon and nitrogen concentrations in sandy loam soils in Georgia, USA. Soil Till Res 63: 167–179.
17. Rochette P, Gregorich E (1998) Dynamics of soil microbial biomass C, soluble organic C and CO_2 evolution after three years of manure application. Can J Soil Sci 78: 283–290.
18. Collins HP, Rasmussen PE, Douglas CL (1992) Crop rotation and residue management effects on soil carbon and microbial dynamics. Soil Sci Soc Am J 56: 783–788.
19. Fynn RWS, Morris CD, Edwards TJ (2009) Effect of burning and mowing on grass and forb diversity in a long-term grassland experiment. Appl Veg Sci 7: 1–10.
20. Paul EA (1996) Soil microbiology and biochemistry. San Diego, California: Academic Press. 512 p.
21. Liski J, Nissinen A, Erhard M, Taskinen O (2003) Climatic effects on litter decomposition from arctic tundra to tropical rainforest. Global change biol 9: 575–584.
22. Joffre R, Ågren GI (2001) From plant to soil: litter production and decomposition. In: Roy J, Saugier B, Mooney HA, editors. Terrestrial global productivity. San Diego, California, , USA: Academic Press. pp. 83–99.
23. Larkin RP (2003) Characterization of soil microbial communities under different potato cropping systems by microbial population dynamics, substrate utilization, and fatty acid profiles. Soil Biol Biochem 35: 1451–1466.
24. Smith JL, Papendick RI, Bezdicek DF, Lynch JM (1992) Soil organic matter dynamics and crop residue management. In: Metting FB, editor. Soil microbial ecology: applications in agricultural and environmental management. New York: Marcel Dekker Inc. pp. 65–94.
25. Filser J, Fromm H, Nagel RF, Winter K (1995) Effects of previous intensive agricultural management on microorganisms and the biodiversity of soil fauna. Plant Soil 170: 123–129.
26. Sekiguchi H, Kushida A, Takenaka S (2007) Effects of cattle manure and green manure on the microbial community structure in upland soil determined by denaturing gradient gel electrophoresis. Microbes Environ 22: 327–335.
27. Kennedy AC, Smith KL (1995) Soil microbial diversity and the sustainability of agricultural soils. Plant Soil 170: 75–86.
28. Fox CA, MacDonald KB (2003) Challenges related to soil biodiversity research in agroecosystems-issues within the context of scale of observation. Can J Soil Sci 83: 231–244.
29. Zhong WH, Cai ZC (2007) Long-term effects of inorganic fertilizers on microbial biomass and community functional diversity in a paddy soil derived from quaternary red clay. Appl Soil Ecol 36: 84–91.
30. Garland JL, Mills AL (1991) Classification and characterization of heterotrophic microbial communities on the basis of patterns of community-level sole-carbon-source utilization. Appl Environ Microb 57: 2351–2359.
31. Ebhin Masto R, Chhonkar PK, Singh D, Patra AK (2006) Changes in soil biological and biochemical characteristics in a long-term field trial on a sub-tropical inceptisol. Soil Biol Biochem 38: 1577–1582.
32. Chang EH, Chung RS, Tsai YH (2007) Effect of different application rates of organic fertilizer on soil enzyme activity and microbial population. Soil Sci Plant Nutr 53: 132–140.
33. Zhang W, Parker KM, Luo Y, Wan S, Wallace LL, et al. (2005) Soil microbial responses to experimental warming and clipping in a tallgrass prairie. Glob Change Biol 11: 266–277.
34. Wan SQ, Luo YQ (2003) Substrate regulation of soil respiration in a tallgrass prairie: results of a clipping and shading experiment. Global Biogeochem Cy 17: 1054.
35. Cheng XL, Luo YQ, Su B, Wan SQ, Hui DF, et al. (2011) Plant carbon substrate supply regulated soil nitrogen dynamics in a tallgrass prairie in the Great Plains, USA: results of a clipping and shading experiment. J Plant Ecol 4: 228–235.
36. Melillo JM, Prentice IC, Farquhar GD, Schulze ED, Sala OE (1996) Terrestrial biotic responses to environmental change and feedbacks to climate. In: Climate change 1995: the science of climate change. . New York: Cambridge University Press. pp. 445–481.
37. Tans PP, Bakwin PS (1995) Climate change and carbon dioxide. Ambio 24: 376–378.
38. Zhang QF, Zheng Z, Jin YX (1990) Studies on the forest succession in Dalao ridge, Hubei province. Acta Phytoecologica et geobotanica sinica 14: 110–117 (In Chinese with English abstract).
39. Lu RK (2000) Analytical methods of soil agrochemistry. Beijing: China Agricultural Science and Technology Press. 638p (In Chinese).
40. Tian G, Granato TC, Cox AE, Pietz RI, Carlson CR, Jr., et al. (2009) Soil carbon sequestration resulting from long-term application of biosolids for land reclamation. J Environ Qual 38: 61–74.
41. Post WM, Emanuel WR, Zinke PJ, Stangenberger AG (1982) Soil carbon pools and world life zones. Nature 298: 156–159.
42. Zhou J, Guo WH, Wang RQ, Han XM, Wang Q (2008) Microbial community diversity in the profile of an agricultural soil in northern China. J Environ Sci 20: 981–988.
43. Dong M, Wang YF, Kong FZ, Jiang GM, Zhang ZB, et al. (1996) Survey, observation and analysis of terrestrial biocommunities. Beijing: Standards Press. 290 p (In Chinese).

44. Garland JL (1996) Analytical approaches to the characterization of samples of microbial communities using patterns of potential C source utilization. Soil Biol Biochem 28: 213–221.

45. Haack SK, Garchow H, Klug MJ, Forney LJ (1995) Analysis of factors affecting the accuracy, reproducibility, and interpretation of microbial community carbon source utilization patterns. Appl Environ Microb 61: 1458–1468.

46. Magurran AE (1988) Ecological diversity and its measurement. Princeton, NJ: Princeton university press. 179 p.

47. Staddon WJ, Duchesne LC, Trevors JT (1997) Microbial diversity and community structure of postdisturbance forest soils as determined by sole-carbon-source utilization patterns. Microb Ecol 34: 125–130.

48. Ter Braak CJF, Šmilauer P (2002) CANOCO reference manual and CanoDraw for Windows user's guide: software for canonical community ordination (version 4.5). Ithaca, New York: Microcomputer Power. 500 p.

49. Schlesinger WH (1995) Soil respiration and changes in soil carbon stocks. In: Woodwell GM, Mackenzie FT, editors. Biotic feedbacks in the global climatic system: will the warming feed the warming. New York: Oxford University Press. pp. 159–168.

50. Conant RT, Paustian K, Elliott ET (2001) Grassland management and conversion into grassland: Effects on soil carbon. Ecol Appl 11: 343–355.

51. Silveira ML, Liu K, Sollenberger LE, Follett RF, Vendramini J (2012) Short-term effects of grazing intensity and nitrogen fertilization on soil organic carbon pools under perennial grass pastures in the southeastern USA. Soil biol Biochem 58: 42–49.

52. Thornley JHM, Fowler D, Cannell MGR (1991) Terrestrial carbon storage resulting from CO$_2$ and nitrogen fertilization in temperate grasslands. Plant Cell Environ 14: 1007–1011.

53. Kuzyakov Y, Domanski G (2000) Carbon input by plants into the soil. Review. J Plant Nutr Soil Sci 163: 421–431.

54. Chen HM. (2005) Environmental soil science. Beijing: Science Press. 516 p (In Chinese).

55. Pyšek P, Lepš J (2009) Response of a weed community to nitrogen fertilization: a multivariate analysis. J Veg Sci 2: 237–244.

56. Thibodeau L, Raymond P, Camiré C, Munson AD (2000) Impact of precommercial thinning in balsam fir stands on soil nitrogen dynamics, microbial biomass, decomposition, and foliar nutrition. Can J Forest Res 30: 229–238.

57. Porporato A, D'odorico P, Laio F, Rodriguez-Iturbe I (2003) Hydrologic controls on soil carbon and nitrogen cycles. I. Modeling scheme. Adv Water Resour 26: 45–58.

58. Ruess RW, McNaughton SJ, Coughenour MB (1983) The effects of clipping, nitrogen source and nitrogen concentration on the growth responses and nitrogen uptake of an East African sedge. Oecologia 59: 253–261.

59. Nicolardot B, Recous S, Mary B (2001) Simulation of C and N mineralisation during crop residue decomposition: a simple dynamic model based on the C: N ratio of the residues. Plant Soil 228: 83–103.

60. Magill AH, Aber JD (2000) Variation in soil net mineralization rates with dissolved organic carbon additions. Soil biol Biochem 32: 597–601.

61. Zak DR, Tilman D, Parmenter RR, Rice CW, Fisher FM, et al. (1994) Plant production and soil microorganisms in late-successional ecosystems: a continental-scale study. Ecology 75: 2333–2347.

62. Palm CA, Myers RJK, Nandwa SM (1997) Combined use of organic and inorganic nutrient sources for soil fertility maintenance and replenishment. In: Buresh RJ, Sanchez PA, Calhoun F, editors. Replenishing soil fertility in Africa. Madison, Wisconsin: SSSA special publication. pp. 193–217.

63. Chen Q, Hooper DU, Lin S (2011) Shifts in species composition constrain restoration of overgrazed grassland using nitrogen fertilization in Inner Mongolian steppe, China. PloS One 6: e16909.

64. Goetz H (1969) Composition and yields of native grassland sites fertilized at different rates of nitrogen. J Range Manage 22: 384–390.

65. Dumanski J, Desjardins RL, Tarnocai C, Monreal C, Gregorich EG, et al. (1998) Possibilities for future carbon sequestration in Canadian agriculture in relation to land use changes. Climatic Change 40: 81–103.

66. Smith P, Powlson DS, Glendining MJ, Smith JU (1998) Preliminary estimates of the potential for carbon mitigation in European soils through no-till farming. Global change biol 4: 679–685.

20

Potential of Global Cropland Phytolith Carbon Sink from Optimization of Cropping System and Fertilization

Zhaoliang Song[1,2,3]*, **Jeffrey F. Parr**[4], **Fengshan Guo**[2]

1 Zhejiang Provincial Key Laboratory of C Cycling in Forest Ecosystems and C Sequestration, Zhejiang Agricultural and Forestry University, Lin'an, Zhejiang, China, **2** School of Environment and Resources, Zhejiang Agricultural and Forestry University, Lin'an, Zhejiang, China, **3** State Key Laboratory of Environmental Geochemistry, Institute of Geochemistry, Chinese Academy of Sciences, Guiyang, Guizhou, China, **4** Southern Cross GeoScience, Southern Cross University, Lismore, New South Wales, Australia

Abstract

The occlusion of carbon (C) by phytoliths, the recalcitrant silicified structures deposited within plant tissues, is an important persistent C sink mechanism for croplands and other grass-dominated ecosystems. By constructing a silica content-phytolith content transfer function and calculating the magnitude of phytolith C sink in global croplands with relevant crop production data, this study investigated the present and potential of phytolith C sinks in global croplands and its contribution to the cropland C balance to understand the cropland C cycle and enhance long-term C sequestration in croplands. Our results indicate that the phytolith sink annually sequesters 26.35 ± 10.22 Tg of carbon dioxide (CO_2) and may contribute $40\pm18\%$ of the global net cropland soil C sink for 1961–2100. Rice (25%), wheat (19%) and maize (23%) are the dominant contributing crop species to this phytolith C sink. Continentally, the main contributors are Asia (49%), North America (17%) and Europe (16%). The sink has tripled since 1961, mainly due to fertilizer application and irrigation. Cropland phytolith C sinks may be further enhanced by adopting cropland management practices such as optimization of cropping system and fertilization.

Editor: Senjie Lin, University of Connecticut, United States of America

Funding: The authors are grateful for support from the National Natural Science Foundation of China (Grant No. 41103042), the Zhejiang Province Key Science and Technology Innovation Team (No. 2010R50030), and the Opening Project of State Key Laboratory of Environmental Geochemistry (SKLEG9011). The funders had no role in study design, data collection and analysis, decision to publish, or preparation of the manuscript.

Competing Interests: The authors have declared that no competing interests exist.

* E-mail: songzhaoliang78@163.com

Introduction

Present understanding of the global carbon (C) cycle and climate feedbacks is limited by uncertainty over terrestrial C balance [1–5]. As one of the largest terrestrial ecosystems deeply influenced by human activities, the croplands cover an area of 15.33×10^8 hm^2 globally and may play a significant role in terrestrial C balance [3,6]. Although croplands were traditionally considered to be the largest biospheric source of C lost to the atmosphere in most areas of the world [7–12], they may also be significant C sinks under proper management [3,6,13–15].

Phytolith-occluded C (PhytOC), where C is entrapped within recalcitrant silicified structures when they are deposited within plant tissues [16–18], is particularly prolific in many crops such as rice [19], wheat [20], millet [21] and sugarcane [22]. PhytOC is highly resistant against decomposition [18,23–25] and may accumulate in soil for several thousands of years after plant decomposition [18], demonstrating the potential of phytoliths in the long-term biogeochemical sequestration of atmospheric carbon dioxide (CO_2) [5,26]. Soil PhytOC accumulation is an important persistent C sink mechanism for croplands [18–22] and other grass-dominated ecosystems [26,27]. Moreover, Jansson et. al. [28] suggest that the production of PhytOC in croplands could be greatly enhanced through crop breeding. However, the present and potential of global cropland phytolith C sink have not been revealed.

In the present study, we quantifed the present and potential of phytolith carbon sink and its contribution to the global cropland C balance by constructing a silica content-phytolith content transfer function and calculating the magnitude of the phytolith C sink in global croplands with relevant crop data including the PhytOC and silica content, farm crop output, the Si-rich organ ratio (mass ratios of the Si-rich organ: crop output) and the PhytOC stability factor. The purposes of the study are to guide the management of cropland ecosystems to maximize phytolith C sequestration and mitigate climate change.

Materials and Methods

Ethics Statements

No specific permits were required for the described field studies, because the experimental field is owned by Zhejiang Agricultural and Forestry University, and the School of Environment and Resources performs the management. No specific permits were required for these locations/activities, because the location is not privately-owned or protected in any way, the field studies did not involve endangered or protected species, and each sample consisted of no more than 500 grams (fresh weight).

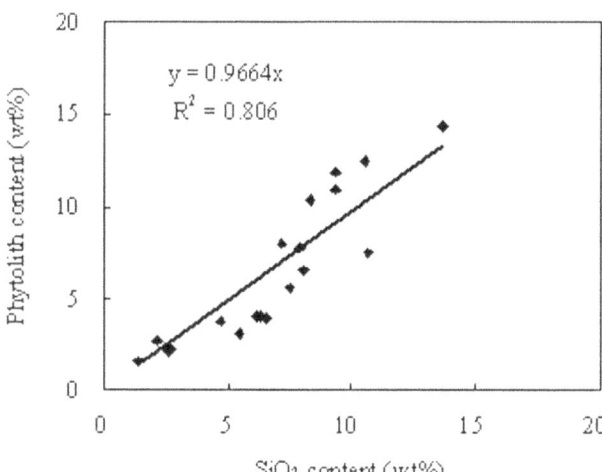

Figure 1. The correlation of phytolith content (%) to SiO$_2$ (%) content in different crop species (p<0.01).

Constructing the Transfer Function for the Phytolith:Silica Content

Plant phytolith content may be estimated from plant silica content data using the transfer function for the phytolith:silica content [26]. To construct the silica content- phytolith content transfer function, mature crop organ samples were collected– each sample consisted of approximately 500 g of composite plant material.

Plant samples were oven-dried at 65°C to a constant mass and cut into small pieces (<5 mm). They were ashed at 500°C to remove organic matter, fused with lithium metaborate, dissolved in dilute nitric acid and analyzed for silica content using inductively coupled plasma-optical emission spectroscopy (ICP-OES; Optima 7000 DV, Perkin Elmer, Massachusetts, USA). Plant phytoliths were isolated using a microwave digestion process followed by a Walkley–Black type digestion to ensure the removal of extraneous organic material [19,27]. The isolated phytoliths were dried to a constant mass at 75°C for 24 h in a fan-forced oven and weighed to determine the plant phytolith content. The occluded-C content within phytoliths was also determined [19,27]. The error was <5% in phytolith and silica measurements and <10% in PhytOC measurements using plant standards (GSV-1) and triplicate analyses.

The plant silica content- phytolith content transfer function was constructed using regression analysis based on the phytolith and silica contents determined for the samples (Figure 1). Silica content was converted to phytolith content using the following equation (R^2 = 0.806, p<0.01):

$$\text{Phytolith content (wt\%)} = 0.9664 \times \text{silica content (wt\%)} \quad (1)$$

Data Collection, Phytolith and PhytOC Content Estimation

Farm productivity data was obtained from Food and Agriculture Organization of the United Nations (FAO) Statistics [29]. Silica content data was obtained from published monographs [30,31], papers [19–21,32,33] and also determined in the present study. Silica content of crop species was used to estimate phytolith content using a conversion factor of 0.9664; see equation (1). The

Table 1. General information and PhytOC content of the dominant arable crops.

Farm crops	Area (10^6 hm^2)[a]	Plant Si-rich organs	PhytOC (%)[b]	
			mean	SE
Crops (total)	1532.6		0.13	0.05
Cereals (total)	697.7	Stem, sheath and leaf	0.19	0.07
Rice	164.1	Stem, sheath and leaf	0.25	0.07
Wheat	220.4	Stem, sheath and leaf	0.16	0.08
Maize	170.4	Stem, sheath and leaf	0.16	0.05
Soybeans	103	Stem and leaf	0.02	0.01
Roots and Tubers	54.3	Stem and leaf	0.02	0.01
Oil-bearing crops	62	Stem and leaf	0.08	0.08
Seed cotton	35.2	Stem and leaf	0.02	0.01
Sugar cane	25.4	Sheath and leaf	0.25	0.07

[a]values from FAO (2012).
[b]estimated from phytolith and silica content data (This study; ref. [19–21,30–33]) using equation (1) and occluded-C content in phytolith of 3±1% (This study; ref. [19–21,27]).

PhytOC content in plant organs was estimated from phytolith content data using an occluded-C content in phytolith of 2~4% (average 3%) according to the present study and references [19–21,27].

Estimating PhytOC Production and the Phytolith C Sink

The production of PhytOC is primarily affected by plant PhytOC concentration and aboveground net primary productivity (ANPP) of Si-rich organs [34], where plant PhytOC is mainly determined by PhytOC content in phytoliths [26] and plant Si content [30,33]. This allowed the crop PhytOC production rate to be estimated from the PhytOC content and total ANPP of the Si-rich organs of an area as:

$$\begin{aligned} \text{PhytOC production rate} &= \text{PhytOC content} \\ &\times \text{ANPP of Si-rich organs} \times 44/12 \end{aligned} \quad (2)$$

where PhytOC production rate is the PhytOC production by a particular crop's Si-rich organs per year (Tg CO$_2$ yr^{-1}), PhytOC content is the concentration of PhytOC in a crop's Si-rich organs (wt %) and ANPP is the total aboveground net primary productivity of Si-rich crop organs (Tg yr^{-1}) of an area estimated from Si-rich organ factor [35,36] and crop output [29].

As the PhytOC sequestration rate is controlled by the PhytOC production rate in plants and the stability of phytolith in environments, the phytolith C sink rate can be estimated from data of PhytOC production rate and phytolith stability factor as:

$$\begin{aligned} \text{Phytolith C sink rate} &= \text{PhytOC production rate} \\ &\times \text{phytolith stability factor} \end{aligned} \quad (3)$$

where PhytOC production rate may be estimated from equation (2) and the phytolith stability factor is assumed to be 0.9±0.05 as most phytoliths have been proved stable for thousands of years though some small phytolith particles containing little carbon may be partly dissolved depending on formation sites and chemical

Table 2. Estimated phytolith C sink produced by global farm crops in 2011.

Farm crops	Crop output (Tg yr^{-1})[a]	Si-rich organ factor[b]	ANPP[c] of Si-rich organs (Tg yr^{-1})[c]	Phytolith C sink (Tg CO$_2$ yr^{-1})[d]	
				Mean	SE
Crops (total)		1.43	8091	26.35	10.22
Cereals (total)	2587	1.37	3557	22.39	8.41
Rice	723	1.1	795	6.60	1.99
Wheat	704	1.29	906	4.93	2.30
Maize	883	1.35	1194	6.14	2.46
Soybeans	261	1.5	391	0.27	0.14
Roots and Tubers	807	0.58	468	0.3	0.17
Oil-bearing crops	105	2.2	231	0.59	0.70
Seed cotton	77	2.91	225	0.18	0.07
Sugar cane[e]	1794	0.18	323	2.63	0.74

[a]values from FAO [29];
[b]mass ratios of the Si-rich organ: crop output from Huang et al. [35] and Zhu et al. [36];
[c]ANPP: above-ground net primary productivity;
[d]estimated from the crop output and Si-rich organ factor and PhytOC content in Table 1 using equations (2, 3);
[e]The crop output of sugar cane is fresh cane weight.

composition of phytoliths in plant organs, and deposition environments of phytoliths after plant decay [18,25].

Results

Distribution of PhytOC in Dominant Arable Crops

The global area of croplands is 1532.6 10^6 hm^2, about half of which is covered by cereals (Table 1). The PhytOC content varies greatly among different crops (0.02–0.25%, with an average of 0.13%) (Table 1). Generally, sugar cane and cereals have higher PhytOC contents in dry biomass (0.16–0.25%) than than other crops (0.02–0.08%). Within cereals, rice has higher PhytOC content in dry biomass (0.25±0.07%) than other cereal crops such as wheat (0.16±0.08%) and maize (0.16±0.05%).

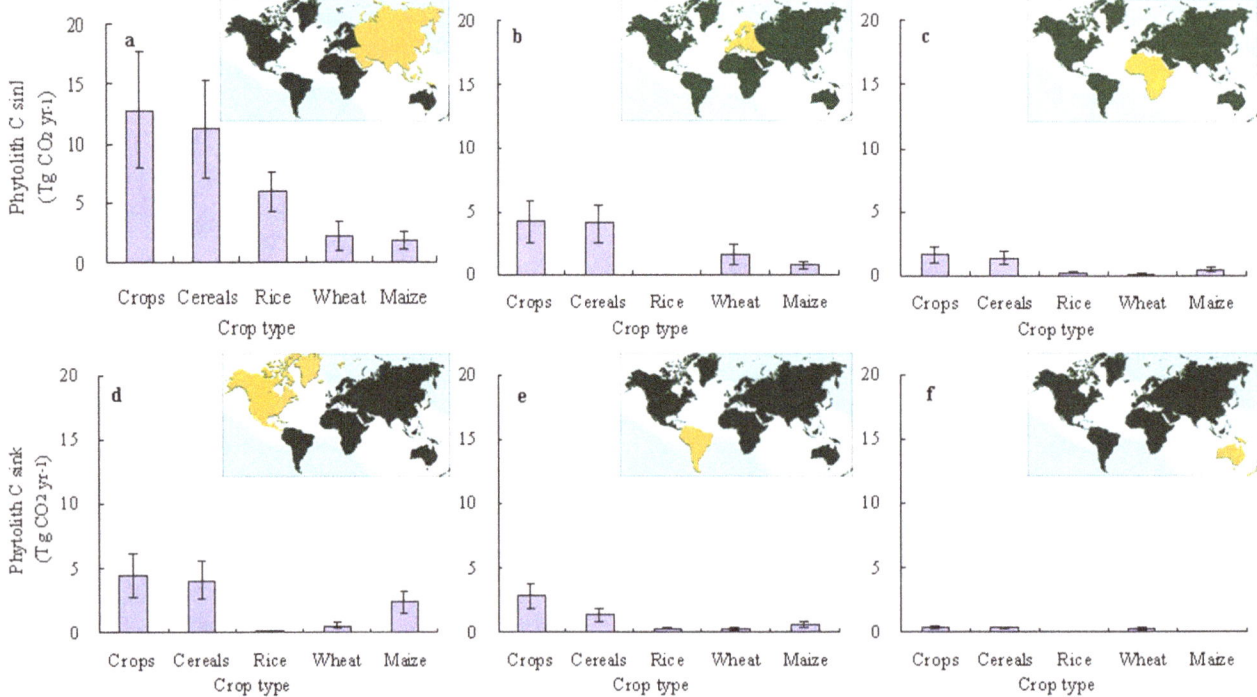

Figure 2. Phytolith carbon sink production by farm crops from different continents in 2011. Where **A:** Asia, **B:** Europe, **C:** Africa, **D:** North America, **E:** South America, **F:** Oceania. 'Crops' represents the sum of all farm crops and "Cereals" represents the sum of all cereal crops including rice, wheat, maize etc.

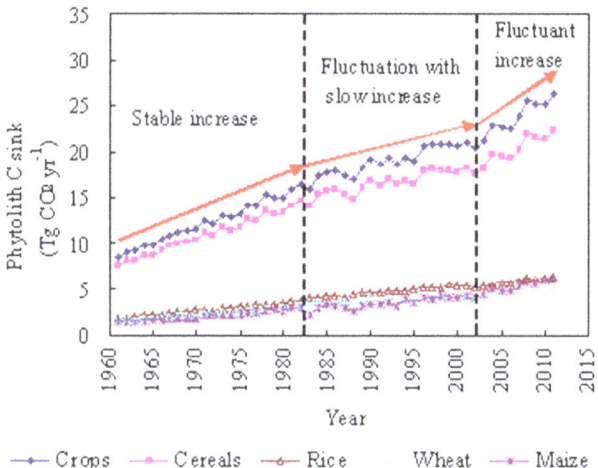

Figure 3. Phytolith carbon sink produced by global farm crops from 1961 to 2011. Crops include all farm crops in Table 1. Cereals include rice, wheat, maize, millet, barley and sorghum.

Phytolith Carbon Sink of Global Croplands

The phytolith C sink varies greatly among different crops (Table 2). The phytolith C sinks generated by rice, wheat and maize (6.60 ± 1.99, 4.93 ± 2.30 and 6.14 ± 2.46 Tg CO_2 yr^{-1}, respectively) are much higher than other crops. The total phytolith C sink produced by global farm crops is around 26.35 ± 10.22 Tg CO_2 yr^{-1}, 85% of which is contributed from cereals, including rice (25%), wheat (19%) and maize (23%).

Figure 2 displays the relative land area of the major continents and the phytolith C sink produced by farm crops in each in 2011. The largest phytolith C sinks occur in Asia (12.80 ± 4.90 Tg CO_2 yr^{-1}), North America (4.50 ± 1.74 Tg CO_2 yr^{-1}) and Europe (4.21 ± 1.66 Tg CO_2 yr^{-1}), which account for 49, 17 and 16% of the total global croplands, respectively.

The total phytolith C sink of global croplands has tripled since 1961 (Figure 3). In general, the evolution of the phytolith C sink since 1961 may be divided into three stages:

(1) 1961–1982: the total phytolith C sink increased steadily from 8.61 to 16.57 Tg CO_2 yr^{-1}.

(2) 1983–2002: the total phytolith C sink fluctuated, with a slow increase from 15.99 to 20.51 Tg CO_2 yr^{-1}.

(3) 2003–2011: the total phytolith C sink increased, with some fluctuation, from 21.14 to 26.35 Tg CO_2 yr^{-1}.

Discussion

Contribution of Phytolith C Sink to Global Cropland C Balance

By comparing phytolith C sink data and the global cropland soil C balance during 1961–2100, the contribution of the phytolith C sink to the net global cropland C balance was estimated (Table 3). Ruddiman [37] estimated that the emission from land-use conversion during the postindustrial era (i.e. 200 years), using 0.8 Gt C yr^{-1} (or 2.93 Pg CO_2 yr^{-1}), at 160 Gt C (or 587 Pg CO_2). Using an average soil C sink rate of -2.93 Pg CO_2 yr^{-1} during 1961–2015, the total soil C sink is about -161.2 Pg CO_2. Lal [6,14] estimated a high and attainable soil C sequestration potential of 0.55 Gt C yr^{-1} (or 2.02 Pg CO_2 yr^{-1}) for global croplands assuming judicious land use and recommended management practices (RMPs) were taken world-wide. Taking an average soil C sink rate of 2.02 Pg CO_2 yr^{-1} during 2016–2100, the total soil C sink is about 171.7 Pg CO_2. The total net soil C sink of global croplands during 1961–2100 is about 10.6 Pg CO_2. Taking an average phytolith C sink rate of 0.03 Pg CO_2 yr^{-1}, the total phytolith C sink of global croplands during 1961 and 2100 is 4.2 ± 1.9 Pg CO_2 yr^{-1}, 40 ± 18% of the total net soil C sink (Table 3).

Enhancing Phytolith Carbon Sink through the Optimization of Cropping System and Fertilization

Carbon sink trading has been carried out in many fields [38]. If phytolith carbon sink can be measured exactly, we believe that the sink may also be traded to increase the income of farmers. Therefore, in the future, farmers will be optimising carbon sequestration besides increasing yields.

Regional analysis of crop structures and farm productivity (FAO, 2012) suggests that the high phytolith C sinks in Asia, North America and Europe are due to the relatively wide production of rice, maize and wheat, respectively. The rapid increase of total phytolith C sink since 1961 has been due to cropland expansion and increase in the cereal yield per unit area as a result of fertilizer application and irrigation.

Although the global cropland area is difficult to increase significantly in the near future, the findings of the study suggest that the present global cropland phytolith carbon sink could be further enhanced through the optimization of cropping system and fertilization (Table 4).

Cropping system optimization measures include enhancement of cereal area percentage in croplands and enhancement of multi-cropping index (Table 4). For example, Parr and Sullivan [20] and Li et al. [19] revealed that the enhancement of rice and wheat area percentage in croplands might significantly increase the total phytolith C sink in croplands because of their higher phytolith contents than other crops with low costs. Enhancement of multi-

Table 3. Contribution of the phytolith C sink to the global cropland C balance for 1961–2100.

	Phytolith C	Soil C			Phytolith C contribution (%)
Period	1961–2100	1961–2015[a]	2016–2100[b]	1961–2100	1961–2100
Sink rate (Pg CO_2 yr^{-1})	0.03 ± 0.01	-2.93	2.02	0.08	40 ± 18
Total sink (Pg CO_2)	4.2 ± 1.9	-161.2	171.7	10.6	40 ± 18

Sinks are positive values and sources are negative values.
[a] the average soil C sink rate data of 1961–2015 are after Ruddiman [37].
[b] the average soil C sink rate data of 2016–2100 are after Lal [6,14] assuming judicious land use and recommended management practices (RMPs) are applied worldwide during 2016–2100.

Table 4. Potential measures to enhance global cropland phytolith carbon sink.

Types	Measures	Mechanisms	Comments
Optimization of cropping system	Enhancement of cereal percentage in croplands	Enhancing crop output and phytolith content	High efficiency in all croplands with low costs
	Enhancement of multi- cropping index	Enhancing crop output	High efficiency in all croplands with low costs
Fertilization	Silicon fertilizer application	Enhancing crop phytolith content	High efficiency in cereal croplands and sugarcane with high costs
	Rock powder amendment	Enhancing crop phytolith content	High efficiency in cereal croplands and sugarcane with low costs
	Organic mulching	Enhancing crop output and phytolith content	High efficiency in cereal croplands and sugarcane with low costs
	Traditional fertilization	Enhancing crop output	High efficiency with high costs

cropping index may significantly increase the total cropland phytolith C sink by enhancing crop output with low costs.

Fertilization measures include silicon fertilizer application, rock powder amendment, organic mulching, and traditional fertilization (Table 4). Silicon fertilizer application, rock powder amendment and organic mulching will increase soil bioavailable silicon input, plant silicon uptake and phytolith content for cereals and sugarcane [5,19]. Traditional fertilization (N, P, K fertilizer application) may also increase total phytolith C sink in croplands by enhancing crop output.

Although the potential measures proposed for promoting cropland phytolith C sink based on the study are meritable, more data is required. The exact efficiency and costs of the proposed measures need further assessment before practical measures may be implemented to sequester globally-significant amounts of atmospheric CO_2.

Conclusions

Relative to the liable biomass C sink, the phytolith C sink in croplands is certain and stable, and can be sustained for several hundreds or thousands of years in most regions of the world. The phytolith sink of global croplands is a stable net sink of

26.35 ± 10.22 Tg CO_2 yr^{-1}, and may play a significant role in global cropland C balance for 1961–2100. The high phytolith sinks in Asia, North America and Europe can be attributed to the relatively high production of rice, maize, and wheat, respectively. The total phytolith C sink of global croplands has tripled since 1961 mainly due to fertilization, irrigation and cropland expansion. Taking an average phytolith C sink rate of 0.03 Pg CO_2 yr^{-1}, the total phytolith C sink of global croplands during 1961 and 2100 is 4.2 ± 1.9 Pg CO_2 yr^{-1}, $40 \pm 18\%$ of the total net soil C sink. Our data suggest that the cropland phytolith C sinks may be further enhanced by adopting cropland management practices such as optimization of cropping system and fertilization.

Acknowledgments

We thank Mr. David Cushley of ISE for language editing.

Author Contributions

Conceived and designed the experiments: ZLS JFP. Performed the experiments: ZLS FSG. Analyzed the data: ZLS. Contributed reagents/materials/analysis tools: ZLS. Wrote the paper: ZLS JFP FSG.

References

1. Cao M, Woodward FI (1998) Dynamic responses of terrestrial ecosystem carbon cycling to global climate change. Nature 393: 249–252.
2. Heimann M, Reichstein M (2008) Terrestrial ecosystem carbon dynamics and climate feedbacks. Nature 451: 289–292.
3. Piao S, Fang J, Ciais P, Peylin P, Huang Y, et al. (2009) The carbon balance of terrestrial ecosystems in China. Nature 458: 1008–1013.
4. Pan Y, Birdsey RA, Fang J, Houghton R, Kauppi PE, et al. (2011) A large and persistent carbon sink in the world's forests. Science 333: 988–993.
5. Song ZL, Wang HL, Strong PJ, Li ZM, Jiang PK (2012a) Plant impact on the coupled terrestrial biogeochemical cycles of silicon and carbon: Implications for biogeochemical carbon sequestration. Earth-Sci Rev 155, 319–331.
6. Lal R (2004a) Soil carbon sequestration impacts on global climate change and food security. Science 304: 1623–1627.
7. Houghton RA (1995) Changes in the storage of terrestrial carbon since 1850. In: Lal R, Kimble JM, Levine E, Stewart BA (Eds.), Soils and Global Change. CRC/Lewis, Boca Raton, FL, pp, 45–65.
8. Houghton RA (1999) The annual net flux of carbon to the atmosphere from changes in land use 1850 to 1990. Tellus 50B: 298–313.
9. Houghton RA, Hackler JL, Lawrence KT (1999) The U.S. carbon budget: contributions from land-use change. Science 285: 574–578.
10. Schimel DS (1995) Terrestrial ecosystems and the carbon cycle. Global Change Biol 1: 77–91.
11. Intergovernmental Panel on Climate Change (IPCC) (2001) Climate Change: The Scientific Basis. Cambridge Univ Press, Cambridge, UK.
12. Smith PC (2004) Carbon sequestration in croplands: the potential in Europe and the global context. Eur J Agron 20: 229–236.
13. Lal R (2001) Managing world soils for food security and environmental quality. Adv Agron 71: 155–192.
14. Lal R (2004b) Soil carbon sequestration to mitigate climate change. Geoderma 123: 1–22.
15. Six J, Ogle SM, Jay breidt F, Conant RT, Mosier AR, et al. (2004) The potential to mitigate global warming with no-tillage management is only realized when practised in the long term. Global Change Biol 10: 155–160.
16. Siever R, Scott RA (1963) Organic geochemistry of silica, in: Berger IA (Ed.), Organic Geochemistry. Pergammon Press, Elmsford, NY, pp, 579–595.
17. Piperno DR (1988) Phytolith Analysis: An Archaeological and Geological Perspective. Academic Press, London.
18. Parr JF, Sullivan LA (2005) Soil carbon sequestration in phytoliths. Soil Biol Biochem 37: 117–124.
19. Li ZM, Song ZL, Parr JF, Wang HL (2013) Occluded C in rice phytoliths: implications to biogeochemical carbon sequestration. Plant Soil. Published online first. DOI: 10.1007s11104-013-1661-9.
20. Parr JF, Sullivan LA (2011) Phytolith occluded carbon and silica variability in wheat cultivars. Plant Soil 342: 165–171.
21. Zuo X, Lü H (2011) Carbon sequestration within millet phytoliths from dry-farming of crops in China. Chinese Sci Bull 56: 3451–3456.
22. Parr JF, Sullivan LA, Quirk R (2009) Sugarcane phytoliths: Encapsulation and sequestration of a long-lived carbon fraction. Sugar Tech 11: 17–21.
23. Wilding LP (1967) Radiocarbon dating of biogenetic opal. Science 156: 66–67.
24. Mulholland SC, Prior C (1992) Processing of phytoliths for radiocarbon dating by AMS. Phytolitharien Newsletter 7: 7–9.

25. Meunier JD, Colin F, Alarcon C (1999) Biogenic silica storage in soils. Geology 27: 835–838.

26. Song ZL, Liu HY, Si Y, Yin Y (2012b) The production of phytoliths in the grasslands of China: implications to biogeochemical sequestration of atmospheric CO_2. Global Change Biol 18: 3647–3653.

27. Parr JF, Sullivan LA, Chen B, Ye G, Zheng W (2010) Carbon bio-sequestration within the phytoliths of economic bamboo species. Global Change Biol 16: 2661–2667.

28. Jansson C, Wullschleger SD, Kalluri UC, Tuskan GA (2010) Phytosequestration: Carbon biosequestration by plants and the prospects of genetic engineering. BioScience 60: 685–696.

29. Food and Agriculture Organization of the United Nations. FAO (2012): Statistics. Available: http://faostat.fao.org/site/567/default.aspx#ancor, from 1961 to 2011.

30. Hou X (1982) Vegetation Geography of China and Chemical Composition of its Dominant Plants, Science Press, Beijing, pp, 188–243.

31. Xu X, Yang L, Dong Y (1998) Rice field ecosystem in China. China Agriculture Press.

32. Zhou Q, Li Y, Lin J, Ye Y, Yang L (2005) Effects of irrigation on the vascular bundle structures and contents of silicon, magnesium and zinc in leaves of two sugarcane varieties. Sugar Crops of China (in Chinese with English abstract) 2: 1–4.

33. Ding TP, Tian SH, Sun L, Wu LH, Zhou JX et al. (2008) Silicon isotope fractionation between rice plants and nutrient solution and its significance to the study of the silicon cycle. Geochim Cosmochim Ac 72: 5600–5615.

34. Blecker SW, McCulley RL, Chadwick OA, Kelly EF (2006) Biologic cycling of silica across a grassland bioclimosequence. Global Biogeochem Cy 20: 1–11.

35. Huang Y, Zhang W, Sun WJ, Zheng XH (2007) Net primary production of Chinese croplands from 1950 to 1999. Ecol Appl 17: 692–701.

36. Zhu J, Li R, Yang X (2012) Spatial and temporal distribution of crop straw resources in 30 years in China. Journal of Northwest A & F University (in Chinese with English abstract) 40: 139–145.

37. Ruddiman WF (2003) The anthropogenic greenhouse era began thousands of years ago. Climate Change 61: 261–293.

38. Fang J, Chen A, Peng C, Zhao S, Ci L (2001) Changes in forest biomass carbon storage in China between 1949 and 1998. Science 292: 2320–2323.

21

Chinese Tallow Trees (*Triadica sebifera*) from the Invasive Range Outperform Those from the Native Range with an Active Soil Community or Phosphorus Fertilization

Ling Zhang[1,2], **Yaojun Zhang**[1], **Hong Wang**[1], **Jianwen Zou**[1]*, **Evan Siemann**[1,2]

1 College of Resources & Environmental Sciences, Nanjing Agricultural University, Nanjing, China, **2** Department of Ecology & Evolutionary Biology, Rice University, Houston, Texas, United States of America

Abstract

Two mechanisms that have been proposed to explain success of invasive plants are unusual biotic interactions, such as enemy release or enhanced mutualisms, and increased resource availability. However, while these mechanisms are usually considered separately, both may be involved in successful invasions. Biotic interactions may be positive or negative and may interact with nutritional resources in determining invasion success. In addition, the effects of different nutrients on invasions may vary. Finally, genetic variation in traits between populations located in introduced versus native ranges may be important for biotic interactions and/or resource use. Here, we investigated the roles of soil biota, resource availability, and plant genetic variation using seedlings of *Triadica sebifera* in an experiment in the native range (China). We manipulated nitrogen (control or 4 g/m^2), phosphorus (control or 0.5 g/m^2), soil biota (untreated or sterilized field soil), and plant origin (4 populations from the invasive range, 4 populations from the native range) in a full factorial experiment. Phosphorus addition increased root, stem, and leaf masses. Leaf mass and height growth depended on population origin and soil sterilization. Invasive populations had higher leaf mass and growth rates than native populations did in fresh soil but they had lower, comparable leaf mass and growth rates in sterilized soil. Invasive populations had higher growth rates with phosphorus addition but native ones did not. Soil sterilization decreased specific leaf area in both native and exotic populations. Negative effects of soil sterilization suggest that soil pathogens may not be as important as soil mutualists for *T. sebifera* performance. Moreover, interactive effects of sterilization and origin suggest that invasive *T. sebifera* may have evolved more beneficial relationships with the soil biota. Overall, seedlings from the invasive range outperformed those from the native range, however, an absence of soil biota or low phosphorus removed this advantage.

Editor: Harald Auge, Helmholtz Centre for Environmental Research - UFZ, Germany

Funding: This work was supported by the National Natural Science Foundation of China (NSFC-41225003), the PADA (the Priority Academic Program Development of Jiangsu Province), the Ministry of Education 111 project (B12009), and US-NSF (DEB 0820560). The funders had no role in study design, data collection and analysis, decision to publish, or preparation of the manuscript.

Competing Interests: The authors have declared that no competing interests exist.

* E-mail: jwzou21@njau.edu.cn

Introduction

Exotic plant invasions threaten ecosystem functions and stability [1–3]. Identifying the mechanisms underlying successful plant invasions will help guide effective invasive plant control and aid in ecosystem restoration. Two mechanisms that have been proposed to explain successful plant invasions are: 1) that exotic plants benefit from greater resource availability (the increased resource availability hypothesis or "IRAH"; [4,5]) and 2) exotic plants benefit from weak effects of natural enemies (the enemy release hypothesis or "ERH"; [6]) and/or strong effects of mutualists (the enhanced mutualists hypothesis or EMH, [7]).

The IRAH posits that the opportunities for invasions increase as resource availability increases in a community [4]. This increased resource availability does not necessarily reflect higher nutrient input because resource availability reflects the balance of resource supply and uptake by resident plants [5]. While most, but not all, exotic invaders may be better adapted to high nutrient conditions

than native species ("ruderals" [8]), pre-adaptation or post-introduction adaptation of exotic plants to high nutrient conditions may confer an advantage to exotic plants compared to less well-adapted native plants. For instance, invasive plants may be favored by increased soil resources (*e.g.* nitrogen [N], phosphorus [P]) that favor plants with low root to shoot ratios [9]. Similarly, plants with high N dependent maximal growth rates will be favored over those with high N use efficiencies when that N availability is high. Because plants with high N demand may not also have high P demand, for instance because of different symbiotic relationships (*e.g.* rhizobial or mycorrhizal) or allocation to high N (proteins) or P (nucleic acids) compounds, soil resources may vary in their impacts on invasions [2,10]. Moreover, nutrient assimilation by invasive plant species may vary due to positive and/or negative biotic interactions with more positive or less negative interactions facilitating nutrient uptake of the host plant.

The ERH posits that exotic plants benefit from introduction to a new range without specialist enemies in combination with not being preferred by generalist enemies [6]. Recent studies suggested that escape from soil pathogens may be at least as important as escape from aboveground specialist insect herbivores in their contribution to successful plant invasions [11–13]. Since soil communities include pathogens, parasites, and herbivores as well as beneficial groups (*e.g.* mycorrhizae, rhizobia) [14,15], the overall impact of soil biota on plant performance will reflect the net effect of both negative and positive interactions [16]. Strong negative impacts of soil microbial communities on invasive plants have mostly been observed in natural population of these plants growing in their native ranges [11,17] indicating that negative interactions are relatively stronger than beneficial ones [16]. This could reflect stronger negative effects or weaker positive effects on plant performance [7,18,19].

Differences in biotic or abiotic factors between the native and invasive ranges of plants can lead to genetic differences in morphological or physiological traits between populations in the native and introduced ranges [20–23]. One example of a shift in morphological traits is a lower root to shoot ratio [24,25]. In general, a lower root to shoot ratio provides an advantage in competition for aboveground resources and a disadvantage in competition for belowground resources [9]. In addition, escape from natural enemies, in particular specialists, in the invasive range may lead to a reallocation from defense to growth [26–28]. Moreover, more beneficial soil mutualisms in the invasive range [11] may lead to genetic differences in plant traits relevant to these interactions. However, resource requirements and biotic interactions are not independent [29,30]. In addition, shifts in traits of invasive plants may lead to altered soil microbial communities [25,31], which may in turn impact soil N and P use [32–35]. However, the dependence of invasive plant performance on genetic variation in plant traits, interactions with the soil biota, and availability of N and P is poorly understood.

Here, we examined effects of interactions between soil nutrients (N and P), soil microbial communities (active or sterilized), and population origin (native or invasive range) using Chinese tallow tree (*Triadica sebifera* (L.) Small, henceforth *T. sebifera*) as a model plant. *T. sebifera* is native to China and was first introduced into the USA in 1772 to Savannah, GA then subsequently to several sites along the Gulf Coast and is now invasive in grasslands, forests, and disturbed habitats throughout the southeastern USA, converting them to monospecific forests [36–39]. Previous studies have demonstrated that invasive *T. sebifera* had unusually positive interactions with the soil biota relative to native tree species in the

introduced range [40]. Conducting studies in the native range with populations from the native and introduced ranges provides additional insights into how genetic differences in *T. sebifera* populations may influence the net effects of the soil biota on *T. sebifera* performance. In an experiment conducted in the native range, we addressed the following questions: (1) Do *T. sebifera* seedlings perform better with N and/or P addition? (2) What are the net effects of the soil biota in the native range? (3) Do *T. sebifera* seedling responses to nutrient additions and soil biota manipulations differ between population origins?

Materials and Methods

Focal Species

T. sebifera is native to China where it has been cultivated for 14 centuries and is now an aggressive invader in the southeastern USA [36,41]. Studies demonstrated *T. sebifera* in the invasive range (invasive populations) are faster-growing relative to native conspecifics (native populations) or non-invasive co-occurring plant species [25,27,42]. Invasive *T. sebifera* rapidly accumulates soil pathogens in the invasive range relative to co-occurring native resident species which decreases the performance of *T. sebifera* seedlings under conspecifics [43,44]. However, *T. sebifera* has also been shown to be more mycorrhizal dependent in its invasive range compared with native trees [40,44]. In addition, *T. sebifera* seedlings from the invasive range have stronger responses to N addition than ones from the native range perhaps partly due to facilitation of N mineralization [34].

Seeds and Seedlings

In November 2009, we hand collected seeds of naturalized *T. sebifera* in China and the USA (Table 1). All seed collections were from public areas where no permission was required for collection. *T. sebifera* is not an endangered or protected species in either country. All seeds were collected from at least five haphazardly selected trees. Seeds used for planting were weighed by populations to evaluate the potential impacts of seed provisioning on seedling performance. Results of an ANOVA showed that seed masses of populations were independent of population origin ($F_{1,6} = 3.99$, $P = 0.09$). Also, seedling height ($F_{1,6} = 0.25$, $P = 0.64$) and number of leaves ($F_{1,6} = 2.59$, $P = 0.16$) at the time of transplanting were independent of population origin. Together these results suggest that there were no strong maternal effects due to differences in seed provisioning. In January, we treated seeds in a 10% bleach rinse and then soaked seeds in water with lab detergent to remove the waxy seed coat [43]. All seeds were then surface sterilized by 0.5% potassium permanganate and planted in 100 ml Conetainers (Stuewe & Sons, Corvallis, OR, USA) filled with sterilized field soil (see below). Seeds germinated in early April, 2010. After seedlings had secondary leaves, seedlings of similar heights were transplanted into pots (1.5 L). Pots received three soil treatments in a full-factorial design (N = 256, 2 population origins×4 populations×2 soil sterilization×2 N×2 P×4 replicates). To coincide with the growing season of *T. sebifera* in this area seeds were grown for 4 months in a non-heated greenhouse from June 2010 to November 2010 at Nanjing Agricultural University, Nanjing, China.

Soil Treatments

Soil was collected from the top 20 cm in a fallow agricultural field. *T. sebifera* trees were at least 200 m away from where soil was collected to reduce the potential buildup of specific soil organisms [16]. Soils characteristics were: carbon % = 2.32±0.11; nitrogen % = 0.22±0.007; C:N = 10.53±1.65 (means ±1 se). Previous

Table 1. Native (China) and invasive (USA) *T. sebifera* populations used in this experiment.

Source population	Latitude	Longitude
China		
Hefei, Anhui	31°38~39′N	117°50~51′E
Bengbu, Anhui	32°57~58′N	117°20~21′E
Nanjing, Jiangsu	32°02~03′N	118°50~51′E
Shanghai	31°31~32′N	121°52~53′E
USA		
Limehouse, SC	32°09~10′N	81°05~07′W
Hutchinson Island, GA	31°23~24′N	81°15~16′W
Houston, TX	29°41~42′N	95°25~26′W
Gainesville, FL	29°34~35′N	82°21~22′W

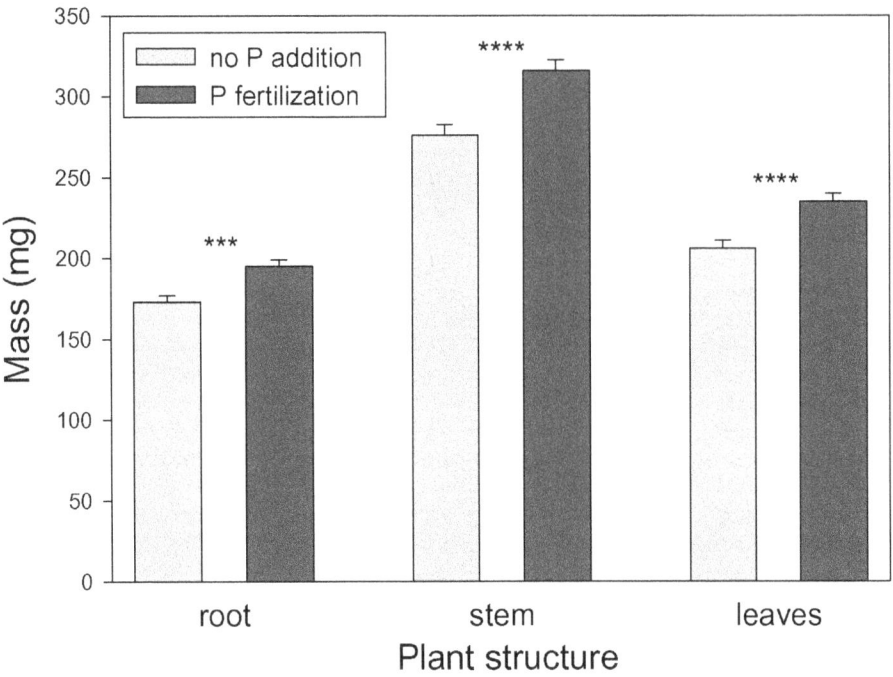

Figure 1. The dependence of root, stem and leaf masses of *T. sebifera* **seedlings on P addition.** Means+1 SE. ***: $P < 0.001$; ****: $P < 0.0001$.

studies focused on home- and away-soil effects indicate buildup of negative soil organisms in conspecific (home) soils in both the native and introduced ranges [43,44]. The soils used here are suited for making inferences about the effects of soil nutrients and the soil biota during the process of colonization in the native range and spread in the introduced range. Half of the soil was autoclaved

at $121°C$ for 40 minutes ("sterilized soil") and the other half was left untreated ("fresh soil").

Pots that were in the N fertilizer treatment received 4 g m^{-2} of N as KNO_3 (equivalent to 15.1 mg/L of soil). Plants in the control (no addition) N treatment received an equivalent volume of deionized water. Pots in the P fertilizer treatments received P at a

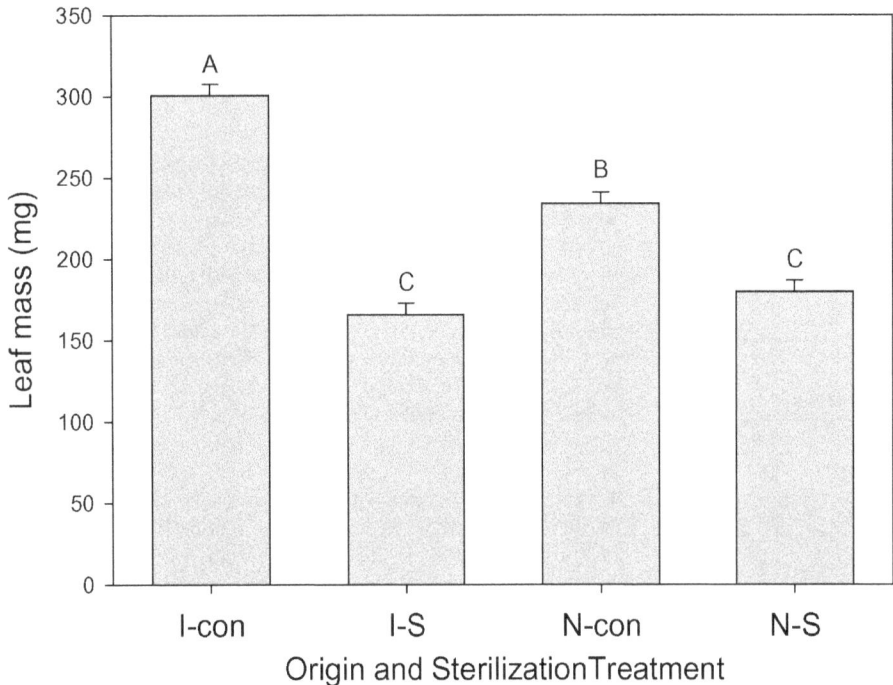

Figure 2. The dependence of leaf mass on population origin ("I" is invasive, "N" is native) and soil sterilization treatments ("con" is control, "S" is sterilization). Means+1 SE. Means with the same letter were not significantly different in post-hoc tests ($P < 0.05$).

Table 2. The dependence of root, stem, leaf biomass on origin, N addition, P addition and soil sterilization and their interactions in a MANOVA and follow-up ANOVAs.

Effect	MANOVA			Root			Stem			Leaf		
	DF	F	P	DF	F	P	DF	F	P	DF	F	P
Origin	3,4	2.52	0.1971	1,6	1.09	0.3366	1,6	0.12	0.7446	1,6	5.35	0.0600
N	3,189	1.01	0.3911	1,191	2.40	0.1227	1,191	0.25	0.6167	1,191	0.18	0.6723
P	**3,189**	**7.15**	**<0.0001**	**1,191**	**12.20**	**0.0006**	**1,191**	**16.79**	**<0.0001**	**1,191**	**16.21**	**<0.0001**
Sterilization	**3,189**	**83.84**	**<0.0001**	1,191	0.23	0.6339	1,191	0.02	0.8918	**1,191**	**171.29**	**<0.0001**
Origin×N	3,4	0.48	0.7147	1,6	0.44	0.5323	1,6	0.42	0.5402	1,6	0.01	0.9915
Origin×P	3,4	3.95	0.1087	1,6	1.46	0.2717	1,6	0.09	0.7773	1,6	0.99	0.3589
Origin×Sterilization	**3,4**	**7.16**	**0.0437**	1,6	0.69	0.4376	1,6	0.48	0.5142	**1,6**	**14.52**	**0.0089**
N×P	3,189	0.59	0.6228	1,191	0.06	0.8117	1,191	0.25	0.6205	1,191	0.53	0.4684
N×Sterilization	3,189	1.75	0.1574	1,191	0.07	0.7855	1,191	0.32	0.5716	1,191	2.26	0.1348
P×Sterilization	3,189	1.26	0.2910	1,191	2.30	0.1311	1,191	3.56	0.0609	1,191	0.57	0.4509
Origin×N×P	3,4	0.22	0.8789	1,6	0.43	0.5385	1,6	0.42	0.5420	1,6	0.80	0.4056
Origin×N×Sterilization	3,4	0.39	0.7676	1,6	1.34	0.2914	1,6	0.88	0.3850	1,6	0.05	0.8314
Origin×P×Sterilization	3,4	0.81	0.5514	1,6	0.15	0.7111	1,6	0.06	0.8211	1,6	1.36	0.2872
N×P×Sterilization	3,189	0.57	0.6342	1,191	0.64	0.4243	1,191	0.01	0.9176	1,191	0.59	0.4445
Origin×N×P×Sterilization	**3,4**	**8.45**	**0.0332**	1,6	0.43	0.5385	1,6	1.39	0.2834	1,6	3.85	0.0973

Significant results shown in bold.

rate of 0.5 g m^{-2} as KH_2PO_4 (equivalent to 1.9 mg/L of soil) and control (no addition) P pots received an equivalent volume of deionized water. Fertilizer additions were made one month after seedlings were transplanted.

Data Collection

We measured stem height of each seedling from ground surface to terminal bud at both the beginning and the end of the experiment. We thoroughly cleaned equipment between measurements. At the end of the experiment (4 months), seedlings were clipped at ground level (then separated into leaves and stems) and roots were gently washed from the soil. Total leaf area (cm^2) was obtained by scanning fresh leaves and analyzing them with SCNIMAGE (Scion Corporation; www.scioncorp.com). Seedling roots, stems, and leaves were then dried at 60°C to constant mass and weighed. We calculated height growth rates (HGR, mm cm^{-1} day^{-1}) as: HGR = ln (harvest stem height/initial stem height at transplanting)/days. Specific leaf area (SLA, leaf area per unit leaf dry mass, cm^2 g^{-1}) was calculated dividing leaf area by leaf biomass.

Statistical Analyses

We first conducted a MANOVA to examine the effects of seedling origin, N treatment, P treatment, soil treatment, and their interactions on *T. sebifera* root mass, stem mass, and leaf mass. We used variation among populations to test for differences between population origins (and corresponding interaction terms with population to examine interactive effects with origin). Because there were significant MANOVA results, we then conducted ANOVAs for each of the biomass variables. We also conducted ANOVAs to examine the dependence of height growth rate and specific leaf area on our treatments. We used partial difference adjusted means contrast tests to examine differences among treatment means for significant interactive effects. Data did not need to be transformed to meet the assumptions of ANOVA.

Differences at $\alpha = 0.05$ level are reported as significant. All statistical analyses were carried out in SAS (SAS Institute, Cary, NC, USA).

Results

Plant Biomass

In the MANOVA, P addition, sterilization, origin×sterilization, and origin×N×P×sterilization all had significant effects on root, stem and leaf biomass (Table 2). In follow-up ANOVAs, P addition significantly increased biomass of roots, stems and leaves (Table 2; Fig. 1). In addition, leaf biomass depended on sterilization and origin×sterilization with greater increases in leaf biomass in fresh soil compared to sterilized soil for seedlings from invasive populations versus native populations (Fig. 2).

Plant Growth Rate and Specific Leaf Area

Height growth rate depended on origin, P addition, soil sterilization, origin×P, and origin×sterilization (Table 3). Seedlings from invasive populations had significantly higher growth rates with P addition but ones from native populations did not (Fig. 3A).

In addition, the height growth increases in fresh soil compared to sterilized soil were significantly larger for invasive populations (Fig. 3B). Specific leaf area was significantly higher in fresh soil (Fig. 4A) and SLA also depended on N addition, P addition, and origin×N×P (Table 3). This interactive effect reflected significantly higher SLA for seedlings from invasive populations but significantly lower SLA for those from native populations when both N and P were added (Fig. 4B).

Discussion

Root, stem and leaf biomass of both origins were increased with P addition. In previous studies of plant invasions and soil P, most reported increased P availability in invaded areas [32,33,45–47]

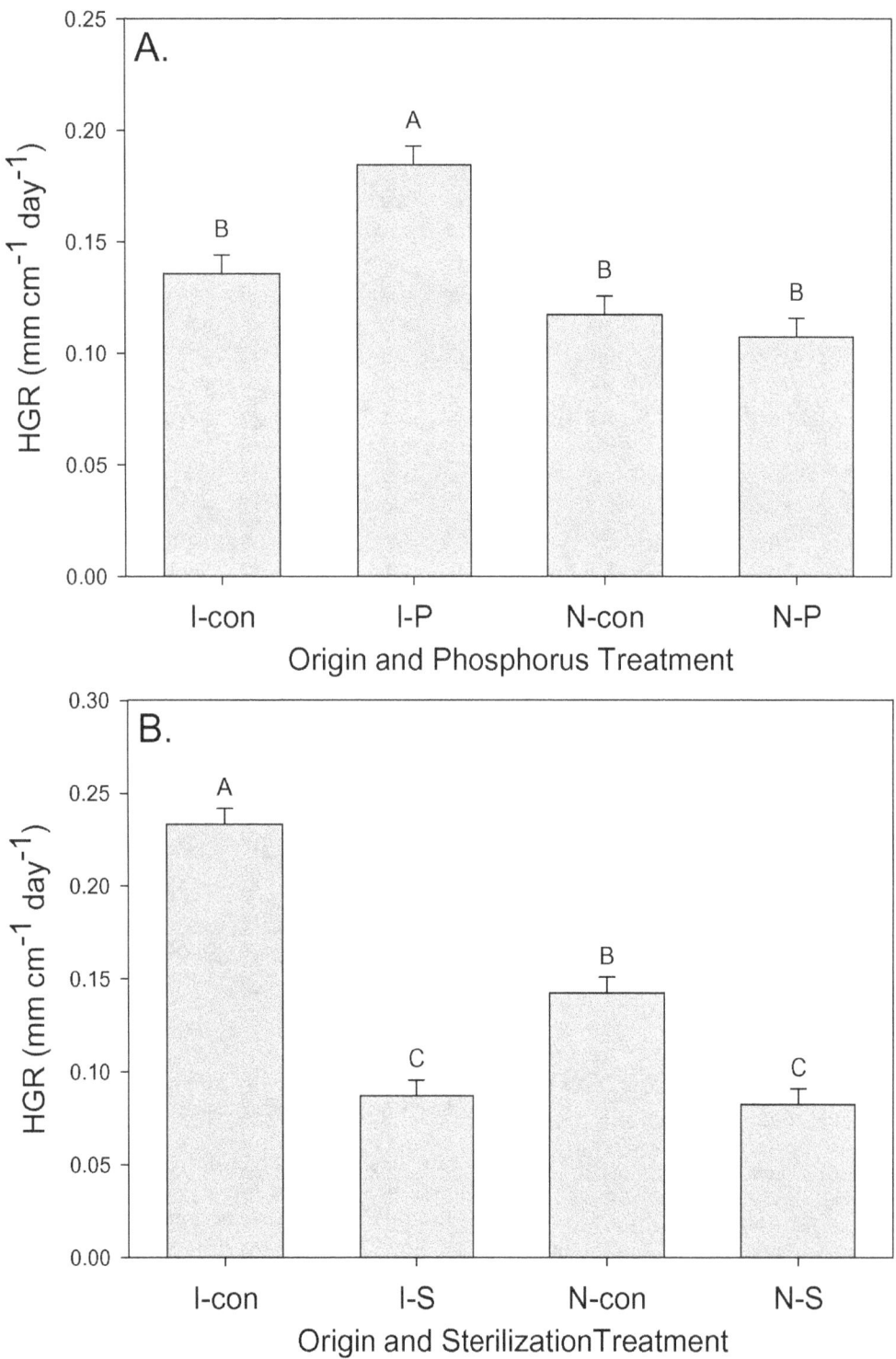

Figure 3. The dependence of height growth (HGR) on A) population origin ("I" is invasive, "N" is native) and P treatment ("con" is control, "P" is addition) and B) population origin and soil sterilization treatments ("con" is control, "S" is sterilization). Means+1 SE. Means with the same letter were not significantly different in post-hoc tests (P<0.05).

suggesting that invasive species may have evolved to mineralize soil P at a higher efficiency relative to native ones. Additional studies have demonstrated the importance of P availability for competitive ability and range expansion for invasive plant species [48,49]. Our results indicated that seedlings from both native and invasive origins were P limited since each responded positively to P addition, but had no response to N addition. However, N is another important soil nutrient that may limit plant growth and range expansion of *T. sebifera*. Zou *et al.* [34] found higher soil organic N mineralization in soils associated with *T. sebifera* of

Table 3. The dependence of height growth rate (HGR) and specific leaf area (SLA) on origin, N addition, P addition and soil sterilization and their interactions in ANOVAs.

Effect	DF	HGR		SLA	
		F	P	F	P
Origin	1,6	**33.78**	**0.0011**	3.70	0.1026
N	1,233	0.02	0.8999	**5.67**	**0.0181**
P	1,233	**5.94**	**0.0156**	4.65	0.0322
Sterilization	1,233	**146.88**	**<0.0001**	150.86	<0.0001
Origin×N	1,6	2.91	0.1388	0.52	0.4979
Origin×P	1,6	**12.36**	**0.0126**	0.74	0.4218
Origin×Sterilization	1,6	**27.72**	**0.0019**	2.44	0.1691
N×P	1,233	0.34	0.5618	0.02	0.8884
N×Sterilization	1,233	1.11	0.2940	1.05	0.3074
P×Sterilization	1,233	0.82	0.3664	1.55	0.2145
Origin×N×P	1,6	0.09	0.7728	**8.51**	**0.0267**
Origin×N×Sterilization	1,6	0.35	0.5749	1.01	0.3528
Origin×P×Sterilization	1,6	1.03	0.3494	0.02	0.8811
N×P×Sterilization	1,233	0.18	0.6687	3.03	0.0832
Origin×N×P×Sterilization	1,6	0.04	0.8530	0.01	0.9739

Significant results shown in bold.

invasive origin, which might lead to increased soil N availability. In addition, invasive *T. sebifera* plants have been shown to have a stronger positive response to inorganic N levels relative to those from native populations [34]. However, growth of *T. sebifera* seedlings from invasive populations invading coastal prairies in the introduced range responded significantly to N and K addition

alone but only responded positively to P addition when N was also added [48]. The strong positive response to P addition but not N addition we found here may reflect the extremely high levels of N deposition in the native range of *T. sebifera* [50].

The negative effects of soil sterilization on leaf biomass and height growth rate suggested *T. sebifera* seedlings had net positive interactions with the soil biota in the native range. Specific leaf area also decreased with soil sterilization (Table 3; Fig. 4A). Higher SLA is usually associated with lower leaf construction cost and higher N use efficiency in invasive plants [51–53]. One interaction that is important for P assimilation by plant species is arbuscular mycorrhizal fungi [54]. The higher arbuscular mycorrhizal colonization level observed for invasive *T. sebifera* relative to the native tree species in the introduced range is evidence that *T. sebifera* is arbuscular mycorrhizae dependent [40]. In our study, soil sterilization interacted with seedling origin to impact leaf biomass, with invasive origin seedlings more strongly inhibited by soil sterilization relative to ones of native origin. Thus, it appears that *T. sebifera* from both origins have overall positive interactions with the soil microbial communities but that those interactions are more beneficial for those of invasive origin relative to those of native origin. Although our populations spanned a broad geographical range and included descendants of both introductions, including a larger number of populations may have increased the number of population origin effects that were significant.

Assuming the negative effect of soil sterilization was simply the removal of mutualists important in P or N uptake [10,55], the negative effect of soil sterilization on growth might be weakened when N and/or P were added. However, there was not such a significant interactive effect on the mass of leaves, stems, or roots or on height growth rate. Perhaps uptake was so poor in sterilized soils that additional nutrients were not available to plants. The greater decline in leaf biomass for invasive origin plants relative those of native origin indicated a greater net beneficial interaction

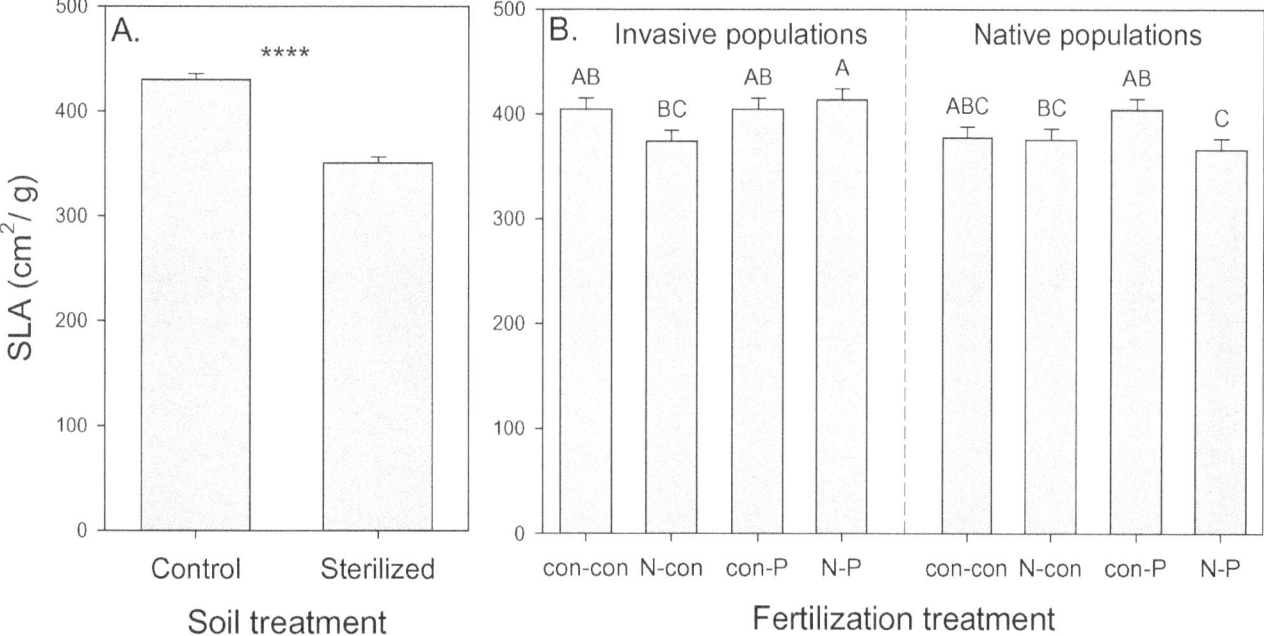

Figure 4. The dependence of specific leaf area (SLA) on A) soil treatment and B) population origin ("I" is invasive, "N" is native) and fertilization treatment ("con-con" is no fertilization, "N-con" is N addition, "con-P" is P addition, and "N-P" is N and P addition). Means+1 SE. Means with the same letter were not significantly different in post-hoc tests (P<0.05). ****: P<0.0001.

with the soil biota [40]. This could reflect greater positive interactions or weaker negative interactions but these possibilities cannot be evaluated in this study. If a similar pattern exists in the introduced range, it might be a mechanism contributing to its successful invasion.

Height growth rate of seedlings from the invasive range significantly increased with P addition but those from the native range did not respond to P addition (Table 3; Fig. 3). Generally, in a high resource, low stress environment, plants with a higher growth rate would be more successful when competing for light [56,57]. There was a significant interactive effect of origin, N addition, and P addition in which seedlings from invasive populations had especially high SLA and seedlings from native populations had especially low SLA (Fig. 4B). This is consistent with seedlings from invasive populations being more responsive to increased resources. Overall, the strong P response of seedlings from the invasive range together with comparable performance of seedlings without P addition suggests that seedlings from invasive populations may only have a competitive advantage in high P conditions [58].

It should be noted that this study focused on interactions with generalists in the native range since we collected soil more than 200 m away from any *T. sebifera* trees [59]. It is possible that we would have observed overall more negative effects of interactions with the soil biota had we used soil collected near conspecifics [60]. The interactions of *T. sebifera* seedlings of different origins might also differ if the soil community included more specialists [21]. If *T. sebifera* interacts with few specialists in the introduced range, the results of this study may help to understand the role of plant-soil interactions and soil resources in invasions. Research conducted on *Robinia pseudoacacia* by collecting soil from native, expanded (naturalized), and invasive ranges indicated that invasive plants are successful due to acquiring mutualisms and meanwhile, escaping from pathogens to gain a net positive effect of soil biota [11]. Further studies conducted in areas where *T. sebifera* is naturalized but not invasive [41,61] would increase our knowledge of the role soil communities play in range expansion of *T. sebifera*.

Acknowledgments

We thank: Chunxiao Wang, Hao Wang, Mengyue Wang and Rui Qu for their assistance in collecting and processing soil; Nannan Chen and Xiaomei Yang for their help in data collection; colleagues at Rice University, Wuhan Botanical Garden and Fudan University for assistance in collecting seeds. Further thanks should be addressed to the anonymous reviewers for their considerate comments that have greatly improved the earlier version of this manuscript.

Author Contributions

Conceived and designed the experiments: LZ JWZ. Performed the experiments: LZ YJZ HW. Analyzed the data: LZ ES. Wrote the paper: LZ JWZ ES.

References

1. Mack RN, Simberloff D, Lonsdale WM, Evans H, Clout M, et al. (2000) Biotic invasions: Causes, epidemiology, global consequences, and control. Ecol Appl 10: 689–710.
2. Ehrenfeld JG (2010) Ecosystem consequences of biological invasions. Annu Rev Ecol Evol Syst 41: 59–80.
3. Vitousek PM, DAntonio CM, Loope LL, Rejmanek M, Westbrooks R (1997) Introduced species: A significant component of human-caused global change. New Zeal J Ecol 21: 1–16.
4. Davis MA, Grime JP, Thompson K (2000) Fluctuating resources in plant communities: a general theory of invasibility. J Ecol 88: 528–534.
5. Shea K, Chesson P (2002) Community ecology theory as a framework for biological invasions. Trends Ecol Evol 17: 170–176.
6. Keane RM, Crawley MJ (2002) Exotic plant invasions and the enemy release hypothesis. Trends Ecol Evol 17: 164–170.
7. Reinhart KO, Callaway RM (2004) Soil biota facilitate exotic *Acer* invasions in Europe and North America. Ecol Appl 14: 1737–1745.
8. Grime JP (2001) Plant strategies, vegetation processes, and ecosystem properties. Chichester: Wiley. 417 p.
9. Reynolds HL, Pacala SW (1993) An analytical treatment of root-to-shoot ratio and plant competition for soil nutrient and light. Am Nat 141: 51–70.
10. Richardson DM, Allsopp N, D'Antonio CM, Milton SJ, Rejmanek M (2000) Plant invasions - the role of mutualisms. Biol Rev 75: 65–93.
11. Callaway RM, Bedmar EJ, Reinhart KO, Silvan CG, Klironomos J (2011) Effects of soil biota from different ranges on *Robinia* invasion: acquiring mutualists and escaping pathogens. Ecology 92: 1027–1035.
12. Wardle DA, Bardgett RD, Klironomos JN, Setala H, van der Putten WH, et al. (2004) Ecological linkages between aboveground and belowground biota. Science 304: 1629–1633.
13. van der Putten WH, Klironomos JN, Wardle DA (2007) Microbial ecology of biological invasions. ISME J 1: 28–37.
14. Kempel A, Schmidt AK, Brandl R, Schadler M (2010) Support from the underground: Induced plant resistance depends on arbuscular mycorrhizal fungi. Funct Ecol 24: 293–300.
15. Reinhart KO, Johnson D, Clay K (2012) Conspecific plant-soil feedbacks of temperate tree species in the southern Appalachians, USA. PloS ONE 7: e40680.
16. Reinhart KO, Callaway RM (2006) Soil biota and invasive plants. New Phytol 170: 445–457.
17. Inderjit, van der Putten WH (2010) Impacts of soil microbial communities on exotic plant invasions. Trends Ecol Evol 25: 512–519.
18. Parker MA, Wurtz AK, Paynter Q (2007) Nodule symbiosis of invasive *Mimosa pigra* in Australia and in ancestral habitats: a comparative analysis. Biol Invasions 9: 127–138.
19. Callaway RM, Newingham B, Zabinski CA, Mahall BE (2001) Compensatory growth and competitive ability of an invasive weed are enhanced by soil fungi and native neighbours. Ecol Lett 4: 429–433.
20. Bossdorf O, Auge H, Lafuma L, Rogers WE, Siemann E, et al. (2005) Phenotypic and genetic differentiation between native and introduced plant populations. Oecologia 144: 1–11.
21. Joshi J, Vrieling K (2005) The enemy release and EICA hypothesis revisited: incorporating the fundamental difference between specialist and generalist herbivores. Ecol Lett 8: 704–714.
22. Orians CM, Ward D (2010) Evolution of plant defenses in nonindigenous environments. Annu Rev Entomol 55: 439–459.
23. Fukano Y, Yahara T (2012) Changes in defense of an alien plant *Ambrosia artemisiifolia* before and after the invasion of a native specialist enemy *Ophraella communa*. PLoS ONE 7: e49114.
24. Allred BW, Fuhlendorf SD, Monaco TA, Will RE (2010) Morphological and physiological traits in the success of the invasive plant *Lespedeza cuneata*. Biol Invasions 12: 739–749.
25. Zou J, Rogers WE, Siemann E (2007) Differences in morphological and physiological traits between native and invasive populations of *Sapium sebiferum*. Funct Ecol 21: 721–730.
26. Huang W, Siemann E, Wheeler GS, Zou JW, Carrillo J, et al. (2010) Resource allocation to defence and growth are driven by different responses to generalist and specialist herbivory in an invasive plant. J Ecol 98: 1157–1167.
27. Zou JW, Siemann E, Rogers WE, DeWalt SJ (2008) Decreased resistance and increased tolerance to native herbivores of the invasive plant *Sapium sebiferum*. Ecography 31: 663–671.
28. Blossey B, Notzold R (1995) Evolution of increased competitive ability in invasive nonindigenous plants - a hypothesis. J Ecol 83: 887–889.
29. Blumenthal D, Mitchell CE, Pysek P, Jarosik V (2009) Synergy between pathogen release and resource availability in plant invasion. P Natl Acad Sci USA 106: 7899–7904.
30. Bozzolo F, Lipson D (2013) Differential responses of native and exotic coastal sage scrub plant species to N additions and the soil microbial community. Plant Soil doi:10.1007/s11104-013-1668-2.
31. Norton JB, Monaco TA, Norton U (2007) Mediterranean annual grasses in western North America: kids in a candy store. Plant Soil 298: 1–5.
32. Vanderhoeven S, Dassonville N, Chapuis-Lardy L, Hayez M, Meerts P (2006) Impact of the invasive alien plant *Solidago gigantea* on primary productivity, plant nutrient content and soil mineral nutrient concentrations. Plant Soil 286: 259–268.
33. Thorpe AS, Archer V, DeLuca TH (2006) The invasive forb, *Centaurea maculosa*, increases phosphorus availability in Montana grasslands. Appl Soil Ecol 32: 118–122.
34. Zou JW, Rogers WE, DeWalt SJ, Siemann E (2006) The effect of Chinese tallow tree (*Sapium sebiferum*) ecotype on soil-plant system carbon and nitrogen processes. Oecologia 150: 272–281.
35. Hawkes CV, Wren IF, Herman DJ, Firestone MK (2005) Plant invasion alters nitrogen cycling by modifying the soil nitrifying community. Ecol Lett 8: 976–985.

36. Bruce KA, Cameron GN, Harcombe PA, Jubinsky G (1997) Introduction, impact on native habitats, and management of a woody invader, the Chinese tallow tree, *Sapium sebiferum* (L) Roxb. Nat Area J 17: 255–260.

37. Dewalt SJ, Siemann E, Rogers WE (2006) Microsatellite markers for an invasive tetraploid tree, Chinese tallow (*Triadica sebifera*). Mol Ecol Notes 6: 505–507.

38. DeWalt SJ, Siemann E, Rogers WE (2011) Geographic distribution of genetic variation among native and introduced populations of Chinese tallow tree, *Triadica sebifera* (Euphorbiaceae). Ann Bot 98: 1128–1138.

39. Siemann E, Rogers WE (2006) Recruitment limitation, seedling performance and persistence of exotic tree monocultures. Biol Invasions 8: 979–991.

40. Nijjer S, Rogers WE, Lee CTA, Siemann E (2008) The effects of soil biota and fertilization on the success of *Sapium sebiferum*. Appl Soil Ecol 38: 1–11.

41. Wang HH, Grant WE, Gan JB, Rogers WE, Swannack TM, et al. (2012) Integrating spread dynamics and economics of timber production to manage Chinese tallow invasions in southern US forestlands. PloS ONE 7: e33877.

42. Siemann E, Rogers WE (2003) Herbivory, disease, recruitment limitation, and success of alien and native tree species. Ecology 84: 1489–1505.

43. Nijjer S, Rogers WE, Siemann E (2007) Negative plant-soil feedbacks may limit persistence of an invasive tree due to rapid accumulation of soil pathogens. Proc Biol Sci 274: 2621–2627.

44. Yang Q, Carrillo J, Jin HY, Shang L, Hovick SM, et al. (2013) Plant–soil biota interactions of an invasive species in its native and introduced ranges: Implications for invasion success. Soil Biol Biochem 65: 78–85.

45. Chapuis-Lardy L, Vanderhoeven S, Dassonville N, Koutika LS, Meerts P (2006) Effect of the exotic invasive plant *Solidago gigantea* on soil phosphorus status. Biol Fert Soils 42: 481–489.

46. Herr C, Chapuis-Lardy L, Dassonville N, Vanderhoeven S, Meerts P (2007) Seasonal effect of the exotic invasive plant *Solidago gigantea* on soil pH and P fractions. J Plant Nutr Soil Sc 170: 729–738.

47. Kueffer C, Klingler G, Zirfass K, Schumacher E, Edwards PJ, et al. (2008) Invasive trees show only weak potential to impact nutrient dynamics in phosphorus-poor tropical forests in the Seychelles. Funct Ecol 22: 359–366.

48. Siemann E, Rogers WE (2007) The role of soil resources in an exotic tree invasion in Texas coastal prairie. J Ecol 95: 689–697.

49. Suding KN, LeJeune KD, Seastedt TR (2004) Competitive impacts and responses of an invasive weed: dependencies on nitrogen and phosphorus availability. Oecologia 141: 526–535.

50. Liu XJ, Zhang Y, Han WX, Tang AH, Shen JL, et al. (2013) Enhanced nitrogen deposition over China. Nature 494: 459–462.

51. Baruch Z, Goldstein G (1999) Leaf construction cost, nutrient concentration, and net CO_2 assimilation of native and invasive species in Hawaii. Oecologia 121: 183–192.

52. Grotkopp E, Rejmanek M (2007) High seedling relative growth rate and specific leaf area are traits of invasive species: Phylogenetically independent contrasts of woody angiospernis. Ann Bot 94: 526–532.

53. Feng YL, Fu GL, Zheng YL (2008) Specific leaf area relates to the differences in leaf construction cost, photosynthesis, nitrogen allocation, and use efficiencies between invasive and noninvasive alien congeners. Planta 228: 383–390.

54. Schweiger PF, Thingstrup I, Jakobsen I (1999) Comparison of two test systems for measuring plant phosphorus uptake via arbuscular mycorrhizal fungi. Mycorrhiza 8: 207–213.

55. de Groot CC, Marcelis LFM, van den Boogaard R, Kaiser WM, Lambers H (2003) Interaction of nitrogen and phosphorus nutrition in determining growth. Plant Soil 248: 257–268.

56. Poorter H, Niklas KJ, Reich PB, Oleksyn J, Poot P, et al. (2012) Biomass allocation to leaves, stems and roots: meta-analyses of interspecific variation and environmental control. New Phytol 193: 30–50.

57. Valladares F, Wright SJ, Lasso E, Kitajima K, Pearcy RW (2000) Plastic phenotypic response to light of 16 congeneric shrubs from a Panamanian rainforest. Ecology 81: 1925–1936.

58. te Beest M, Stevens N, Olff H, van der Putten WH (2009) Plant-soil feedback induces shifts in biomass allocation in the invasive plant *Chromolaena odorata*. J Ecol 97: 1281–1290.

59. Johnson DJ, Beaulieu WT, Bever JD, Clay K (2012) Conspecific negative density dependence and forest diversity. Science 336: 904–907.

60. McCarthy-Neumann S, Kobe RK (2010) Conspecific plant-soil feedbacks reduce survivorship and growth of tropical tree seedlings. J Ecol 98: 396–407.

61. Bower MJ, Aslan CE, Rejmanek M (2009) Invasion potential of Chinese tallow tree (*Triadica sebifera*) in California's central valley. Invasive Plant Sci Manag 2: 386–395.

22

Nutrient Limitation of Native and Invasive N₂-Fixing Plants in Northwest Prairies

Andrea S. Thorpe[1]*¤, **Steven Perakis**[2], **Christina Catricala**[2], **Thomas N. Kaye**[1]

1 Institute for Applied Ecology, Corvallis, Oregon, United States of America, 2 United States Geological Survey, Forest and Rangeland Ecosystem Science Center, Corvallis, Oregon, United States of America

Abstract

Nutrient rich conditions often promote plant invasions, yet additions of non-nitrogen (N) nutrients may provide a novel approach for conserving native symbiotic N-fixing plants in otherwise N-limited ecosystems. *Lupinus oreganus* is a threatened N-fixing plant endemic to prairies in western Oregon and southwest Washington (USA). We tested the effect of non-N fertilizers on the growth, reproduction, tissue N content, and stable isotope $\delta^{15}N$ composition of *Lupinus* at three sites that differed in soil phosphorus (P) and N availability. We also examined changes in other Fabaceae (primarily *Vicia sativa* and *V. hirsuta*) and cover of all plant species. Variation in background soil P and N availability shaped patterns of nutrient limitation across sites. Where soil P and N were low, P additions increased *Lupinus* tissue N and altered foliar $\delta^{15}N$, suggesting P limitation of N fixation. Where soil P was low but N was high, P addition stimulated growth and reproduction in *Lupinus*. At a third site, with higher soil P, only micro- and macronutrient fertilization without N and P increased *Lupinus* growth and tissue N. *Lupinus* foliar $\delta^{15}N$ averaged $-0.010‰$ across all treatments and varied little with tissue N, suggesting consistent use of fixed N. In contrast, foliar $\delta^{15}N$ of *Vicia* spp. shifted towards $0‰$ as tissue N increased, suggesting that conditions fostering N fixation may benefit these exotic species. Fertilization increased cover, N fixation, and tissue N of non-target, exotic Fabaceae, but overall plant community structure shifted at only one site, and only after the dominant *Lupinus* was excluded from analyses. Our finding that non-N fertilization increased the performance of *Lupinus* with few community effects suggests a potential strategy to aid populations of threatened legume species. The increase in exotic Fabaceae species that occurred with fertilization further suggests that monitoring and adaptive management should accompany any large scale applications.

Editor: Daniel Ballhorn, Portland State University, United States of America

Funding: Funding for this project was provided by the US Geological Survey, Forest and Rangeland Ecosystem Science Center. The funders had no role in study design, data collection and analysis, decision to publish, or preparation of the manuscript.

Competing Interests: The authors have declared that no competing interests exist.

* E-mail: athorpe@neoninc.org

¤ Current address: National Ecological Observatory Network, Inc., Boulder, Colorado, United States of America

Introduction

Conservation of endangered species often depends on active management of critical habitats and ecosystems [1]. However, management of remnant habitats to support native species is frequently complicated by the overlap in plant traits (for example, phenology, life history, or tolerance to fire or grazing) between native and exotic species [2], [3]. In the Pacific Northwest (USA), one of the primary targets for restoration is *Lupinus oreganus* A. Heller (Kincaid's lupine, Fabaceae; hereafter '*Lupinus*'), a federally threatened species. This species is endemic to upland prairies in western Oregon and southwest Washington. In addition to habitat loss, habitat degradation by invasive plants poses one of the greatest threats to this species [4]. Identifying effective management techniques for *Lupinus* is of particular concern as it is the primary larval host plant for the endangered butterfly, *Plebejus icarioides fenderi* Macy (Fender's blue butterfly, Lepidoptera: Lycaenidae).

Plant growth in most terrestrial ecosystems is nitrogen (N) limited [5], [6], but some plants, including those in the Fabaceae family, form symbioses with bacteria that permit fixation of atmospheric N₂ gas into plant-available forms. Nitrogen fixation,

in turn, can be limited by the high phosphorus (P) requirements of symbiotic N-fixing species, or by micronutrients such as molybdenum (Mo) that are essential to the nitrogenase enzyme that fixes atmospheric N₂ into ammonia [7–10]. In agricultural settings, P fertilization has been shown to increase the success of legumes [11–13], although there can be great variability in response to fertilization across species [14], [15] and genotypes [16]. While less is known from natural systems, legumes there have also generally been found to have a positive response to P fertilization [17–20]. Comparatively fewer studies report evidence of Mo limitation of legumes, though as with P, Mo limitation responses can be species- and site-specific [21]. Molybdenum deficiency can occur in agricultural soils of western Oregon, particularly on acidic hill soils surrounding the floor of the Willamette Valley [22].

Almost 10% of the invasive species in North America are members of the Fabaceae [23]. These species are typically associated with low-nutrient and/or disturbed ecosystems, and their invasion success is linked to their ability to alter nutrient cycling and ecosystem dynamics via N-fixation [20], [24–27]. The majority of research on invasive N-fixers has been conducted on woody perennial legumes. Less is known about the effects of herbaceous invasive legumes, though it is likely that their effects differ from those of woody species [28].

The objectives of our study were to determine if P or other non-N nutrients limit the growth of native *Lupinus*, whether these nutrients affect abundance of non-native N-fixers, and whether changes in N fixation can explain these responses. Non-native herbaceous legumes are also common in our study region, but tend to only dominate recently disturbed habitats (e.g. roadsides and recently cleared fields) and are not usually considered significant competitors of native *Lupinus* in remnant and restored prairies. We did not include N fertilization among our experimental treatments out of concern that non-N-fixing species would outcompete these remnant federally threatened populations of *Lupinus*. Previous N fertilization studies have observed losses of diversity and reduction in legume abundance in mixed communities containing symbiotic N fixers [19], [29]. We hypothesized that non-N fertilizers would increase the growth of *Lupinus* and other N-fixing species, but have no effect on non-N-fixing species in these communities. Although sites were initially selected with the intent of being replicates of each other, we hypothesized that there would be differences in soil nutrient status and species responses between the sites due to differences in historic and current management.

Materials and Methods

Study sites and treatments

We established three study sites in remnant patches of upland prairie habitat in the Willamette Valley in western Oregon, USA. This region has a seasonal quasi-Mediterranean climate with cool wet winters and warm dry summers. Experimental plots were established in *Lupinus* populations at Baskett Butte (near Dallas; research permission from US Fish and Wildlife Service, Willamette Valley Refuges), Lupine Meadows (near Philomath; research permission from Greenbelt Land Trust), and Wren (research permission from a private landowner) in summer 2006 (See Figure S1 for a map of site locations). Baskett Butte is located on the Baskett Slough National Wildlife Refuge. Although it is likely that this area was at one time grazed by sheep and/or cattle, it has been more than 45 years since this activity occurred. Current management consists of occasional fall mowing. Since 2004, Lupine Meadows (owned and managed by the Greenbelt Land Trust) has been managed primarily for conservation purposes. Prior to entering into a conservation easement, the property was managed as a low-utilization horse pasture and private family recreation area for about 20 years. The Wren population is located on private property and has a history of light grazing and occasional mowing; however, these activities have not occurred for greater than 10 years. All sites are located on Mollisols and differ in surface (0–10 cm) mineral soil carbon (C), nitrogen (N), C:N, and available P (phosphorus) (Table 1).

At each site, large areas containing *Lupinus* were mapped onto a grid of 1 m × 1 m cells. Cells with either no *Lupinus* or greater than 80% cover of *Lupinus* were rejected as potential treatment plots. From the remaining cells, we selected 20 as treatment plots, with a 1 m buffer surrounding each plot, and staggered the plots to further minimize nutrient movement among them. We used a fully factorial design of two levels each (present/not present) of two fertilizer amendments consisting of phosphorus-only (P) and other non-nitrogen and non-phosphorus macronutrients and micronutrients (M), for a total of four treatments (control, P, M, M+P). Each treatment was replicated five times at each of the three sites. The P treatment was added as super triple phosphate (10 g P m^{-2} yr^{-1}). The M treatment was added as potassium chloride (10 g K

m^{-2} yr^{-1}), calcium chloride (10 g Ca m^{-2} yr^{-1}), magnesium sulfate (5 g Mg m^{-2} yr^{-1}, 6.75 g S m^{-2} yr^{-1}), sodium molybdate (0.02 g Mo m^{-2} yr^{-1}), and the commercial nutrient blend Granusol (0.168 g Mn m^{-2} yr^{-1}, 0.168 g Zn m^{-2} yr^{-1}, 0.168 g Fe m^{-2} yr^{-1}, 0.05 g Cu m^{-2} yr^{-1}, and 0.017 g B m^{-2} yr^{-1}); Granusol was crushed into a powder and all M fertilizers were mixed prior to application Fertilization levels were chosen to largely follow [30]. Each annual fertilization was split evenly into early winter (November/December) and spring (March) applications for 4 years, from November 2006 to March 2009.

Plant community responses

Measurements of plant cover and composition occurred in May or June of each year, during peak flowering of *Lupinus*, and before the onset of summer drought. In early June 2006, prior to initial fertilization, we measured *Lupinus* cover and counted the number of leaves, and mature and aborted inflorescences. In 2009, we repeated these measurements and also estimated the areal cover of all vascular plant species, as well as mosses/lichens, bare ground, and rock, in all plots. We were not able to sample biomass due to the protected status of *Lupinus* and presence of the endangered Fender's blue butterfly in our plots. For analyses of *Lupinus* responses to treatments, cover was determined by measuring the approximate rectangular area (length and width to the nearest cm) occupied by a clump. For analysis of community responses to treatments, cover of all other species was visually estimated in June 2009 to the nearest 1%; species with less than 1% cover were given a value of 0.1%. Total cover for a plot was at least 100%, and often exceeded that if many layers of vegetation were present. Species nomenclatures, provenance, and duration followed the USDA Plants database [31].

Lupinus analyses

Due to the differences in past and current management at our sites, we tested for fertilizer treatment effects at each site separately. Phosphorus and M were used as fixed factors in all analyses; with two levels of each factor (present/not present). Cover of *Lupinus* was arcsin transformed and analyzed using a General Linear Model (SPSS 17.0, 2008) with cover in 2006 as a covariate. We also used a General Linear Model (SPSS 17.0, 2008) to analyze treatment effects on the number of inflorescences m^{-2} of *Lupinus*, with inflorescences m^{-2} in 2006 as a covariate (except at Baskett Butte, where no inflorescences were produced in 2006). Due to the non-normality of the data from Baskett Butte, at this site, we tested for differences in the rank abundances of the number of inflorescences m^{-2} using a General Linear Model (SPSS 17.0, 2008). We used a Generalized Linear Model (SPSS 17.0, 2008) with a negative binomial distribution to analyze treatment effects on the count of leaves, using the number of leaves in 2006 as a covariate; the response of leaves generally followed that of cover (see Tables S1, S2) and thus is not reported here.

Community analyses

Although relatively minor components of the community initially, over time we observed increases in several non-native Fabaceae species, including *Vicia sativa* L., *V. hirsuta* (L.) Gray and *Trifolium dubium* Sibth. In our analyses, we considered all non-*Lupinus* legumes together, hereafter termed 'other Fabaceae'. No other native Fabaceae species were present in the plots. We used a General Linear Model (SPSS 17.0, 2008) to test for a fertilizer effect on arcsin transformed cover of other Fabaceae.

We used multi-response permutation procedure (MRPP) to test for community responses to fertilizer treatments using PC-ORD 6 [32]. MRPP is a nonparametric permutation test that compares

Table 1. Soil characteristics in control plots at the study sites.

Site	Soil Type (USDA)	Soil C (%)	Soil N (%)	Soil C:N	NH$_4^+$ (mg/kg)	NO$_3^-$ (mg/kg)	Net N min (mg/ kg−28d)	Bray-P (mg/kg)	pH (H$_2$O)
Lupine Meadows	Ultic Argixerolls	3.43^a (0.05)	0.23^a (0.002)	15.2^a (0.78)	1.03 (0.15)	0.00 (0.00)	−1.4^c (0.20)	2.86^a (0.18)	6.4 (0.1)
Baskett Butte	Ultic Haploxeroll	3.41^a (0.18)	0.29^b (0.010)	11.6^b (0.29)	1.62 (0.62)	0.03 (0.03)	4.3^a (2.08)	1.97^b (0.14)	6.6 (0.3)
Wren	Ultic Argixerolls	4.23^b (0.11)	0.32^b (0.007)	13.3^c (0.57)	1.73 (0.31)	0.01 (0.01)	−0.1^b (0.98)	1.84 ^{b} (0.15)	6.5 (0.1)

Notes: Soil type data from NRCS Soil Surveys for Polk and Benton Counties. Soil chemical variables reflect means of five control plots sampled to 10 cm depth, with standard errors in parentheses. Site differences in chemical variables were analyzed using one-way ANOVA. Values in columns followed by different letters are significantly different (p<0.05) by Tukey post-hoc comparison.

groups based on plot dissimilarity. The effect size, or A-statistic, reflects the degree to which groups differ and is independent of sample size [33]. We used Sørensen distances to calculate community dissimilarity. Analyses utilized 1000 Monte Carlo simulations in PC-ORD 6 [32]. Due to the potentially strong effect of the cover of *Lupinus* on these analyses (plots were selected to have relatively high cover of *Lupinus*), we conducted two analyses, first using the absolute cover of all species within the plots, and second using the relative cover of all species except *Lupinus* (thus removing the potential effect of *Lupinus*). When a significant treatment effect was found, we used an uncorrected multiple comparison test for differences between groups and indicator species analysis (utilizing 5,000 Monte Carlo simulations) to test association of species within treatment groups.

Plant and soil nutrients

We determined plant tissue N concentrations and natural abundance $\delta^{15}N$ stable isotope composition on four species common to all sites at the end of the 2008 growing season. In each plot, we collected entire aboveground tissues of 3–4 individuals each of *Fragaria virginiana*, *Plantago lanceolata* and *Vicia* spp, (*V. sativa* and *V. hirsuta* were sampled together as at the time of sampling, the species were indistinguishable) and leaves from 3–4 *Lupinus*, and composited samples by individual species. Samples were dried at 65°C for 48 hrs, ground to fine powder, and analyzed for %N and $\delta^{15}N$ on a Thermo Electron Delta plus Advantage mass spectrometer with a Costech ECS-4010 elemental analyzer combustion at the Colorado Plateau Stable Isotope Laboratory. We used a fully-factorial ANOVA (SPSS 17.0; 2008) to evaluate effects of fertilization on plant tissue N and $\delta^{15}N$ separately for each species. We also used least-squares linear regression (SPSS 17.0; 2008) of plot-level data to assess species-specific relationships between tissue N and $\delta^{15}N$.

We hoped to use $\delta^{15}N$ data of putative N-fixers (*Lupinus* and *Vicia* spp) and the average $\delta^{15}N$ of non-fixers (*Fragaria virginiana*, *Plantago lanceolata*) to calculate the percent of N derived from fixation (NDFA) by assuming that N-fixation occurred at the $\delta^{15}N = 0‰$ value of atmospheric N$_2$ gas [34]. However, across all treatment and site combinations, our calculated NDFA for *Lupinus* violated method assumptions by significantly exceeding 100% in 1 of 12 possible cases, and the NDFA for *Vicia* spp. was significantly less than 0% in 1 out of 10 possible cases for which this species was present (one-sample t-tests). *Lupinus* spp. grown on N-free media in the greenhouse and forced to obtain N solely from fixation can display leaf and shoot $\delta^{15}N$ ranging from −1.6‰ to + 4.2‰ depending on growth conditions, growth stage, degree of nodulation, and specific strain of rhizobial inoculum [35–38], and this range suggests that the assumption of fixed N = 0‰ could be violated. Likewise, assuming that non-fixing reference species

adequately sample soil N available for putative N-fixers is always contentious, and our finding that *Vicia* spp. in one control plot were significantly more depleted than reference species suggestions this assumption may have been violated. We therefore rely on $\delta^{15}N$ values alone to examine potential differences in N fixation among species.

We collected three mineral soil samples in each plot by coring (2 cm diameter, 0–10 cm depth) and created one composite soil sample per plot at the end of the 2008 growing season. Soils were sieved to 2 mm in the laboratory, and extracted within 24 hours of collection for inorganic N and P and within 48 hours for pH. Inorganic N was extracted by adding 35 mL of 2M KCl to 7 g field-moist soil, shaking for 1 h, then filtering by pouring through pre-rinsed Whatman 42 filter paper, followed by analysis for ammonium (NH$_4^+$) and nitrate (NO$_3^-$) on a Lachat QuikChem 8000 flow-injection autoanalyzer using the salicylate and cadmium reduction methods, respectively (QuikChem Methods 12-107-06-2-A and 12-107-04-1-F, Lachat Instruments, Milwaukee, WI, USA). Soil inorganic P was extracted by adding 25 mL of Bray-1 extracting solution (0.03 N NH$_4$F-0.025 N HCl) to 5 g field-moist soil in 50 mL centrifuge tubes, shaking vigorously for 1 min, centrifuging at 3400 rpm for 5 min, filtering through Whatman 42 filter paper, followed by analysis using the molybdenum blue method (QuikChem Method 12-115-01-1-A, Lachat Instruments, Milwaukee, WI, USA). Soil solution pH was determined by mixing 20 mL deionized water with 10 g field-moist soil (2 water: 1 soil), allowing the mixture to equilibrate for 30 min, then measuring pH of the supernatant with an Accumet pH meter (Fisher Scientific, Hampton, NH, USA). We assayed for potential N mineralization in mineral soil by 28 day laboratory incubation of 10 g soil at 25 °C and 60% water holding capacity, followed by extraction and analysis for ammonium and nitrate as above, corrected for initial inorganic N. Gravimetric soil moisture content of these samples was determined by comparing the weight of a 10 g subsample before and after drying at 105°C for 48 hr. We also determined total C and N concentrations in control plots. 20 g of sieved soil were dried at 65°C for 48 hr, ground to a fine powder, then analyzed on a Costech ECS-4010 elemental combustion analyzer (Costech Analytical, Valencia, CA, USA). We used factorial ANOVA (SPSS 17.0; 2008) to evaluate effects of fertilization treatments on soil chemical variables.

Results

Response of *Lupinus*

At Baskett Butte, P addition significantly increased *Lupinus* inflorescences (Figure 1; $F_P = 3.846$, df = 1,16, $P_P = 0.068$; subscripts denote factors/levels); no inflorescences were produced in plots that received neither M or P. There was also an increase in

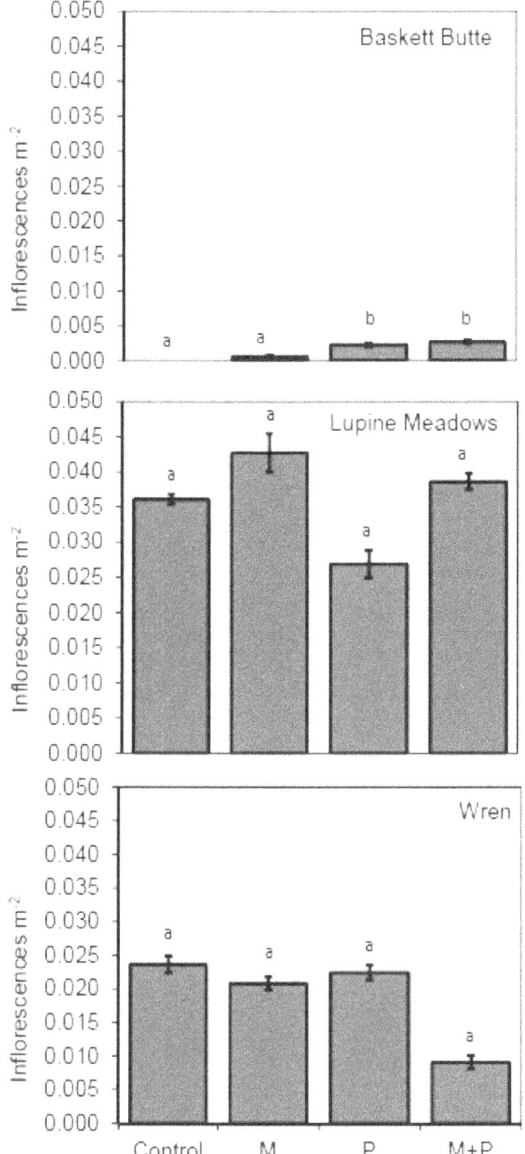

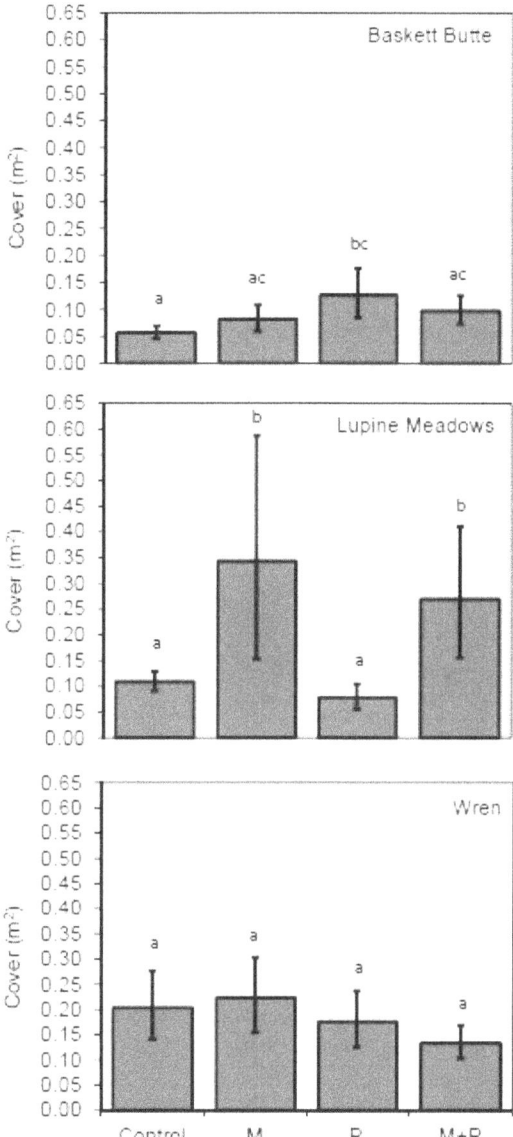

Figure 1. Mean number of *Lupinus* inflorescences m⁻² lupine cover in plots treated with micronutrients and/or phosphorus. Treatments were applied at 3 sites in western Oregon. Bars are means±1 S.E.; n = 5 per treatment. There were effects of P at Baskett Butte ($P = 0.063$). Letters denote differences between treatments within species.

Figure 2. Cover of *Lupinus* in plots treated with micronutrients and/or phosphorus. Treatments were applied at 3 sites in western Oregon. Bars are means±1 S.E.; n = 5 per treatment. There were effects of P at Baskett Butte ($P = 0.064$) and M at Lupine Meadows ($P = 0.057$). Letters denote differences between treatments within species.

cover (Figure 2; $F_P = 3.989$, df = 1,15, $P_P = 0.063$) of *Lupinus* in plots treated with P. There were no effects of the M treatments, alone or in combination with P, on any of the *Lupinus* variables at this site (Figure 1, 2; see Tables S2, S3).

Treatment with M increased *Lupinus* cover (Figure 2; $F_M = 0.318$, df = 1,15, $P_M = 0.057$) at Lupine Meadows. While there was a trend for increased inflorescence production in plots treated with M, this effect was not significant (Figure 1; $F_M = 0.476$, df 1,15, $P_M = 0.501$). There were no effects on *Lupinus* of P addition and no significant interactions between P and M at this site (Figures 1, 2; see Tables S2, S3).

There were no treatment effects on cover or inflorescence production at Wren (Figures 1, 2, see Tables S2, S3).

Community Response

The response of other Fabaceae to fertilizer treatments differed by site (Figure 3). Phosphorus alone increased the cover of other Fabaceae ($F_P = 42.76$, df = 1,16, $P_P < 0.0005$) at Baskett Butte. At Lupine Meadows, both M and P fertilizer treatments increased cover of other Fabaceae ($F_M = 5.222$, df = 1,16, $P_M = 0.036$; $F_P = 9.90$, df = 1,16, $P_P = 0.006$). Finally, there was no effect of fertilizer treatments on other Fabaceae cover at Wren, and there were no treatment interactions on other Fabaceae at any of the sites (Table S2).

Using MRPP, we found no fertilization effects on community composition at our sites when all species were included in the analyses (Baskett Butte, A = −0.0555, $P = 0.786$; Lupine Meadows, A = −0.0064, $P = 0.901$; Wren, A = 0.0262, $P = 0.139$). However, when we tested for treatment effects on the relative

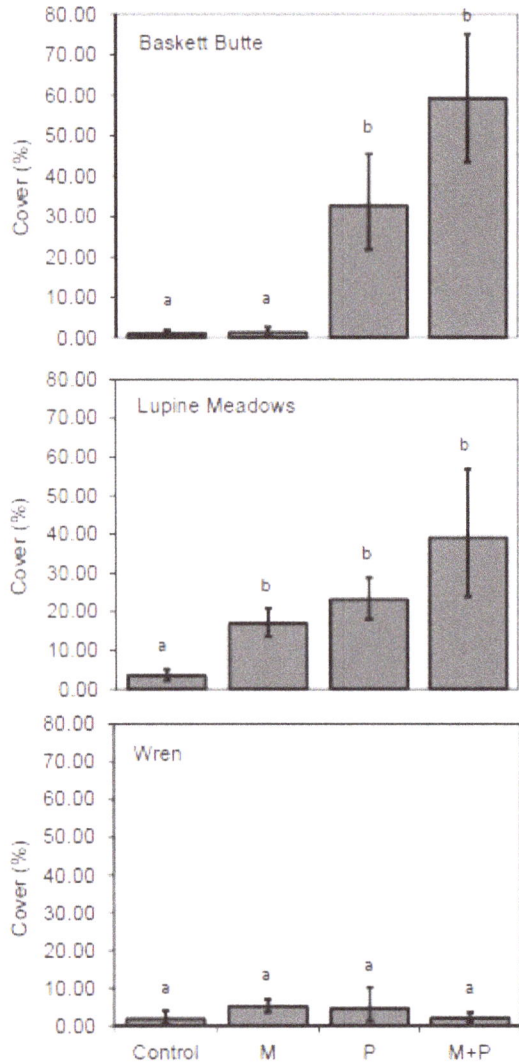

Figure 3. Cover of other Fabaceae in plots treated with micronutrients and/or phosphorus. Bars are means±1 S.E.; n=5 per treatment. Treatment effects were significant for P at Baskett Butte (P <0.0005) and both M (P=0.036) and P (P=0.006) at Lupine Meadows. Letters denote differences between treatments within species.

cover of all species except *Lupinus*, we did not find a treatment effect on community structure at Baskett Butte (A = 0.0382, P = 0.115) or Wren (A = −0.0020, P = 0.498), but there was a significant effect at Lupine Meadows (A = 0.0847, P = 0.004). Control plots differed from plots treated with M (A = 0.0983, P = 0.008), P (A = 0.1147, P = 0.007), and M+P (A = 0.1498, P = 0.004). *Leucanthemum vulgare* was associated with plots treated with P (IV = 36.2, P = 0.0116) and *Vicia hirsuta* was associated with plots treated with M+P (IV = 49.3, P = 0.029). There were no other significant indicator species.

Plant and soil nutrients

In control plots, we found consistent differences among N-fixing and non-N-fixing species in tissue %N and $\delta^{15}N$ at all sites (see Table S4). *Lupinus* had significantly higher tissue %N and more enriched $\delta^{15}N$ than other species. *Vicia* spp. $\delta^{15}N$ differed, but tissue %N did not, from non-N-fixing species at Wren and Lupine Meadows (*Vicia* spp. were not present in control plots at Baskett Butte thus precluding comparisons). Non-N-fixing *Plantago* and *Fragaria* did not differ significantly from one another in either tissue %N or $\delta^{15}N$ at any site.

Across all possible sites and treatments (n = 12), *Lupinus* $\delta^{15}N$ did not differ significantly from 0‰ in 11 out of 12 cases, but control plots at Wren (0.3693‰) were significantly higher than 0‰ (p = 0.02, one-sample t-test). *Vicia* spp. were present in 10 of 12 possible site and treatment combinations, and displayed $\delta^{15}N$ significantly less than 0‰ in 9 of 10 possible cases (except P treatment at Baskett Butte, P = 0.09). At Baskett Butte, we did not detect a significant effect of fertilization on $\delta^{15}N$ or tissue %N in *Lupinus* (Table 2; Figures 4, 5). As *Vicia* spp. were present only in plots treated with P alone and in combination with M at this site, this precluded an analysis of fertilization effects on tissue %N and $\delta^{15}N$.

At Lupine Meadows, we detected a significant effect of M fertilization on *Lupinus* tissue %N (F_M = 16.54, df = 1,16, P_M<0.001; Table 2), which increased from 2.18% N in control plots to 2.68% N in treated plots (Figure 5). We detected significant increases in *Vicia* spp. tissue %N in M and P fertilized plots (F_M = 16.16, df = 1,16, P_M<0.001; F_P = 5.39, df = 1,16, P_P = 0.034) that corresponded to shifts in *Vicia* spp. $\delta^{15}N$ towards more atmospheric values (F_M = 4.27, df = 1,16, P_M = 0.055; F_P = 4.67, df = 1,16, P_P = 0.046). *Fragaria* tissue %N increased significantly in P and M treatments (F_M = 5.90, df = 1,16, P_M = 0.027; F_P = 9.65 df = 1,16, P_P = 0.007). *Plantago* responded slightly to M fertilization in both tissue %N (F_M = 3.75, df = 1,16, P_M = 0.073) and $\delta^{15}N$ (F_M = 3.95, df = 1,16, P_M = 0.067), with values increasing in %N towards more atmospheric $\delta^{15}N$.

Table 2. Tissue %N and $\delta^{15}N$ of N$_2$-fixing (*Lupinus* and *Vicia*) and non-N$_2$-fixing (*Fragaria* and *Plantago*) species in control plots.

Site	Tissue %N				Tissue $\delta^{15}N$			
	Lupinus	*Vicia*	*Fragaria*	*Plantago*	*Lupinus*	*Vicia*	*Fragaria*	*Plantago*
Lupine Meadows	2.18ᵃ (0.11)	1.29ᵇ (0.14)	1.19ᵇ (0.06)	0.99ᵃ (0.07)	−0.08ᵃ (0.098)	−1.50ᵇ (0.287)	−2.76ᶜ (0.24)	−3.25ᶜ (0.30)
Baskett Butte	1.82ᵃ (0.12)	–	1.49ᵇ (0.08)	1.23ᵇ (0.09)	−0.13ᵃ (0.09)	–	−3.85ᵇ (0.28)	−3.58ᵇ (0.33)
Wren	2.75ᵃ (0.12)	1.60ᵇ (0.13)	1.33ᵇ (0.07)	1.36ᵇ (0.04)	0.37ᵃ (0.10)	−1.87ᵇ (0.25)	−1.08ᶜ (0.05)	−0.74ᶜ (0.22)

Notes: Data are averages with standard errors in parentheses. Dashes indicates that *Vicia* was absent from control plots at Baskett Butte. Significant differences among species within sites were analyzed using one-way ANOVA. Values in rows for tissue %N and $\delta^{15}N$ that are followed by different letters are significantly different (p<0.10) by Tukey post-hoc comparison.

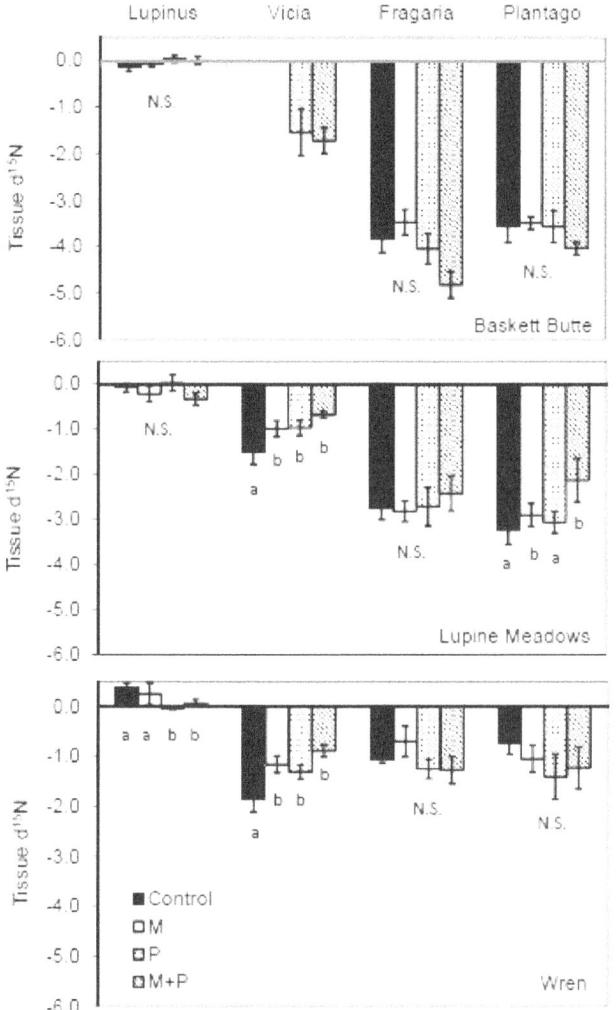

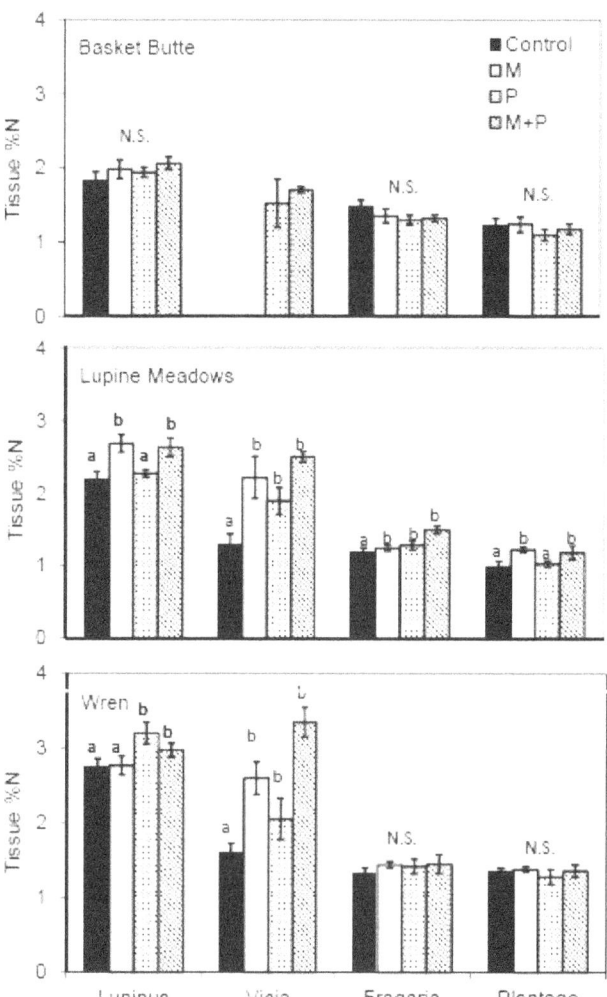

Figure 4. $\delta^{15}N$ of *Lupinus, Vicia, Frgaria,* and *Plantago* in plots treated with micronutrients and/or phosphorus. Bars are means±1 S.E.; n=5 per treatment (No *Vicia* were present in control and M plots at Baskett Butte). Letters denote differences between treatments within species. N.S. indicates where no differences were found. Treatment effects were not tested on *Vicia* at Baskett Butte as we were not able to obtain tissue samples within the control and M plots.

Figure 5. Tissue %N of *Lupinus, Vicia, Frgaria,* **and** *Plantago* **in plots treated with micronutrients and/or phosphorus.** Bars are means±s1 S.E.; n=5 per treatment (No *Vicia* were present in control and M plots at Baskett Butte). Letters denote differences between treatments within species. N.S. indicates where no differences were found. Treatment effects were not tested on *Vicia* at Baskett Butte as we were not able to obtain tissue samples within the control and M plots.

At Wren, significant shifts in tissue%N and $\delta^{15}N$ due to fertilization occurred only in N-fixing species (Table 2; Figures 4, 5). For *Lupinus*, P fertilization significantly increased tissue%N ($F_P = 7.75$, df=1,16, $P_P = 0.014$) and slightly decreased $\delta^{15}N$ towards more atmospheric values ($F_P = 4.77$, df=1,16, $P_P = 0.045$). In *Vicia* spp., both M and P significantly increased tissue%N ($F_M = 29.04$, df=1,16, $P_M<0.001$; $F_P = 7.97$, df=1,16, $P_P = 0.012$) and increased $\delta^{15}N$ towards more atmospheric values ($F_M = 10.31$, df=1,16, $P_M = 0.005$; $F_P = 5.44$, df=1,16, $P_P=0.033$).

When plot-level tissue data were evaluated by site, we found that *Vicia* spp. tissue%N consistently increased as $\delta^{15}N$ shifted from depleted values towards 0‰ (Figure 6). For *Lupinus*, $\delta^{15}N$ values overall were near 0‰, and there was only a weak linear relationship observed between tissue %N and $\delta^{15}N$ at one site.

We observed very few fertilization effects on soil N and pH (data not shown). As expected, P fertilization increased soil extractable P at all sites, on average 10-fold over control and M-only plots. Soil

NH_4^+ was marginally higher ($F_P = 3.8$, df=1,16, $P_P = 0.069$) in P treated plots only at Lupine Meadows, and NO_3^- did not differ among any treatments. Laboratory potential N mineralization was significantly lower ($F_P = 12.1$, df=1,16, $P_P = 0.003$) in P treated plots only at Baskett Butte. Soil pH was ~0.2–0.6 units lower in M treated plots at all sites.

Discussion

Our results demonstrate that non-N nutrients can limit N fixation and growth of threatened *Lupinus*. At sites where P or M promoted *Lupinus* reproduction or growth, the same nutrient generally increased *Lupinus* tissue N. *Lupinus* $\delta^{15}N$ was consistently near the atmospheric value of 0‰ across all sites in both control and fertilized plots (0.03‰±0.14‰; mean±SE, n=3 sites), whereas non-N-fixing reference species displayed depleted $\delta^{15}N$ (*Fragaria*: –2.56‰±0.81; *Plantago*: –2.52‰±0.90), suggesting strong reliance of *Lupinus* on N-fixation as a source of N. Other *Lupinus* species growing on low-N soils can also show strong

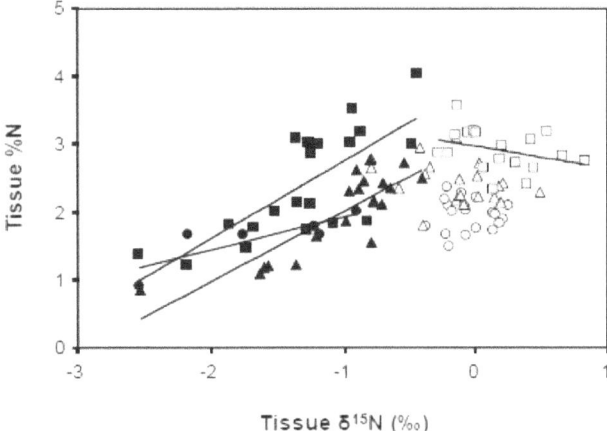

Figure 6. Relationship between tissue %N and $\delta^{15}N$ for *Lupinus* and *Vicia* in all treatment plots. Filled symbols indicate *Vicia*, open symbols indicate *Lupinus*. Linear regressions are significant for *Vicia* at Baskett Butte (filled circles: $R^2 = 0.69$, $P = 0.04$), Lupine Meadows (filled triangles, $R^2 = 0.73$, $P < 0.001$) and Wren (filled squares $R^2 = 0.56$, $P < 0.001$), and for *Lupinus* at Wren (open squares $R^2 = 0.50$, $P < 0.001$).

reliance on atmospheric N [39], although elevated N availability can shift *Lupinus* N acquisition towards soil N [40]. Due to the relative constant and small differences in $\delta^{15}N$ values of *Lupinus*, these data can provide only suggestive evidence of N fixation patterns. Additional data on nodule biomass and nitrogenase enzyme activity would have been helpful, but unfortunately the federally threatened status of *Lupinus* at our field sites precluded more destructive soil sampling required for such measurements.

The apparent reliance of *Lupinus* on fixed N in our study plots likely reflects low ambient soil N availability, as suggested by low concentrations of total N, low potential net N mineralization, and low extractable NO_3^- relative to NH_4^+ in surface mineral soils (Table 1). These low-N soils contrast sharply with the unusually N-rich soils of the coastal forest province bordering the Willamette Valley [41], and suggest that aboriginal burning of prairies [42], [43] likely maintained low-N soil conditions that favored N-fixers such as *Lupinus* in upland prairies. It is unclear whether recent increases in atmospheric N deposition in the Willamette Valley have altered N dynamics in vascular plant communities that include *Lupinus*, as has been observed for arboreal lichen communities in the area, which display increased thallus N, shifts in community composition towards nitrophilous species, and loss of N-fixing cyanolichens [44].

Non-native annual *Vicia* spp. also showed nutrient limitation of N fixation and growth, but in ways that differed from native, perennial *Lupinus*. At two of the sites (Baskett Butte and Lupine Meadows), *Vicia* spp. displayed $\delta^{15}N$ values intermediate between atmospheric N_2 and non-fixing species, and at the third site (Wren) $\delta^{15}N$ overlapped with non-fixers. *Vicia* spp. also showed less site-specificity in response to fertilization treatments than *Lupinus*, responded to both M and P in the two sites where it was present in all plots, and was present only in P and M+P plots at the third site. In addition, the concomitant increase in tissue N with a shift towards atmospheric $\delta^{15}N$ across all study plots (Figure 6) provides evidence for N limitation of *Vicia* spp., and suggests that constraints on N-fixation limit *Vicia* spp. tissue N to low values in these low-N sites. Thus, both N and non-N nutrients may be more important in shaping success of non-native *Vicia* spp. than native *Lupinus* in these sites. The relatively plastic reliance of *Vicia* spp. on soil versus atmospheric sources of N further suggests a potential to respond

rapidly to variation in factors controlling N supply. This is consistent with evidence from agricultural settings showing that high N additions can suppress N-fixation in this genus [45]. Thus, non-N fertilization that increases N-fixation in invasive *Vicia* species is likely to lead to significant increases in growth and reproduction.

We found evidence that background levels of soil P, both alone and interactively with soil N, shaped patterns of nutrient limitation observed in *Lupinus*. At two sites with low soil P - Baskett Butte and Wren – we found significant effects of P fertilization on *Lupinus*. However, responses to P fertilization varied with background soil N availability. At Wren, where both soil P and soil N mineralization were low, P fertilization significantly increased *Lupinus* tissue N and changed tissue $\delta^{15}N$ towards atmospheric values. At Baskett Butte, where soil P was low and soil N mineralization was high, P fertilization did not change tissue N or $\delta^{15}N$ but did significantly increase *Lupinus* growth and reproduction. It is thought that P may limit either N-fixation or growth in symbiotic N fixing organisms [46]. Our results raise the possibility that the mode of P limitation may depend on N supply, whereby P primarily limits N fixation when soil N is scarce, but limits N-fixer growth when soil N is abundant.

Higher background soil P at Lupine Meadows appeared to alleviate P limitation, and only micronutrient fertilization increased *Lupinus* growth and tissue N. The higher levels of soil P and resulting micronutrient limitation of *Lupinus* at Lupine Meadows may reflect site management history; Lupine Meadows is the most recently grazed site, and grazing can preferentially increase the availability of P relative to other nutrients [47], [48]. Slightly lower soil pH at Lupine Meadows may have also decreased soil Mo availability [21], leading to stronger Mo limitation at this site, although future Mo-only fertilization is needed to conclusively identify which specific micronutrient stimulated *Lupinus* growth and tissue N at this relatively P-rich site. Indeed, species-specific access to and requirements for Mo, P and other non-N nutrients may contribute to the differential response of *Lupinus*, *Vicia* spp. and other Fabaceae to fertilization, and may constitute an important yet poorly understood control of symbiotic N_2-fixation in many ecosystems worldwide [15], [49], [50].

The response of *Vicia* spp. to both M and P at our sites was surprising, as these elements are not substitutable in the biochemistry of N-fixation [51]. While this response may indicate co-limitation by these nutrients, we cannot exclude the possibility that the response of *Vicia* spp. to P fertilizer may also indicate a response to trace contamination of Mo in the P fertilizer. Analysis of our P fertilizer shows trace contamination with Mo of 9.2 mg Mo/kg (Julie Pett-Ridge, *personal communication*), within the range of typical triple-superphosphate fertilizers (2.4–18.5 mg Mo/kg) [52], resulting in low-levels of inadvertent Mo addition of 0.35 mg Mo $m^{-2}yr^{-1}$ in the P-only treatment. This represents only 1.7% of the Mo added deliberately in the M treatments and would be quite a sensitive response given that surface soil organic matter has high potential to immobilize Mo [53], whereas the symbiotic N_2-fixing plants that we studied root and produce nodules in mineral soil. However, asymbiotic N fixation in surface organic soil can be stimulated by trace levels of Mo addition [10] and the soils that we studied are generally considered low in available Mo [22], so we cannot exclude it as a possibility.

The observed changes in the plant community suggest that increases in N fixation are primarily benefiting N-fixing species. At the site where fertilization significantly altered plant community composition (Lupine Meadows), the indicator species for this effect were the non-native non-N-fixing forb, *Leucanthemum vulgare*, and non-native N-fixer, *V. hirsuta*. The increase in *Leucanthemum* at Lupine Meadows was unexpected; future year's analyses should consider whether this species continues to respond to phosphorus fertilization. While we observed increased cover of non-native N-fixers in response to P fertilization at Baskett Butte, the lack of overall plant community response at this site is likely due to the relatively high variability in community composition between plots within the same treatment group (as indicated by an effect size close to 0; A = 0.0382). Greater shifts in plant community composition may not be apparent unless there is significant mortality of *Lupinus*, thus decreasing competition for light [54], [55]. While changes in plant species composition may take several years to occur, plant biomass may respond relatively quickly in response to fertilization [54]. Several treatment plots appeared visually to have higher biomass levels than control plots, and although we were not able to sample biomass due to the presence of the endangered Fender's blue butterfly, future measurements could include a surrogate for biomass, such as vegetation height.

The difference in the ability of the non-native *Vicia* spp. and native *Lupinus* to increase nitrogen fixation in response to nutrients may be related to differences in these species' life histories. *Lupinus* is a native, long-lived perennial forb that is generally found in relatively undisturbed remnant prairies. These prairies are expected to be relatively low in N, thus this species' long-term persistence in these communities likely depends on N-fixation. In contrast, the other Fabaceae species in our plots were exotic annual species. As these species are short-lived, they would benefit from being flexible and opportunistic in their strategy to obtain N. Our work adds to prior field studies that have demonstrated increased growth in leguminous species in natural ecosystems in response to P fertilization [17–20].

Five years of experimental fertilization with non-nitrogen fertilizers succeeded in increasing the growth and reproductive capacity of threatened *Lupinus* with few non-target plant-community effects. However, our results also suggest some caution in use of fertilization for recovery of rare N-fixing species, as we found both an apparently greater benefit to invasive nitrogen fixers to fertilization, as well as some site- and species-specificity of whether micronutrients or phosphorus fertilization were most effective. Such site specificity due to differences in management history and local abiotic conditions may contribute to the difficulties in identifying a generalized approach for successful restoration [56]. Background increases in atmospheric N deposition across the Pacific Northwest are poised to further increase the dominance of non-native grasses over native forbs, as has occurred in shrubland-grasslands of California [57]. It is unknown how such N inputs may interact with restorative non-N fertilization to shape community composition and other dynamics of prairie ecosystems. To the extent that changes in background N inputs, climate and other factors further threaten native plant communities in this region, it is possible that non-N fertilization may prove to be useful to maintain population levels of threatened legume species.

Supporting Information

Figure S1 Locations of experimental sites

Table S1 Summary of tests of effects of micronutrient and phosphorus fertilizers on the number of *L. oreganus* leaves.

Table S2 Summary of two-way ANOVA of effects of micronutrient and phosphorus fertilizers on cover and inflorescences of *L. oreganus* and cover of other Fabaceae species.

Table S3 Summary of significant ANOVA tests of effects of micronutrient (M) and phosphorus (P) fertilizers on 2008 tissue %N and δ15N of *L. oreganus*, *Vicia* spp., *Fragaria*, and *Plantago*.

Acknowledgments

The authors are grateful for the cooperation of the US Fish and Wildlife Service, Willamette Valley Refuges, and Greenbelt Land Trust. Any use of trade, product, or firm names is for descriptive purposes only and does not imply endorsement by the U.S. Government.

Author Contributions

Conceived and designed the experiments: AT SP TK. Performed the experiments: AT SP CC TK. Analyzed the data: AT SP TK. Contributed reagents/materials/analysis tools: AT SP CC. Wrote the paper: AT SP CC TK.

References

1. Franklin JF (1993) Preserving biodiversity: species, ecosystems, or landscapes? Ecol Appl 3:202–205.
2. Sheley RL, Krueger-Mangold J (2003) Principles for restoring invasive plant-infested rangeland. Weed Sci 51:260–265.
3. Stanley AG, Kaye TN, Dunwiddie PW (2008) Regional strategies for restoring native prairies: observations from a multisite collaborative research project. Native Plants J 9:255–266.
4. US Fish and Wildlife Service (2010) Recovery plan for the prairie species of western Oregon and southwestern Washington. U.S. Fish and Wildlife Service, Portland, Oregon. pp xi +241.
5. Vitousek PM, Howarth RW (1991) Nitrogen limitation on land and in the sea-How can it occur? Biogeochemistry 13: 87–115.
6. LeBauer DS, Treseder KK (2008) Nitrogen limitation of net primary productivity in terrestrial ecosystems is globally distributed. Ecology 89:371–379.
7. Schindler DW (1977) Evolution of phosphorus limitation in lakes. Science 195:260–262.
8. Howarth RW, Cole JJ (1985) Molybdenum availability, nitrogen limitation, and phytoplankton growth in natural waters. Science 229:653.
9. Smith VH (1992) Effects of nitrogen: phosphorus supply ratios on nitrogen fixation in agricultural and pastoral ecosystems. Biogeochemistry 18:19–35.
10. Barron AR, Wurzburger N, Bellinger J, Wright SJ, Kraepiel AML, et al. (2009) Molybdenum limitation of asymbiotic nitrogen fixation in tropical forest soils. Nat Geosci 2:42–45.
11. Almeida JPF, Hartwig UA, Frehner M, Nösberger J, Lüscher A. (2000) Evidence that P deficiency induces N feedback regulation of symbiotic N₂ fixation in white clover (*Trifolium repens* L.). J Exp Bot 51:1289.
12. Leidi E, Rodrigues-Navarro D (2000) Nitrogen and phosphorus availability limit N₂ fixation in bean. New Phytol 147:337–346.
13. Wall LG, Hellsten A, Huss-Danell K (2000) Nitrogen, phosphorus, and the ratio between them affect nodulation in *Alnus incana* and *Trifolium pratense*. Symbiosis 29:91–106.
14. Tang CR, Fang Y, Raphael C (1998) Factors affecting soil acidification under legumes: II. Effect of phosphorus supply. Aust J Agr Res 49:657–664.
15. Roscher C, Thein S, Weigelt A, Temperton VM, Buchmann N, et al. (2010) N₂ fixation and performance of 12 legume species in a 6-year grassland biodiversity experiment. Plant Soil 341: 333–348.
16. Lotscher M, Hay MJM (1997) Genotypic differences in physiological integration, morphological plasticity and utilization of phosphorus induced by variation in phosphate supply in *Trifolium repens*. J Ecol 85:341–350.
17. Haubensak KA (2001) Invasion and Impacts of Nitrogen-fixing Shrubs Genista Monspessulana and Cytisus Scoparius in Grasslands of Washington and Coastal California. Ph.D. dissertation University of California, Berkeley, California, USA348 pp.
18. Harpole WS, Goldstein L, Alcher R (2007) Resource limitation. In: D'Antonio CM, Corbin J, Stromberg M (eds) Ecology and Management of California Grassland University of California Press Berkeley, California, USA119–227.
19. Finzi AC, Rodgers VL (2009) Bottom-up rather than top-down processes regulate the abundance and activity of nitrogen fixing plants in two Connecticut old-field ecosystems. Biogeochemistry 95:309–321.
20. Haubensak KA, D'Antonio CM (2010) The importance of nitrogen-fixation for an invader of a coastal California grassland. Biol Invasions 13:1275–1282.

21. Johansen C, Kerridge PC, Sultana A (1997) Responses of forage legumes and grasses to molybdenum. In: Gupta CU (ed) Molybedenum in Agriculture Cambridge University PressCambridge202–228.

22. Oregon State College Experiment Station. (1958) The Molybdenum Problem in Oregon. Agricultural Experiment Station.COregon State College, orvallis, Oregon, USA

23. Ehrenfeld JG (2003) Effects of exotic plant invasions on soil nutrient cycling processes. Ecosystems 6:503–523.

24. Vitousek PM, Walker LR (1989) Biological invasion by *Myrica faya* in Hawai'i: plant demography, nitrogen fixation, ecosystem effects. Ecol Monog 59:247–265.

25. Witkowski E (1991) Effects of invasive alien acacias on nutrient cycling in the coastal lowlands of the Cape fynbos. J Appl Ecol:1–15.

26. Lonsdale W, Miller I (1993) Fire as a management tool for a tropical woody weed: *Mimosa pigra* in Northern Australia. J Environ Manage 39:77–87.

27. Van Auken OW, Bush JK (1997) The importance of neighbors, soil pH, phosphorus, and nitrogen for the growth of two C_4 grasses. Plant Sci 158:325–331.

28. Yelenik SG, Stock WD, Richardson DM (2006) Functional group identity does not predict invader impacts: Differential effects of nitrogen-fixing exotic plants on ecosystem function. Biol Invasions 9:117–125.

29. Suding KN, Collins SL, Gough L, Clark C, Cleland EE, et al. (2005) Functional- and abundance-based mechanisms explain diversity loss due to N fertilization. P Natl Acad Sci 102: 4387–4392.

30. Huenneke LJ, Hamburg SP, Koide R, Mooney HA, Vitousek PM (1990) Effects of soil resources on plant invasion and community structure in Californian serpentine grassland. Ecology 71:478.

31. USDA (2012) The PLANTS Database (http://plants.usda.gov). National Plant Data Team, Greensboro, NC 27401-4901 USA.

32. McCune B, Mefford MJ (2005) Multivariate Analysis on the PC-ORD System. MjM Software Design, Glenedon Beach, OR, USA.

33. McCune B, Grace JB (2002) Analysis of Ecological Communities. MjM Software Design, Glenedon Beach, OR, USA.

34. Högberg P (1997) ^{15}N natural abundance in soil-plant systems. New Phytol 37:179–203.

35. Turner GL, Bergersen FJ (1983) Natural abundance of ^{15}N in root nodules of soybean, lupine, subterranean clover and lucerne. Soil BiolBiochem 15:525–530.

36. Bergersen FJ, Turner GL, Amarger N, Mariotti F, Marrioti A (1986). Strain of *Rhizobium lupini* determines natural abundance of ^{15}N in root nodules of *Lupinus* spp. Soil Biol Biochem 18: 97–101.

37. Evans J, Turner GL, O'Connor GE, Bergersen FJ (1987). Nitrogen fixation and accretion of soil nitrogen by field-grown lupins (*Lupinus angustifolius*). Field Crops Res 17:109–120.

38. Unkovich M, Pate JS (2001) Assessing N_2 fixation in annual legumes using ^{15}N natural abundance. In: Unkovich M, Pate NcNeill A, Gibbs DJ (eds) Stable isotope techniques in the study of biological processes and functioning of Ecosystem. Kluwer Academic, Dordercht 103–118.

39. Myrold DD, Huss-Danell K. (2003). Alder and lupine enhance nitrogen cycling in a degraded forest soil in northern Sweden. Plant Soil 254:47–56.

40. Goergen E, Chambers JC, Blank R (2009) Effects of water and nitrogen availability on nitrogen contribution by the legume, *Lupinus argenteus* Pursh. Appl Soil Ecol 42:200–208.

41. Perakis SS, Matkins JJ, Hibbs DE (2011) Interactions of tissue and fertilizer nitrogen on decomposition dynamics of lignin-rich conifer litter. Ecosphere 3:1–12.

42. Boyd R (1999) Indians, Fire, and the Land in the Pacific Northwest. Oregon State University Press, Corvallis, Oregon, USA. 313 pp.

43. Walsh MK, Pearl CA, Whitlock C, Bartlein PJ, Worona MA (2010) An 11 000-year-long record of fire and vegetation history at Beaver Lake, Oregon, central Willamette Valley. Quaternary Sci Rev 29:1093–1106.

44. Geiser LH, Jovan SE, Glavich DA, Porter M (2010). Lichen-based critical loads for atmospheric nitrogen deposition in Western Oregon and Washington Forests, USA. Environ Pollut 158:2412–2421.

45. Hardarson G, Danso SKA, Zapata F, Reichardt K (1991) Measreuments of nitrogen fixation in fababean at different N fertilizer rates using the ^{15}N isotope dilution and 'A-value' methods. Plant Soil 131:161–168.

46. Vitousek PM, Menge DNL, Reed SC, Cleveland CC (2013) Biological nitrogen fixation: Rates, patterns, and ecological controls in terrestrial ecosystems. PhilosT RS B.

47. Haynes RJ, Williams PH (1993) Nutrient cycling and soil fertility in grazed pasture ecosystem. Adv Agron 49:119–199.

48. Brown JR (1996) Fertility management of harvested forages in the northern states. Nutrient cycling in forage systems; In: Joost E, Roberts CA (eds) Proceedings Symposium, 7–8 Mar. 1996. Columbia, Missouri, USA.

49. Ritchie ME, Tilman D. (1995) Responses of legumes to herbivores and nutrients during succession on a nitrogen-poor soil. Ecology 76: 2648–2655.

50. West JB, Hille Ris Lambers J, Lee TD, Hobbie SE, Reich PB (2005) Legume species identity and soil nitrogen supply determine symbiotic nitrogen-fixation responses to elevated atmospheric CO_2. New Phytol 167:523–530.

51. Fisher K, Newton WE (2002). Nitrogen fixation – a general overview. In: Leigh GJ (ed) Nitrogen Fixation at the Millennium. Elsevier, Amsterdam, The Netherlands, pp 1–27.

52. Charter RA, Tabatabai MA, Schafer JW (1995) Arsenic, molybdenum, selenium, and tungsten contents of fertilizers and phosphate rocks. Commun Soil Sci Plan 26:17–18.

53. Wurzburger N, Bellenger JP, Kraepiel AML, Hedin LO (2012) Molybdenum and phosphorus interact to constrain asymbiotic nitrogen fixation in tropical forests. PLoS ONE 7(3):e33710. doi:10.1371/journal.pone.0033710.

54. Maron JL, Jefferies RL (1999) Bush lupine mortality, altered resource availability, and alternative vegetation states. Ecology 80:443–454.

55. Gosling P (2005) Facilitation of *Urtica dioica* colonisation by *Lupinus arboreus* on a nutrient-poor mining spoil. Plant Ecol 178:141–148.

56. Stanley AG, Dunwiddie PW, Kaye TN (2011) Restoring invaded northwest prairies: management recommendations for a region wide experiment. Northwest Sci 85:233–246.

57. Pardo LH, Fenn ME, Goodale CL, Geiser LH, Driscoll CT, et al. (2011) Effects of nitrogen deposition and empirical critical loads for nitrogen for ecoregions of the United States. Ecol Appl 21:3049–3082.

Permissions

List of Contributors

Qiang Yu, Honghui Wu, Xiaotao Lü and Xingguo Han
State Key Laboratory of Forest and Soil Ecology, Institute of Applied Ecology, Chinese Academy of Sciences, Shenyang, China
State Key Laboratory of Vegetation and Environmental Change, Institute of Botany, Chinese Academy of Sciences, Beijing, China

Zhiping Wang
State Key Laboratory of Vegetation and Environmental Change, Institute of Botany, Chinese Academy of Sciences, Beijing, China

Nianpeng He
Institute of Geographic Sciences and Natural Resources Research, Chinese Academy of Sciences, Beijing, China

James J. Elser
School of Life Sciences, Arizona State University, Tempe, Arizona, United States of America

Jianguo Wu
School of Life Sciences, Arizona State University, Tempe, Arizona, United States of America
Sino-US Center for Conservation, Energy and Sustainability Science (SUCCESS), Inner Mongolia University, Hohhot, Inner Mongolia, China

Sandra Barantal, Nathalie Fromin, Patrick Schevin and Stephan Hättenschwiler
Centre d'Ecologie Fonctionnelle et Evolutive (CEFE), CNRS, Montpellier, France

Nicolas Fanin
Centre d'Ecologie Fonctionnelle et Evolutive (CEFE), CNRS, Montpellier, France
Université of Montpellier II, Montpellier, France

Heidy Schimann
UMR Ecologie (EcoFoG), Campus Agronomique, Kourou, French Guiana

Jennifer Firn
School of Earth, Environment and Biological Sciences, Queensland University of Technology, Brisbane, Queensland, Australia

Suzanne M. Prober
Ecosystem Sciences, CSIRO, Wembley, Western Australia, Australia

Yvonne M. Buckley
School of Biological Sciences, The University of Queensland, St. Lucia, Queensland, Australia
Ecosystem Sciences CSIRO, Dutton Park, Queensland, Australia

Yanyan Wei, M. J. I. Shohag and Xiaoe Yang
Ministry of Education (MOE) Key Laboratory of Environmental Remediation and Ecosystem Health, College of Environmental and Resources Science, Zhejiang University, Hangzhou, People's Republic of China

Jürgen Homeier, Dietrich Hertel and Christoph Leuschner
Albrecht von Haller Institute of Plant Sciences, Georg August University Göttingen, Göttingen, Germany

Tessa Camenzind and Matthias C. Rillig
Institute of Biology, Freie Universität Berlin, Berlin, Germany

Nixon L. Cumbicus
Instituto de Ecología, Universidad Técnica Particular de Loja, San Cayetano Alto, Loja, Ecuador

Dorothee Sandmann, Stefan Scheu and Mark Maraun
J.F. Blumenbach Intitute of Zoology and Anthropology, Georg August University Göttingen, Göttingen, Germany

Guntars O. Martinson
Max Planck Institute for Terrestrial Microbiology, Marburg, Germany
Buesgen Institute - Soil Science of Tropical and Subtropical Ecosystems, Georg August University Göttingen, Göttingen, Germany

Edzo Veldkamp
Buesgen Institute - Soil Science of Tropical and Subtropical Ecosystems, Georg August University Göttingen, Göttingen, Germany

L. Nohemy Poma
Universidad National de Loja, Ciudadela Universitaria Guillermo Falconí sector La Argelia, Loja, Ecuador

Wolfgang Wilcke
Geographic Institute, University of Berne, Berne, Switzerland

Hans Wullaert
Geographic Institute, University of Mainz, Mainz, Germany

Guirui Yu
Key Laboratory of Ecosystem Network Observation and Modeling, Institute of Geographic Sciences and Natural Resources Research, Chinese Academy of Sciences, Beijing, China

Chunyan Luo
Institute of Agricultural Resources and Regional Planning, CAAS, Beijing, China

Yang Gao
Key Laboratory of Ecosystem Network Observation and Modeling, Institute of Geographic Sciences and Natural Resources Research, Chinese Academy of Sciences, Beijing, China
School of Agriculture and Biology, Shanghai Jiaotong University, Shanghai, China

Pei Zhou
School of Agriculture and Biology, Shanghai Jiaotong University, Shanghai, China

Dazhi Wen
Key Laboratory of Vegetation Restoration and Management of Degraded Ecosystems, South China Botanical Garden, Chinese Academy of Sciences, Guangzhou, China
University of Chinese Academy of Sciences, Beijing, China

Enqing Hou
Key Laboratory of Vegetation Restoration and Management of Degraded Ecosystems, South China Botanical Garden, Chinese Academy of Sciences, Guangzhou, China,
Environmental Futures Centre, Griffith School of Environment, Griffith University, Nathan, Queensland, Australia
University of Chinese Academy of Sciences, Beijing, China

Chengrong Chen
Environmental Futures Centre, Griffith School of Environment, Griffith University, Nathan, Queensland, Australia

Megan E. McGroddy
Department of Environmental Sciences, NASA/University of Virginia, Charlottesville, Virginia, United States of America
University of Chinese Academy of Sciences, Beijing, China

Jinyang Wang, Xiaolin Zhang, Yinglie Liu, Xiaojian Pan, Zhaozhi Chen, Taiqing Huang and Zhengqin Xiong
Jiangsu Key Laboratory of Low Carbon Agriculture and GHGs Mitigation, College of Resources and Environmental Sciences, Nanjing Agricultural University, Nanjing, China

Pingli Liu
Jiangsu Key Laboratory of Low Carbon Agriculture and GHGs Mitigation, College of Resources and Environmental Sciences, Nanjing Agricultural University, Nanjing, China
Hebi Academy of Agricultural Sciences, Hebi, Henan, China

Junxia Yan and Hongjian Li
Institute of Loess Plateau, Shanxi University, Taiyuan Shanxi, China

Junjian Li
Institute of Loess Plateau, Shanxi University, Taiyuan Shanxi, China
State Key Laboratory of Urban and Regional Ecology, Research Centre for Eco-environmental Sciences, Chinese Academy of Sciences, Beijing, China

Yuanming Zheng and Jizheng He
State Key Laboratory of Urban and Regional Ecology, Research Centre for Eco-environmental Sciences, Chinese Academy of Sciences, Beijing, China

Xiang Wang
Biology Institute of Shanxi, Taiyuan Shanxi, China

Guangwei Ding
Chemistry Department, Northern State University, Aberdeen, South Dakota, United States of America

Andreas P. Mamolos and Demetrios S. Veresoglou
Laboratory of Ecology and Environmental Protection, Faculty of Agriculture, Aristotle University of Thessaloniki, Thessaloniki, Greece

Barry Thornton
Environmental and Biochemical Sciences Group, The James Hutton Institute, Craigiebuckler, Aberdeen, United Kingdom

George Menexes
Laboratory of Agronomy, School of Agriculture, Aristotle University of Thessaloniki, Thessaloniki, Greece

Stavros D. Veresoglou
Laboratory of Ecology and Environmental Protection, Faculty of Agriculture, Aristotle University of Thessaloniki, Thessaloniki, Greece
Plant Ecology, Freie Universität Berlin, Berlin, Germany

Paul Szpak, Jean-François Millaire and Christine D. White
Department of Anthropology, The University of Western Ontario, London, Ontario, Canada

Fred J. Longstaffe
Department of Earth Sciences, The University of Western Ontario, London, Ontario, Canada

Christian Jessen
Coral Reef Ecology Group (CORE), Leibniz Center for Tropical Marine Ecology (ZMT), Bremen, Germany

Cornelia Roder, Javier Felipe Villa Lizcano and Christian R. Voolstra
Red Sea Research Center, King Abdullah University of Science and Technology (KAUST), Thuwal, Saudi Arabia

Christian Wild
Coral Reef Ecology Group (CORE), Leibniz Center for Tropical Marine Ecology (ZMT), Bremen, Germany
Faculty of Biology and Chemistry, University of Bremen, Bremen, Germany

Jinling Meng
National Key Laboratory of Crop Genetic Improvement, Huazhong Agricultural University, Wuhan, Hubei, China

Zunkang Zhao, Likun Wu, Guangda Ding, Taoxiong Shi, Didi Zhang, Lei Shi and Fangsen Xu
National Key Laboratory of Crop Genetic Improvement, Huazhong Agricultural University, Wuhan, Hubei, China
Microelement Research Centre, Huazhong Agricultural University, Wuhan, Hubei, China

Fuzhao Nian
Microelement Research Centre, Huazhong Agricultural University, Wuhan, Hubei, China

Li Cheng-Fang, Kou Zhi-Kui, Zhang Zhi-Sheng, Wang Jin-Ping, Cai Ming-Li and Cao Cou-Gui
College of Plant Science and Technology, Huazhong Agricultural University, Wuhan, Hubei, China

Zhou Dan-Na
Institute of Animal Husbandry and Veterinary Science, Hubei Academy of Aguicultural Sciences, Wuhan, Hubei, China

W. H. Gera Hol
Department of Terrestrial Ecology, Netherlands Institute of Ecology NIOO-KNAW, Wageningen, The Netherlands

Katrin M. Meyer
Department of Terrestrial Ecology, Netherlands Institute of Ecology NIOO-KNAW, Wageningen, The Netherlands
Ecosystem Modelling, Faculty of Forest Sciences and Forest Ecology, University of Goettingen, Göttingen, Germany

Matthijs Vos
Department of Ecology & Ecosystem Modeling, Institute of Biochemistry and Biology, University of Potsdam, Potsdam, Germany
Department of Aquatic Ecology, Netherlands Institute of Ecology NIOO-KNAW, Wageningen, The Netherlands

Wolf M. Mooij
Department of Aquatic Ecology, Netherlands Institute of Ecology NIOO-KNAW, Wageningen, The Netherlands
Aquatic Ecology and Water Quality Management, Department of Environmental Sciences Group, Wageningen University, Wageningen, The Netherlands

Aad J. Termorshuizen
BLGG Research, Wageningen, The Netherlands

WimH. Van der Putten
Department of Terrestrial Ecology, Netherlands Institute of Ecology NIOO-KNAW, Wageningen, The Netherlands
Laboratory of Nematology, Wageningen University, Wageningen, The Netherlands

Diana Pauly, Sylvia Worbs, Sebastian Kirchner, Olena Shatohina, Martin B. Dorner and Brigitte G. Dorner
Center for Biological Security - Microbial Toxins, Robert Koch-Institut, Berlin, Germany

Qingfeng Meng, Shanchao Yue, Xinping Chen, Zhenling Cui and Fusuo Zhang
Center for Resources, Environment and Food Security, China Agricultural University, Beijing, China

Youliang Ye
College of Resources and Environmental Sciences, Henan Agricultural University, Zhengzhou, China

Wenqi Ma
College of Resources and Environmental Sciences, Hebei Agricultural University, Baoding, China

Yanan Tong
College of Resources and Environmental Sciences, Northwest Sci-Tech University of Agriculture and Forestry, Yangling, China

Hao Shen, Zhang-Ming Wang and Wan-Hui Ye
Key Laboratory of Vegetation Restoration and Management of Degraded Ecosystems, South China Botanical Garden, Chinese Academy of Sciences, Guangzhou, Guangdong, PR China

Shu-Jun Xu
Key Laboratory of Vegetation Restoration and Management of Degraded Ecosystems, South China Botanical Garden, Chinese Academy of Sciences, Guangzhou, Guangdong, PR China
College of Life Sciences, University of Chinese Academy of Sciences, Beijing, PR China

Lan Hong
College of Horticulture and Landscape Architecture, Zhongkai University of Agriculture and Engineering, Guangzhou, Guangdong, PR China

Itxaso Barberia, Fernando Blanco, Carmelo P. Cubillas and Helena Matute
Departamento de Fundamentos y Me´todos de la Psicología, Universidad de Deusto, Bilbao, Spain

Zhaoliang Song
Zhejiang Provincial Key Laboratory of C Cycling in Forest Ecosystems and C Sequestration, Zhejiang Agricultural and Forestry University, Lin'an, Zhejiang, China
School of Environment and Resources, Zhejiang Agricultural and Forestry University, Lin'an, Zhejiang, China
State Key Laboratory of Environmental Geochemistry, Institute of Geochemistry, Chinese Academy of Sciences, Guiyang, Guizhou, China

Fengshan Guo
School of Environment and Resources, Zhejiang Agricultural and Forestry University, Lin'an, Zhejiang, China

Jeffrey F. Parr
Southern Cross GeoScience, Southern Cross University, Lismore, New South Wales, Australia

Yaojun Zhang, Hong Wang and Jianwen Zou
College of Resources & Environmental Sciences, Nanjing Agricultural University, Nanjing, China

Ling Zhang and Evan Siemann
College of Resources & Environmental Sciences, Nanjing Agricultural University, Nanjing, China
Department of Ecology & Evolutionary Biology, Rice University, Houston, Texas, United States of America

Thomas N. Kaye
Institute for Applied Ecology, Corvallis, Oregon, United States of America

Andrea S. Thorpe
Institute for Applied Ecology, Corvallis, Oregon, United States of America
National Ecological Observatory Network, Inc., Boulder, Colorado, United States of America

Steven Perakis and Christina Catricala
United States Geological Survey, Forest and Rangeland Ecosystem Science Center, Corvallis, Oregon, United States of America

Index

www.ingramcontent.com/pod-product-compliance
Lightning Source LLC
Chambersburg PA
CBHW080413190526
45161CB00003B/221